Comprehending the Complexity of Countries

Hans Kuijper

Comprehending the Complexity of Countries

The Way Ahead

Springer

Hans Kuijper
Rotterdam, The Netherlands

ISBN 978-981-16-4708-6 ISBN 978-981-16-4709-3 (eBook)
https://doi.org/10.1007/978-981-16-4709-3

This Springer imprint is published by the registered company Springer Nature Singapore Pte Ltd.
The registered company address is: 152 Beach Road, #21-01/04 Gateway East, Singapore 189721, Singapore

Ṛta/order (Rig-Veda)

Indrajāla/Indra's net (Atharva-Veda)

Lǐ/order (Lǐjì)

The great Interdependence of all differences (Yìjīng)

All things are one (Heraclitus)

Everything in everything (Anaxagoras)

The ten thousand things are all one (Zhuangzi)

The sage tries to comprehend the whole (Liezi)

ἕν τὸ πᾶν/the all is one (Cleopatra the Alchemist)

The ultimate is the one (Wang Bi)

Pratītyasamutpāda/no dharma is self-sufficient (Buddhism)

One is all and all is one (Zhiyan)

Unity and diversity are interrelated (Śaṃkara)

There is nothing isolated (Zhu Xi)

Tout tient à tout (Antoine de la Sale)

D'ogni cosa la parte ritiene in sé la natura del tutto (Leonardo da Vinci)

Qui n'entend qu'une partie n'entend rien (Alain-René Lesage)

In uno omnia (Athanasius Kircher)

Singulae in omnibus, omnes in singulis (Friedrich Oetinger)

Das Ganze im Kleinen erblicken (Goethe)

Immer strebe zum Ganzen (Schiller)

Alles ist vernetzt (Alexander von Humboldt)

Das Ganze ist das Wesen (Hegel)

Le tout est dans tout (Joseph Jacotot)

All X are all Y (William Hamilton)

In the particular is contained the universe (James Joyce)

In the whole, all is one and one is all (Feng Youlan)

To Mary and Billy

Preface

Abstract The gestation period is described, the perspective into which the book is to be read is provided, and acknowledgements are made.

This book argues for a radically new kind of area/country studies, because the current ones are fundamentally flawed. Seemingly a matter of little concern to most people, the issue is quite serious, for politicians who are one-sidedly, poorly or downright misleadingly informed about a foreign country are likely to pursue the wrong policy toward it, which is not conducive to better understanding among nations. Although practitioners of Iranology, Turkology, Japanology, Koreanology, Sinology, Tibetology, (Latin) American, Caribbean, Russian, Slavic, European, Central Asian, South Asian, Southeast Asian, African or Arabic studies have an *explanandum* (a thing [area or country] to be explained), they do not have a distinctive *explanans*, that is, a tested or testable theory that does the explaining and is different from theories developed in disciplines such as sociology, anthropology, demography, economics and political science. On the one hand, they claim to be knowledgeable about everything concerning the area/country of their interest or predilection, cold-shouldering anyone who is not an Iranologist, Turkologist, etc. and speaks not the country's language. On the other hand, they admit (not always readily) that it is impossible to be an expert on, say, the economy, polity and society of a country, given the prodigious amount of information available, the conflicting schools of thought in economics, political science and sociology, and the increasing awareness of the complexity of the world. Of course, the *explanandum* of Iranologists differs from the study object of Japanologists, but the point I will try to make is that, while Iranologists, Japanologists and similar scholars are giving detailed descriptions of somebody or something in the country they are interested in, they do this without a country *theory*, ignorant of the marked and noticeable difference between describing, analyzing and synthesizing. The conclusions of the book are: (a) proper understanding of a country is only possible if a truly interdisciplinary (as distinct from multidisciplinary) approach is taken, and (b) the insights of systems and computer scientists can be profitably used in doing so. Redesigning the study of countries and emphasizing that, as geographical areas, they are parts of a whole (the world), I argue for a radically new kind of area/country studies, having profound implications for both the social and human sciences. As the subject is of great importance for international peace and security, it should be high on the agenda of the higher education discussion that has been dragging on for quite some time but has not yet resulted in practical ways to deal with the problems universities are facing.

The gestation period of the book has been a long one. As freshman at Leiden's Sinological Institute (now confusingly called Institute for Chinese Studies [not: Studies about China]) and sitting at the feet of a professor who in his previous life had been subservient to the colonial administration in the Dutch East Indies, I ventured to ask him how to digest the huge chunk of China information collected in the Institute's library, how to find the main avenues between the disciplines that had produced the thousands of volumes stacked up in bookcases almost reaching the ceiling of the large hall, and which books to read in what order. The answer I got was surprisingly simple: "Read, read, and read". However, an order in what appeared to me a chaos of information on China was not provided, confirming my suspicion that I was dealing with somebody "deep versed in books, but shallow in himself" (Milton). Vainly searching among lecturers and staff members for a more conclusive response in the subsequent years, and noticing that it would be better not to raise such painful questions (I incurred the displeasure and wrath a high-ranking teacher by openly questioning the scientific content of Sinology during a coffee break), I finally decided to find out for myself about something that seemed to be addressing the very foundations of the study I had chosen, with the naivety of a young man not expecting the cowardice of his classmates and the dogged resistance he would encounter on his odyssey.

"Graduated in Sinology" and more than ever convinced of the importance of being scientifically disciplined, I declined the generous offer (made by the Institute's director) of being librarian of the prestigious Chinese library in Leiden. Instead, being lucky enough to have the necessary financial resources, I enrolled for economics at Erasmus University Rotterdam, where the great and humble Nobel Laureate Jan Tinbergen lectured on the convergence of economic systems. Confused about the concepts of market and money (at Leiden University, I had been laughed out for asking about it), I decided to do so at the time that China was making the paradigm shift from Mao Zedong's to Deng Xiaoping's economic strategy. In Rotterdam, I was glad to be disciplined in economics but realized how important it also is not to look at a country one-sidedly. I tried to move Professor Tinbergen to travel to China to promote his ideas, but his frail health did not allow him to do so. Having obtained a Bachelor's degree in economics, but finding my financial resources exhausted, I had to look for a full time job. I joined the Dutch Ministry of Economic Affairs, where I became responsible for doing in depth research and internal reporting on the political-economic developments in East Asian countries, particularly in the People's Republic of China. I felt (and showed to feel) uncomfortable, however, in being called "China expert". Seconded to the Ministry of Foreign Affairs, I was appointed as Commercial Attaché at the Netherlands Embassy in Singapore, where, besides other preoccupations, I got practically involved in forming a Dutch consortium that would tender to dig the tunnel for, and to perform the construction of, the Metro (MRT), which—eventually built by French and Japanese contractors who had been wise enough to pull their resources together—is now successfully operated. Meanwhile, my search for the foundations of Sinology had to be put on the back-burner. Four years later, I was appointed as Trade Representative of the Netherlands in Taiwan, where I learned about the details of the vexed "Taiwan-China Problem" and began to realize how complicated life can be. Retired from Civil Service in 2000, I resumed studying philosophy of the social sciences and became increasingly interested in the science of (complex) systems and computer science, with the question of the scientific status of the study of China anew on my agenda but broadened to quality of country studies.

At Leiden University, I had seen and bluntly stated that the study of China lacked disciplinary depth. At Erasmus University Rotterdam, I had seen and silently noticed that the study of any country whatsoever would be unbalanced and lopsided when conducted from an economic point of view only. At the dusk of my life, with the wine being aged for many years, it dawned on me that the study of each and every country should be interdisciplinary and international in character, and that, in such a study, depth and width could and should be combined by forming a scientific team which—aided by ever more powerful computers—takes a systemic view and focuses its attention on the overall pattern and underlying principle.

Deeply unsatisfied with higher education, I left Leiden, but would never forget the main ideas of the great thinkers in ancient China which I had become acquainted with almost accidentally, for my Chinese language instructors laid more stress on philological *Akribie* than on philosophical wisdom. It was only later that I saw the interconnectedness of the ideas. Particularly, two of them have kept buzzing round in my head: the concepts of 道 (*dào* [way, process]) and 理 (*lǐ* [structure, pattern]). At the time when the revolutionary language-ideas of Ludwig Wittgenstein (1889–1951) became all the rage at Dutch universities, I ventured to write a long, unpublished pre-doctoral paper on the latter concept entitled "The character *lǐ*: an attempt to define a Chinese key concept". Unfortunately, the paper, in which I expressed my wondering and musings on the relationships between language, consciousness and reality, was judged by China scholars, who only looked philologically at my translation of passages from the Chinese Classics. Pondering over the difference between a nonlinear trail that gradually emerges and a road built after its drawing on a drawing board, I began to understand the meaning of *dào*. The concepts of pattern and process, I learned after retirement, are central to cosmology, geology, biology, demography, ecology and geopolitics. Recently, Chinese and Japanese physicists have shown that the relationship between the structure and dynamics of a network is of crucial importance for understanding the behavior of complex systems. "Pattern" and "process" refer to the arrangement of different things into order, the former concept emphasizing (but not overemphasizing) the synchronic and, concomitantly, the latter accentuating (but not overaccentuating) the diachronic. In a nutshell, "*dào*" stands for the continous dynamic and "*lǐ*" for the sructural unity of complexity. Like process (diachronic order) and pattern (synchronic order), *dào* and *lǐ* are inextricably interlinked. The usual translation of "*dàolǐ*" with "reason" is misleading. It has led to the tragical misunderstanding between China and the Western world, where the word reason connotes rationality. "Reason", having many different meanings, should be replaced by "principle" or "basic idea". "*Dàolǐ*" refers to what explains all things. The cardinal concept of order (the struggle for the truth of which being, according to Max Weber, "*die eigentliche Substanz der Geschichte*") is etymologically related to the Vedic word *Ṛta* (meaning what is right and properly joined, the truth—literally: the course and order of things in nature). "Order" refers to the Cosmos, which is essentially a huge dynamic system, a vertiginously complex whole evolving in time. *Seeing* order is the essence of intelligence, the capacity for comprehending or the ability to take and see things together, orderly.

Interestingly, 氣 (*qì* [vital energy, *élan vital*]), that which does the ordering, i.e., that without which there can be neither *dào* nor *lǐ*, is *der Dritte im Bunde*. *Qì* (originally the vapor above boiling rice, hence vaporization, condensation) is the nothing (*wú*) which enables the emergence of all that is (*yǒu*), the emptiness (*xū*) on which both *dào* and *lǐ* are predicated. Studying it may settle the fundamental disagreement between (the followers of) Albert Einstein and Niels Bohr, a disunity that seems to be only a conceptual one.[1] *Qì* shall not be equated with the Holy Spirit, understood differently in the competing Abrahamic religions. A critical comparison between *dào*, *lǐ* and *qì* on the one side and the Trinity dogma on the other side[2] would therefore reveal the most basic difference between the Chinese and Western way of thought, a difference which the Jesuits who, in the sixteenth and seventeenth centuries, were the pioneers and trailblazers of Western Sinology did not see[3]; a difference which Scott Austin in his last work, *Tao and Trinity* (Palgrave Macmillan, 2014), and Catherine Klancer in her book, *Embracing Our Complexity: Thomas Aquinas and Zhu Xi on Power and the Common Good* (SUNY, 2015), did not see either, but a difference of paramount importance in this age of globalization.

The *dào—lǐ—qì* thesis is mathematically interesting. In 1859, Bernhard Riemann famously conjectured that the prime numbers are not randomly (without pattern) scattered over the other numbers. If his hypothesis, one of the Clay Mathematics Institute's "million dollar problems", could be confirmed, the intuition of the ancient Chinese (and Indians) would have been correct and we should admire their *esprit de finesse*. Basically, they taught, there is unity, no duality [*bùèr*, *advaita*]). According to them, the fabric of the cosmos has no breaks, cracks, or cuts, and *Ṛta* is universal. It would be rather easy to expand the series $1 + 3 + 5$, i.e., to see the pattern of the sequence. To tell which primes come after the (thus far) last calculated one, however,[4] has turned out to be a hard nut to crack, because it seems to be impossible to know exactly what happens when proceeding from one prime number to another. The word digit refers to a finger (the hand's limb used to point at or indicate) and counting on the ten fingers is said to be the beginning of mathematics. However, going from one finger to the next one implies different things differently.[5]

Whereas most Westerners believe that God (male or female?) created the world, and the order of things was imposed on the chaos by "somebody" *outside* of it, leaving the theological problem "Who created God?", or "Where did God come from?", unresolved,[6] the Chinese have tended to believe that things are ordered spontaneously, arranged by themselves, dictated from the *inside* rather than from the outside. Modern Chinese thinkers are thus not surprised by the works of Hilary Putnam and Saul

Kripke, who stopped thinking of God as the source of all order, challenged the metaphysics of David Hume (1711–1776) and "made belief in essences or essential natures once more respectable". Nor are they astonished at the idea of self-regulation or self-organization Western scientists such as Hermann Haken, Humberto Maturana and Stuart Kauffman now vigorously advance. Those familiar with the *Dàodéjīng* (道德經), a Chinese classic text traditionally credited to Laozi (sixth-century BC), would even take a step further. The wisdom of *Dào* is: after appearing and emerging (birth) come disappearing and fading away (death) and vice versa; sunset follows sunrise, and the other way round. Asking (religious) leaders in the West to admit publicly that the old Chinese worldview is more plausible than theirs, however, would be asking the impossible, because it would set the herd adrift, and things would get completely out of control—control that these leaders, exhibiting the characteristics of a narcissist,[7] will try to keep at all costs, for "*der Wille zur Macht*" is the only thing that drives them and will drive them until death. That is why the myth of God, the Creator, is maintained and persistently clung to.

Describing the political issues faced by contemporary China, Paul Gewirtz, professor of constitutional law at Yale's Law School and director of Yale's China Center, writes:

> China today places great value on making money and on self-interested material success, long denied to the Chinese. But values in addition to individual materialism are needed to hold a country together and make it a good country. Where will these values continue to come from in China? The announced ideology of China's Communist Party no longer seems to be a source of moral values for Chinese society. Indeed, it is no longer clear what that ideology really is. China certainly has no equivalent to the United States' faith in its Constitution as a continuing source of our country's values, as almost a civic religion. Moreover, China does not have a strong conventional religious tradition that can be the source of values. Furthermore, the close family structures that were a traditional forum for the generational transfer of values have been weakened as Chinese society has become more mobile and, yes, more free. (https://www.chinafile.com/xi-mao-and-chinas-search-usable-past)

In an article entitled "Zhuangzi's Ironic Detachment and Political Commitment" and published in the journal *Dao* (15:1 [March 2016], pp. 1–17), Bryan W. Van Norden, an American translator of Chinese philosophical texts and celebrated "scholar of Chinese and comparative philosophy", comments on the passage. However, I am not quite sure whether Gewirtz and Van Norden fully understand the deep marks left by ancient Chinese thinkers on contemporary China.

The difference between the Western and Chinese ways of thought (a subject on which I elaborate in an article entitled "The Foundation of Chinese Systems Thinking" and to be published in *Systems Research & Behavioral Science*) has grave ethical consequences. In the West, it is often said that without God there can be no morality, that the very meaning of "good" and "bad" stems from God's will, that only belief in God can serve to motivate man to act morally, and that the "Death of God" (secularism) is the root cause of the mounting problems of Western society. Marin Heidegger (1889–1976), who is regarded as one of the most important philosophers of the 20th century, said: "*Nur noch ein Gott kann uns retten*". Hugo Grotius (1583–1645), however, was one of the very few who dared to advance the idea that there could be moral obligation without reference to God. Choosing his words carefully, he famously wrote: "*Haec quidem quae iam diximus, locum aliquem haberent etiamsi daremus, quod sine summo scelere dari nequit, non esse Deum, aut non curari ab eo negotia humana* (what we have been saying would have a degree of validity even if we conceded that which cannot be conceded without the utmost wickedness: that there is no God, or that the affairs of men are of no concern to Him). However, Grotius' natural rights theory originated in his attempt to justify the colonial claims of Dutch traders. The dominant view that without God there can be no morality would mean that the Chinese are completely devoid of any sense of morality and that they are mentally ill, an insulting and immoral conclusion that flies in the face of reality. The Chinese have considered "attuning to nature" (naturalness) to be their most fundamental duty, and "going against the grain of things" has been anathema to them. For example, Mou Zongsan (1909–1995), a prominent modern Chinese philosopher, highlights the moral metaphysics of Confucianism (its fundamental linkage between morality and reality) without assuming the existence of God. Unfortunately, his masterpiece, *Yuánshàn Lùn* (on the highest good), is still untranslated into English. It would thus be interesting to compare the basic tenets of Christian and Chinese ethics, keeping the words of Plato in mind: "The good may be said to be not only the author of knowledge to all things known, but of their being and essence, and yet the good is not essence, but far exceeds essence in dignity and power" (*The Republic*, 509).

The fundamental difference between the Western and Chinese worldview has also far-reaching juridical and political consequences. For a country that, in principle, is ruled on the premise that things are ordered spontaneously looks different from countries holding the view that God, as transcendent being, is the ultimate lawgiver and prime judge—a view contested by the adherents of positive law (the law promulgated and imposed by a central government). Whereas Western civilization is predicated on the assumption that man, being by nature bad ("original sin"), needs to be ruled and controlled (an assumption that runs counter to the idea of liberty), the culture of China is predicated on the assumption that man, being by nature good, is in no need of being ruled and controlled, at least not in principle. "Whoever sees a small child poised to fall into the water will spontaneously rush to rescue it", Mencius (372–289 BC) perceptively remarks, implying that no external decree, order or law for rescuing a child in imminent danger of drowning needs to be obeyed or consented to. The rescuing just happens, not because the rescuer regards it as his/her moral duty. The Western sets of values, norms, aspirations and expectations that constitute its cultural system are fundamentally different from the sets having formed the Chinese culture.

Another (but not an entirely other) idea imprinted on my mind was the concept of law upheld by ancient Chinese thinkers. They read the rules to govern man from nature rather than from books not belonging to the venerated Classics (*jīng*). They were opposed to the idea that society should be subjected to the will of man, because man is short-sighted and inclined to prejudice and partiality. Ancient Chinese thinkers trusted nature, but were deeply suspicious of everything that is unnatural, man-made. Confucius (551–479 BC), who differentiated between the *jūnzǐ* (superior, noble man) and de *xiǎorén* (inferior, ignoble man), used to say: "*Jūnzǐ bú dǎng*" (a noble man [not: nobleman!] does not join a party), a simple saying that should be food for thought to our sneaky, double-tongued politicians and their zealous supporters. I never lost my interest in the Chinese concept of law, as clearly distinct from the Western idea but remarkably concurrent with the Indian concept of law. Particularly the sensitive issue of human rights has kept drawing my attention.[8]

Though the idea of reducing the length of the book was appealing, I resisted the suggestion that I should shorten it, because I wanted the book to cover the entire field of area and country studies. If I had restricted myself to criticizing the work of Sinologists only, it might have looked as if I excused people like Japanologists, Mongolists, Koreanologists, Tibetologists, Indologists, Iranologists, Eurpeanists, Russianists, Africanists, Americanists and Latin Americanists for engaging in studies which I claim (and will argue) to be fundamentally flawed. I do not want the book to be repetitious, but saying the same differently (*eadem sed aliter*), or varying on a theme, may augment the power of my argument.

It would be inappropriate to mention the names of those who formed the current against which I had to swim and attempted to frustrate my efforts to uplift the study of countries to a *scientific* level. Suffice it to say that without them this book would probably not have been written. Since the book is scathingly critical about area and country studies, I could not find any of their practitioners to be kind enough to read the manuscript and to make suggestions how to improve on it. However, I wish to express my deepest gratitude for the help and encouragement of the following persons. My lovely wife Mary has been and will always remain the only person who knows what it takes to write a book like this one. She is gifted by nature with, and conditioned by her culture toward, the attitude that is essential to attune harmoniously to the vagaries of life. Zhu Xi (1130–1200) and Wang Yangming (1472–1529), the main protagonists of Neo-Confucianism, opened my eyes for the difference between the universe outside and the universe inside, a difference by which our universities, where scientists of nature and scientists of culture have become alienated from each other, are characterized. Surendranath Dasgupta's *magnum opus*, *A History of Indian Philosophy* (1922–1955) may be slightly dated, but the multi-volume book unlocked to me the treasure trove of Indian philosophy and got me to reading Sarvepalli Radhakrishnan and Charles Moore (eds.), *A Source Book in Indian Philosophy* (Princeton University Press, 1957), Mysore Hiriyanna, *The Essentials of Indian Philosophy* (Allen & Unwin, 1978), K. Satchidananda Murty, *Philosophy in India* (Motilal Banarsidass, 1985), Chandradhar Sharma, *A Critical Survey of Indian Philosophy* (Motilal Banarsidass, 1987), Jitendra Mohanty, *Between Two Worlds: East and West* (OUP, 2002), Deepak Sarma, *Classical Indian Philosophy: A Reader* (Columbia University Press, 2011), Irina Kuznetsova et al. (eds.), *Hindu and Buddhist Ideas in Dialogue: Self and No-Self* (Routledge, 2016) and Jonathan Ganeri (ed.), *Handbook of Indian Philosophy* (OUP, 2017). The work of Ludwig von Bertalanffy, the largely neglected founding father of systems science, has been a great source of inspiration. Unfortunately, I wasn't able to personally communicate with Murray Gell-Mann (1929–2019), who used to admonish his students to be aware of life's highly intricate nonlinearities and to "take a crude look at the whole", no matter their specialism (https://www.santafe.edu/news-center/news/murray-gell-mann-passes-away-89). The book in which he brilliantly elaborates on this great idea, *The Quark and The Jaguar: Adventures in the Simple and the Complex* (Henry Holt, 1994), left a deep impression on me. On February 25, 2017, my dear friend Elart von Collani, professor of mathematics and statistics at Würzburg University, passed away. He was a noble man and a true and truthful scientist, who, at the time that I was struggling with the idea of uncertainty, drew my attention to the much neglected masterpiece of Jacob Bernoulli, *Ars Conjectandi*. I am greatly indebted to Tan Sri Andrew Sheng, who sits on the Advisory Board of Para Limes, a renowned institute established in 2006 by scientists in Europe and the USA "to invigorate explorations in a complex world". Dr. Sheng kindly read the manuscript carefully and provided me with invaluable advice on how to improve its content. Needless to say, I take full responsibility for any error or imperfection the critical reader may discover. I did my utmost to put forward my ideas about current and future area/country studies by arguing convincingly and writing clearly, without being superfluous. However, as the Dutch saying goes: there is no spotted cow without a stain.

I am also grateful to the following publishers:

- *Sage Publications*, in New Delhi, permitted me to use, or to elaborate on, parts of my article "Is Sinology a Science?", published in *China Report*, 36:3 (July-September 2000), pp. 331–354;
- *Znanstvena založba Filozofske fakultete Univerze v Ljubljani* (Ljubjana University) generously allowed me to draw on the article "What's Wrong with the Study of China/Countries", published in *Asian Studies*, 2:1 (May 2014), pp. 151–185; and
- *Uchitel*, in Volgograd, kindly permitted me to reuse parts of the article "Redesigning Area/Country Studies", published in *Journal of Globalization Studies*, 8:2 (November 2017), pp. 42–67.

Finally, I am grateful for the help and guidance given and the patience exercised by Nick Melchior and his editorial team at Springer Publishing Company.

Rotterdam, The Netherlands Hans Kuijper

Notes and References

1. See Abraham Pais, *Subtle is the Lord* (OUP, 1982), Carsten Held, *Die Bohr-Einstein-Debatte* (Mentis, 1998), Edmund Blair Bolles, *Einstein Defiant* (Henry Press, 2004), Danian Hu, *China and Albert Einstein* (HUP, 2005) and Suk Gabriel Choi and Jung-Yeup Kim (eds.), *The Idea of Qi/Gi* (Lexington, 2019). Also visit https://press.princeton.edu/search?search=einstein.
2. See Melville Stewart (ed.), *The Trinity* (Springer, 2003). Visit https://plato.stanford.edu/entries/trinity, https://www.iep.utm.edu/trinity and https://philpapers.org/s/trinity.
3. See Jacques Gernet, *Chine et Christianisme* (Gallimard, 1991, Chap. 5).
4. Go to http://www.prime-numbers.org.

5. Visit https://theconversation.com/has-one-of-maths-greatest-mysteries-the-riemann-hypothesis-finally-been-solved-103974 and https://www.quantamagazine.org/mathematicians-will-never-stop-proving-the-prime-number-theorem-20200722. I leave the interested reader in the hands of Robert Langlands, Abel Prize Laureate 2018.
6. See Jan Assmann, *From Akhenaten to Moses* (AUC Press, 2014). Visit https://plato.stanford.edu/entries/monotheism and https://aeon.co/essays/why-did-an-ancient-egyptian-king-erase-all-gods-but-aten.
7. See Michael Maccoby, *Narcissistic Leaders* (Harvard Business School, 2007).
8. See Marina Svensson, *Debating Human Rights in China: A Conceptual and Political History* (Rowman & Littlefield, 2002) and Sarah Biddulph and Joshua Rosenzweig (eds.), *Handbook on Human Rights in China* (Edward Elgar, 2019). Chang Weijen, *An Annotated Bibliography of Chinese Legal Philosophy* (Academia Sinica, 1976) is indispensable.

Introduction

Abstract The research method followed is justified, a summary of the chapters is given, the difference between science and *philosophy* of science is emphasized and a comprehensive dimensional lay-out of the country to be studied is presented, the important point being the intimate relationship between the research-dimensions. The worldviews and lifestyles of, e.g., desert (or mountain) people, forest dwellers, miners, fishermen, peasants, factory workers, herdsmen, monks, prostitutes, captains of industry, bankers, ministers, parliamentarians, generals, soldiers, lawyers, academics, villagers, city dwellers, yougsters and adults greatly differ, but the geological, geographic, biological, demographic, environmental, archeological, political, legal, military, economic, financial, social, psychological, linguistic, literary, educational, medical, religious and cultural dimension of a country are inseparably intertwined, though analytically distinguishable. The *points of view* may and for practical reasons should be different, but the subject being considered (country) is a whole—not a "*Gesamtheit*", but a "*Ganzheit*". Ontology should not get mixed up with epistemology.

The School of Oriental and African Studies, in London, Oxford's School of Global and Area Studies, the Institute for Area Studies, in Leiden, the National Institute for Oriental Languages and Civilizations, in Paris, the French School of the Far East, in Paris, the German Institute of Global and Area Studies, in Hamburg, the Centre for Modern East Asian Studies, in Göttingen, the Center for East and South-East Asian Studies, in Lund, the Institute of Oriental Studies of the Russian Academy of Sciences, in Moscow, and the Japan Consortium for Area Studies, in Tokyo, are some of the institutions entirely dedicated to area or country studies. Dedicated to the study of Africa, China, Latin America, the Middle East, Japan, Russia and Eastern Europe, and South Asia, the School of Global and Area Studies at the University of Oxford even boasts "the largest community of area studies scholars anywhere in the world". Since 1952, Brill publishes the *Handbook of Oriental Studies.*[1] Since 1956, Gallimard publishes, under the auspices of UNESCO, the series *Connaissance de l'Orient*. The European Alliance for Asian Studies, a co-operative platform of European institutions specializing in the study of Asia, was established in 1997 and aims to "bring together the varied competences and expertise on Asia and Asian Studies found in Europe". Claiming "to strengthen this rich and unique tapestry of academic endeavor found in the region", the Alliance "facilitates a multilevel communication channel between its members, while it sets out to raise public awareness on the scholarly excellence they offer to the benefit of their constituencies and national and transnational environments". From its inception in 1998, the biennial International Convention of Asia Scholars (ICAS) has become "the largest gathering of its kind in the world", and ICAS 11, whose much-trumpeted academic program was "bigger than ever" (including "close to 550 panel sessions, roundtables and book/Ph.D. presentations"), was held in July 1919. In 2016, the three-year research project Rethinking Asian Studies in a Global Context, funded by the Andrew W. Mellon Foundation and coordinated by the International Institute for Asian Studies, was "successfully completed", but, my unsolicited advice notwithstanding, there had been no *fundamental* rethinking.

Over the past few decades, there have been heated debates about area or country studies.[2] However, the critical participants in the discussion have not been critical enough and the defendants have (deliberately?) ignored the elephant in the room, lacking in "*Reflexion auf eigenes Tun*". Neither party has gone to the very heart of the matter. Neither the plaintiffs nor the defendants have raised the fundamental question of the scientific status of area/country studies, a question no insider seemingly dares or wants to burn his/her fingers on. The crucial difference between *explanandum* and *explanans* is not mentioned. Even more remarkably, none of them have pointed out the enormous potential of systems and computer science to realize genuinely interdisciplinary and truly international area or country research. Admitting the importance of multidisciplinary research at best, they do not see the difference between multi and interdisciplinary research, between juxtaposition and composition. They look at the parts but forget about the whole (or the other way around), not realizing that (*a*) parts that do not belong to a whole are not parts at all and (*b*) a whole without parts is not a whole. There are very few if any area/country studies which are truly international. Ethnocentrism is extremely difficult to eradicate.

The area/country debate has been actively conducted or passively followed by academics affiliated with an association related to area/country studies. For a list, which is by no means exhaustive, see Appendix C.

To justifiably reject the work of a single so-called area/country expert would require the correct quoting of every statement made in it and thought by me to be scientifically wanting. This would be a difficult but not impossible job if he/she is not a prolific writer. However, to fairly reject the work of thousands of practitioners of area/country studies would require the correct quoting of every written statement made by them and considered by me to be scientifically unacceptable, a task nobody can be reasonably asked to carry out. My thesis that so-called area or country experts have no theory of their own (distinctive *explanans*) and consequently cannot be considered scientists (even if the word scientist is taken *sensu lato*, i.e., in the sense of *Wissenschaftler*) bears thus the character of a null hypothesis; it is not proved but is possibly disproved. In other words, the thesis advanced in this book stands as long as it has not been refuted. My claim can, *pace* Popper, be upheld so long as it has not been rebutted, repudiated or overthrown by mentioning the name of only one, single area/country scholar who really has a distinctive *explanans* related to the area or country he/she is paying attention to, that is to say, an *explanans* which is compatible with a general, well-established country theory. So rather than addressing numerous debatable issues discussed in the literature, I will try to justify my claim that whoever is called, or advertises himself or herself as, area/country expert is in fact a pseudo-scientist, who does not belong in a university, research institute or think tank. In this book, I attempt to break new ground in a field of academic research where, apart from those who fancy themselves area or country experts *tout court*, social scientists are rubbing shoulder with human scientists but fail to reach consensus on the right country approach. Whether I succeed in doing so is up to the unbiased and fair reader to judge.

The book is to be subsumed under the category of philosophy of science. To get an idea about its logical structure, the reader is advised to see the abstract at the beginning of every chapter. The features of the book, the chapters of which are followed by three indispensable appendices, are:

- Readability
  Abstruse language, jargon and vagueness of expression are avoided. Wherever needed a clear definition is given.
- Coherence
  Special attention is paid to the logic and coherence of the book. The occasional meandering serves its main message.
- Relevance
  At universities, research institutes and think tanks across the world, tens of thousands of scholars are lecturing on a certain country. The book presents a challenge they cannot afford not to take on. The study of countries is pre-eminently a field of research which complex systems scientists should be concerned with. It would be the consummation of their ambition, because countries, nonlinearly related to other countries, are complex systems *par excellence*. The book will be relevant as long as "country experts" refuse to quit their habit of writing and lecturing about a country without being thoroughly schooled in the discipline applicable to the field they cover.
- Timeliness
  The staggering *Afghanistan Papers* (published by *The Washington Post*) and the announced withdrawal of American troops are proof of it that a wrong policy based on poor country scholarship can cause widespread sufferings and bring about great costs.
- Richness of information
  When the book is well read, it will provide much food for thought.
- Originality
  The approach to the study of countries is innovative. Criticizing current country studies and suggesting a new way to conduct future ones, I am not repeating or paraphrasing what has already been said or written. By calling for *computer-aidedcross-disciplinaryandinternationalcountry studies based on*

systems science, I attempt to create order, "to put Humpty together again", so to speak. In close conjunction with advancing the book's main theme, I emphasize the importance of the difference between the Western (dichotomizing, excluding, separating) and Chinese (complementary, including, unifying) way of thought. A unique attempt is made to combine them and to forge a bridge between the natural and cultural sciences.

Overall, I do not confine myself to a scathing attack on current country studies. By taking the step from the real to the ideal and by changing from the critical to the constructive, I think I added the sweet to the sour, trying to serve a more palatable dish. If the reader has a tight agenda, he/she can suffice with reading the main text, but to the consecutively numbered "Notes and References" at the chapter ends will then not be given due attention. Since the book covers an extraordinary wide field of academic research and education (studies of Africa, America [USA], Canada, Latin America, the Caribbean, the Middle East, East Asia, Central Asia, South Asia, Southeast Asia, Australia, the Slavic world, and Europe), I cannot deal with all the individual cases and would like to recall the research method justified above. To some cases, however, I do call attention by way of illustrating. In Chap. 12, the reader will find a *summa summarum*.

It is important to make a distinction between science and *philosophy* of science; confusion of the two results in misunderstanding and miscommunication. In the fourth edition of his standard work, *A Historical Introduction to the Philosophy of Science* (Oxford University Press, 2001, pp. 1–4), John Losee sketches four points of view on philosophy of science. The fourth viewpoint, which he adopted (and I adopt in this book), is that philosophy of science is a second-order discipline. That is to say, a distinction is to be made between doing science and thinking about how science is, or ought to be, done. The subject matter of philosophy of science is the procedure, structure, foundations, concepts and implications of science, which is supposed to start from "facts". Since science is divided into natural, social, human, and formal science (and subsequently into physics, chemistry, astronomy, earth science, botany, zoology, ecology, demography, linguistics, geography, anthropology, sociology, psychology, political science, law, economics, medicine, education, literary science, science of religion, science of art, mathematics, statistics, and computer science [the list is by no means exhaustive]), philosophy of science is usually classified under philosophy of physical, biological, social, human, or formal science. The foundations of these philosophies are in turn subjects of critical investigation.[3] Many books have been written on the different sciences, but what science precisely is remains a mystery. The English word science (French *science*, German *Wissenschaft*, Spanish *ciencia*, Italian *scienze*) is cognate with the word conscience (French *conscience*, German *Gewissen*, Spanish *conciencia*, Italian *coscienza*). Though the word science has thus a moral/social connotation, this is often forgotten or (purposely?) neglected.

If Imre Lakatos (1922–1974), a Hungarian philosopher of science known for his thesis on the fallibility of mathematics, was right in saying that philosophy of science without history of science is empty, and history of science without philosophy of science is blind, his words should be extended by including that philosophy and history of science are always conditioned, i.e., deeply rooted in a society soaked with culture. Sociology of scientific knowledge (SSK), which has become the main theme of postmodernity,[4] is also an important object of study, "context" being its core concept. SSK raises the question as to whether or not reality exists independently of our consciousness or conceptual scheme of it. In other words, sociology of science poses the perennial problem of the relationship between epistemology and ontology.[5] "Reality" is an elusive concept. So the nature of knowledge (science) is unknown.[6] For the vexed problems are: knowledge of what and is it then real knowledge? Since the issue of the extent and limits of human knowledge is a perplexing one,[7] the argument set forth in this book is essentially speculative, not based on "hard facts". This statement is not meant to be a disclaimer. I do not intend to shirk my responsibility for the statements made in the book, but in the end we all have to acknowledge that we groping around in the darkness and that our knowledge, however impressive, is and will always remain severely restricted.

The country to be investigated can be outlined by a dimensional lay-out, representing its articulated or detailed geological, geographic, paleontological, biological, demographic, environmental, archeological, political, legal, military, economic, financial, social, psychological, linguistic, literary, educational, medical and cultural dimension. The well-managed team of disciplinarily collaborating, internationally recruited and highly seasoned researchers should bear in mind that (*a*) the country dimensions are closely interrelated aspects or parts of the whole country, (*b*) cross-related happenings can take place along every dimension and (*c*) events and actions are fundamentally different happenings (for explanation, see below).

If whatever happens on or between the dimensions would be measurable, vector analysis could be applied to the study of a country. Unfortunately, that is not the case. Measuring, or counting, is only possible in regard to things that can be measured, or counted. In close conjunction with advancing the main theme of the book (a plea for computer-aided cross-disciplinary and international country studies based on the science of complex systems), I will argue that nature and culture are intimately interrelated, for which reason the sciences of nature cannot ignore the studies of culture and the latter should be trying hard to be as exact as the former. Though it tends to be forgotten, both the sciences of nature and the studies of culture begin with, and are based on, *experience*. Without empirical research the sciences of nature are nothing but speculations and the studies of culture are essentially studies of experience expressed, of the countless manifestations of the immeasurable human mind.

Writing this book, I took my cue from Kwang-chih Chang (commonly know an as K. C. Chang), a Taiwanese-American archeologist who introduced modern, Western methods of archeology to the study of ancient Chinese history. In *Kwang-chih Chang 1931–2001: A Biographical Memoir* (The National Academy of Sciences, 2012), Robert Murowchick writes:

> The emphasis on collaboration—between disciplines and between countries—was the hallmark of K. C.'s work throughout his life. (p. 36)

Notes and References

1. For "oriental studies" or "orientalism", terms smacking of the era of colonialism and European expansion, search the Internet. Also go to http://www.orientalismstudies.com (> bibliographies) and explore http://orientalismo.blogspot.com. In addition, see Donald Lach, *Asia in the Making of Europe* (Chicago University Press, 1965–1993), Joachim Gentz, *Keywords Re-Oriented* (Universitätsverlag Göttingen, 2009, pp. 41–47), David Mungello, *The Great Encounter of China and the West, 1500–1800* (Rowman & Littlefield, 2013), Riley Quinn, *An Analysis of Edward Said's Orientalism* (Macat, 2017), Dawid Rogacz, "The Birth of Enlightenment Secularism: From the Spirit of Confucianism", *Asian Philosophy*, 28:1 (2018), pp. 68–83, Laura De Giorgi and Federico Greselin (eds.), *150 Years of Oriental Studies at Ca'Foscari* (Ca'Foscari, 2018), André Bueno et al., *Orientalismo Conectado* (Sobreontens, 2019) and *Korea Observer*, 50:2 (2019), pp. 287–303. The University of Naples "*L'Orientale*" and the School of Oriental and African Studies (SOAS), in London, stick to their guns, and Brill (Leiden), Vandenhoeck & Ruprecht (Göttingen) and Harrassowitz (Wiesbaden) are some of the publishers still bringing out oriental books. The flawed core dogma of orientalism, typical example of the Enlightenment's dark side, is that East and West are to be measured on different scales.
2. See Neil Waters (ed.), *Beyond the Area Studies Wars* (University Press of New England, 2001), Masao Miyoshi and Harry Harootunian (eds.), *Learning Places: The Afterlives of Area Studies* (Duke University Press, 2002), Walter Mignolo, *The Idea of Latin America* (Blackwell, 2005), William Moseley, "Area Studies in a Global Context", *The Chronicle of Higher Education*, November 29, 2009, Lesley Pitman, *Supporing Research in Area Studies* (Chandos, 2015), Guido Franzinetti, "The Strange Death of Area Studies and the Normative Turn", *Quaderni Storici*, December 2015, pp. 834-872, Valentina Pedone and Ikuko Sagiyama (eds.), *Transcending Borders* (Florence University Press, 2016), Ivan Kurilla and Victoria Zhuravleva (eds.), *Russian/Soviet Studies in the United States, Amerikanistaka in Russia* (Lexington, 2016), Edith Clowes and Shelly Bromberg (eds.), *Area Studies in the Global Age* (Northern Illinois University Press, 2016), Torsten Loschke, *Area Studies Revisited* (V&R, 2017), Katja Mielke and Anna-Katharina Hornidge (eds.), *Area Studies at the Crossroads* (Palgrave Macmillan, 2017) and Ariel Ahram et al. (eds.), *Comparative Area Studies* (OUP, 2018).
3. See the *Minnesota Studies in the Philosophy of Science* (1956ff), *the Boston Studies in the Philosophy of Science* (1963ff), the series *Handbook of the Philosophy of Science* (2006ff), Stathis Psillos and Martin Curd (eds.), *Companion to Philosophy of Science* (Routledge, 2013), Matthias Unterhuber et al., "Philosophy of Science in Germany, 1992–2012", *Journal for General Philosophy of Science*, 45:1 (October 2014), pp. 71–160, Paul Humphreys (ed.), *Handbook of Philosophy of Science* (OUP, 2016), Thomas Hickey, *Twentieth-Century of Philosophy of* Science (Thomas Hickey, 2016), Anouk Baberousse et al. (eds.), *The Philosophy of Science: A Companion* (OUP, 2018) and Wolfgang Balzer and Karl Brendel, *Theorie der Wissenschaften* (Springer, 2019). The journal *Philosophy of Science* (1934ff) is sponsored by the Philosophy of Science Association at Baltimore's University (MD). Visit https://philpapers.org/browse/general-philosophy-of-science. The German philosopher Jürgen Mittelstraß is editor of the eight-volume *Enzyklopädie Philosophie und Wissenschaftstheorie* (Springer, 2005–2018). Philosophy of science interfaces with the formal sciences (see Chap. 5, Under Systems Science).
4. See Perry Anderson, *The Origins of Postmodernity* (Verso, 1998) and Lawrence Cahoone (ed.), *From Modernism to Postmodernism: An Anthology* (Blackwell, 2003). Barry Barnes, David Bloor, Sal Restivo, Randall Collins, Gaston Bachelard, Harry Collins, Karin Knorr Cetina, Paul Feyerabend, Steve Fuller, Martin Kusch, Bruno Latour, Mike Mulkay, Derek J. de Solla Price, Lucy Suchman and Anselm Strauss have theorized about SSK.
5. See Peter Dybjer et al., *Epistemology versus Ontology* (Springer, 2012).
6. See Duncan Pritchard et al., *The Nature and Value of Knowledge* (OUP, 2010) and Kevin McCain, *The Nature of Scientific Knowledge* (Springer, 2016).
7. See Nicholas Rescher, *The Limits of Science* (University of Pittsburgh Press, 1999) and Marcelo Gleiser, *The Island of Knowledge* (Basic Books, 2014).

Contents

Chapter 1
The Enduring Confusion

Abstract The enduring confusion about area or country studies is set out, and the fundamental problem dealt with in the book is described. The often misused word expert is explained, and a distinction is made between "country experts" *tout court* (mainly found In Europe) and scientists who, interested in a country, cannot resist overstepping the bounderies of their discipline (largely found in the U.S.). Country studies are not the same as cultural studies. In this chapter, bold claims are made, but they are backed up with data.

"Area study" refers to a field of scholarship pertaining to a particular geographical, national or cultural region. The term is a general description of many fields of research. It tends to be overlooked or glossed over, but "area/country experts" such as Africanists, Americanists, Russianists, Sinologists, Arabists, Hispanists and Mongolists have the ingrained habit of trespassing on fields of study belonging to social or other kinds of science. Since the main idea advanced and developed in this book hinges on the meaning of "expert", I offer here my understanding of the word expert.

An expert (from *expertus* [sound, valid, adequate]) has broad and deep competence in terms of knowledge or experience in a certain field. Particularly, a *scientific* expert has broad and deep competence in terms of knowledge in a certain field of research (discipline). I do not know of anyone having this kind of competence in research fields such as study of the Middle East, Europe, Africa, Latin America, Southeast Asia, the USA, Russia, China, India and Japan. However, there is a myriad of "area/country experts" affiliated with universities, research institutes or think tanks across the world. A scientific expert is thus, first and foremost, educated and disciplined. He or she has acquired well-organized knowledge. A person knowing all kinds of things in a particular field or in regard to a country is not an expert yet. What he/she misses is scientifically based, well-grounded knowledge. If knowing all kinds of things concerning a country would be a sufficient condition for being rightly called country expert, every visitor of the place would qualify.

An America-expert (Americanist), for example, should have broad and deep competence in terms of well-organized knowledge regarding the USA. In other words, he should not only be knowledgeable about the Great Depression, New

H. Kuijper, *Comprehending the Complexity of Countries*,
https://doi.org/10.1007/978-981-16-4709-3_1

Deal, Intel, IBM, Microsoft, Facebook, AT&T, HP, Ford, GE, GM, Yahoo, Twitter, Uber, Airbnb, Netflix, Tesla, ExxonMobil, Greyhound, Walmart, Colgate-Palmolive, Johnson & Johnson, Procter & Gamble, Boeing, Lockheed Martin, McDonald's, Burger King, KFC, Coca-Cola, Pepsi, Starbucks, Disney, Nike, Sears, Chevron, Pfizer, the Indian removal, John Wayne, Afro-Americans, Latino-Americans, New York, Chicago, New Orleans, San Francisco, Los Angeles, Detroit, Houston, Dust Bowl, Rust Belt, Bible Belt, Las Vegas, show business, ABC, NBC, CNN, MTV, Broadway, Hollywood, Tom Hanks, Brad Pitt, Al Pacino, Tom Cruise, Nicolas Cage, jazz, the Great Lakes, the National Parks, the Grand Canyon, Kennedy Space Center, civil rights movement, gun politics, capital punishment, federalism, slavery, racism, Republican Party, Democratic Party, elections, constitution, presidential powers, Mount Rushmore, the White House, George Washington, Thomas Jefferson, Theodore Roosevelt, Abraham Lincoln, Franklin Roosevelt, Woodrow Wilson, Dwight Eisenhower, John F. Kennedy, Watergate, Ronald Reagan, George Marshall, Norman Schwarzkopf, Thomas Edison, Alexander Graham Bell, Henry Ford, the Getty family, Edgar Allan Poe, Walt Whitman, Emily Dickinson, Fenimore Cooper, Herman Melville, Mark Twain, Thomas Paine, Henry James, Ezra Pound, Eugene P'Neill, Tennessee Williams, Arthur Miller, Theodore Dreiser, Sinclair Lewis, Dos Passos, F. Scott Fitzgerald, William Faulkner, Ernest Hemingway, John Steinbeck, Kurt Vonnegut, Norman Mailer, John Updike, Jerry Salinger, Saul Bellow, Tom Wolfe, Philip Roth, John Dewey, William James, C.S. Peirce, John Rawls, Robert Nozick, Allan Bloom, Daniel Bell, Lionel Trilling, Walter Lippmann, Noam Chomsky, Thomas Nagel, Oliver Wendell Holmes, Ronald Dworkin, Joel Feinberg, Saul Kripke, Richard Rorty, Hilary Putnam, Frank Lloyd Wright, Lewis Mumford, Leonard Bernstein, George Balanchine, Martha Graham, Elvis Presley, Dolly Parton, Jim Reeves, John Denver, Benny Goodman, Glenn Miller, Louis Armstrong, Duke Ellington, Nat King Cole, Mahalia Jackson, Frank Sinatra, Beyoncé, Madonna, Whitney Houston, Michael Jackson, Prince, James Brown, Oprah Winfrey, Bobby Fischer, Pete Sampras, Muhammad Ali, Michael Jordan, Tiger Woods, Michael Phelps, the Founding Fathers, Thomas Edison, FBI, CIA, Rockefeller, Carnegie, Wall Street, JPMorgan Chase, the Citigroup, Pentagon, NASA, skyscrapers, health care, higher education, the MET, the MoMA, the Guggenheim Museum, the LOC, the Mob, the G.I.s, the Ivy League, Science and technology, Silicon Valley, Neil Armstrong, Paul Samuelson, Milton Friedman, James Buchanan, Paul Krugman, transportation, pragmatism, the American Revolution, the Civil War, 9/11, baseball, apple pie, the American Nobel Prize laureates and a host of other topics related to the geology, geography, demography, economy, society, polity, military, technology, history or culture of the USA. He should also have *organized* knowledge of the subjects, that is to say, knowledge grounded in, and firmly based on, science. He should be able to put all these items into a unified scientific framework, to see the pattern of things, to factor in all the actors and factors that are relevant in relation to the USA. In other words, he should have a tested or testable *theory* about the country. If he is unable to develop such a theory, he is but a cabinet of curiosities, a storehouse of unconnected issues, just a repository of interesting things, in brief, he is nothing but a toppled-over bookcase about the USA. So-called America-experts, whose number multiplies when presidential elections are forthcoming, easily deflate, because, wanting to dazzle their audience, they are eager to disclose what they know about the USA. Sorely lacking in theoretical foundation and empirical validation, however, they do not really understand what is happening in the country. That is why Americanists, Russianists, Iranologists, Indologists, Koreanologists, Japanologists, Sinologists and other kindred "experts" have difficulty in finding shelter in higher education, the difficulty being glaring testimony of their dubious scientific status.

The word science has two related meanings to be distinguished: (*a*) a subjective and (*b*) an objective meaning. *Subjectively*, science is organized knowledge. It is not only knowledge (a subject with which epistemologists have been concerned); it is also knowledge that is organized or systematized. Subjectively, science is not only a mental ability; it is also the skill to engage in or carry out the scientific activity in an ordered, systematic way. Knowing instead of believing something is thus not enough; it should also be the ability to see something in concection with other things, to put something into perspective, to understand not only the scheme or pattern of things, but also to see the *way* of things, i.e. how they are changing, unceasingly. *Objectively*, science is a system of statements made by scientists, the reflection of science in subjective sense. Objectively, science is largely the result of the scientific activity of many people, ideally the scientific activity of a group of interdisciplinarily and internationally collaborating people. Think of a collective and well-structured scientific report on the economy of, or environmental situation in, a country. Though the ongoing discussions about knowledge and science are related, they should be distinguished from each other, because science is a special kind of knowledge. Philosophers of science are concerned with the foundations, structure, concepts, diversity, methods, social relevance and technical implementation of science. See Stathis Psillos and Martin Curd (eds.), *Companion to Philosophy of Science* (Routledge 2013) or Dov Gabbay et al. (eds.), *Handbook of the Philosophy of Science* (Elsevier, 2006ff), "a comprehensive, multi-volume research collection with contributions from leading figures in all areas of the philosophy of science". Visit https://www.iep.utm.edu/category/s-l-m/science and see note 11 in Chap. 3. Lists of epistemologists and philosophers of science can be found on the Internet.

Area/country study programs in (mainly) Europe cover a bit of geography, demography, history, language, literature, art, religion, politics, law, economics, sociology, anthropology, archeology, psychology, education, science and technology and philosophy, inasmuch as the subjects of these disciplines are thought to be relevant to the region or country concerned. Such programs lead the graduate to be "*un touche à tout sans profondeur*" (someone who has only a smattering of knowledge about whatever is brought up in regard to the area or country), for the courses given are targeted at students who do not want to be disciplined in a specific scientific, or philosophical, domain but seek to be knowledgeable about everything concerning the area or country of their predilection, i.e., at the student who does not want to become expert in a particular discipline, and yet—after graduation—considers him/herself to be, and wants to be introduced as, a country connoisseur or area expert.

In the USA, Canada and Australia—where higher education authorities who realize the harmful consequences of blurring academic borders and stress the advantages of labor division are critical of the policies pursuit by their European counterparts—enrolling students interested in a particular country are required to narrow down their interest and make a discipline choice. However, graduated in, say, economics and publicly asked about their opinion on the country, they are generally quick to leave their home turf (in the present case, to relinquish their job of testing economic theories on, or applying them critically to, the country's economy). They trespass on somebody else's property, without being arrested by the interviewer, who thereby adds to the confusion and misleading of the audience. Doing so, the expert, whose scope of inquiry is restricted and thus not comprehensive, becomes a dabbler; the professional becomes an amateur. At American, Canadian and Australian universities, there are many experts in all kinds of fields of research, but there are no country experts, let alone area specialists. Whoever consents to being introduced as such, on the television or in other media, is wearing no clothes.

For the sake of politeness, I should abstain from mentioning names, confident that a good listener needs no words (*sapienti sat*). For the sake of good order, however, it is important to make a clear distinction from the outset, to cull the herd, so to speak. I restrict myself to the study of China.

On the one hand, there were/are

- anthropologists like Brigitte Baptandier, Davidson Black, Francesca Bray, Myron Cohen, Elisabeth Croll, Stephan Feuchtwang, Maurice Freedman, Susan Greenhalgh, Stevan Harrell, Charlotte Ikels, Andrew Kipnis, Frank Pieke, Jack Potter, Steven Sangren, William Skinner, Leon Stover, Lothar von Falkenhausen, Franz Weidenreich and Arthur Wolf;
- archeologists like Johan Andersson, Robert Bagley, Katheryn Linduff, Robert Thorp, Judith Treistman and Anne Underhill;
- art historians like Anneliese Bulling, James Cahill, Craig Clunas, Max Loehr, Martin Powers, Osvald Sirén, Peter Sturman, Michael Sullivan, Mary Tregear, William Watson, William Willetts and Susan Bush;
- demographers like John Aird, Isabelle Attané, Judith Banister and Irene Taeuber;
- ecologists like Vaclav Smil;
- economists like Loren Brandt, Robert Dernberger, David Dollar, Audrey Donnithorne, Peter Drysdale, Alexander Eckstein, Ross Garnaut, Jane Golley, Robert Gregory, John Gurley, Carsten Herrmann-Pillath, Bert Hofman, Nicholas Lardy, Barry Naughton, Peter Navarro, Dwight Perkins, Louis Putterman, Thomas Rawski, Carl Riskin, Scott Rozelle, Terry Sicular, Dani Rodrik and Hilton Root;
- ethnologists like Marcel Granet and Adeline Herrou;
- geographers like George Cressey, Frank Leeming, Clifton Pannell, Theodore Shabad and Thomas Treager;
- legal experts like William Alford, Maria Carrai, Donald Clarke, Jerome Cohen, Randle Edwards, Jean Escarra, Benjamin Liebman, Stanley Lubman, Robin Munro, Marc van der Valk and Harro von Senger;
- linguists like William Baxter, John DeFrancis, Paul Kratochvil, John McCoy, Jerry Norman, Robert Ramsey and Rint Sybesma;
- paleontologists like Walter Granger and Pierre Teilhard de Chardin;
- philosophers like Philip Ivanhoe, Donald Munro and Bryan Van Norden;

- political scientists like Mary Gallagher, Harry Harding, Sebastian Heilmann, Scott Kennedy, Kenneth Lieberthal, Werner Meissner, Andrew Nathan, Michel Oksenberg, Lawrence Reardon, Gunter Schubert, David Shambauch, Ross Terrill and Luigi Tomba;
- psychologists like Michael Bond;
- sociologists like Deborah Davis, William Parish, Jean-Louis Rocca, Gilbert Rozman, Jonathan Unger, Ezra Vogel, Andrew Walder and Martin Whyte.

These people were/are scientists or philosophers specialized in studying the aspect of China related to their discipline. Occasionally and unfortunately, however, they were/are amateurs by overstepping the boundaries of their discipline, wrongly believing that being highly educated means being knowledgeable about everything. As if an economist has expertise in the field of immune diseases, an ecologist could replace a criminologist, a biochemist would be able to take over the job of a historian of technology, an anthropologist is disciplined in literary studies, or a linguist is qualified to talk about demographic transition. Forgetting that a virologist knows next to nothing about a country's macroeconomy or jurisprudence is the general tendency among academics engaged with China (Taiwan or the Prople's Republic of China). Surely, all things hang together and form one system, but this does not mean that no distinctions can and have to be made. The complexity of a country is beyond anybody's comprehension. Scientific collaboration is absolutely required to comprehend this complexity.

There is another *caveat*. In *China's 40 Years of Reform and Development, 1978–2018*, a collection of essays from scholars around the world edited by Ross Garnaut, Ligang Song and Cai Fang and published by the Australian National University (ANU) Press, in July 2018, "not a single chapter broaches the isues of corruption or asset-stripping, the creation of an oligarchy, the linkages between 'private' sector and the Communist Party, the fortunes held abroad by PRC leaders, the profiteering that was part and parcel of the dual-track price system, the use of China's economic prowess for political purposes abroad, the social credit system and its far-reaching consequences, the censorship of the internet, the behavior of the petroleum gang and its impact on the energy sector, environmental pollution and censorship of those who talk about it too loudly (see the film *Under the Dome* by Chai Jing), the arrest of labor activists and lawyers, and the economic inequality and its implication" (book reviewer Carsten Holz's critical remarks in *The China Journal* [January 2020, pp. 179-189]). Noticeable: the volume (in the ANU book series *China Update*) was published *before* the Wuhan coronavirus outbreak, which exposed the weakness of political systems and caused stock markets to plummet all over the world. I concur with virologist Jaap Goudsmit, who thinks that the war against COVID-19 cannot be won without interdisciplinary and international research. Visit https://council.science/covid19.

On the other hand, and I am confining myself again to Western academics interested in China, there were/are countless people calling themselves or wanting to be introduced as, Sinologists.[1] They were/are not thoroughly schooled in any scientific discipline whatsoever (in chapter eight, I will express my doubts as to whether historians are scientists). Prone to escapism and obscurantism, but eager to be part of things, they were/are dabbling in many skills, rather than gaining expertise by focusing their mind on one. Like butterflies, they were/are fluttering around in the vast garden of science, staying at one place or another according to their whims and fancies, or to the issues of the day. Hooked by the country, they were/are caught by the "China bug". Even if they did/do focus their mind on a particular aspect of the country, they were/are doing so without being well-grounded in the basics of, or having a Ph.D. degree in, the discipline concerned, apparently unaware of, or not embarrassed by, the richness of the geological, biological, geographical, demographic, archeological, political, jurisprudential, military, economic, financial, social, anthropological, psychological, educational, technological, medical, linguistic, literary, religious, cultural, ecological or philosophical domain.

The scholarly output of China-connoisseurs is staggering. It is piled up and gathering dust in big libraries, the collections of which are so large that one better browse them by riding a bicycle. *The Cambridge History of China*, mainly written by more than hundred "leading scholars" in the Anglo-Saxon world (almost none of whom was/is schooled in a scientific discipline), counts fifteen hefty volumes, two of which published in two books (see Michael Loewe and Edward Shaughnessy [eds.], *The Cambridge History of Ancient China* [CUP, 1999, pp. 1–7]). However, thus far "China experts" have not been able to pull their resources together and to provide a professional, integral and annotated translation of China's Standard Histories (*zhèngshǐ*). Only a tiny fraction of (*a*) the Chinese Classics and the indispensable commentaries on them (*shísānjīng zhùshū*), (*b*) the treasure trove of the Collected Philosophers (*zhūzǐjíchèng*), (*c*) the Confucian Canon (*rúzàng*), (*d*) the Daoist Canon (*dàozàng*), (*e*) the Buddhist Canon (*dà zàngjīng*) and (*f*) the Local/Provincial Monographs (*dìfāngzhì*) has been translated into English — after more than 400 years of Western Sinology and with countless Chinese scholars being available! It is a shame that the Great Books of the Chinese World, comparable to the series of Great Books of the Western World (chief-edited by Mortimer Adler and published by Encyclopædia Britannica, Inc.), still cry out to be professionally translated, annotated, edited and published, and to be supplemented with a "syntopicon" (collection of topics). It would make systematic comparison between the Western and Chinese ways of thought much easier, an undertaking ugently needed in the Age of Globalization. Geoffrey Lloyd, "senior scholar in residence" at the famous Needham Research Institute, in Cambridge, seems to have been doing this in *Ancient Greece and China Compared* (CUP, 1918), but he claims to be knowledgeable not only about one, but about two countries.

In order to avoid any misunderstanding, I leave the countless pro- or anti-China columnists, commentators, journalists or reporters employed by news agencies or television broadcasting stations out of account, as they are below category. I leave also the numerous Chinese, Korean or Japanese "China experts" out of account, as this book is primarily addressed to *Western* "country experts". Being fluent in Chinese (certainly not a trifle when one is not a native speaker) does not mean having mastered a discipline (which is difficult enough), let alone several disciplines (which is absolutely impossible). To believe that language mastery qualifies for being scientist is the big hoax at centers for East Asian Studies. These centers are often maintained and reinforced by lavish Chinese grants and the concomitant employment of cohorts of native speakers as language instructors, who are *assumed* to be loyal to their host country and not to have a hidden agenda. The country given attention to may be different, but the distinction being made (between professionals and amateurs) remains the same.

The growing presence of Chinese on the campus of American, Canadian, European and Australian universities should be a cause of concern. Apart from the proliferation of so-called Confucius Institutes (a subject on which *éminence grise* Marshall Sahlins wrote a highly commendable booklet) and the relentless implementation of Xi Jinping's "Belt and Road Initiative" (see below), that presence provides ample proof of China's growing soft power—a power which "open societies" do not really know how to counterbalance. Count the Chinese affiliated with the Columbia-Harvard China and the World Program and read the online article "How Should Universities Respond to China's Growing Presence on Their Campuses?".

The way of finding out whether, say, Sinologists really are what they (tacitly) pretend to be (experts on China) is to make inquiries about how comfortable they are with quantitative reasoning and information technology, about their familiarity with the "mixed methods research", about their nomenclature (the key terms in their debate), about the property of the relations between their basic concepts, about their scientific model of China, about the underlying, unstated assumptions of their argument, about the kind and number of successfully tested hypotheses framed in their line of work, about the *Grundriss*, *Grundstein* or *Gipfel* of their conceptual *Gebäude*, about the core subject (*problématique*) of their profession, about the landmarks or milestones in the history of Sinology or about the central point that assures its unity.

Like other "country experts", "China experts" have a keen eye for details but do not let them speak as parts of a whole, as biologists do when they put rice, bamboo, pakchoi, soya, garlic, jasmine, litchi, peony or mulberry in the category of angiosperms. "China experts" take keenness wrongly for profundity. Ignoring the good advice of Pliny the Younger: "*non multa sed multum*" (not many but much), there is multitude rather than plenitude in their writings. To be convinced of it, the reader should browse through *T'oung Pao*, "the leading scholarly journal on all aspects of traditional China", or *The China Quarterly*, "the leading scholarly journal in its field, covering all aspects of contemporary China including Taiwan". "Its interdisciplinary approach", the editorial board of the SOAS publication boasts, "covers a range of subjects including anthropology/sociology, literature and the arts, business/economics, geography, history, international affairs, law and politics". "China experts" do not provide us with insight into the coherence and depth of their study object. They afford us no view of the country's architecture. The beat, barcode, style or idiom of it are not given. *Běn bù guàn yú mò; yī bù xiǎn yú duō* (the root is not connected with the twigs; the one is not shown in the many). Indeed, the *plures* presented are insignificant so long as the *unum* is elusive, for it is only against the background of the general that the specific acquires any meaning. Having no theory of their own and mistaking the cramming of facts for taste and discernment in selecting the important ones, "country experts", however, they may be defined or classified, do not deserve their intellectual status.

Geologists use words such as "syncline", "outcrop", "uplift", "crust", "shearing", "fault", "slump", "isostasy", "subduction", "cyclothem", "erosion" "varve", "cuesta", "guyot" and "hot spot". Linguists are familiar with words like "phoneme", "syllable", "discourse", "syntax", "lexicon", "turn", "copula", "performative", "segment", "parataxis", "morpheme", "subjunctive", "clause", "triphthong" and "glottal stop". Economists communicate with each other in terms like "opportunity costs", "indifference curve", "managed float", "crowding out", "returns to scale", "invisible hand", "adjusted present value", "externalities", "dumping", "versioning", "inflation", "employment", "production function", "current account" and "purchasing power parity". Ask Turkologists. Iranologists, Russianists, Sinologists, Japanologists or Koreanologists for a glossary (list of technical terms used in country studies), and the justifiable and legitimate request cannot be granted, because they do not have a theoretically determinate set of designations or categories that are particular of, and specific to, the study of any country. In short, "country experts" are nothing but unreliable talking heads, unreliable because they never tell from which discipline they are borrowing and whether/how the borrowed theory has been tested.

There are different plant groups: archaeplastida, bryophytes, non-vascular plants, vascular plants, spermatophytes, pteridophytes, gymnosperms and angiosperms. Their studies are subsumed under botany, a branch of biology. There are 60 different groups of animals: apes, bears, bees, camels, cats, crocodiles, elephants, falcons, kangaroos, otters, pigs, sharks, snakes, swans, whales, wolves, etc. Their studies are classified under zoology, another branch of biology that studies the structure, evolution and habits of animals and their interactions with the ecosystem in which they live. Human beings can be differently classified. It depends on the main characteristic a researcher wants to focus his/her mind on. Anthropology is the study of man, the building block of any social or human science. It has branched off into everal subdisiplines, spilling over into sociology, the discipline anthropology is closely related to, but jealously kept separated from. Social anthropology is concerned with the ways in which people live together, economic anthropology with the ways in which people who live together make a living, political anthropology with the ways in which people who live together are legally related to each other, linguistic anthropology with the ways in which people in a community communicate with each other and cultural anthropology with the norms, values, aspirations and expectations prevalent in a community or society. Biological anthropology (aka physical anthropology) is about the physical traits and development of human beings, which are taken to be a particular kind of animals. Botany, zoology and anthropology have competing theories by which plants, animals and *ἄνθρωποι*, respectively, are explained. The study of no country, however, has a *country*-theory. That is why country studies are fundamentally flawed.

Much more interesting things could be written on, for example, the concept or practice of law in China, if the authors were well up in the writings of Plato, Cicero, Aquinas, Althusius, Grotius, Hobbes, Montesquieu, Jeremy Bentham, John Austin, Oliver Wendell Holmes, François Gény, Gustav Radbruch, Hans Kelsen, Herman Dooyeweerd, Lon Fuller, Harold Berman, John Rawls, Joel Feinberg, Joseph Raz, Richard Posner and Ernest Weinrib. Similarly, books or articles about Chinese art would gain in importance, if the writers thereof were acquainted with the aesthetic views of Plato, Aristotle, Brunelleschi, Baumgarten, Winckelmann, Kant, Lessing, Schiller, Hegel, John Ruskin, Heinrich Wölfflin, Benedetto Croce, Clive Bell, Walter Benjamin, Roman Ingarden, Susanne Langer, Mikel Dufrenne, Monroe Beardsley, Richard Wollheim, Stanley Cavell and Arthur Danto. Even if the author of books or articles about "Chinese law" or "Chinese art" were well-informed about the subject, he/she would be, respectively, a *juridical* and *artistic* expert paying his/her professional attention to law, or art, in China—but certainly not a *China* expert or Sinologist.

In 1990, Sidgwick and Jackson, in London, published a book written by Michael Loewe. It is entitled *The Pride that was China.* The chapters of the book are on "the gifts of nature and their problems", "kingdoms and empires", "man and his neighbours", "the spoken and the written word", "beliefs, hopes and fears", "the organs of government", "literature and scholarship", "tombs and their treasures", "capital cities", "the arts", "commercial practice" and "the advancement of science and its applications". Notes for further reading are given on pp. 181–296. Unfortunately, Professor Loewe, who was President of the European Association for Chinese Studies, is not a geologist, geographer, archeologist, demographer, political scientist, anthropologist, sociologist, linguist, literary scientist, economist or art historian. Science and technology (S&T) is a vast field of research he cannot but acknowledge that he is only an amateur in. It encompasses the natural, agricultural, medical and social sciences, the humanities and engineering and technology. Loewe's book is, therefore, interesting but not scientifically compelling.

A thorough study of the modernization of China is arguably the most important research project that scientists could possibly embark upon. In his 876 pages counting bestseller *The Search for Modern China* (W.W. Norton, 1990), being a survey of the last several hundred years of Chinese history, Jonathan Spence, emeritus professor at Yale University, heroically tried to tackle the problem single-handedly, bypassing the available literature on the subject of modernity and "forgetting" to mention the names of Knight Biggerstaff, Joseph Levenson, Marion Levy, Jack Gray, Rhoads Murphey, Immanuel Hsü, Kung-chuan Hsiao, Richard Baum, Richard Suttmeier and Gilbert Rozman, who had already tried their hand at, but underestimated the difficulty of, the subject of modernity in China. *Mirabile dictu*, Spence, who received eight honorary degrees in America, Great Britain and China, did not discuss the issue of modernity that in particular Robert Palmer (1909–2002) and Joel Colton (1918–2011) had treated in their landmark book *A History of the Modern World* (first edition in 1950, thirteenth edition in 2017). The lemmas "enlightenment", "counter-enlightenment", "modernism", "post-modernism", "modernity", "post-modernity", "modernization", "Renaissance", "Humanism", "Reformation", "state", "nation-state", "John Locke", "liberty", "liberalism", "equality", "justice", "John Rawls", "citizenship", "civil society", "privacy", "property", "(human) rights", "rule of law", "*trias politica*", "legitimacy", "(popular) sovereignty", "mass media", "scientific revolution", "industrial revolution", "electric age", "electronic age", "information age", "commercialization", "capitalism", "welfare", "progress", "rationality", "secularization", "emancipation", "estrangement", "alienation", "decadence", "European miracle", "disenchantment", "westernization" and "globalization" are conspicuous by their absence from the index of Spence's blockbuster book. Perhaps most importantly, Spence did not engage in a critical review of Barrington Moore's *Social Origins of Dictatorship and Democracy: Lord and Peasant in the Making of the Modern World* (Penguin, 1967, Chap. IV), John Dunn's *Modern Revolutions: An Introduction to the Analysis of a Political Phenomenon* (CUP, 1972, second edition in 1989) or Theda Skocpol's *States and Social Revolutions: A Comparative Analysis of France, Russia and China* (CUP, 1979).

The Western debate about modernity has become ferociously complex (and so-called China experts know how to spin a yarn from it). The heart of the matter seems to be that modernity is understood as the attempt to find the secular and rational principle of order (see Preface). There is disagreement as to when this attempt began: with the Milesian cosmologists, the Renaissance–Humanism, the Age of Discovery/Explorations, the Protestant Reformation or the rise of the Nation-State? No one disagrees, however, that the revolutions instigated by Copernicus, Galilei, Vesalius, Newton and Darwin were crucial catalysts. Understanding the world as potentially comprehensible, and subject to the human will, these men radically changed the intellectual discourse. Rather than to give the awesome power of directing the fate of individuals to God or the Pope ("God's representative on Earth"), or to a monarch claiming to rule "by the grace of God", they posed the revolutionary question: what if society (always proclaimed and thought to be a great organism) is made up of *individual* organisms, each with the power to shape its own destiny? Society's future would then be what its members decide to make of it.[2] Unfortunately, professor Spence has nothing to say about these important matters in *The Search for Modern China* (W.W. Norton, 1990; the second, expanded edition, appeared in 1999, and the third edition rolled in 2013 from the press). He deems it even unnecessary to make a single reference to them in his well-marketed book. Even in the book's third edition, he fails to mention that a history of modernity denying its metaphysical or theological dimension is superficial, a point powerfully made by Charles Taylor in *A Secular Age* (Harvard University Press, 2007) and Michael Gillepsie in *The Theological Origins of Modernity* (University of Chicago Press, 2009).

The idea of order is of central importance not only in politics, IR (theory of international order), economics (theory of economic order) and organizational studies but also in astronomy (cosmology), physics (entropy) and informatics. It should be the first subject on any agenda for an East–West Dialogue, and it should certainly be dwelt upon in a book about the modernization of China. Under "God's Eye", the phrase "*novus ordo seclorum*" (new order of the ages) appears on the reverse of the Great Seal of the United States (the words "*e pluribus unum*" [out of many, one] appearing on the obverse). In addition, the phrase is printed on each American one-dollar bill. However, what that order should be, how it relates to social contract theories (https://www.iep.utm.edu/soc-cont, https://plato.stanford.edu/entries/contractarianism-contemporary) and what the word God stands for are points worthy to be discussed in a frank and genuine dialogue, without taking a biased position, particularly in a time when we are witnessing the emergence of a New World Order.

Area/country studies have been strongly influenced by, but shall not be confused with, cultural studies, a very broad, extremely varied and highly contested field of research tied up with literary theory, Globalization Studies and World History. Their practitioners are brave enough to take the whole shebang into account but do not get down to the nitty–gritty work of investigating the particulars; they are preoccupied with the general (abstract) but neglect the particular (concrete), forgetting that the former is related to the latter, and vice versa. Rather than investigating a specific area of the world, or a specific cultural practice, and putting the specific into perspective, cultural studies concentrate on how a medium or message is related to ideology, social class, gender, sexuality, ethnicity or race. In 1996, Alan Sokal, professor of physics and mathematics, submitted an article to *Social Text*, an academic journal of postmodern cultural studies. The submission was an experiment to investigate whether a leading journal of cultural studies would publish an article peppered with sheer nonsense if the paper would flatter the ideology of the editors. The article was accepted and published. On the day of publication, the author revealed that the paper was a hoax. The painfully embarrassed editors did not fail to attempt to justify their decision. Fairly acknowledging that they had been terribly wrong, were rightly blamed for it and would quickly change their bad behavior was too much for them, just like the owner of a dog that bites into the leg of an unsuspecting passerby saying (instead of excusing himself) "My dog never does such things", thereby implying that the passerby was somehow guilty of the animal's wounding bite. The shameful behavior of the editors of *Social Text* and of cultural scholars at large is not exceptional; it represents a general practice among academics, people said to be using their brains and to be reasonable.[3] Whoever disagrees with me should read the arguments pro and contra abortion, Brexit, death penalty, euthanasia, globalization, gun control, immigration or impeachment. Academic discussions without advocacy of personal interests do not exist. No matter the subject matter, political discussions are inherently infected with partisanship. *Dialogues* (as distinct from debates) are rare commodities.

Area/country studies are akin to cultural anthropology but, whereas cultural anthropology is the scientific study of different human communities (and shares with *philosophical* anthropology and with humanism the interest in human beings), area/country studies take also the location and place/territory where people live into account, thereby sitting close to, but not being identical with, geography. Though they differ in the region being studied, area/country studies are not, in contrast to physical or cultural anthropology, disciplined.

History, defined as the study of the past as described in documents (in contrast to archeology, aka "study of the spade") and differentiated in more than a dozen subfields, is concerned with things, events, persons or actions *sub specie temporis*.[4] Like geography, area/country studies spatialize, and like history, they temporalize. However, neither spatializing nor temporalizing is their quintessence. "Area study" and "country study" are container concepts, not referring to any science whatsoever, not referring to a defining scientific model, not referring to an analytical point of view, "research program" (Imre Lakatos) or "research tradition" (Larry Laudan). To drive this point home, let me give three examples.

There is much confusing talk about the problems of population growth and massive immigration. Demography, which can be traced back to antiquity, is the statistical study of human populations. It encompasses the study not only of their size and composition but also of their dynamics, their changes in response to birth, migration and death. Demography is to be distinguished (but not separated) from the still young population biology, a combination of population ecology and population genetics.

In recent decades, there has been "a rapid development of demographic models and methods. There is an explosive growth in the range of applications of population analysis" (see the *Springer series on Demographic Methods and Population Analysis* [1988ff] or visit http://www.cicred.org). Though demography is at the crossroads of disciplines such as biology, ecology, sociology, economics, politics, anthropology and epidemiology, it offers the tools to approach a wide range of population issues. Acknowledging that both the determinants and the effects of demographics are to be ultimately explained by other disciplines, demographers have a clear and distinct analytical point of view, a unique disciplinary matrix, a scientifically distinctive sound or beat. Similar statements can be made about geography, the core business of which is studying, in close cooperation with geology, biology, ecology and the social and human sciences, the surface of (a part of) the Earth, that is to say, the spatial distribution of its physical and human characteristics (see Derek Gregory et al. [eds.], *The Dictionary of Human Geography* [Wiley, 2009] or Douglas Richardson et al. [eds.] *The International Encyclopedia of Geography* [Wiley, 2017]).

Archeology, the study of the human past from material remains (artifacts or biofacts), draws heavily on physics, chemistry, geology, geography, biology and computer science. Its dependence on these disciplines is even a key feature of archeological research. Nevertheless, its methodology is characteristic and defining: survey, excavation, post-excavation analysis, dating methods and preservation. Archeologists have also advanced and developed different theories (models). See Chris Scarre (ed.), *The Human Past* (Thames & Hudson, 2018). The book is "absolutely essential reading for all students in archaeology" (Barry Cunliffe). Archeology is to be distinguished from paleontology, the study of life prior to the Holocene, the current geological epoch. There are archelogists focusing their mind on archeological sites in Egypt, Ethiopia, Somalia, Kenya, Mauritania, Mali, Morocco, Angola, Botwana, South Africa, Tanzania, Turkey, Armenia, Israel, Jordan, Syria, Saudi Arabia, Afghanistan, Uzbekistan, Iran, Iraq, India, Pakistan, Bangladesh, Indonesia, Easter Island, Australia, Samoa, China, Taiwan, Korea, Japan, Vietnam, Kazakhstan, Mongolia, Mexico, Belize, Bolivia, Ecuador, Peru, Ireland, Denmark, Finland, Germany, Romania, Bulgaria, Greece, Cyprus, Spain, Italy, Hungary, Poland, Ukraine or Russia.

Dimensional/disciplinary specification is a prerequisite for comprehending the complexity of countries, but a country's dimension can only be the subject of serious research when its other dimensions are taken into account by the scientists concerned. A linguistic study of the Chinese language, a study of the Japanese literature conducted by the holder of a Ph.D. in literary science, or a political study of Iran's political system: yes, provided that other dimensions (being the subjects of research not carried out by linguists, literary scientists or political scientists, repectively) are paid due attention to. "The study of a country" *tout court*, however, is a shameless brutality. Unless scientists from different departments work together and resort is taken to systems science (see Chap. 5), country studies such as Egyptology, Turkology, Iranology, Indology, Sinology, Koreanology and Japanology should, therefore, be removed from the higher education curriculum.

Ethnoarcheology is an interdisciplinary field of research. Pioneered by Jesse Fewkes (1850–1930), Gordon Childe (1892–1957) and André Leroi-Gourhan (1911–1986), it combines archeology and anthropology, aiming at understanding how current facts, actions or events could fossilize, i.e. become archeological evidence. Ethnoarcheology takes both present processes and their appearance in past processes into account. It represents the inclusion of multiple methods and viewpoints in a distinguishable research program. For the sake of better understanding, the reader should visit the *Musée de l'Homme*, in Paris. The avowed objective of this wonderful museum is to collect and orderly display everything related to (*a*) "*l'Homme dans sa chaîne évolutive*", (*b*) "*l'Homme dans son unité et sa variété biologique*" and (*c*) "*l'Homme dans sa diversité culturelle et sociale*". For current ethnoarcheological research, see *World Archaeology*, 48:5 (2016), pp. 605–713.

A mature science consists of subdisciplines. The workers in these vineyards occupy themselves with a part without losing sight of the whole it belongs to. For example, biology deals with living things at different levels in the biosphere. The growth of the discipline was triggered by division of labor. Zoologists are interested in animals, ethologists in the behavior of these creatures, botanists in plants, mycologists in fungi, phycologists in algae and microbiologists in bacteria and viruses. Here, the ramifying does not stop. When we restrict ourselves to the animal kingdom, mammalogists are concerned with mammals, entomologists with insects, carcinologists with crustaceans, arachnologists with spiders and their relatives, ornithologists with birds, ichthyologists with fishes, malacologists with molluscs and herpetologists with reptiles and amphibians. Despite their differences, all the divisions and subdivisions are interrelated; mother, daughters and granddaughters belong to one family. The splitting of biology into specialisms has been guided by the same principles. The dialects are different, but the language spoken is the language of biologists, "cell" being the key concept. *Mutatis mutandis*, the same goes for geology, ecology, psychology, sociology, anthropology, archeology, pedagogy, medicine, economics, politics, law and musicology.

After World War II, area/country studies started to diversify, an academic process fueled by America's foreign policy. By any stretch of the imagination, though, I cannot see how the diverse groups being formed belong to a family. There is no intellectual kinship, scientific lineage or academic genealogy. For example, postwar "China experts" have—scientifically speaking (in terms of *explanans*)—nothing in common, neither among themselves nor in relation to those occupying themselves with another country. They still have no command of a tight-knit network of basic notions applicable to China, their *explanandum*. There is an endless stream of books and articles about China, but the images given of the country are incoherent and inconsistent. The picture is jumbled. Humpty Dumpty is not put together. The situation in, or development of, China, said to be moving fast to the center stage of world politics, is the subject of many private and public discussions, but there is no *Sinological* debate, no *Sinological* perspective, no *Sinological* research method, no *Sinological* body of organized knowledge, no *Sinological* way of formulation and presentation. Nothing, *nada*! Sure enough, Sinologists do have a domain. It is carefully guarded against being entered into by "people who do not know anything about China" or, worse still, "do not speak Chinese". The obfuscation is that they do not have a distinctive *explanans*, a theory of their own that explains the country by or of itself. After all, Sinologists are supposed to be scientists who are concerned with a country called *China*, just like economists who study the way in which money and goods are produced and used.

Theories (see Chap. 2) are explanatory *systems* (see Chap. 5). An explanation renders something, in this case China, intelligible. In other words, it gives information that makes the *explanandum* clear and understandable. An explanation is a system of interrelated statements intended to clarify how something works. For details, the reader may see Johannes Persson and Petri Ylikoski (eds.), *Rethinking Explanation* (Springer, 2007) and visit https://www.iep.utm.edu/explanat and https://plato.stanford.edu/entries/scientific-explanation.

Sinologists do not have concepts that are foundational in the study of China. Every discipline aims at foundational concepts to organize a seemingly chaos of observations. Indeed, historically, a discipline has proved to be legitimate in so far as it can define at least one foundational concept that is distinctive from any concept in another discipline, for example, "atom" in physics, "molecule" in chemistry, "cell" in biology, "neuron" in neuroscience, "location" in geography, "population" in demography, "phoneme" in linguistics, "text" in literary science, "psyche" in psychology, "family" in sociology, "man" (as building block, agent or vertex) in anthropology, "state" in political science and "market" in economics. In glaring contrast, no "China expert" can tell what the foundational concept of Sinology or Chinese Studies is.

The change from "Sinology" to "Chinese Studies" is misleading on two accounts. By definition, "*Chinese* Studies" are studies carried out by Chinese scientists or scholars on some subject (not necessarily a subject related to China). If "Chinese Studies" is a substitute for "Sinology", the question rises what kind of study in regard to China (PRC or ROC?) is meant: a geographic, demographic, linguistic, literary, economic, political, legal, military, social, ecological or another kind of study? "Professor of Chinese, or Russian, Studies" sounds like "Professor of Social, or Human, sciences", almost empty designations, for there are many and widely different social, or human, sciences. Nobody is or will ever be a professional in all of them. "Professor of Middle Eastern Studies", "Professor of African Studies", "Professor of Latin American Studies" or "Professor of European/Easteuropean Studies" sounds strange and inadequate. Europe, for example, consists of 50 sovereign but strikingly dissimilar countries. Budgetary constraints at universities may make it necessary to lump some countries together for investigation, but this kind of lumping together runs counter to the Western tendency toward scientific analysis.

Having metaphorically broken their *explanandum* up into various parts (e.g., the Chinese geology, geography, population, language, literature, mindset, society, polity/politics and economy), Sinologists cannot do anything but to rely upon or to borrow from the natural, social or human sciences. Presumably aware of the fundamental problem they are beset with, they have dealt with it half-heartedly. They have split up Sinology into "Chinese/China Studies", but have no common, scientific language. There is no systematic way of studying China as such, *an und für sich*. One looks in vain for the *Principia Sinologica*. Ask a Sinologist for the Sinological counterpart of Adam Smith, Émile Durkheim, Max Weber, Mary Parker Follet, Carl Ritter, Henri Fayol, John Finnis, Franz Boas, Ferdinand de Saussure, Adolphe Quetelet, Herbert Hart, Ronald Dworkin, Leonid Kantorovich, Jean Piaget,

Kurt Lewin, Yuri Lotman, Michael Goodchild or Hugo Riemann, and you will get a stare. That is the reason why I argue that area studies in general and China studies in particular are highly questionable. Clearly, something is basically wrong in the domain of country or area studies.

The result has proved disastrous. America's staunch support of the Nanjing Government after 1927, its friendship with Generalissimo Chiang Kai-shek and his nefarious wife during the Second World War and its military backup of the nationalists during and after the Chinese Civil War (1945–1949) were driven by the specter of a communist behemoth in Eurasia, a specter fueled by the secret reports of "respected and reliable China experts": If Communist China (Mao Zedong) would make common cause with the Soviet Union (as it seemed to do at the time), Mackinder's Heartland would be lost, and America's role in the "Far East" (the designation speaks volumes) would be over and done with forever. This turned out to be a strategic blunder of the first order when the Sino-Soviet Split surfaced in the years after 1956, but nobody thought of it to hold the "respected and reliable China experts" accountable for the dire miscalculation (Joseph McCarthy's support and popularity were rapidly fading). The "experts" seized the opportunity and changed the signs smartly, but who was (and is) really well informed about the Soviet Union or the People's Republic of China is still an open question. History has taken many twists and turns since the eventful decades in the middle of the twentieth century. Whether the Administration in Washington has learned the lessons-from-the-past, remains to be seen. My spirit sinks when I read about the current China whisperers of the President of the United States of America. Perhaps I am too skeptical. Fact is that "China experts" such as David Lampton and David Shambaugh are welcome guests at the premises of the U.S. Department of State. I can only hope that they know how to evaluate the "friendly relations" between Xi Jinping and Vladimir Putin correctly and understand the geopolitics of the current Beijing–Moscow Consensus (read Andrew Korybko, "Is Russia backtracking on China's 'Belt and Road' [BRI]?", *Global Research*, August 31, 2020). Russia and China, being allies of convenience, have a long common border. They are economic unequals (go to https://doi.org/10.1057/s41311-020-00251-7 and read Jonathan Hillman, "China and Russia: Economic Unequals", July 2020). With a growing population (and little arable land, encroached upon by urban developers), China is or will be seeking "*Lebensraum*". Sooner or later, a conflict between the two giants will be unavoidable. If "*historia docet*", one lesson to be learned is that the Russian bear and the Chinese dragon do not tolerate each other. Whether the American President had the nerve to hold his fire, only future can tell.

In April 2019, the Russian government approved the construction of a 2000-km highway, linking Europe to Central Asia. The four-lane highway, planned to run from the Russian border with Belarus to the Russian border with Kazakhstan and budgeted at $ 9.5 billion, will take about fourteen years to be completed. Being part of a 5000-km planned road-project which will eventually connect Hamburg with Shanghai, the highway will be an integral part of China's Belt and Road Initiative (BRI), a much-publicized, perhaps ill-conceived, but sometimes falsely and maliciously represented project. For an analysis of the image of Russia and the Soviet Union in Chinese history textbooks of the 2000s, visit https://doi.org/10.21638/11701/spbu13.2017.410.

In *Eurasianism: An Ideology for the Multipolar World* (Rowman & Littlefield. 2020), Paolo Pizzolo focuses his mind on Eurasianism, a political doctrine mainly developed in Russia. He describes Halford Mackinder's "Heartland Theory", Nicholas Spykman's "Rimland Theory", Karl Haushofer's "*Kontinentalblock* Theory" and Aleksandr Dugin's "Fourth Political Theory (*Chetvertaya Politicheskaya Teoriya*"), pointing out the incompatible dichotomy and irreconcilable dualism between thalassocracy (sea-based hegemony) and tellurocracy (land-based hegemony). The author, who takes Russia and China to be telluric powers, explores the BRI, considering it an initiative "aimed at integrating the Eurasian space", but he underestimates the importance of the "Maritime Silk Road", China's response to the challenge posed by *naval* powers. Pizzolo forgets that China's strategic response to the West, apart from its digital silk road, is *two*-pronged: one by land (telluric) and another by sea (thalassic). Explore https://www.globalaffairs.ch. In addition, visit https://www.vox.com/2018/4/6/17206230/china-trade-belt-road-economy and www.brookings.edu/globalchina.

In August 2019, the Washington-based Center for Strategic and International Studies (CSIS), a "bipartisan, nonprofit policy research organization dedicated to providing strategic insights and policy solutions to help decision-makers chart a course toward a better world", released a report, entitled *China and the United States: Cooperation, Competition, and/or Conflict*. One of its key points is that "China's economic progress and regional economic outreach will often be more of the central focus of its grand strategy than the modernization and expansion of its military forces". The implication of this finding seems to be that an analysis only focusing on either side of the civil-military developments in China cannot fully describe the evolution of China's national strategy. Unfortunately, Anthony Cordesman, the compiler of the report who claims to be an expert on—believe it or not—Afghanistan, China, counter-terrorism and homeland security, defense industry, acquisition and Innovation, defense strategy and capabilities, defense and security, energy and sustainability, geopolitics and international security, international development, Korea, the Middle East, NATO, North Africa, Pakistan and Turkey, provides us no insight into what he means by "grand strategy" and whether or to what extent this noun phrase is the correct translation of the Chinese core concept of *mǒulüè*. Clearly, comprehending another country is impossible without recourse to the science of complex systems, i.e., without thorough and collaborative research on its current situation put into historical perspective.

In the margin of the 35th ASEAN summit in Bangkok, in November 2019, the USA, Japan and Australia announced a plan to establish a trilateral "Blue Dot Network", meant to help develop an infrastructure "in the Indo-Pacific and around the world". It is the Western response to the BRI challenge. No details are available yet, but Professor Peter McCawley, of the Australian National University, noted, with the usual hedging: "It seems clear that one of the main aims of establishing the new network is to counter China's ambitious international investment program in Asia and Africa". According to Chris Devonshire-Ellis, the "Mandarin, Cantonese, English,

Greek, Latin, Mongolian and Russian speaking" chairman and founding Partner of the "Russia & China Belt Road Initiative Office", "If the Blue Dot Network is not actually in the business of providing guarantees, then it is destined to be just a piece of anti-China, Western PR hot air, and another stick which Washington in particular can use to beat up China about 'global quality standards' along its Belt and Road".

In January 2020, the nineteenth volume in the NBR series *Strategic Asia* entitled *Strategic Asia 2020: U.S.-China Competition for Global Influence* was released. It offers "a forward-looking assessment of how the rivalry between China and the United States is playing out around the globe". Each chapter "examines how a country or region is coping with the consequences of Sino-U.S. competition and draws implications for U.S. policymakers". The book, released at the George Washington University's Elliott School of International Affairs, is edited by Ashley J. Tellis (Counselor at the National Bureau of Asian Research and Senior Fellow at the Carnegie Endowment for International Peace), Alison Szalwinski (NBR's Vice President of Research) and Michael Wills (NBR's Executive Vice President), all "first-class China experts". Interesting is also the book review roundtable at https://www.nbr.org/publication/daniel-s-markeys-chinas-western-horizon-beijing-and-the-new-geopolitics-of-eurasia.

At the present time, an intense debate is going on in America about what the U.S. stance toward China should be. Dozens of self-styled China scholars, thinking of their next visa application for the PRC, argue that China is not an enemy. The backlash against these panda huggers came from John Pomfret, author of the "exhaustively researched" book *The Beautiful Country and the Middle Kingdom: America and China, 1776 to the Present* (Henry Holt, 2016). Reading James Thomson, Jr. et al., *Sentimental Imperialists: The American Experience in East Asia* (Harper, 1981) could also have a sobering effect. The essence of the Carsten Holz's article "Have China Scholars All Been Bought?" in the *Far Eastern Economic Review* (170:3 [April 2007], pp. 36–40) still holds true, and the online article of Elizabeth Redden, "Censorship in a Chinese Studies Journal" (April 19, 2019), is worrisome.

In their foreword to *Essays on Chinese Civilization*, a book written by Derk Bodde (Princeton University Press, 1981), the editors Charles Le Blanc and Dorothy Borei, two of Bodde's former students, write

> Sinology implies an integrated, interdisciplinary approach to the study of China as a coherent whole. This approach is largely necessitated by the close interrelatedness of the many aspects of China's culture–language, social structure, political and economic institutions, the arts, religious and philosophical beliefs, and so on. A Sinologist is one who has grasped the fundamental 'organic' character of Chinese civilization and whose study of any particular element of that culture is integrated into a larger vision of the society. Derk Bodde's scholarship is a prime example of Sinology at its best, for it provides detailed knowledge of various facets of the Chinese experience at the same time that it expresses an acute awareness of their interconnectedness. (p. xiii)

If these words were misleading (because they do not reflect the real situation) nearly forty years ago, when the literature on China was already overwhelming, they sound hollow and hypocritical at the time that the volume of publications about "any particular element" of the country has become enormous. Derk Bodde (1900–2003), who translated Feng Youlan's massive *Zhōnggúo Zhxué Shǐ* (a history of Chinese philosophy), wrote extensively about law in Imperial China without any degree in law. In a book entitled *Chinese Thought, Society, and Science: The Intellectual and Social Background of Science and Technology in Pre-modern China* (University of Hawaii Press, 1991), he attempted to elucidate why modern science did not arise in China, without defining or analyzing the words science, technology and modernity. Reading the book, which has been heavily criticized by Shigehisa Kuriyama,[5] one wonders whether professor Bodde really "grasped China's fundamental character".

Oxford Bibliographies in Chinese Studies (OBCS) is an enterprise that covers the study of China across many disciplines. Emulating *Das große China-Lexikon* (edited by Brunhild Staiger†, Stefan Friedrich, Hans Schütte and Reinhard Emmerich, and published in Darmstadt by WBG, in 2000), OBCS developed from two sources: (a) the long-standing tradition of Sinology (still strong in Europe) that uses philological and literary tools to study mainly "pre-modern" China, and (b) the "area study" approach (dominant in the USA, Canada and Australia) that focuses on "modern China". More recently, China's rapid economic growth has led to the expansion of "Chinese Studies", with scholars from the PRC or Taiwan strengthening the ranks of Western "China experts" who fall short of being able to read (classical or modern) Chinese. However, the editor-in-chief of OBCS seems not to care about the theoretical framework that would provide the other members of the editorial board with scientific guidelines for conducting integrative research. In fact, each editor, only asked to fulfill some editorial requirements, is allowed to operate autonomously and independently, the result once again being an enormous grab bag for anyone interested in things Chinese, an impressive patchwork of brief narratives or short and half-baked analyses not linked up with an overall model or overarching theory related to China in and of itself.

The problem discussed is compounded when books or articles about "Chinese philosophy" are subsumed under the category "Chinese Studies". Though the literature on philosophy of science is vast and expanding, the domains of philosophy and science differ from each other in many respects (some would even say: fundamentally). Science is a matter of (a) identifying, prioritizing and formulating questions that relate to a particular piece of reality, (b) designing, prioritizing and executing experiments meant to answer chosen question, (c) drawing inferences from the evidence thus provided, (d) evaluating the results of c and (e) formulating new questions, in a closed-loop fashion. Philosophy, on the other hand, studies anything and everything.[6] In contrast to the bittiness of science, it strives for an integrated account of the world. Philosophy is concerned with matters such as reality, cause, change, experience, knowledge, truth, mind, reason, meaning and value. The core of its domain is conventionally divided into five interrelated subdomains: metaphysics, epistemology, logic, ethics and aesthetics, with metaphysics, epistemology and logic usually separated from ethics and aesthetics, particularly in the USA. Each subdomain is articulated or segmented. For instance, there are many kinds of logic (visit https://plato.stanford.edu/contents.html [logic]), and ethics is divided into four parts: descriptive ethics (aka comparative ethics), prescriptive/normative ethics, meta-ethics and applied ethics. The last part consists of the subparts bioethics, business ethics, environmental ethics and professional ethics, each of them being covered by a large and growing literature. In *The Mind of Science: From Aristotle to Einstein* (CreateSpace, 2015), Michael Sidiropoulos explores the intersection of science and philosophy in the pursuit of knowledge. Reading books or articles about "Chinese philosophy" (a phrase that—like "Chinese mathematics", "Chinese logic", or "Chinese physics"—may raise eyebrows),[7] I very much doubt whether their authors are able to speak (and read!) Chinese and are schooled in any of the subdomains, parts or subparts of philosophy. Let me give a few examples:

(1) Professor Christoph Harbsmeier (University of Oslo) is author of the first part of the final volume of Joseph Needham's well-known *Science & Civilisation in China*. The book was published in1998 and deals with language and logic in traditional China, but Harbsmeier never graduated in linguistics or logic.
(2) In *Philosophy Compass* (10:5 [May 2015], pp. 301–309), professor Jana Rošker (University of Llubljana) "provides an introduction to Classical Chinese Logic" and "addresses the close link between Chinese logic and the Chinese language" without any degree in logic or linguistics.
(3) Sinologists like Chad Hansen, Roger Ames, Franklin Perkins, Chris Fraser, Michael Puett, John Berthrong, Rudolf Wagner and John Makeham come out to inform us about "Chinese metaphysics" without making any reference to current work in metaphysics and ontology.

Being a logician (someone who graduated in logic) is a necessary condition for being able to say or write something sensible about logic in China (not: Chinese logic). However, someone schooled and skilled in logic and focusing his mind on logic in China or on the work of a Chinese logician may be accused of not speaking Chinese or knowing next to nothing about China. Chinese speaking Sinologists have the obnoxious habit of playing up this point to cover up their own lack of education in philosophy, or in science. *Mutatis mutandis*, the same applies to "country experts" like Japanologists, Koreanists, Mongolists, Tibetologists, Indologists, Iranologists and Turkologists. There are myriads of books on philosophy, science (and technology), religion and art (particularly literature), but the relationships between these domains should not be neglected. Philosophers disagree among each other; the fragmentation of science is a fact of life; religious fighting continues unabated, and the proliferation of arts is unending. Nobody seems to be able to provide an overview. Nobody seems to be able to show a way out of this jungle. Sinologists use this as an excuse for selling sheer nonsense about "Chinese philosophy". They succeed in doing so, because there are enough people who buy it.

There are no Sinological schools comparable to the schools of thought in political science, law, psychology, sociology, linguistics, literary theory, or economics, simply because there is no Sinological language other than the undisciplined speaking or writing about China. Rather than speaking a Sinological language different from the language of, say, archeologists, sociologists, economists or political scientists, "China experts" do not speak a scientific language at all. The emperor is wearing no clothes. Since the state of affairs in other area/country studies is the same, the claimed postwar split of these studies is a case of deceptive appearance. The weakness indeed the Achilles' heel of so-called area/country experts was and still is: they have an *explanandum sine explanans.* It is for this reason that they are not qualified to wear the sacred mantle of science (sensu stricto *or* sensu lato).

Whoever believes that the all-rounders in respect of an area or country are dead and gone is mistaken. The proverbial jack-of-all-trades-but-master-of-none/one is still alive and kicking. Open a newspaper or listen to a talk show on the television, and the risk of being informed about a country by a non-scientist is real. Some of these all-purpose pundits do not even shrink from predicting the future of the country, apparently unaware of the nonlinear revolution of the 1970s that emphasized the certainty of uncertainty and led to a redefinition of causality.[8] If pretending to be, or making no objection to being introduced as, expert on some aspect of, say, the Middle East or Russia without holding a degree in the discipline concerned is reprehensible; downright unforgivable is it to make no bones about changing bonnets and masquerade as connoisseur of the area or country *tout court*. Those who are guilty of doing so corroborate the words of Alexander Pope ("fools rush in where angels fear to tread"), and those who lend these pseudoscientists an ear should be more critical. Let me give just one example.

In 2009, a book entitled *Understanding Iran: Everything you need to know, from Persia to the Islamic Republic and from Cyrus to Khamenei* was published by Palgrave Macmillan. The author, a former professor of history at Harvard University and the University of Chicago, is William Polk, who (*a*) is related to James Polk, the 11th President of the United States, and to Frank Polk, a prominent American lawyer and diplomat, (*b*) served as advisor to McGeorge Bundy, National Security Advisor to John F. Kennedy and Lyndon B. Johnson and (*c*) was foreign policy advisor to Dennis Kucinich, candidate for the Democratic nomination for president of the United States in 2004 and 2008. Rouhoullah ("Ruhi") Ramazami (1928–2016), "the experts' expert on Iran" who had been consulted by Jimmy Carter during the Iranian hostage crisis in 1979–1981, eulogized the book in *The Middle East Journal* (64:2 [2010], pp. 307–309), calling it "an extraordinary book by an extraordinary author". Former Senator George McGovern wrote: "This marvelous book promises to be the definitive account of Iran today in historical context – a fitting companion to the author's brilliant *Understanding Iraq*". Graham Fuller, vice-chair of the National Intelligence Council (NIC), joined the chorus by writing: "[Polk's] sure grasp of the social, political and psychological culture of Iran is invaluable to understanding this complex land". Unfortunately, neither Polk nor Ramazami, and neither McGovern nor Fuller predicted any important event in the Middle East. They were unable to do so, because none of them presented a plausible, i.e., tested or testable theory about Iran, or about any other country in that tumultuous area. The same applies to "Middle East experts" like Ali Banuazizi, Farhad Kazemi, Garry Sick, Mohsen Milani, Yudith Yaphe, Trita Parsi, Rola El-Husseini and Hillary Clinton's political adviser Kim Ghattas.

Area/country studies are sometimes subsumed under the category philology. Many people are unfamiliar with the word, but "philology" (derived from *φιλολογία* [love of logos] and referring to textual scholarship, knowledge of Arts and Letters, familiarity with the world of words, and acquaintance with arguments [the art of discourse]) was for centuries almost synonymous with the heterogeneous group of studies encompassing not only the study of Graeco-Roman literature and the Bible but also studies such as Egyptology, Hittitology, Assyriology, Celtic, Slavic, Arabic, Mongolian, and Semitic Studies. In brief, philology was queen of the human sciences and humanities, standing for knowledge about human (as distinct from natural) affairs.[9] Some of the philological specialisms (*Studienfächer*) have been as follows:

- Classical philology: Greek philology, Latin philology, Byzantine Studies, Bible Studies and Medieval Latin philology;
- Neo-philology: English Studies, American Studies, Baltic Studies, Bohemistics, Hungarian Studies, Finnish Studies, German Studies, Celtic Studies, Albanology, Modern Greek Studies, Spanish Studies, Lusitanic Studies, Catalan Studies, French Studies, Italian Studies, Romani Studies, Scandinavian Studies, Slavic Studiesand Russian Studies;
- Asian philology: Sumerology, Hittitology, Assyriology, Jewish Studies, Dravidian Studies, Indology (Indian Studies), Tamil Studies, Iranology (Iranian Studies), Japanology (Japanese Studies), Koreanology (Korean Studies), Mongology (Mongolian Studies), Sinology (Chinese Studies), Hong Kong Studies, Taiwan Studies, Thai Studies, Tibetology, Turkology, Vietnamese Studies, Myanmar (Burma) Studies and Indonesian Studies;
- African philology: Egyptology, Ethiopian Studies and Southern African Studies;
- Latin American philology: Mexican Studies, Andean Studies, Peruvian Studies; and
- Overarching philology: Oriental Studies, Indo-European Studies, Native American Studies, Arabic Studies, African Studies, Asian Studies, East Asian Studies, South Asian Studies, Southeast Asian Studies, Central Asian Studies, Middle Eastern Studies, Caucasian Studies, Austronesian Studies, Latin American Studies, Caribbean Studies, Mesoamerican Studies, Pre-Columbian Studies, Maghrib Studies, Bantu Studies, Ibero-Caucasian Studies and Semitic Studies.

Lists of notable philologists can be found on the Internet.

Today, "philology" is an archaic word. The vast domain referred to has been taken over by, and divided among, archeology, linguistics, literary science, art history, economics, business administration, agronomy, religious studies, mythology, health science, sociology, psychology, educational studies, anthropology, demography, geography, law, political science and military science, which are distinctive sciences that—with the new "noetic revolution" (a global-scale paradigm shift comparable to the important change which took place in the Axial Age and in the Age of Enlightenment) gathering momentum—are becoming increasingly mathematical and computational in character. Ancient or exotic philosophies are now taken care of by professional philosophers, who leave the patchwork and potpourri of philologists far behind, but, being equally ignorant of the drawbacks and pitfalls of rendering a text from one language into another,[10] usually have no other choice than to rely on philological translations.

At the end of his overview of the entire contemporary discourse on philology, "The Many Returns of Philology" (see note 9), Andrew Hui rightly writes

While one might decry the rise of the expert as a fragmentation, a loss of encyclopaedic knowledge, another way would be to celebrate such perspectival and professional diversity. If we adopt the Baconian notion of the advancement of science, this division of labor is inevitable, but so is its circulation: as each expert deepens her own field of knowledge, she contributes to the greater world republic of letters. The danger, of course, is that increasing specialization can lead to narrowness. Turner's sweeping account of Western humanities, Pollock's global kaleidoscope and Hamacher's heady theorizing attempt to resist such parochialism.

The mistake is often made that those who are (believed to be!) fluent in, say, Chinese can be expected to be expert on the Chinese language (classical-literary Chinese [*wényán*] as well as modern-standard Chinese [*pǔtōnghuà*]) and to have thorough knowledge of the Chinese literature (traditional as well as modern), or of anything else related to China, as if a Dutchman proficient in English could deliver a scientific lecture on the old or new rules of English grammar, or on the English literature (regardless period, region, or genre), and could be reliably consulted about the economic, financial, social, educational, environmental, political, juridical, military or religious situation in the United Kingdom or (why not!) the United States. I cannot conceive of making a bigger mistake. The point is important, for it has been argued that to study a country is to study its literature, and to study the literature of a country is to study the language in which it is written, the implication being that those who studied the language of a country (no matter whether the study was disciplinary or not) is an expert on the country. For the current status of Sinological linguistics, see Weldon South Coblin, "Jerry Norman: Reflections on the Development and Current Status of Sinological Linguistics", *Bulletin of Linguistic Studies*, 21 (2018), pp. 56–78, particularly the sobering conclusion on pp. 77–78.

Whether all Sinologists master the Chinese language is highly questionable. In 2018, I attempted to mobilize moral support for the English translation of modern masterpieces of philosophical or scientific thinking in China, so that Western decision-makers would be less dependent on secondary literature regarding the fast resurrecting country. The electronic invitation was sent to hundreds of Sinologists associated with universities or research institutes throughout the world, emphasizing that only moral support of the project was sought, no financial commitment whatsoever would be involved and offerings to translate one of the books mentioned on the attached list would be accepted with great delight. Since not enough moral supporters could be found and very few qualified translators stepped forward, Brill Publishers, though initially enthusiastic about my initiative, deemed the project, meant to be a Western response to the strategic challenge posed by China's ambitious "Belt and Road" project, not viable.

Meanwhile, the latter project, consistently underestimated and widely disparaged in the West, is making steady progress. Barring unforeseen circumstances, it has a fair chance to be completed and to become a game changer of the first order. Visit https://prochetmoyen-orient.ch/routes-de-la-soie-maritimes-et-armees! Probably due to lack of China-expertise, my attempts to get the European Commission interested in the translation idea miserably failed, a remarkable fact, given the Commission's strategic report concerning China's initiative released on 19

September, 2018. The information gathered from this report should be compared with Nadège Rolland (ed.), *Securing the Belt and Road Initiative: China's Evolving Military Engagement along the Silk Roads* (NBR, September 3, 2019), a report subsequently discussed at a conference co-hosted by the National Bureau of Asian Research (NBR) and the Daniel Morgan Graduate School, on November 13.

In a speech delivered on September 3, 2019, Xi Jinping pointed out that under Mao Zedong the Chinese people had stood up (*zhàn qǐlái*), under Deng Xiaoping and his successors, they became rich (*fù qǐlái*), and under his rule, they would become powerful (*qiáng qǐlái*). The speech, warming over ideas already advanced by Yan Fu (1854–1921), recapitulated what Xi had said in his (downloadable) address to the 19th National Congress of the Communist Party of China, in October 18, 2017 ("Secure a Decisive Victory in Building a Moderately Prosperous Society in All Respects and Strive for the Great Success of Socialism with Chinese Characteristics for a New Era"). Whether the Chinese will ever be able to raise their heads proudly because they are free from political overlordship and have finally become citizens rather than subjects (from *sub* [under] and *iacere* [to throw]) remains to be seen, but the West would do better to take Xi's words seriously.

On 27 January, 2020, the NBR released a special report entitled *China's Visions for a New World Order*. In it, Nadège Rolland, senior fellow political and security affairs, lays out "the intellectual and ideological underpinnings that inform China's vision for a new world order and examines the process of transition from thought into concrete policymaking". The reader of this downloadable report should judge for himself whether she really comprehends China and the vision of its leadership. Explore https://www.newamerica.org/cybersecurity-initiative/digichina and https://www.datainnovation.org (fill in: China). In *Mǒulüè: Unerkannte Denkdimensionen aus der Reich der Mitte* (Munich: Hanser, 2018), Harro von Senger argues that so-called China experts, having no English equivalent of the earthbound Chinese concept of 谋略 (*Supraplanung*), are uninformed about China's foreign policy. If this claim can be substantiated, a great deal of the analyses of that policy and of Xi Jinping's historic Belt and Road Initiative has to be taken with a pinch of salt.

The claims made in this chapter are bold. They may upset or even enrage many professors of, or lecturers on, Asian, African, South American, Middle American, Middle Eastern, Russian, European, American, Canadian or Australian studies. Yet inconvenient truths should be exposed for correction without delay, because they are the ugly reality which casts a shadow on our universities.

Notes and References

1. A complese list of them would be only a small subset of the surprisingly large set of "area or country experts". Currently, there are at least 4000 "China experts" active in Europe and North America and 1200 in Japan. See José Frèches, *La Sinologie* (PUF, 1975), Martin Kern, "The emigration of German Sinologists 1930–1945", *Journal of American Oriental Society*, 118:4 (October–December 1998), pp. 507–529, Helmut Martin and Christiane Hammer (eds.), *Chinawissenschaften* (Institut für Asienkunde, 1999),

Zbigniew Wesołowski, *Le détour et l'accès: la sinologie en tant que nouvelle herméneutique pour l'Occident* (Monumenta Serica Sinological Research Center, 2004, pp. 145–202, 285–359), Ming Wilson and John Cayley (eds.), *Europe Studies China* (Floating World, 2006), Yves Camus, "Jesuits' Journeys in Chinese Studies", March, 2007 (online), Chia-Mei Jane Coughlan, *The Study of China in Universities* (Cambria Press, 2008), Thomas Kampen, *Sinologie im 20. Jahrhundert* (Mattes, 2011), Fabio Lanza, "Do we need to rethink Sinology?", *The China Review*, 14:2 (October 2014), pp. 155–158, Eyüp Saritaş, "On the General Characteristics of Sinology Researches in Turkey", *Humanities and Social Sciences*, 3:6 (November 2015), pp. 293–298, Mechthild Leutner and Hauke Neddermann (eds.), *Challenging Narratives: Blind Spots of Sinology* (LIT, 2015), Jennifer Hargrave, "Marco Polo and the Emergence of British Sinology", *SEL*, 56:3 (Summer 2016), pp. 515–537 (online), Kate Rose (ed.), *China From Where We Stand: Readings in Comparative Sinology* (CSP, 2016), Chih-yu Shih (ed.), *Post-Communist Sinology in Transformation* (The Chinese University Press, 2016), Chih-yu Shih et al. (eds.), *From Sinology to Post-Chineseness* (Chinese Social Science Press, 2017), Luisa Paternicò and Chih-yu Shih (eds.), *Oral History of China Studies in Italy* (Fabrizio Serra, 2018), Ren Xiao and Liu Ming, *Chinese Perspectives on International Relations in the Xi Jinping Era* (NBR Special Report, June 2020, pp. 1–12) and André Bueno et al., *Sinologia Hoje* (Universidade do Estado do Rio de Janeiro, 2020). *Issues & Studies* (38:4/39:1, December 2002/March 2003) dedicated a special issue to "The State of the China Studies Field", *The China Review* (14:2, Fall 2014) dedicated a special issue to "Doing Sinology in Former Socialist States" and *Contemporary Chinese Thought* (49:1, 2018) dedicated a special issue to "Sinologism and New Sinology: Discussions and Debate on China-West Studies". After Marco Polo and other medieval voyagers, the Portuguese were the first who sent exploratory missions to "China". The Sinology Research Unit at the Catholic University of Leuven (Belgium), calling itself "Seat of Wisdom", is not concerned with Sinology (the study of China) but with China, or things Chinese. So the Unit's flag does not cover its load. The same applies to the *European Journal of Sinology* (2010ff), edited by Martin Woesler, and to the European Research Centre for Chinese Studies, in Beijing. Also, visit http://sinologystudy.com and http://china-studies.com. It would be unfair to deny that some of the observations or comments made by "area or country experts" were/are highly perceptive. My claim and complaint, however, are that none of these "experts" has ever advanced a tested or testable area or country-theory. We are flooded with opinions only.

2. See Ellis Cashmore and Chris Rojek (eds.), *Dictionary of Cultural Theorists* (Arnold, 1999, pp. 2–4). It should be noted that "modern science" is greatly indebted to China. See Dawid Rogacz, "How China Created Europe", *Diametros*, 54 (2017), pp. 138–160.
3. See Pierre Bourdieu, *Homo academicus* (Minuit, 1984). For cultural studies, see Toby Miller (ed.), *A Companion to Cultural Studies* (Blackwell, 2001),

Pepi Leistyna (ed.), *Cultural Studies* (Blackwell, 2005), Simon During, *The Cultural Studies Reader* (Routledge, 2007), Lawrence Grossberg, *Cultural Studies in the Future Tense* (Duke University Press, 2010), Jeff Lewis, *Cultural Studies* (Sage, 2011), Brian Longhurst et al., *Introducing Cultural Studies* (Routledge, 2016), Kieran Connell and Matthew Hilton (eds.), *Cultural Studies 50 Years On* (R&L, 2016), Chris Barker and Emma Jane, *Cultural Studies* (Sage, 2016), Jaafar Aksikas, *Cultural Studies in Practice* (R&L, 2017), *Cultural Studies* (1987ff), the *International Journal of Cultural Studies* (1998ff) and the book series *Radical Cultural Studies* (2015ff). In addition, visit the Web site of the Cultural Studies Association.

4. In Chap. 8, the question as to whether historians rightfully claim to have a theory of their own will be dealt with.
5. Visit https://muse.jhu.edu/article/8008.
6. See Roger Scruton, *Modern Philosophy: An Introduction and Survey* (Penguin, 1996, pp. 1–15).
7. There are Chinese mathematicians and Chinese physicists, but there is no such thing as Chinese mathematics, or Chinese physics. For the Chinese concept of philosophy in Asia, see Raji Steineck et al. (eds.), *Concepts of Philosophy in Asia and the Islamic World, Vol. I: China and Japan* (Brill, 2018, pp. 53–457). Explore https://utcp.c.u-tokyo.ac.jp/index_en.php.
8. See Alwyn Scott, *The Nonlinear Universe* (Springer, 2007, pp. 86–100).
9. See Sheldon Pollock et al. (eds.), *World Philology* (Harvard University Press, 2015), James Turner, *Philology* (PUP, 2015), *Philological Encounters* (Brill, 2016ff), Andrew Hui, "The Many Returns of Philology: A State of the Field report", *Journal of the History of Ideas*, 78:1 (2017), pp. 137–156 and Gerhard Richter and Ann Smock (eds.), *Give the Word: Responses to Werner Hamacher's 95 Theses on Philology* (University of Nebraska Press, 2019). The Web site to explore is https://lukemcmullan.org. According to W.V.O. Quine, "philology has been recapulated by epistemology" (*Word and Object* [MIT Press, 1960, p. vi]). The human sciences and humanities, the history of which can be traced back to ancient Greece and a major shift of which occurred in the Renaissance, are gripped by particularism (knowing more about less). See Michel Foucault, *Les Mots et les Choses: Une archéologie des sciences humaines* (Gallimard, 1966), Georges Gusdorf, *Les sciences humaines et la pensée occidentale* (Payot, 1966–1988), Toby Miller, *Blow Up the Humanities* (Temple University Press, 2012) and Rens Bod et al. (eds.), *The Making of the Humanities* (Amsterdam University Press, 2010–2014). *International Journal of Humanities and Social Sciences* (2011ff), *Humanities* (2012ff), *Annals of Studies in Science and Humanities* (2015ff) and *Social Sciences & Humanities* (2019ff) are open-access journals. *History of Humanities* (2016ff), not to be confused with *History of the Human Sciences* (1988ff), is sponsored by the Society for the History of the Humanities. Sylvie Mesure and Patrick Savidan (eds.), *Le dictionnaire des sciences humaines* (PUF, 2006) and Helmut Reinalter and Peter Brenner (eds.), *Lexikon der Geisteswissenschaften* (Böhlau, 2011) are indispensable reference works. For "New Directions in the

Humanities", visit http://thehumanities.com. The crisis of the human sciences and humanities is only understandable if "the phenomenological movement" (Spiegelberg) is taken into account. Visit https://humanitiesacrossborders.org, https://www.oxfordbibliographies.com (search for humanities), http://cipsh.net/htm and https://doi.org/10.3167/choc.2020.150102.

10. See Harald Kittel et al. (eds.), *Übersetzung • Translation • Traduction* (De Gruyter, 2011), Kirsten Malmkjaer and Kevin Windle (eds.), *Handbook of Translation Studies* (OUP, 2011), Carmen Millán and Francesca Bartrina (eds.), *Handbook of Translation Studies* (Routledge, 2013), Ives Gambier and Luc van Doorslaer (eds.), *Handbook of Translation Studies* (John Benjamins, 2014), Laurence Wong, *Where Theory and Practice Meet: Understanding Translation through Translation* (Cambridge Scholars Publishing, 2016), Manfred Frühauf, "Tour d'horizon über translatorische Theorien und Beobachtungen aus der Praxis des Übersetzens", *Bochumer Jahrbuch zur Ostasienforschung*, 40 (2017), pp. 5–32 and Sue-Ann Harding and Ovidi Cortés (eds.), *Handbook of Translation and Culture* (Routledge, 2018). See also the journals *Approaches to Translation Studies* (1975ff), *Target* (1989ff), *Translation and Interpreting Studies* (2006ff), *Translation Studies* (2008ff), *Compilation and Translation Review* (2008ff) and *Translation Spaces* (2012ff). Visit http://www.atisa.org, https://www.esist.org, https://est-translationstudies.org, https://www.iatis.org and http://societefrancaisedetraductologie-soft.fr. However, first explore the Web on "translation".

Chapter 2
Theories and Models

Abstract The nature of scientific theories is discussed, and the difference between scientific theories and scientific models is explained. Inasmuch as scientific theories are based on axioms, from which theorems are deduced or inferred, the idea is advanced that science is not completely different from religion.

Since my allegation that "country experts" cannot provide the essentials of a country *theory* and have no distinctive *explanans* would be vague without further clarification, I dwell in this chapter upon the issue as to what are scientific theories and models.

Generally, a theory is taken to be an explanation of how or why a piece of reality is the way it is or works the way it works. In the day-to-day use of the word, a theory is a guess, an untested idea, something that might be true or not-true; it is just an opinion, i.e., my, your, his, her or anybody else's opinion; anything goes. In the sciences, however, this is not what "theory" means. In science, a theory is considered an explanation of something (not: anything or everything) when based on repeatedly tested and widely established ideas. It is a *scientific* theory. For science, visit https://undsci.berkeley.edu. Explaining is clarifying, telling someone clearly about something; scientifically explaining is clarifying in a scientific way.[1] In order for any explanation to be a scientific theory, it has to meet a number of criteria:

(1) **Generality**

A theory needs to be built on lots of different bits of evidence,[2] not on just a single observation or perception. This makes sure that the theory applies in all cases, within the normal statistical variation, the logical consequence of it being that the theory does not apply when the statistical variation is not normal. The criterion of generality obviates the criticism leveled at logical positivism by Norwood Russell Hanson, who in his book *Patterns of Discovery* (Cambridge University Press, 1958, p. 19) maintained that seeing is "theory-laden" (observation of x is shaped by prior knowledge of x).[3] It needs to be emphasized that the criterion of generality cannot be the requirement or requisite of many people supporting the theory, which can still be based on the alleged observations of a single person. If the support of many people were sufficient, the vision of the founder of a religion would qualify for being called a

H. Kuijper, *Comprehending the Complexity of Countries*,
https://doi.org/10.1007/978-981-16-4709-3_2

scientific theory. Though Moses, Jesus/Paulus and Muhammad managed to assemble many believers, they were visionaries, not scientists.

(2) **Falsifiability**

While being general in nature, a theory has to make predictions and be testable in a way that could prove the theory wrong. In science, it is impossible to prove anything right. Instead, scientists run experiments to try and get results that are not predicted to prove the theory wrong. Performing an experiment is not a piece of cake.[4] A theory holds as long as it is not contradicted, falsified, overthrown, refuted or repudiated. There is science only when there are systematic attempts (successful or unsuccessful) to refute. If a hypothesis (a supposition or proposed explanation made on the basis of limited evidence as a starting point for further investigation) stands up to severe tests, it has only been corroborated. In other words, whereas the verificationist maintains that an assertion is correct if and only if it is verified, the falsificationist holds that an assertion is correct if and only if it is not falsified. A theory that is not testable is not a scientific theory. In his landmark study *The Logic of Scientific Discovery* (1959), Karl Popper worked out a calculus of corroboration, famously proposing that the falsifiability or refutability of a statement, hypothesis or theory should be the criterion for demarcating science from nonscience and religion. Whether such a statement, hypothesis or theory is the demarcation criterion remains to be seen, however, for even the most rigorous scientific theory contains a religious element (has an a priori element) and requires what Søren Kierkegaard called "a leap of faith" (accepting something outside the boundaries of reason). Google "Philosophy of the Humanities–YouTube" and see below (axioms).

In his book *Popper, Otto Selz and the Rise of Evolutionary Epistemology* (CUP, 2004), Michel ter Hark claims that Karl Popper (1902–1994), who had earned a doctorate in psychology under the supervision of Karl Bühler (one of the founders of the Würzburg School), failed to acknowledge his indebtedness to the German psychologist Otto Selz (1881–1943), who also inspired Herbert Simon to develop his famous problem-solving computer algorithms. Popper could thus be suspected of having been dishonest and having committed plagiarism? Unfortunately, the concept of plagiarism is contested. Plagiarism overlaps with, but is not the same as infringement of copyright, a form of intellectual property, which did not become an important issue until the nineteenth century.[5] Nobody can be charged with breaking a law that has not been made and promulgated. However, plagiarism has not only a legal but also a moral dimension; it touches on the intricate relationship between law and morality, two normative systems.[6] Are Isaac Newton, Adam Smith and Charles Darwin accusable of plagiarism or are the works of these British icons par excellence immune from control on such wrongdoing? Are they sacrosanct? How about the hundreds of Nobel laureates, and how about historians? The line between old and new is blurred; the latter grows out of and is not a sudden, unexpected change of the former. The new does not fall out of the blue. One climbs a tree by holding with one hand tight on one of its branches. Scientists are always standing on the shoulders of other scientists. The question could, therefore, be raised whether the names of those belonging to the latter group should be mentioned at relevant and pertinent places in the books or articles written by those belonging to the former group, or whether it would be sufficient to mention the names in the ritual "acknowledgments", the short and obligated text at the beginning or end of a book. Scientists who feel called to throw stones to a colleague should ask whether they themselves have been honest and have room to accuse others. Ultimately, today's physicists are heavily indebted to pre-Socratic thinkers. If scientists were strictly required to come out with ideas that are absolutely original, or to name all their direct or indirect predecessors in order not be accused of any form of plagiarism, their productivity in terms of writing would be drastically reduced, and many publishing houses would go bust. If plagiarism is a matter of degree, the waiting is still for a clear and unambiguous criterion. A woman cannot be a bit pregnant.

(3) **Consistency**

A theory has to be able to explain all relevant existing data and predict outcomes with as good accuracy as, or better accuracy than, existing theories on the subject, because in science, you do not get to pick and choose your facts, a fraud committed by many researchers. The question, however, is: how do we know that all "existing data" are taken into account and who decides whether or to what extent the existing data are "relevant"? The concept of relevance, crucial to epistemology, is hard to define.[7]

These three points are the most important when establishing a scientific theory. However, strong theories usually have two additional attributes.

(4) **Adaptability**

No theory is perfect at the time of its creation, but strong theories are able to adapt to new information with only small changes, while maintaining their core idea. Bayesian statistics, used to update probabilities (which are degrees of belief) after obtaining new data, presents the possibility of adapting a theory, and non-monotonic logic, according to which scientists reserve the right to retract drawn conclusions in light of further information, does the same thing differently.[8] However, where is the line to be drawn between big and small "changes" and what is meant by "maintaining their core idea"?" More generally, reasoning in or about uncertain situations has become a subject of intense discussion.[9] Since the adaptability of a strong theory may not leave its core idea unchanged, the adaptability criterion is debatable.

(5) **Parsimony**

Strong theories are parsimonious. They uphold the principle of the Razor of Occam. This can be interpreted as follows: Among competing theories, the one with the fewest assumptions should be selected.[10] The parsimony criterion is weak, for one can never be sure that no assumption has been overlooked. Scientists are finding themselves in a space of uncertainty; their light is shining in deep darkness.

There are hundreds of scientific theories, some of them having been revolutionary (paradigm-shifting). For example, the heliocentric theory of Copernicus, the invisible-hand theory of Adam Smith (now shown to be a weak theory), the oxygen theory of combustion of Antoine Lavoisier, the set theory of Georg Cantor, the evolution theory of Charles Darwin, the genetic theory of Gregor Mendel, the Gestalt theory of Christian von Ehrenfels, the linguistic theory of Ferdinand de Saussure, the relativity theory of Albert Einstein, the quantum theory of Max Planck and Niels Bohr, the game theory of John von Neumann and Oskar Morgenstern, the information theory of Claude Shannon, the general systems theory of Ludwig von Bertalanffy, the plate theory of Alfred Wegener, the symbolic interaction theory of

George Herbert Mead, the social theory of Kenneth Arrow and the computational mind theory of Alan Turing. For a list of alternative theories and critics in the field of physics or natural philosophy, visit http://editionsassailly.com/dissidents_english.htm.

Many people discount the value of a theory in practical situations, but practice cannot do without theory, and vice versa. As the Romans said, "*Praxis sine teoria sicut currus sine* via *sed teoria sine praxi sicut currus sine axi*" (practice without theory is like a chariot without road, but theory without practice is like a chariot without axle). "Theory" comes from *θεωρία* (to look at, to behold), "practice" from *πράττω* (to do).[11] According to Aristotle, "the best way of life is devotion to the appreciation of the truth that organizes the universe, to the realization of our nature and to the full development of our mind". This activity, the finest and most excellent we can be engaged in, he called *θεωρία* (theoretical wisdom), which he distinguishes but not separated from *φρόνησις* (practical wisdom). Aristotle, who holds a position of unparalleled importance in the history of philosophy (and is an excellent philosopher), explicated in *Politics* (the continuation of his *Nicomachean Ethics*) that the best form of *φρόνησις* is directed toward the practical promotion of the good life of the whole community (*πόλις*); the essence of *θεωρία* is "seeing things together and contemplating the whole". Knowing and doing, he emphasized, shall not be uncoupled.[12] If the Stagirit (Plato's best pupil) was right, one wonders what will happen to a society where the two are separated. Is it doomed to collapse?

Modern scientists do not embrace the views of Aristotle. They take scientific theories to be explanations of only parts of the whole, that is to say, theories based on ideas which are not universally tested. The essence of their theories is seeing *some* things together but not contemplating the whole, the togetherness of *all* things. Generally speaking, and with due respect, modern scientists are thus narrow-minded. Their theories are consistent with a pretty long time of scientific inquiry and have so far weathered every test designed to prove them wrong, whether people subscribe or appeal to them or not. They are claimed to be close to the facts, leaving room for more learning and new understanding (adaptability criterion)—room not always taken up, however.

Theories of the solar system, based on Newtonian gravity, not only explain but also allow us to predict where a planet will be on some day and even hour 100 years from now. Darwin's theory of natural selection is an unsurpassed explanation, but it barely predicts and allows experiments of only a quite limited sort. Social theories explain, predict and permit experimentation, all to a degree, not with perfect confidence, but with enough confidence to be more useful than just winging it. Permitting social experimentation, however, does not mean that there is an opportunity for it or that the experiment is actually carried out. Running social experiments can be severely restricted or even prohibited. Country pundits, on the other hand, conjure up mystifying explanations without predicting accurately or even approximately. Not meeting any of the criteria mentioned, they have only stories, narratives and opinions to offer, not *scientific* theories. Barring the possibility of borrowing a theory from social scientists, who never speak *unisono* and rarely if ever step out of their Western box, country pundits only hazard guesses and express their belief. They just hedge, leaving their audience or readership in the dark: It might freeze, it might thaw. No Sinologist predicted the "Cultural Revolution" launched by squeaky-voiced Mao Zedong, the big economic turnaround initiated by little Deng Xiaoping, or the "Belt and Road Initiative" taken by enigmatically smiling Xi Jinping. No Iranologist predicted Ayatollah Khomeini's return to Iran, no Arabist predicted the "Arab Spring" (or its untimely demise), and no Americanist predicted Kennedy's assassination, Nixon's resignation or Trump's election. As the theories of "area/country experts" have not been *scientific* theories, one begins to wonder what they are doing at universities and what they are paid for.

If a scientific theory is an explanatory set of statement, the question may be asked how these statements are established. Basically, there are three possibilities (by perception, inference or reference), one of them (perception) belonging to the domain of direct knowledge, the others to the field of indirect knowledge. The literature on each possibility is overwhelming. The camps representing them talk at cross-purpose, because they fail to really understand each other. Epistemologists (philosophers who are concerned with the nature, scope, means or source[s] of knowledge) have so far not been able to reconcile the camps. The knowledge distilled into a theory can be established by using one, or a combination, of the possibilities indicated. Knowledge that is based on transmitted statements made by other people (knowledge by reference or testimony) raises not only questions about credibility (the visions of people like Moses, Paulus and Muhammad, who lived a very long time ago, do not count as scientific knowledge) but also questions belonging to the field of linguistic analysis[13] Ultimately, knowledge by reference or testimony is based on knowledge gained by perception or inference. Only two possibilities are then left. If Aristotle was right in saying that "*nihil est in intellectu quod non erat in* sensu", then only one possibility remains: direct knowledge, that is, knowledge by perception. Indian scholars call this *pratyakṣa*, that which is perceptible to the eye or visible.[14] The notion of perception, however, is problematic. In *Les Mots et les Choses: Une archéology des* sciences *humaines* (Gallimard, 1966, *Préface*), Michel Foucault speaks of "*le regard déjà codé*" (the encoded eye).[15] The notion of perception is even *highly* problematic, as scientists in the field of quantum mechanics can explain. With "eyes in eyes" (Shakespeare, *Venus and Adonis*, line 120), what do lovers *see*? The literature on perception is very large, and the dispute over it is far from settled.[16] The foundations of scientific knowledge are thus unstable. No scientist knows the fundamental nature of reality, the most basic relationship between mind and matter, or how to classify classifications.[17] The rules of a universal logic have not been found yet.[18] The 7th World Congress on the Square of Opposition is scheduled for September 2021, in Leuven (visit http://www.square-of-opposition.org).

Scientific theories are based on axioms, statements that are *assumed* to be true and serve as starting point for further reasoning. Axiomatization is the process of reducing a number of scientific statements to a system of axioms.[19] Scientific theories or statements used to be mathematically expressed. However, if all mathematics is based on axioms, mathematically expressed scientific theories are based on assumptions, on uncertainty. In other words, they smacks of religious dogmatism. In the so-called hard sciences, theorems are deduced from axioms, which are based on beliefs, which are thought to belong exclusively to the domain of theologians.The book to be read is Nancy Frankenberry (ed.), *The Faith of Scientists* (PUP, 2008). Science is thus fundamentally shaky for a second reason: Its axioms may be "self-evident" and in no need of further argument or investigation, they are *assumed and believed* to be true all the same. If axioms are self-evident, the axiomatic method is pretty much the same as the phenomenological method.[20] Is this the missing link between natural and cultural sciences? It was the Swiss mathematician Jacob Bernoulli (1654–1705) who proclaimed that axioms and beliefs (no matter their degree) should not be the foundation of science.[21] The word axiom comes from ἀξίωμα (what is thought to be worthy). It has an evaluative flavor, in the sense of being subjective rather than objective. Thought or believed to be worthy is not the same as proven to be worthy. What is the word worthy meaning anyway? Whereas "true" belongs to the ontological domain, "worthy" is, like "valuable", an axiological concept. Are what *is* and what *ought to be* reconcilable with each other? Axioms are believed to be true, leaving no room for being false (not-true) and excluding the possibility of being true and not-true at the same time or place. This possibility (*tertium datur*), ignored by the vast majority of scientists, is far from imaginary, as the following examples may show. A door ajar is a door that is neither closed nor open; it is both closed and open. At twilight, it is neither bright nor dark. Gray things are neither black nor white. Bicycling or driving at the middle of the road, one is moving neither on its left nor on its right side, neither of which stands alone! Playing football is much more than kicking a ball ($\bar{A}$) after obtaining it (A), illogical activities when taken out of their context. If male and female are totally different, why have they been driven toward one another? Speaking of somebody's wife or husband, people from Britain usually say: "your other/better half". Zero, without which no calculation is possible, is neither a positive nor a negative number;[22] at the saddle point, a plane *begins* to slope up and down, but is not yet sloping up or down; and while the state of a bit can only be either 0 or 1, the state of a *qubit* can be a coherently both.[23]

The proposition "$1 + 1 = 2$" seems to be evidently true, but it took Russell and Whitehead many pages to prove this in their three-volume book *Principia Mathematica* (1910–1913). In fact, the proposition is *not* evidently true. Only two basic operations can be performed on natural numbers: addition and multiplication. Addition is concerned with aggregating or totaling *individual* numbers, as in $2 + 3 = 1 + 1 + 1 + 1 + 1$. Multiplication is a matter of adding up *groups* of numbers. Four times five (4×5) means four groups of five. Subtraction and division are inverse operations of addition and multiplication, respectively, i.e., subtraction is performed by undoing addition and division by undoing multiplication. So far so good as long we are engaged in arithmetic, which is—let us not forget it—a human invention. However, if it is true that nothing stands alone, it is impossible to add up what belongs together in the first place, and if it is undeniable that all things hang together, reality can only be divided into parts artificially. Reality is one and undivided. Distinctions and separations are made by man (and the *decoupages* vary from culture to culture). Cutting an apple pie into equal slices is not easy. Somebody has to be satisfied with a lesser part of the delicacy.

In the field of complex systems (to be distinguished from complicated systems!), the old idea that we will be able to understand a whole system by adding up its (analyzed) parts is seriously challenged. One and one may well make two, but in order to really understand two, we have to understand (a) the nature of the first one, (b) the nature of the second one and (c) the meaning of that innocent looking word and. Making or forming an *ensemble* should not be the lumping of different (groups of) things together and thinking about, or dealing with, them in the same way, like putting children together in one class, regardless of their quality. This would be a tall order, seldom if ever duly carried out. The proposition "$1 + 1 = 2$" cannot be taken universally true. In the biological realm, for example, one person and another person may add up to more than two persons. Over time, Adam and Eve (one person and another person) have multiplied themselves into more than 7.8 billion different children (go to https://www.prb.org/2020-world-population-data-sheet). All of them can be counted, one by one, but surprisingly few of them *really* count. Those who belong to "the happy few", the *élite*, or the class of "the High and the Mighty" consider themselves as being "*au-dessus de la mêlée*". They are not democrats, not really. *They* are the ones who really count; the rest is "*vulgus*", the "plebs", "*die Viel-zu-Vielen*" (Nietzsche). The relentless struggle of people for a higher place on the social ladder is proof of it: "democracy" is a pacifier.

Modern democracy (the system of gvernment in which, as officially stated, every citizen in the country has only one vote to elect its officials and votes are not weighed but counted) is a myth which the ruling elite likes to uphold. According to Paul Valéry, democracy means "*l'ensemble des non-chefs est un chef*". No everyday man should delude himself into believing that his vote really counts and that a king, president or party leader considers himself to be just like him: an ordinary person. Think of it: a member of the royal family waiting in the queue, like everybody else, or wearing *prêt-à-porter* clothes! In Lilliput, "all animals are equal, but some animals are more equal than others", George Orwell wryly wrote in his allegorical novella *Animal Farm* (1945).

The point here is that ordinary people avoid committing *lèse-majesté*. Having a deep, ineradicable desire to look up to and revere a person (personality cult) and being ready to prostrate themselves in front of the king or to risk their own life for him in battle, they do not want their king thinking that he is not above the fray and behaving as not being special, which raises the painful question how a gold digging commoner with red blood streaming in her veins suddenly transforms into a blue-blooded person enjoying certain privileges and prerogatives.[24] Everybody knows that a hand's thumb and little finger are equally important in playing the piano. In the eyes of his underlings, however, the king and I (one person and another person) are not just two persons to be equally treated with respect. Though presidents and communist party leaders ousted the king from power and did not allow him to regain the throne (actions about which countless books have been written), they have not renounced his godlike behavior and privileges. The *Ancien Régime* has not really been removed. It is just old wine in new bottles. Petronius, said to be the author of *Satyricon*, knew it already: "*Mundus vult decipi, ergo decipiatur*".

In 1832, on the eve of his fatal duel, twenty year old Évariste Galois, scribbling down the solution to a problem that had been standing for 350 years, laid the foundations of group theory, a major branch of abstract or modern algebra. He made the theory of equations into a part of the theory of groups. According to this theory, a group is a non-empty set of elements, which can be binarily combined in such a way that the combination also belongs to the set, provided three group-*axioms* are satisfied. One of them is the associative law saying that it doesn't matter how we group the numbers. For example, $(6 + 3) + 4 = 6 + (3 + 4)$, because

$9+4=6+7=13$. Whereas a set is a collection of entities, a group is a special set, namely a set that has to obey a number of rules in regard to the group operations carried out. There are many applications of the theory of groups, *e.g.*, in mathematics, physics, chemistry, cryptography, virology, economics and music. [25] Group theory and information theory are interrelated.[26] Group theory explains the phenomenon of symmetry.[27] There seems to be even a group theoretical approach to string theory, which is said to point towards a theory of everything. "The theory of groups is, as it were, the whole of mathematics stripped of its matter and reduced to pure form" (Henri Poincaré). The problem, however, is that group theory cannot really be applied to a society or culture, because the associative law does not hold where human minds are involved. The way in which individuals are grouped certainly does matter, and sometimes very much so; to assume that it doesn't would be utterly unrealistic. Though most people know nothing but "*boire, bouffer et baiser*", they cannot be aggregated like cattle. Contravening this directive is the poverty of economics and politics. Each man is unique and two unique human beings cannot be added together without violating the uniqueness which makes them fundamentally different. It would be trivial to remark that human beings have many things in common. Everybody has to go to the toilet and has a bad breath when waking up in the morning. A thermometer under the tongue of a healthy Eskimo or pygmy gives the same reading.

A man wearing a uniform degrades himself. For, by wearing this set of clothes, he shows and even shows off that he is *not* unique, hence replaceable like any material thing (battery, brick, cartridge, paving stone, etc.). The general in his uniform is proud of being general, not of being special. On the other hand, he is proud of the pinned-up decorations making him special, not general. Being the same means not being different. Uniformity is the negation of existence (from "*ex*" and "*sistere*" [to step out of the row]). The soldier who was summoned by his superior because he had not marched in time during the parade understood it very well by promptly answering that it was not he, but the rest of the platoon who marched out of time. Another anecdote, carrying the same message, is the one about the alleged meeting of Diogenes (of Sinope) and Alexander the Great. The story speaks volumes about the uniqueness and incompatibility of human beings. Who is more important (and to whom?): the king, the queen, Vladimir Poetin, Xi Jinping, the Pope, Jeff Bezos, Bill Gates, Warren Buffet, or I? What is to be praised and stimulated: conformity (uniformity) or non-conformity? A dictator can only maintain his position if people, particularly his hangmen, cronies and sycophants, allow him to be their dictator; a billionaire can only stay being billionaire as long as he manages to hide his way of becoming ultra-rich and succeeds in mesmerizing people who want to be as (or even more) wealthy; and a member of the royal family can only be parasitic as long as the credulous crowd tolerates it. Better is it, I think, to stay aloof from the crowd, to wear no uniform and never to degrade oneself in front of any other person.

Theorems are inferences from and logical consequences of axioms, and in his book *99 Variations on a Proof* (PUP, 2019) mathematician Philip Ording demonstrates that different proofs of the same theorem may be delivered. A long list of theorems can be found on the Internet. Perhaps the most famous one are Kurt Gödel's "incompleteness theorems". The first one, formulated in 1931, ended a half-century of attempts (beginning with the work of Frege and culminating in Whitehead and Russell's landmark book and David Hilbert's formalism) to find a set of axioms sufficient for all mathematics. From 1931 until later in his life, Gödel (1906–1978) vainly attempted to overcome the problem he had presented his perplexed opponents with by calling for the pursuit of a new set of axioms. Man, he thought, would never be able to find a formal system general enough to apply to any kind of classification. In other words, a classification of all classifications (meta-classification) would be beyond man's reach.[28] It should be noted that Gödel's theorems too are statements proven on the basis of axioms, that is, statements deduced from some "*idées claires et distinctes*" (Descartes). Gödel was essentially a believer, not a scientist in the sense Bernoulli had given to "science".

Whereas a scientific theory is an explanatory set of statements about specific phenomena based on repeatedly tested and established ideas (but ultimately on something "self-evidently true"), a model is a tool helpful to understand these phenomena.[29] The word model derives from the Italian "*modello*", a word that originally referred to the drawing, painting, or (clay or wax) sculpture made for a patron either in the hope of gaining a commission to make a piece of art (for which the artist would be paid) or to give a visual idea of the commission already placed. The *modello* was a more elaborate and accurate representation of the final work (still to be made) than the *bozzetto* (sketch), but both provided only intentions of performing a piece of art. Sculpted *modelli* also served as guides for assistants in the atelier or studio, as models after which they could study problems of anatomy or drapery. So-called pattern or model books, which were collections of pages filled with pen and ink drawings, served the function of giving examples. The Western tradition of drawing from the nude goes back to the Renaissance. The models chosen were generally tall and slender. The first drawings of professional models belong to the latter part of the fifteenth century. However, the search for the nature of reality and its concomitant retreat from naturalism (aka realism) in the twentieth century have made drawing from the life, which often means from the nude, problematic, though it is still part of the curriculum in most visual arts schools.[30]

The practice of science is a messy business full of jealousy, gossip, fraud, deception, corruption, abuse of power and frustration, but it is essentially building a model. Scientists usually ask: "What are the equations?", but they are less concerned with the relatedness of everything. Equations can be differently classified. Important types are algebraic, difference, differential and functional equations. Mathematical models (see below, under symbolic models) are systems of equations. Whereas scientific theories are sets of fundamentally shaky statement about a piece of reality, scientific models are incomplete representations of it. A scientific model is thus not an explanation but a representation of a part of reality.

Ernst Mach (1838–1916) famously said:

> *Alle Wissenschaft hat Erfahrungen zu ersetzen oder zu ersparen durch Nachbildung oder Vorbildung von Tatsachen in Gedanken, welche Nachbildungen leichter zur Hand sind als die Erfahrung selbst und dieselbe in mancher Beziehung vertreten können.*[31]

A scientific model is a human construct, a simplification meant to demonstrate, or stimulate research on, particular features of a phenomenon or group of phenomena. Models are interpretative representations, always afflicted with the human factor, never free of the subject(s) having built the model. In *Περὶ Ἑρμηνείας* (on interpretation), Aristotle dealt already with the problem of interpretation (assignment of meaning). A model, having only some of the characteristics of what is modeled, is selective. In model building (modeling), two elements are involved: concepts of something and the relationships thought to prevail between them. Defining the concepts and depicting the form of the relationships among them can be quite formidable tasks. I will not deal with the intricate and convoluted problem of concepts, which seem to be a kind of systems.[32] I only deal, albeit briefly, with the problem of expressing (the form of) the relationships found or thought to prevail between the concepts of the model.[33] The importance of doing so can be seen when explaining the operation of an organization. It would be inadequate just to list the various officers. Even an organogram would not be helpful, for it indicates lines of authority at the most, but tells little, if anything, about working relationships, and absolutely nothing about organization-to-organization interactions. Basically, there are three kinds of models.

(1) **Iconic Models**

Iconic models look like what they represent. Examples are: a scale model (of a car, train, ship, aircraft, engine, building, piece of furniture, garden, town, mountain range, river system, organ, animal or human body), in fact, almost anything that can be visualized. Not all iconic models, however, are three-dimensional. Photographs and X-rays are two-dimensional iconic models. An iconic model is similar in appearance to what is represented. For example, a model train has some of the characteristics of a real train. It has similar parts and possibly moves in a similar manner, but it does not have all the features of a real train. By examining the scale model, one can learn something about a real train's size, shape and design, but we cannot tell much about its horsepower, speed, capacity or mechanical dependability. The word icon derives from εἰκών (image).

According to the Bible (on which the American Constitution is based), man is created in/as the image of God (Gn. 1: 26-27). Consequently, all men are equal, an inference which Chinese steeped in the Confucian tradition will not make and to which privileged people in the West only pay lip service. *Iconostasis* is the name given to the screen separating the nave from the sanctuary in a Byzantine church. The word *iconoclasm* refers to the destruction of icons in Byzantium between 726 and 843. At that time, the worship of icons was considered idolatrous. Generally, Islam tends to be more iconoclastic than Christianity, with Sunni Islam being more iconoclastic than Shia Islam. The name of God has been a controversial issue, indeed an irremovable stumbling block in the Abrahamic religions.[34] Initially, the word *iconography* referred to a discipline auxiliary to archaeology. From the 19th century onwards, however, the meaning of the term was extended to cover the descriptive and classificatory investigation of a piece of art. *Iconology* aims to contextualize works of art culturally and explores their possible meanings.[35] At his home page, W.J.T. Mitchell, speaking of the "jungle of imaginary", defines iconology as "the study of images across the media." For *mimesis* (imitation), see Erich Auerbach, *Mimesis: Dargestellte Wirklichkeit in der abendländischen Literatur* (Francke, 1946) and Saija Isomaa et al (eds.), *Rethinking Mimesis* (Cambridge Scholars Pyblishing, 2012). In 1981, *Simulacres et Simulations*, a groundbreaking book written by Jean Baudrillard (1929-2007), was published in Paris. For the philosophy of imagination, see Amy Kind (ed.), *Handbook of Philosophy of Imagination* (Routledge, 2016), and for scientific imagination, see Arnon Levy and Peter Godfrey-Smith (eds.), *The Scientific Imagination* (OUP, 2020). *Imagology* (aka intercultural hermeneutics) studies cultural manifestations such as books, articles, movies, press releases and theater plays, assessing their imaging of peoples, cultures or societies. It problematizes ethnocentrism and national/ethnic prejudice. Being a mid-20th century offshoot of comparative literature, imagology overlaps with art history and social psychology.[36] Today, images are more important than words. A comparative study of the history and cultural impact of images, words, names and numbers is still a *desideratum*.

An icon is (a) a picture or figure of a person considered by adherents of the Greek or Russian Church to be holy, (b) someone or something thought by many people to represent an important idea (each country has its icons), or (c) a sign or picture on a computer screen that is used to start a particular operation. Photo models are a special category of iconic models. They also look like something, not like a physical object but like an ideal or dream maliciously manipulated by people waiting in the wings and eager to line their pockets, that is to say, by people behaving like pimps.[37] Photo models, i.e., representations of what the viewer would like to have or to be, are often said to be victims of the caprices of fashion. However, flattered by the lustful photo agent or camera operator, these "innocent" models easily sign away all the rights to their image when they pose for stock photos.

Verbal models have advantages comparable with those of iconic models. They can be rich in content and their descriptions can be graphic, or "thick" (a term coined by Gilbert Ryle and used by Clifford Geertz). They can deal with extremely complicated phenomena (*explananda*), *e.g.,* the Protestant Reformation, the French, American, or Russian Revolution, the First, or the Second, World War, or the Rise of China (and the concomitant Decline of the West). Iconic or analog models (see below) are not capable of doing this. However, due to the inherent ambiguity of human language, the advantages do not outweigh the disadvantages of verbal models. For what is to be understood by a "case": a briefcase, bookcase, suitcase, glasses case, pencil case, violin case, packing case, test case, criminal case, civil case or situation? What is meant by "to clean": to wipe (up), brush, dust, mob, sweep, scrub, flush, rinse, vacuum, or wash (up)? What by "to fasten": to buckle, tie, zip, pin, lock or bolt? What by "shape": a square, quadrilateral, rhombus, rectangle, trapezium, parallelogram, triangle, pentagon, hexagon, octagon, circle, sphere, pyramid, prism, cylinder, cone or cube? What "measure" is meant: the thermometer, stopwatch, timer, tape measure, kitchen scale, or bathroom scale? On which chair do you want me to sit: the table chair, armchair, director's chair, captain's chair, office chair, deckchair, lounge chair, swivel chair, ball chair, barber's chair, rocking chair, bubble chair, cantilever chair, folding chair, kneeling chair, shower chair, barrel chair, beach chair, tulip chair, lawn chair, club chair, butterfly chair, carver chair, wing chair, x-chair, zigzag chair, Adirondack chair, Egg chair, Tongue chair, Windsor chair, Glastonbury chair, Wassily chair, Swan chair, stool, wheelchair, bikini chair, throne or electric chair? The context or further specifications may clarify what is meant by "pipe": a tube through which a liquid or gas flows, an organ pipe or a thing used for smoking tobacco. Meaning something that is used to keep something in, we can be more specific by saying "sardine tin", "cheese packet", "match box", "cigar box", "toolbox", "shoebox", "cola can", "milk carton", "butter tub", "toothpaste tube", "shampoo sachet", "beer keg", "baby, beer, milk, perfume, plastic, water, or wine bottle", "potato sack", "shopping bag", "oil barrel", "salt shaker", "coal scuttle", "suitcase", "bookcase", "dustbin", "glass", "flask", "flacon", "tumbler", "flute", "water tank", "crate", "carafe" "container", "jerrycan", "vase", "pail", "carboy", "cup", "mug", "growler", "pitcher", "jar", "tankard", "safe", "pot ", "prison", "cellar", "dungeon", "satchel", "pouch", "atomizer", "chest", "trunk", "decanter", "canteen" or "wallet".

The Occitan vocabulary counts 62 words for swamp, 75 words for lightning and 128 words to designate cultivated land. The fact remains, however, that a word can have more than one meaning, a linguistic phenomenon indicated by "polysemy" (for "colexification", a term used by linguistic anthropologists to avoid the distinction between polysemy and vagueness, see *Science*, 399:6472 [December 20, 2019], pp. 1517–1522). The human language swarms with hypernyms, hyponyms, co-hyponyms, homonyms, meronyms and synonyms, the result often being a breakdown in communications, if not worse. There is no universal definition of "word". What precisely is the link between words and their information content (the aboutness of words in sentences), and has the social embeddedness of natural language (language use) anything to do with that connection? In his landmark paper *Über Sinn und Bedeutung* (1892), Gottlob Frege set the ball rolling—a ball that is still rolling. Verbal models can deal with extremely complicated phenomena or situations, but whether they can deal with them successfully is a matter of dispute. Misunderstanding lurks. The descriptions provided in natural or ordinary languages are not yet (some would say: cannot be) normalized into a set of formal sentences. Mathematical formalization will go at the expense of readability.

(2) **Analog Models**

Whereas iconic models visually depict what they represent, an analog model does not resemble or look like what it models. It is not similar to it, in appearance. An orchard could be planned by constructing an iconic model. One could make a box that had proportionally the same boundary design as the orchard. The box could then be filled with soil to represent the surface area of the land involved, and the potential tree location could be analyzed by placing tree models in various places in the box. However, the process of building such an iconic model may be more difficult than is necessary to solve the problem of tree location. An alternative would be to represent the orchard boundaries and the land contours by lines drawn to scale on a piece of paper. The potential tree locations could be represented by dots with possibly a circle drawn around each point to represent the average area required by a mature tree. Such a model would be a blueprint and could be useful for planning an orchard, but it would certainly not look like an orchard. The builder of this analog model has only indicated the concepts and relationships he feels are relevant to the planning. The model does not provide as much information as an iconic model, but might be more useful. There are blueprints of buildings, seaports, airports, towns and even cities. Comparing the blueprints and searching for their common feature(s) might be useful. Sometimes it is impossible to build an iconic model because the thing to be represented is not visible. Showing the relationship between temperature and time cannot be done with an iconic model. With an analog model, however, it can be done. One could design a chart with time on the horizontal and temperature on the vertical axis, and plot the data as points, connecting these points by a line to represent the relationship between temperature and time. Temperature and time could be replaced by concepts used in biology, ecology, demography, geography, economics, finance, psychology, sociology or political science. An atlas is a collection of maps, which are examples of an analog model.

Sometimes one may think one has an analog model when in fact it is not. The organization chart is a case in point: Lines connect different positions at different levels. For the chart to be an analog model, something about it must be analogous to the relationships in the real situation being represented. If the lines are the critical elements, their thickness could indicate the form of the relationship. The lines could also be dashed or dotted. For the most part, however, traditional organograms are very limited models at best. A similar problem arises with maps depicting the relationship between seaports by connecting them with lines of equal thickness. Very little is revealed by such lines other than that products move between the ports connected. If the width of the lines is made proportional to either the weight or the value of the goods being moved, the relationship between the ports is depicted more significantly.

Foucault (*op. cit.* in this chapter) identifies four kinds of *similitudes*, analogy being one of them. Analogy (from ἀναλογία [proportion]) is the process of transferring meaning from a particular subject (the source) to another (the target). The word analogy can also refer to the linguistic expression that corresponds to such a process. Analogy has been studied since antiquity by philosophers, theologians and lawyers. Over the last few decades, renewed interest in the phenomenon has been shown, most notably in the cognitive sciences, which study the mind and its processes.[38] Some Indian philosophers have been considering analogy (*upamāna*) as a means or source of knowledge in its own right.[39] Analogy is to be distinguished from isomorphism and metaphor. Two mathematical objects are isomorphic if an isomorphism (a one-to-one correspondence) exists between them; sociologists speak of isomorphism when there is a similarity of the processes or structure of one organization to those of another. While analogies compare things so that one can see a relationship between them ("Finding an honest person is like finding a needle in a haystack"), metaphors, which are related to myths, do this by saying that something is something else ("the company is a sinking ship").[40]

(3) **Symbolic Models**

These models use symbols to represent variables (instead of concepts) and their relationships. Symbols stand for, or take the place of, other things; they indicate or signify something. For the sake of convenience, I take the words symbol and sign to be interchangeable, contravening the teachings of Charles Sanders Peirce who argued for three kinds of signs (icons, indexes and symbols). Semantics, a subfield of linguistics, is concerned with the relationship between signifiers and what they stand for, their denotation; it is the study of the meaning of words and phrases and spills over into philosophy of language, which explores the relationship between language and reality. Semiotics, the "science of signs", casts a much wider net than linguistics; it is subsumed under social psychology.[41] Variables are operational concepts, i.e., concepts that can be handled, concepts that are—so to speak—tamed and domesticated. Concepts such as culture, art, wisdom, love, honor, education, experience, happiness, religion, magic, freedom, progress, tradition, war and reality cannot be handled yet; they are too clumsy, too unwieldy, but they are concepts all the same, meaning that things are taken together. For a long time, the concept of heat was also unwieldy, until Sadi Carnot, William Thomson, Rudolf Clausius, James Maxwell,

Ludwig Bolzmann, Josiah Gibbs, Gustav Zeuner, and Johannes van der Waals came along. Their scientific bang is still reverberating.[42]

The World Health Organization (WHO) defines human health as "a state of complete physical, mental and social well-being" and not merely as the absence of disease or infirmity, but the definition lacks operational value, and the word "complete" is problematic, for it makes health practically impossible to achieve. However, physicians, having an arsenal of medical equipment at their disposal, know how to operationalize the concept of health physically, while dozens of rating scales are used in psychology, psychiatry and social psychology. A surgeon, who is supposed to have sufficient knowledge of the human body's anatomy and physiology, knows where and how to cut. Medicines, of which there is a great many of types, consist of biotic or abiotic ingredients that are combined by the pharmaceutical industry with the greatest possible care (whether poor patients can afford to buy them is another matter). Medicines are to be taken in a certain dose at a certain time.

Behavior had been another unwieldy concept until it was realized that (a) human behavior is a range of actions being the concern of scientists from different disciplines and (b) actions are fundamentally different from events. Whereas reasons, motives or interests are involved in actions, causes lay at the root of events.[43] Readers of a textbook in micro-economics will further notice that the concepts of consumer and producer behavior are ingeniously made operational.

"Law", the concept/word that refers to the set of rules regulating human behavior, was operationalized as soon as a distinction was made between natural law and positive law. Philosophy of law, however, is a much debated subject matter. Go to https://iep.utm.edu/law-phil and https://philpapers.org (fill in: philosophy of law). Finally, whoever looks at the list of Olympic sports and knows (a) how individual sports(wo)men are selected, trained, coached and supported by a technical staff, (b) how technologically advanced sports equipment are, and (c) which sporting rules have to be observed will realize to what extent the once vague concept of sport has been operationalized.

Every object, thing or person possesses attributes, properties by which it is characterized. Postponing the dealing with the philosophical problem of the relationship between essential properties, that a thing must have for existing, and accidental properties, that it could exist without, I take a variable to be a set of attributes. A variable is something that can take different values; in other words, it is something (set of attributes) the value of which may vary or change, either over time or in relation to the value of another variable. The value of a variable may be different under different conditions, in different situations or circumstances. The word value is both a verb and a noun. Its root meaning involves human judgment. Axiology (from Greek *ἀξία* [value, worth] and *λόγος* [discourse]) is the intricate study of value.[44] Examples of variables (and their attendant values) are gender (male, or female), social class (upper class, middle class, or lower class), marital status (married, divorced/separated, widowed or single), drinking habits (drinking a lot, drinking socially or not drinking at all), type of car owned (European, Japanese or car made elsewhere), opinion about gay marriage (strongly against, against, in favor, strongly in favor, no opinion), country of birth (Italy, Canada, Argentina, Russia,

Japan, …, or Kenya), eye color (brown, blue, green, hazel, amber, red or "difficult to decide") and age (contains numerous values, because it may be anything from 0 upwards). The value of income may vary over time (it can steadily or abruptly go up or down); it may also differ between men and women or/and between professions practiced in a certain country.

"Variable" should be distinguished from "data" (the word, derived from "*datum*" [what is given], is used in the singular or in the plural). Data are pieces of information that can be collected, stored, updated, transferred, analyzed and processed. Entropy, as understood in "communication theory" (Shannon's term),[45] is the uncertainty of a system behavior, the variety of information or message that a system may convey. Physical (thermodynamic) entropy, interpreted as the degree of disorder of a system, is a special case of *information* entropy. The bit is a measure for the amount of information. Bits are basic units of information, and *communication* entropy is measured in bits. It may be called sort of scandal that in our Information Age a unified and generally accepted definition of information is still lacking. Information theory, being the study of the quantification, storage, retrieval and transmission of information and information science, being concerned with the analysis, classification, manipulation and dissemination of information, should not be muddled up. They are differently related to complexity. A bit (or binary digit) can have only one of two values, commonly represented by either 0 or 1. A byte consists of eight bits. A kilobyte is one thousand bytes, one megabyte one thousand kilobytes, one gigabyte one thousand megabytes of information, etc. Big data are datasets so voluminous and complex that the traditional data-processing application software is inadequate to deal with them. Volume, velocity, variety, veracity and volatility are the 5 Vs of big data. "Volume" refers to the vast amounts of data generated every second (emails, twitter, messages, photos, video clips, sensor data, etc.). "Velocity" refers to the speed at which data are generated and moved around (social media, credit card transactions, etc.). "Variety" refers to the different sources and types of structured or unstructured data. "Veracity" refers to the messiness and trustworthiness of data. "Volatility" refers to the term of data validity and the period of time they should be stored.[46]

There are different types of variables: quantitative, qualitative, dependent, independent, endogenous and exogenous variables. A few words will be said about each of them.

Quantitative Versus Qualitative Variables

Quantitative or numeric variables have values that describe a quantity, so that it makes sense to ask: "How many?", or "How much?" Data collected for a quantitative variable are quantitative data. Quantitative variables may be divided into discrete variables (e.g., the number of registered cars, cows on farms in a certain region, bricks in a wall or children in a family) and continuous variables (e.g., height, weight, time or temperature). A discrete variable cannot take the value of a fraction (a family cannot have two children and half a child); its values have breaks or jumps. The value given to the observation of a continuous variable can be as small as the measuring instrument allows[47]; the values of such a variable have no breaks. Reductionist scientists (see

Chap. 9) have a predilection for discreteness. Thinking of parts (molecules, atoms, elements or members), neglecting or de-emphasizing the relationships between the parts and preferring the concrete to the abstract, they tend to go for the distinct, the recognizable. Asking reductionists to classify a door ajar confuses them. They are inclined to count in the first place. Holistic scientists, however, are inclined to go for continuity, which leads them to the domain of mathematical analysis and its practical applications. Such scientists have an eye for the course of a planet in its orbit, the current of electricity, the fall of a raindrop, the flow of a river, the trajectory of a bullet or rocket. The motion of these objects is smooth, proceeds without jerks or gaps. None of the points on the line representing the motion has the clear-cut individuality of an integer, for between any two points, another point can be found or imagined, and there is no shortest step from one point to the next one; in fact, there is no next point at all (!). A major task of mathematicians is to harmonize the discontinuous and the continuous, i.e., to prove that "the fabric of the cosmos" (Brian Greene) has no holes or cuts.[48] See Preface.

Qualitative or non-numeric variables have values that describe a quality. It makes thus sense to ask: "What category does the variable belong to?" Like quantity, quality has been the subject of intensive and extensive research.[49] Data collected for a qualitative variable are qualitative data. Qualitative variables may be divided into nominal variables (e.g., gender, type of house, brand or religion) and ordinal variables (e.g., academic grade, social class, clothing size or attitude). Nominal variables cannot be assigned any order. Ordinal variables take on values that can be ordered or ranked, with no information about the interval between the things being ordered (e.g., saying "higher/more than" without indicating how much higher/more than).

In building a model, we saw, two elements are involved: concepts and the relations prevailing between them. We also saw that symbolic models use symbols to represent variables (instead of concepts) and their relations. Having explained what a variable is and having indicated the difference between quantitative and qualitative variables, I make a short detour to clarify the concept of relation, because the importance of the relationship between a whole and its constituent parts and between the parts is emphasized in this book.

Like quantity and quality, relation is a basic term, yet its meaning is elusive. Relation is the prototype of an indefinable notion. Bertrand Russell once pointed out, "It seems to be impossible to make any statement of what relation is without using the notion of relation in doing so". There are spatial, temporal, internal, external, physical, chemical, geological, geographic, biological, ecological, personal, sexual, blood/family, public, psychological, social, ontological, epistemological, logical, linguistic, mathematical, economic, political, international, historical, categorical, technical, and computational relations.[50] I take the essence of a relation to be that the *same* is said of at least two *different* things, like a speech act, which can only be accomplished by abstracting, by taking distance from the concrete. Both A (not being B) and B (not being A) are true, though they are different. Apples differ from oranges, but they are the same when talking about fruit. The difference beween apples and oranges is what we abstract from. A higher level of analysis is taken. In a similar vein, fruit differs from cheese, but from the point of view of food, they are the same. White people differ from black people, but they are human beings all the same. A man is not a woman, and ignoring or denying this plain fact would be silly, but finding reason in it to behave towards them differently would be stupid as well. In the relation, A and B are exchanging part of their identity. Going from A to B, A is in the process of losing its identity, and vice versa: becoming (not: being) A, B is losing its identity. In becoming grey "to some degree", black and white are losing their identity. Usually, a relation obtains between two entities (substances, properties, or concepts/ideas).

There is a fundamental difference between the Western and (ancient) Chinese way of thought. Whereas Westerners see the relation between two opposites as being contradictory (the one totally excluding the other), the Chinese consider the relation to be contrary (the one including the other to some degree). So the key question in a constantly changing world is whether "left" and "right", "on" and "off", "day" and "night"", "high" and "low", "big" and "small", or "true" and "false/untrue" are contradictory or contrary concepts. See Chap. 9.

The world is full of relations. Indeed, without relations there would be no world or reality to be known. Heraclitus in Greece and Zhu Xi in China were strikingly *unisono*: there is no autonomy; nothing is isolated; all things are one. "*Alles ist vernetzt*", Alexander von Humboldt famously wrote in his masterpiece *Kosmos: Entwurf einer physischen Weltbeschreibung*. Few people, however, are able to see the overall *Vernetzung*, the universal connectedness, *Indrajāla* (Indra's net). Network scientists seem to have an inkling of it. Yājñavalkya, who probably lived at the end of the 8th century BC and is considered one of the earliest Indian philosophers, had no word for it. He rejected every effort to define the nature of Brahman by saying "*neti neti*" (not this, not this). Śaṃkara (788-820 CE), the genius and great vedāntin who is credited with having unified the different schools of thought in Hinduism, agreed: "*The ultimate truth is one, though appearing as many*".[51] Alexander Grothendieck (1928-2014), considered by many to be the greatest mathematician of the 20th century, wanted, like Stephen Hawking (1942-2018), to see *the truth that organizes the universe* (i.e., *the general structure of being* that Aristotle had been looking for). He wanted to hear *the harmony of the universe* that Pythagoras had been speaking about mysteriously.[52] Like Goethe, he wanted "to know what the world in its innermost core holds together". Perhaps unaware of it, abstracting from the concrete, Grothendieck wished to see the "great interdependence of all differences" eulogized in the *Yìjīng*.

Dependent Versus Independent Variables

The distinction made between dependent and independent (autonomous) variables is about the relation between two variables. For example, one could say that somebody's weight depends on his/her diet, leaving the variables that have a bearing on the diet out of consideration. Physicians, chemists, geneticists, geologists, economists, psychologists and sociologists, appealing to the *ceteris paribus* clause and trying to minimalize the impact of factors that could complicate their problem, search for the variable x causing or influencing variable y and call this exercise "seeing A in relation to B". For the mathematician, however, x depends on y just as y depends on x. Think of using an old-fashioned scale to weigh something; x and y are depending on each other; if y varies, x changes, and vice versa. Should more than two variables be involved, the decision as to what depends on what is more difficult to make. Two seesawing children, for example, are not weighing each other. When two people are riding a tandem bicycle, one of them may push his/her pedals more than the other does. Factors (variables) other than diet may be involved in somebody's weight; they need to be taken into account in order to have the picture right and to judge the situation adequately. The problem, however, is how to track down, identify and evaluate *all* the relevant factors and how to put them into a manageable model.

An equation is a statement asserting the equality of two variables or expressions. Equations express relations. They can be divided into different groups. Two of them may be mentioned:

- Equations can be categorized according to their validity. An equation can be definitional (e.g., profit is defined as the excess of total revenue over total cost), conditional (e.g., the quantity demanded is conditioned by the quantity supplied or the latter is conditioned by the former) or behavioral. The behavioral equation specifies, by means of coefficients, the manner in which a variable behaves in response to the change of another variable. It is through specification that the assumptions adopted for the model are expressed. A coefficient is the number by which an unknown quantity is multiplied, multiplication being, as we saw, a kind of aggregation. In the simple behavioral equations C (cost of production) = 10Q (quantity produced) and $C = Q^2$, 10 and square 2 are the coefficients (co-workers) of Q. Coefficients are statistically estimated (on the basis of samples) and may be written algebraically. There are different types, of which correlation coefficients may be the most familiar one. Correlation coefficients, which vary between 0.00 (no correlation) and 1.00 (perfect correlation), tell us whether and to what degree two sets of data are related. Coefficients may be compared with attributive or predicative adjectives assigned to nouns or substantives.
- Equations can also be classified according to their form. When graphed, a linear equation looks like a straight (not necessarily horizontal) line, and a nonlinear equation looks like a curve, the curvature of which may vary. In the former case, the slope value is constant; in the latter case, the slope value is variable. The linear and the nonlinear relationship, or trend in the data, can be strong or weak and monotonic or non-monotonic. When one variable increases and the other increases all the way, or one variable decreases and the other decreases all the way, the relation is monotonic. Otherwise, we are in the presence of a non-monotonic relationship. Linear and nonlinear relationships are subdivided into different kinds. The degree of a linear equation, being the largest variable exponent, is always 1. The degree of a nonlinear equation is at least 2. Nonlinear relationships/equations, which give a more accurate picture of how things work, play a crucial role in the science of complex systems, the study of patterns which emerge from nonlinearly interrelating parts or interacting humans (see Chap. 5).

Sometimes, there is inequality between (the values of) variables. Whereas the study of equations provides the opportunity to handle situations of equilibrium, the study of inequalities makes it possible to deal with disequilibria. The inequalities, symbolized by < (lesser than) and > (greater than), gave rise to linear programming, which—like game theory—is a special case of mathematical optimization, a subject on which a large literature is available. Input–output analysis, for example, may be thought of as a special case of linear programming. See Chap. 5.

"Function" is a slippery concept. Compare sentences such as "The function of the veins is to carry blood to the heart", "He has a lot of functions to attend", "This room may be hired for wedding functions" and "The search function has been updated". In mathematics, "function" means a variable whose value depends on the value of another variable and changes with that value. In other words, a function indicates or states how a variable depends on another variable. The list of mathematical functions, which can be functions in regard to one or more than one real or complex variable, is long. Functions can be added, subtracted, multiplied and divided. For study of the behavior of functions, the use of mathematical analysis instruments is required.[53]

When more than two variables are involved, graphical representation (visualization) quickly becomes impossible, and the builder of the symbolic model has to resort to a system of equations or inequalities which allows him to make use of advanced mathematical techniques.

Endogenous Versus Exogenous Variables

Endogenous variables are variables the values of which are determined by the values of other model variables. In contrast, variables the values of which are assumed not to be affected by the values of model variables and are taken to be given are called exogenous variables. The two terms have only meaning in the context of a specific symbolic model. Whereas the distinction between dependent and independent variables is about the relationship between two variables, the difference between endogenous and exogenous variables concerns a whole model that may (and often will) contain many variables.[54] Scientists appealing to the *ceteris paribus* clause and attempting to minimalize the impact of factors that could make their problem more difficult to solve ultimately search for the exogenous variable(s) that cause(s) a change of the model variables, calling this exercise "seeing things in relationship". In fact, they operate locally, not globally. Their theories have not been tested internationally/universally. By making variables that were at first considered exogenous endogenous, the piece of reality under investigation will be better represented. To represent/model an entire country (instead of only a part of it), a great number of variables that so far have been taken to be exogenous will thus have to be made endogenous. How many variables the perfect and really realistic country model contains is not yet known. This being the case, I think that (a) Bernoulli's *Ars Conjectandi* (available in excellent translation) will be needed for dealing with stochastic elements[55] and (b) suitable software running on a network of supercomputers will make the building and managing of a country model easier. To this end, I would like to appeal to software engineers who see their profession not as a way to make a living but as a kind of art, as a calling they have to listen to. Our endangered species cannot be saved by robots (physical objects without a soul). It can only be saved by men, i.e., by artists who have the power of limitless imagination, the imagination that, itself being invisible, reveals the true nature of the visible. The late John Lennon is my witness. Albert Einstein reputedly said that imagination is more important than knowledge. Knowledge is and will always remain limited, while imagination, being a product of the mind and potentially unlimited, can embrace the whole world.

A *computer* model is a computer program designed to simulate or mimic the behavior or possible behavior of a system. They are used in astronomy, physics, geology, biology, climatology, and—increasingly—in the social sciences. A computer model consists of algorithms translated into a computer program. Thus, one would "build a computer" and, thereupon, "run a computer program". Computer *simulation*, narrowly defined, is the actual running of the program. More broadly defined, it is a method for studying systems. Computer simulation, enormously facilitated by supercomputers (see Chap. 7), is considered the third pillar of science—next to theory formation and empirical research (experiment). The three are different from, but complementary to, each other.[56] Computer simulation is a technique used in making decisions under uncertainty. A computer program will then be written to simulate *what is likely to happen in the future*. For each variable in the problem, a probability distribution is used. A probability distribution is "a mathematical function providing the probabilities of occurrence of different possible outcomes in an experiment" (Robert Ash, *Basic Probability Theory* [Dover, 2008, pp. 66–69]). There are different kinds of probability distributions, and a list of them can be found on the Internet. Monte Carlo simulation is a method for considering all possible combinations of variables; it enables the investigation of the entire distribution of the outcomes of a project. Sensitivity analysis is a simulation technique less expensive than stochastic simulation. One starts with "best guess" estimates for each variable in the problem and then changes their values (within "reasonable" limits) to see the effects of such changes. Starting in the 1950s, computers have been used for simulating all kinds of systems, for example, cars, trains, planes, spacecraft, financial markets, biomolecular systems and weather systems. Generally speaking, computer simulation allows the building of models and their testing against experimental data. If countries are regarded as dynamical systems with an open and uncertain future, computer simulation would be an important technique for studying them, next to theory formation and empirical research related to systems. Perhaps supercomputers can handle problems with regard to superforecasting. See Philip Tetlock and Dan Gardner, *Superforecasting: The Art and Science of Prediction* (Crown Publishing, 2015).

Before leaving the issue of scientific theory or model, I wish to draw the reader's attention to the nature of, and difference between, <u>de</u>ductive and <u>re</u>ductive reasoning. *Deductive* reasoning (aka top-down logic) works from the general to the specific. Schematically: If A, then B; A is the case; B is, therefore, the case. For example: If it rains, the street is wet; it rains; so the street is wet. A deductive argument is an argument intended to guarantee the truth of the conclusion. On the other hand, *reductive* reasoning (aka bottom-up logic) works from the specific to the general. Schematically: If A, then B; B is the case; A might, therefore, be the case. For example: If it rains, the street is wet; the street is wet; ergo, it is, or was, probably raining. Reductive argument is an argument intended to make it plausible that the conclusion is true, if the conditional statement is true. But plausibility is not the same as guarantee, for the street may be wet for other reasons than raining. In a conditional statement (if A, then B), concluding from premise B that A is the case is logically invalid, and yet many people, even scientists, reason reductively. The justification for reasoning reductively has long been a serious bone of contention among philosophers.[57]

Usually, a distinction is made between inductive reduction (when the antecedent of the conditional statement is a generalization of its subsequent) and noninductive reduction (when the antecedent of such a sentence is not a generalization of its subsequent). When the induction is enumerative, a generalization is inferred from evidence about instances of the generalization. For example: All crows are black, because some crows are black. When the induction is hypothetical, a hypothesis is inferred as an explanation of the evidence provided. For example: concluding that a man is hiding behind the curtain from the perception of the tips of his shoes peeping out from under the curtain.

Using the deductive method, a researcher *tests a theory* by collecting and investigating empirical data from a great variety of sources to see whether the theory he or she constructed is universally (invariably) true. Using the method of inductive reasoning, the researcher *constructs a theory* after collecting and investigating as much empirical data as possible. Usually, the two methods go hand in hand.[58] For more information, the reader should visit https://www.iep.utm.edu/ded-ind, https://plato.stanford.edu/entries/scientific-method, https://plato.stanford.edu/entries/scientific-reductionhttps://plato.stanford.edu/entries/scientific-explanation, https://philpapers.org/s/deduction and https://philpapers.org/s/induction.

Historians proceed reductively in a noninductive way. See Józef Bocheński, *Die zeitgenössischen Denkmethoden* (Francke, 1954, Part V) and Nancy Partner and Sarah Foot (eds.), *Handbook of Historical Theory* (Sage, 2012). Visit https://plato.stanford.edu/entries/history and https://www.iep.utm.edu/history. Compters can be profitably used in the field of historical studies, but whether history (the past) can be studied through machine learning or other data-driven, computational approaches is debatable. Men, having both a body and a mind, cannot be replaced by machines, which can only handle countable or quantifiable things. See Jerome Clubb and Howard Allen, "Computers and Historical Studies", *Journal of American History*, 54: 3 (December 1967), pp. 599–607, Myrddin Lewis and Roger Lloyd-Jones, *Using Computers in History: A Practical Guide* (Routledge, 1996), Johannes Preiser-Kapeller, "Calculating the Middle Ages", *Medieval Worlds: Comparative & Interdisciplinary Studies*, No. 2, 2015, pp. 100–127 and Bojan Bozic et al. (eds.), *Computational History and Data-Driven Humanities* (Springer, 2016), a book in the series *Advances in Information and Communication Technology* published by the International Federation for Information Processing. In addition, visit http://ceur-ws.org/Vol-2461, https://networks.h-net.org/h-digital-history and https://jdmdh.episciences.org/page/histoinformatics.

Notes and References

1. Visit https://www.iep.utm.edu/explanat and https://plato.stanford.edu/entries/scientific-explanation. Readers who want to gain deeper insight into the nature of science should go to the book series *Science and Its Conceptual Foundations* (1985ff), published by The University of Chicago Press, and/or engage themselves with the extensive literature about philosophy of science. They could also visit https://philpapers.org and search for "general philosophy of science".
2. The concept of evidence is contested. See Roderick Munday, *Evidence* (OUP, 2017). Go to https://plato.stanford.edu/entries/evidence and https://www.iep.

utm.edu/evidence. In his book *Objecting to God* (Cambridge University Press, 2011), Colin Howson analyzes the evidence claimed to support belief in God's existence. For "evidence-based reasoning", go to https://foundationalperspectives.wordpress.com.

3. Visit https://plato.stanford.edu/entries/science-theory-observation.
4. See Douglas Montgomery, *Design and Analysis of Experiments* (Wiley, 2017) and visit https://plato.stanford.edu/entries/physics-experiment and https://plato.stanford.edu/entries/biology-experiment.
5. See Judy Anderson, *Plagiarism, Copyright Violation, and Other Thefts of Intellectual Property* (McFarland, 1998, Introduction), Dennis Patterson (ed.), *A Companion to Philosophy of Law and Legal Theory* (Blackwell, 1999, pp. 3–23), Mark Lemley and John Duffy, "Property, Intellectual Property, and Free Riding", *Texas Law Review*, 83:4 (March 2005), pp. 1031–1089, Isabella Alexander and Tomás Gómez-Arostegui (eds.), *Research Handbook on the History of Copyright Law* (Edward Elgar, 2016) and Caterina Sganga, *Propertizing European Copyright* (EE, 2018). Monopolists defend the (intellectual) property rights.
6. See Patterson *op.cit.* pp. 223–260, 436–449. Visit https://plato.stanford.edu/entries/natural-law-theories and https://plato.stanford.edu/entries/legal-positivism.
7. See Victor Lavrenko, *A Generative Theory of Relevance* (Springer, 2009), Ewa Wałaszewska and Agnieszka Piskorska (eds.), *Relevance Theory* (Cambridge Scholars Publishing, 2012), Deirdre Wilson and Dan Sperber, *Meaning and Relevance* (CUP, 2012) and Billy Clark, *Relevance Theory* (CUP, 2013).
8. See Jordi Vallverdú, *Bayesians versus Frequentists* (Springer, 2016).
9. See Joseph Halpern, *Reasoning about Uncertainty* (MIT Press, 2017).
10. See Eliott Sober, *Ockham's Razors: A User's Manual* (Cambridge University Press, 2015).
11. See Herbert Stachowiak, *Pragmatik*(Meiner, 1986–1995, Vol. 1, pp. XIX–L).
12. See Jonathan Barnes (ed.), *Companion to Aristotle* (CUP, 1995, pp. 195–258), Kelvin Knight, *Aristotelian Philosophy* (Polity, 2013, pp. 4–40) and Wolfgang Welsch, *Der Philosoph* (Fink, 2018, Chaps. 7 and 8).
13. See Bernd Heine and Heiko Narrog (eds.), *Handbook of Linguistic Analysis* (OUP, 2015).
14. See Bimal Matilal, *Perception* (Clarendon Press, 1986); https://plato.stanford.edu/entries/epistemology-india.
15. See Robert Vallée, *Cognition et Système* (L'Interdisciplinaire, 1995, pp. 27–103), Stephen Phillips and Ramanuja Tatacharya, *Epistemology of Perception* (Motilal Banarsidass, 2009), Minas Kastanakis and Benjamin Voyer, "The effect of culture on perception", *Journal of Business Research*, 67:4 (2014), pp. 425–433, Adam Geremek et al. (eds.), *Perception Beyond Gestalt* (Psychology Press, 2014), Mohan Matthen (ed.), *Handbook of Philosophy of Perception* (OUP, 2015), Inchan Park and Yeonkoo Hong, "Culture and sensory response to visual stimuli", *International Journal of Psychology*, 53:1 (2018), pp. 77–81, and Bobby Duffy, *The Perils of Perception* (Atlantic, 2018).

16. In a downloadable article entitled “Perception: An Interdisciplinary Essay” (2014), I highlight the problem of perception from a philosophical, physiological and psychological point of view. For the current state of research on sensory categorization, see *Review of Philosophy and Psychology*, 10:3 (September 2019), pp. 419–540.
17. My hope is pinned on developments in category theory. Go to https://arxiv.org/pdf/1803.05316.pdf.
18. Visit https://plato.stanford.edu/contents.html (search for logic) and see Jean-Yves Béziau (ed.), *Studies in Universal Logic* (2008ff).
19. See Frederick Suppe (ed.), *The Structure of Scientific Theory* (University of Illinois Press, 1977, pp. 3–241) and Marvin Greenberg, “Old and New Results in the Foundations of Elementary Plane Euclidean and Non-Euclidean Geometries”, *The American Mathematical Monthly*, 117:3 (March 2010), pp. 198–219. For axiomatization in economics, see Robert Clower, "Axiomatics in Economics”, *Southern Economic Journal*, 62:2 (October 1995), pp. 307–319.
20. See Jaakko Hintikka, “What is the axiomatic method?”, *Synthese*, 183:1 (2011), pp. 69–85, Herbert Spiegelberg, *The Phenomenological Movement* (Kluwer, 1994, pp. 675–719) and Shanti Mishra and Shashi Alok, *Handbook of Research Methodology* (Educreation, 2017). Visit https://plato.stanford.edu/entries/scientific-method.
21. See Elart von Collani, “2005: The Jacob Bernoulli Year”, *Stochastics and Quality Control*, 20:2 (October 2005), pp. 155–169 and Jacob Bernoulli, *The Art of Conjecturing*, translated by Edith Sylla (Johns Hopkins University Press, 2006).
22. The relation between zero and infinity is interesting. See Peter Rowlands, *Zero to Infinity* (World Scientific, 2007) and T.A. Radhakrishnan, *Zero Becomes Infinity* (Educration, 2015).
23. See Michael Nielsen and Isaac Chuang, *Quantum Computation and Quantum Information* (CUP, 2010).
24. See Baldassare Castiglione, *The Book of the Courtier* (1528), Pierre l’Arétin, *La Courtisane* (1534), Étienne de la Boétie, *Discours de la servitude volontaire* (1576), Émile Benveniste, *Le vocabulaire des institutions indo-européennes* (Minuit, Vol. II, 1969, pp. 7–95), Norbert Elias, *Die höfische Geselllschaft* (Luchterhand, 1969), Pierre Bourdieu, *La Distinction* (Minuit, 1979), Gerhard Lenski, *Power & Privilege* (University of North Carolina Press, 1984), Mortimer Adler (ed.), *The Syntopicon* (Encyclopædia Britannica, Inc., 1990, Vol. I, pp. 238–253, 560–575, Vol. II, pp. 163–178, 212–222), Margaret McGlynn, *The Royal Prerogative and the Learning of the Inns of Court* (CUP, 2003), Jörn Leonard and Christian Wieland (eds.), *What Makes the Nobility Noble?* (V&R, 2011), Michael Bush, *Serfdom and Slavery* (Routledge, 2013), Vittorio Cotesta, *Kings Into Gods: How Prostration Shaped Eurasian Civilizations* (Brill, 2015) and Keats Currie, *Royal Prerogative* (Trafford, 2017).
25. Visit https://www.semanticscholar.org (select “mathematics” and search for “group theory”).

26. See Raymond Yeung, *Information Theory and Network Coding* (Springer, 2008, pp. 377–397).
27. See the open access journal *Symmetry* (2009ff).
28. See Daniel Parrochia and Pierre Neuville, *Towards a General Theory of Classifications* (Birkhäuser, 2013, pp. 173–208).
29. See Herbert Stachowiak, *Allgemeine Modelltheorie* (Springer, 1973), Hilary Putnam, "Models and Reality", *Journal of Symbolic Logic*, 45:3 (September 1980), pp. 464–482, Mary Morgan and Margaret Morrison (eds.), *Models as Mediators* (CUP, 1999), Robert Franck (ed.), *The Explanatory Power of Models* (Kluwer, 2002), Ulrich Dirks and Eberhard Knobloch (eds.), *Modelle* (Peter Lang, 2008), Tom Ritchey, "Outline for a Morphology of Modelling Methods", *Acta Morphologica Generalis*, 1:1 (2012), pp. 1–20, Margaret Morrison, *Reconstructing Reality* (OUP, 2015), Franck Varenne, *From Models to Simulations* (Routledge, 2018) and Stephen Downes, *Models and Modelling in the Sciences* (Routledge, 2020). Go to https://www.solvingforpattern.org/gallery-of-models and https://arxiv.org/abs/1612.05627.
30. This information is taken from Hugh Brigstocke (ed.), *Western Art* (OUP, 2001, pp. 483–484).
31. See Karl Heller, *Ernst Mach: Wegbereiter der Modenen Physik* (Springer, 1964, p.48) and Friedrich Stadler (ed.), *Ernst Mach* (Springer, 2019).
32. See Susan Carey, *The Origin of Concepts* (OUP, 2011), Antony Flew, *Essays in Conceptual Analysis* (Creative Media Partners, 2015), Eric Margolis and Stephen Laurence (eds.), *The Conceptual Mind* (MIT Press, 2015) and Sebastian Löbner, *Semantik: Eine Einführung* (De Gruyter, 2015). *The Archiv für Begriffsgeschichte* (1955ff) and its *Sonderhefte* (2000ff) are indispensable. The six-volume *New Dictionary of the History of Ideas*, chief-edited by Maryanne Cline Horowitz and published by Thomson Gale, in 2004, is a treasure trove, and the journal *Contributions to the History of Concepts* (2005ff) "aims to promote the dialogue between the history of concepts and other disciplines, such as intellectual history, history of knowledge and science, linguistics, translation studies, history of political thought and discourse analysis". For systems, see Chap. 5.
33. The following partly draws on Paul Rigby, *Models in Business Analysis* (Charles Merrill, 1969).
34. For the relationship between iconoclasm and idolatry, see David Cooper (ed.), *A Companion to Aesthetics* (Blackwell, 1992, pp. 207–209).
35. See Brigstocke *op. cit.*, pp. 351–354.
36. See Luc van Doorslaer et al. (eds.), *Interconnecting Translation Studies and Imagology* (John Benjamins, 2016, pp. 1–18). The series *Studia Imagologica* (1991ff) is indispensable.
37. One should not underestimate the power of advertising, a form of marketing communication geared to push the sales of products, services or ideas. James Walter Thompson, Arthur Nielsen and David Ogilvy are some of the big names in the advertising business.

38. See Douglas Hofstadter and Emmanuel Sander, *Surfaces and Essences* (Basic books, 2013) and Devin Griffiths, *The Age of Analogy* (The Johns Hopkins University Press, 2016, pp. 10–50). Visit https://plato.stanford.edu/entries/reasoning-analogy.
39. See Uma Chattopadhyay, *Dishonoured by Philosophers: Upamāna in Indian Epistemology* (Printworld, 2009) and idem, *Faultless to a Fault: Gaṅgeśa on Upamāna in Indian Epistemology* (Printworld, 2015).
40. See Bob Hale and Crispin Wright (eds.), *A Companion to the Philosophy of Language* (Blackwell, 1999, pp. 248–268) and Barbara Dancygier and Eve Sweetser, *Figurative Language* (CUP, 2014, pp. 13–72).
41. See Abraham Solomonick, *A Theory of General Semiotics* (Cambridge Scholars Publishing, 2015). For symbolic interactionism, see Larry Reynolds and Nancy Herman-Kinney (eds.), *Handbook of Symbolic Interactionism* (Rowman & Littlefield, 2003) and Norman Denzin (ed.), *40th Anniversary of Studies in Symbolic Interactionism* (Emerald, 2013). For the literary/art movement called symbolism, see Helene Henderson and Jay Pederson (eds.), *Twentieth/Century Literary Movements Dictionary* (Omnigraphics, 2000, pp. 723–747) and David Bradshaw and Kevin Dettmar (eds.), *A Companion to Modernist Literature and Culture* (Blackwell, 2006, pp. 155–162).
42. See Gino Segrè, *A Matter of Degrees* (Penguin, 2002), Ingo Müller, *A History of Thermodynamics* (Springer, 2007), Bernard Lavenda, *A New Perspective on Thermodynamics* (Springer, 2010), Salama Abdelhady, "An Advanced Review of Thermodynamics of Electromagnetism", *International Journal of Research Studies in Science, Engineering and Technology*, 3:6 (June 2016), pp. 10–25, and Claus Borgnakke and Richard Sontag, *Fundamentals of Thermodynamics* (Wiley, 2017). Visit https://arxiv.org/abs/1705.01100, http://iaisae.org/index.php/about-ict20, https://plato.stanford.edu/entries/information-entropy and http://scholarpedia.org/article/Entropy.
43. See Samuel Guttenplan (ed.), *A Companion to the Philosophy of Mind* (Blackwell, 1994, pp. 111–121, 531–535) and Arto Laitinen et al. (eds.), *Reasons and Causes* (Palgrave Macmillan, 2013).
44. See Nicholas Rescher, *Value Matters* (Ontos, 2004), Petrus Handoyo, *Exploring Values* (Book Country, 2015), *The Journal of Value Inquiry* (1967ff) and Brill's *Value Inquiry Book Series* (1993ff).
45. See Stanley Salthe, "Entropy: What does it really mean?", *General Systems Bulletin*, 32 (2003), pp. 5–11, Imre Csiszár et al. (eds.), *Entropy, Search, Complexity* (Springer, 2007) and Evgeni Starikov et al. (eds.), *Entropy-Enthalpy Compensation* (Jenny Stanford, 2021). In addition, go to http://www.scholarpedia.org/article/Entropy and https://entropy2020.sciforum.net.
46. For big data and big data analytics, see Viktor Mayer-Schönberger and Kenneth Cukier, *Big Data* (Houghton Mifflin Harcourt, 2013), Jure Leskovec et al., *Mining of Massive Datasets* (CUP, 2014), Klaus Mainzer, *Die Berechnung der Welt* (Beck, 2014), Ritu Arora (ed.), *Conquering Big Data with High Performance Computing* (Springer, 2016), Lakshmi Prasad, *Big Data Analytics*

Made Easy (Chennai: Notion Press, 2016) and Nataraj Dasgupta, *Practical Big Data Analytics* (Packt, 2018).

47. The standard work on measurement is David Kranz et al., *Foundations of Measurement* (Dover, 2007). Quantum mechanics have problematized the notion of measurement.
48. For a recent attempt, go to https://doi.org/10.1103/PhysRevX.9.011002.
49. See Elart von Collani and Karl Bauer, *Was zum Teufel ist Qualität?* (Heldermann, 2007), Patricia Leavy (ed.), *Handbook of Qualitative Research* (OUP, 2014) and John Beckford, *Quality: A Critical Introduction* (Routledge, 2017). For the nature, scope and methods of, and differences between, quantitative and qualitative research in the social sciences, see Sotirios Sarantakos, *Social Research* (Palgrave Macmillan, 2013, Part IV), Walter Hussy et al., *Forschungsmethoden* (Springer, 2013, Parts I and II), Alan Bryman, *Social Research Methods* (OUP, 2016, Parts II and III), Earl Babbie, *The Practice of Social Research* (Cengage, 2016, Parts II-IV), Kosuke Imai, *Quantitative Social Science* (PUP, 2018) and the journal *Quality & Quantity* (1967ff). Explore www.socialresearchmethods.net and https://www.ncrm.ac.uk. Some qualities admit more or less of them (quantity of quality); see *The Monist*, 84:4 (October 2001), pp. 582–616.
50. See Herbert Hochberg and Kevin Mulligan (eds.), *Relations and Predicates* (Ontos, 2004). Go to https://plato.stanford.edu/entries/relations and https://philpapers.org/s/relation.
51. See Surendranath Dasgupta, *A History of Indian Philosophy* (Motilal Banarsidass, Vol. I, 1975, pp. 429–439).
52. See Eli Maor, *Music by the Numbers: From Pythagoras to Schönberg* (PUP, 2018).
53. See André Giroux, *Initiation à l'Analyse Mathématique* (Ellipses, 2015) and William Dunham, *The Calculus Gallery: Masterpieces from Newton to Lebesgue* (PUP, 2018).
54. The dominant model in macroeconomics is the dynamic stochastic general equilibrium (DSGE) model; the values of the model variables are randomly determined. See Peter Spiegler, *Behind the Model* (CUP, 2015) and Joseph Stiglitz, "Where modern macroeconomics went wrong", *Oxford Review of Economic Policy*, 34:2–4 (January 2018), pp. 70–106.
55. See Anthony Edwards, "Ars Conjectandi: Three Hundred Years On", *Significance*, 10:3 (June 20, 2013), pp. 39–41. Anticipation plays a major role in conjecturing. See Giovanni Pezzulo et al. (eds.), The Challenge of Anticipation (Springer, 2008), Mihai Nadin (ed.), *Anticipation across Disciplines* (Springer, 2016) and Roberto Poli, *Introduction to Anticipation Studies* (Springer, 2017). There is a subtle difference between anticipation (not to be confused with prediction and forecast) and expectation. The concept of expectation was introduced in economics by Gunnar Myrdal (1898–1987) and further developed by George Shackle (1903–1992), John Muth (1930–2005), Edmund Phelps (1933), Robert Lucas (1937) and Thomas Sargent (1943).

56. See Guillaume Dubois, *Modeling and Simulation* (CRC Press, 2018) and visit https://plato.stanford.edu/entries/simulations-science.
57. Visit https://plato.stanford.edu/entries/induction-problem.
58. See Friedrich Stadler (ed.), *Induction and Deduction in the Sciences* (Springer, 2004) and Udo Kuckartz, *Mixed Methods* (VS, 2014).

Chapter 3
The Concept of Country

Abstract The concepts of country and area are differentiated, and the concept of country is further clarified. A distinction between "state" and "nation" is made. The focus of attention shifts from the *explanans* to the *explanandum*, but the importance for scientists to have a distinctive *explanans* is not lost sight of. Attention is also paid to the concept of context, the background against which things or happenings get their meaning. This leads to the book's first conclusion: Proper understanding of a country is only possible if a truly interdisciplinary approach is taken. At universities, research institutes and think tanks, there should thus be no place for "country experts". Since interdisciplinary research implies the involvement of ideas, information and people from different disciplines, the importance but often disregarded linkages between sciences are given due attention. The categories under which the sciences are usually subsumed are explained, and an extensive example of their interconnectedness is given.

Having explained what I mean by the *explanans* I accuse so-called area or country experts of missing, I now turn to their *explanandum*, the nature of which I suspect them of not quite understanding.

"Area" is a vague term, referring to a country (e.g., Israel, Mexico, Russia and South Africa), to a region or part of a country (e.g., the Midwest of the USA, Southeast Brazil, Northeast China and Scotland), to a region of the world (e.g., [sub-Saharan] Africa, the Middle East, [Southeast] Asia and [Middle] America),[1] to a civilization (e.g., the Roman, Egyptian, Indian and Chinese civilization), to a part of a house, hotel or city (e.g., kitchen, dining room, non-smoking area and no-go zone) or to a domain of study (e.g., terrorism, feminism, the EU, the Pitcairn islands marine reserve, Pushkin's poetry of his Lyceum period, climate change and the comma in medieval Hebrew manuscripts). In contrast, "country" (as distinct from "countryside") stands colloquially for a nation-state, where passport holders belong and what may be one's mother– or fatherland. The literature on the nature, (supposed) origin, formation, structure and functioning of the State, or nation-state (which is a type of State), is very large, and the discussion of these topics continues unabated. It would lead me too far afield to explore this field of research thoroughly. Given the subject of this book, however, the concept of state cannot be ignored altogether.

H. Kuijper, *Comprehending the Complexity of Countries*,
https://doi.org/10.1007/978-981-16-4709-3_3

The **state** is a form a human association distinguished from other social groups by its purpose: the establishment of a political order. The second volume of *Internationales Jahrbuch des Deutschen Idealismus* (Walter de Gruyter, 2004) is dedicated to the subject. Few establishers of order know what "order" actually means, but to achieve their objective, they proclaim and enforce laws, which only apply to people belonging to a certain territory. The court that represents the state, which embodies the highest power in a country, has thus no jurisdiction beyond the boundaries of the state's territory. According to Alessandro d'Entrèves (*The Notion of the State* [Clarendon Press, 1967]) and Gianfranco Poggi (*The State: Its Nature, Development, and Prospects* (Stanford University Press, 1990), the word state can be used descriptively (in a non-evaluative way) or prescriptively (in a normative manner). Used descriptively, "state" stands for the body politic or political community, which has existed throughout history in a variety of different forms. Only the nomadic community is excluded from being called a state,[2] though some form of government is intrinsic to human society. "State" implies a fixed relationship between a community and a territory. Used descriptively, the concept of state is the idea that the political community has a universal characteristic. In other words, the state is a kind of activity imposed on man as a necessity. This activity creates a unity among a number of human beings (society) and presupposes a form of rule (a relationship of command and obedience). State members assert themselves in contrast to those who do not belong to the community that consists of people living together in a particual geographical area.

The descriptive definition of a state points to another characteristic. In so far as the state is a kind of activity concerned with ends, it raises the question of the rightness of these ends and the means or methods employed to achieve them. Seen from this point of view, the state is a perennial problem. For the state involves not only the establishment of order; the order should also be the right one. Plato, Aristotle, Augustine, Maimonides, Thomas Aquinas, Niccolò Machiavelli, Jean Bodin, Hugo Grotius, Thomas Hobbes, René Descartes, Benedict Spinoza, John Locke, Montesquieu, David Hume, Rousseau, Immanuel Kant, Adam Smith, Edmund Burke, Thomas Paine, Hegel, John Stuart Mill, Karl Marx, Friedrich Nietzsche, Leo Strauss, Michael Oakeshott, Hannah Arendt, John Rawls, Jürgen Habermas and Robert Nozick occupied themselves with the question as to what the state should be, that is, with the concept of state in a normative way. They had different views about order, the axis of all political thinking. Political *scientists*, legal scientists, economists, sociologists, psychologists, anthropologists, demographers and geographers have greatly contributed to the deepening of political philosophy, which is a proof that scientific disciplines are inextricably interconnected. The approach may be different, but the problem is the same: how to rule a country. "A rose by any other name would smell as sweet", Shakespeare aptly and inimitably said.

In the twentieth century state, the descriptive–prescriptive distinction became blurred. The society as arena of activities was absorbed by the state, which itself was subordinated either to a coalition of political parties forming—*pour le besoin de la cause*—an uneasy alliance, or to a single political party bearing the characteristics of a religious movement. Russia, China, Germany and Italy came under the sway of Stalin, Mao Zedong, Hitler and Mussolini, respectively. In the "free Western world", the cradle-to-grave welfare system flourished in which the state played a key role in the protection and promotion of the economic and social well-being of its citizens, with very few people knowing how to define well-being. This development was counterbalanced by a movement to define the state as an apparatus of rule only, distinguished by its monopoly of coercion. Concomitantly, the idea of minimal state emerged, Robert Nozick (1938–2002) being its main protagonist. See Jonathan Wolff, *Robert Nozick: Property, Justice and the Minimal State* (Stanford University Press, 1991) and visit https://plato.stanford.edu/entries/contractarianism-contemporary. This tendency was related to the rise of positivism, with its insistence that the social sciences must eschew value judgments and devote themselves to the ascertainment of facts. However, the question as to what are and what are not facts has remained unanswered.

Whereas in totalitarian regimes the ruling party falsely claimed to be the voice of the people, in the West the idea was advanced that the rules applying to the body polic were rightfully decided on and constituted by the people. As a result, cynicism became rife in broad sections of the population at both sides. On the other hand, globalization that grew in the wake of advances in transportation and communication technology has reawakened the forces of nationalism. We are now witnessing a decline of globalization ("deglobalization") and neo-nationalism is on the rise. A list of nationalist organizations can be found on the Internet.

Europe has long been recognized as the birthplace, not only of the state, but also of the ideas that have justified it. As heirs apparent of Machiavelli, Hobbes, Locke, Rousseau, Hegel, Mill and Marx, political thinkers wanted to understand how these giants molded our present-day political thoughts. Until the 1970s, that enterprise seemed unproblematic. The works of Friedrich Meinecke (*Die Idee der Staatsräson in der neueren Geschichte*) and George Sabine (*A History Political Theory*) were standard introductions to the subject. They presented great political ideas as passing from one exemplary agent to the next. The focus was on individual writers, on the procession of their classic texts. Over the last forty years or so, a spate of scholarly books, beginning with the publication of John Rawls' *A Theory of Justice* (1971), John Pocock's *The Machiavellian Moment* (1975), Quentin Skinner's two-volume *Foundations of Modern Political Thought* (1978) and John Dunn's *Western Political Theory in the Face of the Future* (1979), has transformed this enterprise. Since then, political philosophy has been a vibrant field in philosophy, one that intersects not only with law and economics, but also with other sections of political science, including just war theory.[3] The attention is not anymore directed to a line of political thinkers but to the development of ideologies within which their writings are more readily intelligible, to thematic continuity and to linkages with broader, previously neglected areas of intellectual history. Rawls, Pocock, Skinner and Dunn were criticized, and the emphasis has undeniably shifted from individual authors to the issues they had been involved in. The political, social, economic and intellectual context of texts is now taken into account. The monumental book series *Cambridge History of Political Thought* (1988ff) is a testimony to this new trend. Yet, the admirable contextualization shown in the series could be furthered by taking *all* the relevant and interrelated aspects or dimensions of history into account. For political, legal, military, social, psychological, educational, economic, financial, demographic, geographic, ecological, religious and cultural affairs, while changing over and developing through time, are interlinked. They are the parts of a whole which should be paid close attention to. This requires interdisciplinary collaboration supported by a network of computers.

Treatment of the intriguing subject Origins of the State must be postponed. The literature on the subject (to be distinguished from *ruling* a country, or state *government*) is steadily growing, as new discoveries generate new hypotheses and empirically based theories, which encourage new generations of researchers to come forward with even more refined hypotheses and theories. Notable names are Julian Steward, André Leroi-Gourhan, Elman Service, Laurence Krader, Morton Fried, Robert Carneiro, Charles Tilly, Henri Claessen, Marshall Sahlins, Pierre Clastres, Bruce Trigger, Richard Blanton, Henry Wright, Kristian Kristiansen, Gary Feinman, Allen Johnson, Timothy Earle, Andrej Korotajev, Nikolay Kradin, and Dmitri Bondarenko. *Social Evolution & History* (2002ff) is a leading journal (online). Hobbes, Locke, Rousseau, Kant and Rawls hypothesized that a social contract laid at the basis of state formation, but Hume, Hegel and Marx disagreed.

The **nation-state** (shortly: nation) is a combination of two elements: the political–territorial and the historical–cultural element. "State" refers to the modern state, which reached fruition after the Renaissance and Reformation.[4] The state is as a set of institutions that may be centralized or decentralized, but possesses the monopoly of taxation and violence in a demarcated and internationally recognized territory.[5] "Nation" refers to a human community with a common ancestry, historical memory, living culture and sense of belonging together.State and nation have a strained relationship. The more pronounced the civic-territorial elements, the less difficult has it been to bring about a fusion of state and nation; conversely, the more prominent the ethnic-cultural elements, the less probable is it that harmony between the two will be reached. A list of national founders can be found on the Internet. Going through it will reveal how difficult nation-building ("the process through which the boundaries of the modern state and those of the national community become congruent") has been. Visit also https://www.oxfordbibliographies.com/view/document/obo-9780199743292/obo-9780199743292-0217.xml.

Tribalism, which (like nationalism) both constrains and enables, is everywhere manifest. It is very much alive, not only in Iraq and Libya (or elsewhere in the Middle East, Asia, or Africa) but also in Australia, Canada and the USA.[7] Broadly defined, tribalism can be recognized in party loyalty, in-group bias, groupthink and the amity–enmity complex, virulent forms that have been omnipresent.[8]

The nation-state goal is the criterion of international legitimacy. Some countries have an ethnic group which distinguishes itself from, and regards itself superior to, minorities, which may be politically active and unwilling to accept the culture and ascendancy of the dominant ethnic group. The Han Chinese, who vastly outnumber the Tibetans and Uyghurs and ruthlessly suppress their culture, is a case in point. One should read the leaked documents which detail China's abuse of Muslims in Xinjiang and whose validity has not been questioned (*The New York Times*, November 16, 2019). See also Sean Roberts, *The War on the Uyghurs* (PUP, 2020).

The prospect of achieving the status of nation-state in the foreseeable future is bleak. Advanced countries like Belgium, Canada, France, Spain and Britain also suffer from deep internal ethnic cleavages. The oppositions of the Walloons, Quebeckers, Bretons, Catalans and Scots, respectively, reveal the error of assuming a nation-state where there is none. Is the United Kingdom a nation-state? The Englishmen, the Scots, the Welsh and the Irish will answer this question differently, reaching for their gun to strengthen their point of view. Only in a loose sense, countries like Belgium, Canada, France, Spain and Britain can be conceived of as nation-states They enjoyed that status internationally before ethnic nationalism (not to be confused with civic nationalism) spread to their ethnic communities. According to Anthony Smith (*Encyclopedia of Political Institutions* [Blackwell, 1987, p. 381]), some 10% of the world's countries can be accorded the status of nation-state (sensu stricto). The number of countries where state territory and institutions are occupied by members of a single, ethnically defined nation with a homogeneous culture, is surprisingly small.

With the rise of globalization, some scholars, particularly those to the political left, have been sounding the death knell of the nation-state, but they may have been rushing things.[9] In *Imagined Communities: Reflections on the Origin and Spread of Nationalism* (Verso, 1983), Benedict Anderson (1936–2015), one of the seminal figures in the field, argued that nationalism had proven to be a most tenacious animal. Addressing a long line of scholars who had prematurely and wishfulthinkingly predicted its demise, he declared: "The reality is quite plain: the end of the era of nationalism, so long prophesied, is not remotely in sight. Indeed, nation-ness is the most universally legitimate value in the political life of our time". Yet since the publication of the book, visions of a globalized "world without borders" have caught the imagination of many scholars. Readers interested in the difference between state and nation should turn to Hans Kohn, *The Idea of Nationalism: A Study in Its Origins and Background* (Routledge, 2017) and make a study of the collapse of the Habsburg monarchy, or of the history of the Kurds. In chapter eight, I will write a few words about the dispute between Europeanists and anti-Europeanists.

An **empire** is nothing but an aggregate or adding together of nations, not an integration of them. Examples are the Akkadian Empire, the Achaemenid Empire, the Egyptian Empire, the Roman Empire, the Holy Roman Empire, the Frankish Empire, the Umayyad Caliphate, the Mongol Empire, the Ottoman Empire, the Spanish Empire, the Portuguese Empire, the Qing Dynasty, the Mughal Empire, the Russian Empire, the Empire of Japan, the British Empire, the French Empire, the German Empire, the Dutch Empire and the American Empire. They considered themselves to be carriers of a universal civilization, but were **de facto** not. The Chinese emperor, for example, saw himself, and was seen by many of his subjects, as the center of the world. Colonial empires, which lasted until the decolonization after the Second World War, often consisted of territorially scattered units. The Austro-Hungarian Empire (dual monarchy) and the Russian Empire, on the other hand, were made of contiguous territories. However, they were constantly afflicted by internal strife or armed conflict. Federalism is a "form of political organization in which unity and regional diversity are accommodated by distributing power among general and regional governments in a manner constitutionally safeguarding the existence and authority of each" (Ronald Watts). Unfortunately, federalism has been and will probably remain a beckoning perspective. If other members of the so-called European Union follow the example of Poland, where the highest judges declared themselves not be bound anymore by any European law, the union, which is actually not a union at all, will be doomed.[10]

So, it could be safely said that a country is a bordered stretch of land with politically or otherwise organized people who, to some extent, share a history and have a sense of belonging. Countless are the soldiers who fell for their homeland ("*Si le nez de Cléopâtre eût été plus court...*"!). On a French, German, American or Russian military cemetery lie the remains of, respectively, *French*, *German*, *American* and *Russian* soldiers, willing executioners of national prejudices. There is still dispute over who reached the North Pole first (Robert Peary or Matthew Henson?), but there is no doubt that it was the *Norwegian* Roald Amundsen who led the first expedition to the South Pole. In 1969, Neil Armstrong planted the *American* flag on the Moon. At the Olympic Games, athletes from different countries compete with each other. We sing a *national* anthem, respect a *national* flag and salute a *national* leader. Nobody seems to think out of the box and to conceive of a world without borders. We are all herd animals, aren't we? The Queen of the United Kingdom is queen of all Britons, including the Scots and Welshmen, and the President of the United States is president of all Americans, including those of Hispanic or Latino origin. The country may be plagued by anomalies, dissent, strife, class or power struggle, even by civil war, as long as it is a sovereign state it is a country.[11] Studies of International Relations (not to be confounded with Global Studies) are studies not pertaining to countries but, as the naming already indicates, to the *relations* between them. These relations can have bearings on a country's internal affairs, and vice versa. Countries can be classified by different criteria (size, site, situation, climate, ethnic or racial composition, level of economic achievement, S&T status, military muscle, political influence, etc.). Ranking countries while taking all these criteria into account is impossible. Some criteria can, others cannot, be quantified.

Meanwhile, it should be borne in mind that applying Western ideas about state or nation-state dynamics to non-Western countries is fraught with the difficulties of imposing a Western framework on a non-Western context, because the legal and political framework which Westerners use to talk about has been defined in terms of a society breaking free from a despotic regime, irrespective of (*a*) the period of absolute monarchy (for which the divine right of kings had been the theological justification), (*b*) the age of enlightened absolutism (aka enlightened despotism) and (*c*) the undeniably totalitarian traits of so-called democratic regimes.

After lengthy discussions of burning questions (behind closed doors), heads of state or government who rely on the advice given or information provided by "area/country experts" sign important documents, thereby determining the fate of tens or hundreds of millions of people. However, looking at official group-photos taken on the occasion of international meetings, one wonders whether each of the stern-looking or big-smiling but carefully lined up figures in the picture really understands his/her colleague's country, that is to say, the complexity and uniqueness of it. For example, at the historic Yalta Conference, in 1945, to what extent were Franklin Roosevelt, Winston Churchill and Joseph Stalin knowledgeable about each other's country? Were their minds only militarily/geo-politically framed? During the sovereign debt crisis faced by Greece in the aftermath of the financial crisis of 2007–2008, what did the "troika" (made up by representatives of the European Commission, the International Monetary Fund and the European Central Bank) really know about the history and culture of Greece, where the cradle of Western civilization stand and "The Great Conversation" began?[12]

Dialogues often degenerate into tug-of-war debates. Whereas a debate is competitive and geared to win, a dialogue is cooperative and intended to exchange information and build a relationship.[13] Game theorists tend to overlook this important difference. When the American and Russian, or Chinese, president sit down and talk face-to-face with each other via interpreters (*traduttore traditore*!), do they only exchange views, each thinking he is absolutely right and only trying to convince the other in the "friendly conversation"? Or do they engage in a genuine dialogue, each interlocutor being ready to bracket his (country's) interests and what he takes to be plain "facts"?[14] Do they open their mind and actively and respectfully listen to each other? Or are they deeply distrustful of one another? Do they take all relevant factors into account, keeping nothing up their sleeve? Or are they not coming clean about their plans? Is there a frank and honest disagreement, each one acknowledging that the other may have good reasons to disagree, each one not only able to convince but also willing to be convinced in a free, untrammeled conversation about, say, the nature and justification of private property, the blessings and curses of liberalism or the desirability and shortcomings of democracy (as discussed by Plato, Aristotle, Montesquieu, Jean-Jacques Rousseau, Alexis de Tocqueville, Friedrich Nietzsche, Robert Michels, Henry Adams, Henry Mencken, Robert Dahl, Joseph Schumpeter, Hélène Landemore, Michael Mann, Roger Scruton, Steven Levitsky, Corey Bretschneider, David Estlund, Frank Ankersmit, Jean-Marie Guéhenno, Claude Lefort, Thomas Christiano, Pierre Rosanvallon, Jacques Rancière, Frédéric Worms, Nancy MacLean, Daniel Bell, Hans-Hermann Hoppe and the contributors to *Rethinking Democracy* [Wiley, 2019])? Visit https://freedomhouse.org and https://www.idea.int.

To the benefit of those whose future is at stake, it would be interesting to know the final verdict on these matters reached by responsible and duty-bound reporters. I suspect many presenters and interviewers of being biased and having compromised their principle of neutrality for the sake of being accepted as interviewer. People like Larry King, Wolf Blitzer, John King, Fareed Zakaria, David Letterman, Jake Tapper, Erin Burnett, Christiane Amanpour, Graham Norton and Victor Blackwell are "professionals", knowing perfectly well what shall not be asked to remain on speaking terms with influential or powerful interviewees. The TV viewer is rarely, if ever, told "the truth, the whole truth and nothing but the truth".

Dubious deals are made in advance, and non-live broadcasts give editors the opportunity to cut out passages found to be unwanted. After all, editing means "making changes to a text or film, deciding what will be removed and what will be kept in, in order to prepare it for being printed or shown" (*Cambridge English Dictionary*). Before appearing on the TV, people, critically assessed by the fashion designer, retouched by the visagist and styled by the hairdresser, are instructed how to behave in front of the camera. The media (meant to be the watchdog of democracy) misinform, spin, frame, fabricate, contrive and are guilty of overt or covert advertising and propaganda. Lying, to some degree, seems to be the profession of many television hosts and newspaper reporters. In this totally commercialized and monetized world, who can be trusted![15]

When heads of state or government meet each other, honor (or what the Chinese call "face") will surely be involved. Honor is a complicated matter, on which Francis Fukuyama has nothing to say in his sensational book, *The End of History and the Last Man* (Free Press, 1992), though he devoted many pages to what Plato called θυμός (spiritedness), Hobbes designated as "pride", Machiavelli called "*gloria*", Alexander Hamilton termed "love of fame", James Maddison denoted with "ambition", Rousseau named "*amour-propre*" and Hegel labeled "*Anerkennung*". Leaving honor out of the discussion, duels and honor killings are difficult to understand.

It would not be easy to define the concept of honor (Homer, Sophocles, Dante, Chaucer, Ariosto, Castiglione, Montaigne, Tasso, Shakespeare, Calderón, Corneille, Voltaire, Schiller, Stendhal, Balzac, Melville, Ibsen, Tolstoy, Shaw, Conrad, Rostand, Joyce, Hemingway and Vargas Llosa dealt with the subject), but, apart from having ethical and psychological dimensions, honor (which, like woman's grace, cannot be measured) has also social, cultural and political aspects. It tells not only something about a person's self-respect ("am I myself a man of honor?"), but also about his social status, cultural clout and political position in the country he is speaking for. When a political leader is given "face", his supporters and possibly the whole country he represents will feel honored. However, giving him "face" is a subtle action, for which a great deal of psychological and cultural knowledge is needed and social skill as well as political and commercial acumen are required. An amicable, well-meant pat on the back may not have the intended result. The same applies to strict formal behavior. The balancing act between familiarity and formalism is difficult and delicate. It is a kind of art, to be performed with naturalness and sincerity. It would be no exaggeration to say that many of today's statesmen do not know this art. A seemingly simple handshake requires intuition.

Though I defined country as a bordered stretch of land with politically organized people who, to some extent, share a history and have a sense of belonging, it may be useful for being able to follow the argument of this book to give a different but essentially similar definition. I would then say that a country has many aspects; it is a huge set of interrelated elements being part of a larger whole, this whole being a part of the world, which in turn is but a tiny part of the universe.

In the long tradition of the *Great Books of the Western World*, the conception of the universe has been intermingled with the ideas of God and man. Cosmology, theology and anthropology have been widely ramifying discourses interpenetrating each other and forming an almost unravelable tangle that is not easy to compare with non-Western civilizations.

A country is thus more than a set: it is a system, indeed it is a complex system. Whereas a set is a collection or group of things (e.g., the vowels of the alphabet, the students enrolled for a class, the rivers in the United States, the capital cities of Europe, or the people living on the Earth),[16] a system is a set of *connected* things or elements (e.g., "Hot water circulates through the heating system", "There appears to be a problem with the patient's respiratory system", "A heavy fall of snow had disrupted the city's transport system", "The accident was caused by the failure of the reactor's cooling system" and "Drink a lot of water to flush the toxins out of your system"). The elements of the system called country can be physical, biological or social in character. Social elements (i.e., the things social and human scientists are concerned with) are dependent on, but not completely reducible to, biological elements; biological elements are dependent on, but not entirely reducible to physical elements. Physical and biological elements are the concern of physicists and biologists, respectively. Their labor division is an ongoing process. If the relations between the elements are nonlinear (one thing does not clearly or directly follow from another), the system is complex. Network science exploits the properties making a system complex. The study of social networks is still in its infancy, but its future holds much in store. If the mind's focus is on, or shifts to, the (relations between) social elements, a further distinction can be made between geographic, demographic, political, legal, military, economic, financial, cultural, linguistic, literary, medical, psychological, educational and environmental phenomena, which, possibly subdivided, can be considered either at a certain time or over a shorter or longer period of time.

A country is thus an enormous complex system (complexity) of complex systems (complexities) delineated by its boundaries, borders or frontiers, and surrounded by other (foreign) countries. It is not an aggregate (*Gesamtheit*) but a ὅλον (*Ganzheit*), composed of (not: added up by) other wholes that can be partitioned and analyzed. A country is a unity in diversity driven by its evolution; it is an intricately structured, geography-shaped, history-molded, culture-soaked and goal-directed whole, a set (*ensemble*) consisting of a multitude of interdependent subsets, which in turn consist of interconnected sub-subsets, etc. This means that a country has different but related layers to be studied. If a country is to be studied thoroughly, all its aspects and all their layers have to be fully taken into account. Distinctively structured, its population is cognitively and emotionally bonded, has a sense of belonging, is proud of her general way of life (internal dissent and strife notwithstanding), and is—perhaps vaguely but unmistakably—aware of the friendly or hostile outside (versus the inside) world. With emergent properties that are irreducible to the properties of its constituent parts,[17] each country is internally characterized by top-down and bottom-up forces, while externally influencing and being affected. Like a living organism, not isolated from its environment, a country is constantly changing, without losing its identity. In its innermost core, a country is held together by information, which is mysteriously related to energy.[18]

The word information comes from *informare* (to form, to make something by combining two or more parts). Information is easy to create but hard to trust, easy to spread but difficult to control. Influencing decision making (the choosing after thinking about possibilities), information is notoriously difficult to define. This is the reason why the basis of information science, – theory, and – technology is unstable. Physicists conjecture information to be a basic property of matter, not an identifiable and measurable thing, but what holds things together. Information would be kind of glue or cement, and the word information would be referring to the very essence of a system, to order (*Ṛta*!).[19] The open-access journal *Entropy* (2000ff) plans a special issue about "creative research on the innovative uses of information theory, and its extensions, to better understand human behavior and social processes". For further information, see the open-access journal *Information* (2010ff), particularly its special issue, "What is Information?" (2020).

A country is an incredibly intricate network or Web, a complexity to be investigated at different, though linked levels by a team of researchers. It is a complex, adaptive and dynamic system of systems to be compared with other complex, adaptive and dynamic systems (countries). All these systems constitute the suprasystem or world-system that is euphemistically called the "Family of Nations" (UN). In fact, the United Nations is disunited. See Peter Zeihan, **Disunited Nations: The Scramble for Power in an Ungoverned World** (Harper Collins, 2020). If the ambition is to—literally—comprehend a country, *i.e.*, to present an adequate summary or *Zusammenfassung* of the different accounts or visions of it, we need scientists with profound knowledge in particular disciplines and proficiency in communicating with neighboring and, figuratively speaking, T-shaped experts. For no "country expert" can be master of all pertinent sciences. A country is like a person, somebody to be respected. It is an *individuum*, something that cannot be divided without losing its identity. Like the elaborate pattern of a carpet that cannot be perceived by an ant, the convoluted pattern of a country cannot be discerned by the political, legal or literary scientist, by the linguist, sociologist or ecologist or by any other scientist looking through his/her stove pipe and working at what is nothing but a department or faculty of a university. Joint effort is needed, scientific collaboration that is well-managed, *i.e.,* guided and directed by an international, respectable steering committee. The insights of the science of complex systems, of network science and of computer/computation science may be useful (as will be explained in Chaps. 5 and 7). For these sciences stress the importance and pervasiveness of connectedness (as the ancient Indians and Chinese did all along), and—more importantly—they have the tools for handling it.

If countries are dynamic, multifaceted, poly-dimensional, multilayered and embedded systems which consist of different, interrelated and divisible subsystems, and if the ambition is to comprehend a country in order to pursue a sound and sensible policy in regard to it, it might be useful to dwell for a while on the concept of context.

"Context" (from *contextus* [consistency, cohesion]) refers to the background or setting of concepts, words, sentences, theories, persons, actions, events, or issues. Only with context can we gain insight and perspective. The context can be different (geographical, historical, political etc.). The context of an open system, being a system that has external interactions, which can take the form of information –, energy – or material transfers into or out of the system, cannot be completely described, because the relations of the inherently composite system with its surroundings are constantly changing. Knowledge is relative. According to Nietzsche, perspectivity is "the fundamental condition of all life". No statement is absolutely true or false, for reality, always in flux and forever incomprehensible, is–here and now–perceived differently (from different, possibly scientific points of view), judged about by someone with a different background (family, education, life, age, career) and a different way of thinking (mentality, attitude, vision, worldview), and described by someone having a different memory and being in a different situation. Features of the situation include not only place and time, but also the role played by the speaker, his relationship to the receiver of his message, the knowledge and experience they share, and what has already been said, in a certain way (stress, loudness, rhythm, intonation, style). Contextualizing a subject means putting it into a certain perspective. Context-dependency is a central characteristic of counterfactual thinking, the human tendency to bring to mind or imagine what is contrary to what actually happened.[20] A sub-context, being a context within another context, is to be distinguished from the super-context, the context to which sub-contexts belong. Science, that "endless cycle of proud proposing and disdainful doubting " (Melanie Mitchell), will only grow if scientists make observations that throw new light on old issues, when they see things from a different point of view or put them in another perspective.[21] People in East Asia are said to be more sensitive to the social context of their behavior. By contrast, Westerners would be more inclined to detach objects or individuals from their context. If this is true, Westerners are more likely to make what social psychologists call the "fundamental attribution error", the unpleasant consequences of which they are or will be experiencing.[22]

The lesson to be learned by those who aspire to comprehend a country is this: Natural, social and human scientists interested in testing their theories on, or applying them critically to, a country should regularly compare notes. They should be searching for a common ground. Collectively reflecting on the country, they should—within the purview of their discipline—zoom in on a single side or aspect of it and, at the same time, be aware not only of the context of the parts they are devoting their professional attention to but also of the context of the country as a whole (its position amidst, its posture toward and its treatment by other countries). In other words, natural, social and human scientists should communicate and genuinely collaborate with each other. For the viewpoint of each of them is a necessary but by itself not sufficient condition for understanding the country. They should focus on one set of things and, by zooming out, keep feeling with colleagues who are focusing on another set. There is no knitting without yarns and no yarn without interlocked fibers. The bunching and twisting of different wires make the strength of the cable. *L'union fait la force*! It is the team rather than the totality, *Gesamtheit* or aggregate of eleven players that wins the football match. "When there is lack of harmony in the army, it cannot win a decisive victory", Sunzi (544–496 BC) famously said.

It is not one-sidedness or partisanship but the intelligent combination of well-founded viewpoints that finally matters. Never ask for the meaning of a word in isolation but for the meaning of a word in the context of a sentence, this being part of an utterance made in a conversation, which—like waltzing or sweep rowing—is a joint activity taking place at a certain place, at a certain time and under certain conditions. It is the *Zusammenhang* of all the parts that counts, not a single part or a restricted number of parts taken out of its/their context according to the *ceteris paribus* clause. There is only offspring through the intimate activities of male and female. We need two legs to walk, two hands to clap (or to wash each other), two jaws to chew, two eyes to see depth, two brain hemispheres to think and speak and act. It takes two parties to negotiate/bargain, two sides to clinch a deal, two lips to kiss and two human beings to make love. What is a man without a woman, or a woman without a man! Without the connecting soundpost, no violin can sing. We inhale and exhale. A *pas de deux* performed by one person is missing somebody essentially. It takes two to tango. A bird needs two wings to fly. Light and darkness make the day, summer and winter solstice make the year complete. Low and high tide alternate. Good things come in pairs (*hǎo shì chéng shuāng*). Without *Yīn* and *Yáng*, no *Dào*.

Godfried Bomans (1913–1971), a Dutch author and comedian, once told he had been totally captivated by the beautiful legs of the German-American actress and singer Marlene Dietriech. He ended the story by telling his audience jokingly that he would have been satisfied if his wife had only one of those gorgeous legs.

A cook knows what and how to blend. One admires the architecture of a building, not the individual stones; the mosaic, not its tesserae; the fabric, not its fibers. To really comprehend a country, its various facets have thus to be seen together, which no single professional is or will ever be able to do. Juxtaposition (multidisciplinary research) will not do; we need *com*position (*inter*disciplinary research). Components have to be seen as components that make up a whole, as musical notes that form a chord. The context is of overriding importance; no identity without difference. With people all over the word eager to know (more) about other countries, one can only deplore that the television medium is not used to its fullest potential to inform the viewers. Instead of chat shows, silly TV games and commercials, the need for which is questionable on several grounds, the provision of country information based on collaboratively conducted research should be developed. The public response to it could serve as feedback that may improve the quality and deepen the content of the information.

Each and every country has aspects or facets that have been historically evolved. For practical purposes, these aspects can be distinguished, but they shall not be isolated. Aspects and facets are parts or dimensions of something that has other aspects or facets as well. All of them constititute the thing that shall not be considered one-sidedly. Let me give a few examples:

- After the crash of two Boeing 737 MAX 8 aircrafts in October 2018 and March 2019, causing hundreds of deaths, the 737 MAX series was grounded. To solve the problem, Boeing Company had to scrutinize each and every aspect of the accident: from the designing of the plane, to its production (in each and every phase), to the training of its pilots, to the loading of the plane and to the weather conditions during its flights.
- Abuses in horse racing are rife in many places. Horses are herd animals. Their needs and emotions should not be ignored for the pleasure or benefit of man. Seeing a horse being whipped, Nietzsche went mad. Usually, horse racing is glamorously presented in the media (fashionably dressed women, gentlemen wearing a top hat, sportsmanship, etc.), but the racing industry, which breeds tens of thousands of horses each year, is horribly corrupt and plagued by all kinds of accusations. To regulate this industry, looking at one aspect only is not enough. Each aspect and every dimension of it is important. All the actors who are involved and all the factors that are relevant should be taken into account: breeders, owners,

sponsors, grooms, stewards, trainers, jockeys, bookmakers, veterinarians, nutrition, medication, whipping, horse surgeons, logistics, stables, horsetack (riding equipment such as saddles, bridles, stirrups, halters, reins, bits and martingales), sulkies and racing tracks.
- Interior designing (decorative arts and furnishings) is the planning and ordering/organization of social space; it is a branch of spatial ordering and to be distinguished from time-related ordering. In interior designing, "total design" is important. There is, or should be, no clear disparity between the elements making up the *Ganzheit*. There are, or should be, neither detonating elements nor clashing colors. The whole shines through its details, which compose the whole, the *Gestalt*. Furniture, carpets, curtains, lamps and artifacts like paintings, sculptures, pottery, basketry and glassware fit harmoniously together, giving the interior its character, which, ideally, correspond with, or match to, the identity/individuality of the person who will use, or will be living in, the space. Interior designers and consultants know this, and "country experts" should learn from them.

The nature, structure, organization, scope, methods, rise, development and impact of science and the relation between science, religion and philosophy (the three intellectual enterprises of mankind), are matters the discussion of which I must postpone. However, a few words may be said about the objects scientists are studying. They fall into five categories and are studied by the physical, biological, social, human and formal sciences.

- The physical sciences, encompassing physics, chemistry, astronomy and earth science (each science having a range of subdisciplines), are concerned with non-living systems; they are elementary and relatively simple sciences.
- The biological sciences, comprising more than fifty subdisciplines, are concerned with living systems (organisms); these systems are based on, but irreducible to, non-living systems. How life emerged from cosmic dust is still a mystery, all the efforts of molecular biologists and biochemists notwithstanding.[23] An impressive index of molecular biology and biochemistry articles can be found on the Internet.
- The social sciences, including but not limited to linguistics, communication studies, economics, politics, polemology, sociology, education and law, are concerned with the relations among, and the concomitant behavior of, human beings; they take for granted the results of research conducted by biological scientists. Usually, social scientists do not trespass on the territory of biological scientists and the other way around, but collaborative research by members of both groups may result in revolutionary discoveries.
- The human sciences, including but not limited to literary studies, art history, religious sciences, psychology and anthropology, are concerned with human beings as conscious and creative individuals. Closely related to the social sciences,[24] they are fundamentally different from the three aforementioned groups of sciences, because object and subject tend to be fused in the human sciences. A stone, a plant or a language can be objectified and measured, but man (the objectifying and measuring subject) cannot. The humanities or *Kulturwissenschaften* threaten

to be pushed aside by the natural sciences. Like a garden that becomes overgrown with weed, mind threatens to be ravaged by matter. In common language: materialism is on the rise. It tends to be forgotten, that creativity, man's prerogative, is the mind set free; it is and will always remain ungraspable.[25]
- The formal sciences will be dealt with in chapter five.

The system of sciences (not to be confused with the science of systems) is not a modular system like a house consisting of separate modules (vestibule, kitchen, storage space, living room, bathroom, bedroom, etc.). Neither is it a scattered group of islands varying in size or number of inhabitants. The system of sciences is like an archipelago, a cluster of islands having things in common; it is a complex, historically evolved system composed of a steadily growing number of subsystems. Whatever the advantages of labor division, science is essentially unitary; it is a body of organized knowledge. Like a cathedral, which is not a pile of stones but a great number of stones arranged/ordered in a meaningful way, science is not an accumulation but an organization, not an aggregate but a whole, not a **Gesamtheit**, but a **Ganzheit**. Scientists from different disciplines should, therefore, at the instigation and under the umbrella or auspices of the university they are associated with, regularly have a joined conversation or brainstorming session in order to recognize the connectedness of the businesses they are in and to decide on the way they should go, separately as well as collectively. Explore https://digitalcommons.unl.edu/sunburst.html and see Katy Börner, *Atlas of Science: Visualizing What We Know* (MIT Press, 2010).

The borders between the sciences are porous and permeable. A look at the following, non-exhaustive list may clarify this: astrophysics, chemical physics, geophysics, geobiology, biophysics, biochemistry, psychophysics, biostatistics, psychopharmacology, biodemography, bioeconomics, biopolitics, biopsychology, medical physics, medical biology, econophysics, sociophysics, health economics, sociobiology, physical geography, physical anthropology, social geography, economic geology, economic geography (aka geographical economics), political geography, military geography, biogeography, molecular biology, evolutionary psychology, evolutionary economics, social psychology, sociolinguistics, social ecology, archaeo-chemistry, psycholinguistics, linguistic anthropology (aka ethnolinguistics), cognitive anthropology, ethnomusicology, linguistic demography, linguistic geography, archaeological anthropology (aka ethnoarchaeology), economic anthropology, social anthropology, educational anthropology, political anthropology, legal anthropology, physical anthropology, cultural anthropology, ecological anthropology, psychological anthropology, medical anthropology, sociology of literature, literary linguistics, social aesthetics, political aesthetics, cultural economics, medical sociology, psychiatric anthropology, legal anthropology, political demography, social demography, political economy, financial economics, demographic economics (aka population economics), ecological economics (aka bioeconomics), ecological psychology, ecological anthropology, education economics, educational psychology, political sociology, political psychology, military psychology, military medicine, legal psychology, legal economics, social economics, psychological economics, population biology, population geography, environmental chemistry, mathematical economics, mathematical sociology, mathematical anthropology, mathematical psychology, mathematical biology, mathematical statistics, computational mathematics, computational statistics, computational biology, computational archeology, computational sociology, computational economics, computational finance and computational anthropology.

Clearly, it is impossible for anybody to master even one of these bilaterally related sciences, let alone more than one. However, there is a common denominator of all of them. This denominator consists of three closely interconnected factors: (*a*) reality, (*b*) knowing reality and (*c*) the means of knowing reality. A country is but a piece of reality. To know or comprehend a country is what some academics vainly attempt to do single-handedly. Language is the means or instrument by which they usually communicate their partial knowledge of the country, with each other or with a wider audience, or readership. The scientific study of language (linguistics) and, more generally, of signs (semiotics) is of crucial and vital importance to comprehending the complexity of countries.

Scientists in such fields as biochemistry, social psychology, sociolinguistics, political economy, legal economics and mathematical economics attempt to abstract from the particulars in one field of research (e.g., law or economics) and to look for patterns in a pair of fields considered to be related. Trying to widen their view and to *com*prehend, they are thus beginning to collaborate scientifically. What they fail to do, however, is to follow through all the way, making a joint effort to comprehend the complexity of Turkey, Iran, Russia, Korea, China or any other county. Trying to understand these countries, they should search for balance and give no primacy to any science. Physicists, for example, are inclined to reduce everything to matter, or to state that everything is physical and is ruled by the laws of nature, leaving no room for metaphysical idealists who, in turn, claim that reality is a form of thought. These are extreme, hence unreliable, visions. In my view, the mental is dependent on, but not completely reducible to, the physical. Both the physical and the mental are real. The physical is real to the followers of "doubting Thomas", immortalized by Caravaggio; the mental is real to the Platonists. The clue is to see unity in diversity and vice versa, to consider the parts of a whole and the whole to which the parts belong. The concrete (*le visible*) and the abstract (*l'invisible*) should not be confused.

The sciences taken together form an enormous matrix. Whereas the colums of the matrix are divided over the physical, biology, social and human sciences (which can be broken down into subdisciplines), the rows are assigned to subjects like sociology of voting, sociology of terrorism, agrarian sociology, industrial sociology, sociology of old age, sociology of slums, electoral geography, religious geography, geography of slums, urban geography, urban economics, economics of slums, economics of transportation, economics of the public sector or economics of war. The impression of science being fragmented and confusing is strengthened by the relentless pairing up of sciences or sub-sciences and the increasing number of scientific applications. Science has strained relations with philosophy on the one side and with religion on the other side. The three domains should be distinguished, but shall not be separated.

There is a large literature on *the history* and *philosophy* of science, and the relationship between science and religion is a matter of hot debate.[26]

The sciences are intimately connected; they form a continuously changing whole, an expanding corpus of developing knowledge, each and every part/element of which has become the object of historical and philosophical investigations. History, philosophy and sociology of science form a trio, the strength of which stems from the confluence of history, philosophy and sociology. The history of a scientific discipline should not be confused with the history of what the discipline has been explaining or is supposed to explain. For example, the history of economics (economic thought) must not be confounded with the history of a country's economy. Furthermore, history of philosophy is not the same as philosophy of history. The word history refers either to a chain of happenings (events or actions) in the past (History I) or to the inquiry into, or account of, these happenings (History II). Mixing these two meanings up has been the cause of much misunderstanding.

Matured sciences go with the adjective "mathematical" or "computational", mathematics having computational and computation mathematical aspects. As matters of political, legal, military, economic, financial, geological, geographical, environmental, demographic, social, educational, psychological or religious importance hang tightly together, I believe it is time for investigating both the unity in the diversity and the diversity in the unity of sciences, that is, for applying the science of systems to the system of sciences. In an excellent, but almost forgotten article, "General Systems Theory – The Skeleton of Science" (*Management Science*, 2:3 [April 1956], pp. 197–208), Kenneth Boulding developed a framework for resolving the dispute over gaps and miscommunication between the sciences. In the early 1930s, Mary Parker Follett (1868–1933), a pioneer in the field of organizational studies, had already pleaded for "undepartmentalized thinking", for a coordinated, interdisciplinary approach to organizations. This was several years after Aleksander Bogdanov's (1873–1928) trailblazing *Tektologiya* had been published, a book now regarded as precursor of systems theory and related aspects of synergetics. In 1959, the English version of *Le Phénomène humain*, written by the French Jesuit and paleontologist Pierre Teilhard de Chardin (1881–1955), was published. Developing Vladimir Vernadsky's idea of noösphere (the sphere of human thought), the book aroused some interest in the complexity-science community but has suffered from its verbosity and Teihard's effort to save the Christian doctrine. In 1978, James Grier Miller picked up Boulding's thread and originated the living systems theory with his book *Living Systems: The Basic Concepts*, a beautiful text (online) that "should be read by all those who intend to put the so-called social sciences on a truly scientific basis in order to reach a unification of knowledge, at least as far as methodology is concerned".[27] While science of science (SciSci), which is to be distinguished from philosophy of science, seeks to offer a "quantitative understanding of the interactions among scientific agents across diverse geographic and temporal scales", I am pleading for genuine collaboration between all academic disciplines, no matter whether they are quantitatively or qualitatively oriented. "One is the ultimate reality", Wang Bi (third century) said. Different views about the One are all we have. It is time to open our eyes and to look at the **whole** elephant. "Unity and diversity are interrelated", Śaṃkara (eighth century) said. Humpty should be put together again. So let us listen to the *Rigveda* and return to the all-pervasive order. Read the list of quotations at the beginning of this book!

The following examples may substantiate my claim that the sciences are interconnected.

- *Financial economics* is occupied with the financial or monetary aspect/side of economic activities, because, in a modern society, financial assets of one type or another (money, stocks, bonds, mortgages, notes, bills, foreign exchange and derivatives) appear at both the supply and the demand side of the financial market. Its concern is thus the relationships between financial variables such as house prices, equity prices, interest rates and exchange rates. Financial economics is closely related to both microeconomics (which looks at economic activities through a microscope, studying the economic behavior of firms or households) and macroeconomics (which studies the economy as a whole, focusing on a nation's production, employment and general price level [inflation or deflation]). Macroeconomics is supposed to be based on microeconomics. Many micro or macroeconomists focus their attention on either the "aggregate demand" or the "aggregate supply" of goods. The question, however, is whether demands/supplies, can be added up.Economic systems are characterized by nonlinear relations. Financial economics is divided into four interrelated parts: personal, corporate, public and international finance. The literature on financial economics is growing by the day.
- *Political economy*, a discipline that originated in moral philosophy, is the vast minefield of studying economic activities (including their financial side) in relation to the rules and regulations established by a government, the forms of government differing by power structure, power source, ideology or other attributes.[29]
- *Legal economics* is a vigorously researched area. In *Texas Law Review* (53 [1975], pp. 757–782), Richard Posner, one of its prolific exponents, defined it as "the application of the theories and empirical methods of economics to the central institutions of a legal system". The bridge between law and economics is a heavily trafficked two-way one.[30]
- *Social economics* (aka economic sociology) is concerned with the social embeddedness of economic activities. More realistic than hard-core economics, it examines how social relations affect, and are affected by, economic relations. Many economists and sociologists are ignorant of each other's work. As a result, the understanding of the world has been seriously impaired. Economic anthropology is the twin sister of social economics.[31]
- *Psychological economics* is closely related to social economics and social psychology. It focuses more on the psychological aspects of economic phenomena and processes. Economists who are uninformed about developments in psychology are blinkered.[32]
- *Education economics* is concerned with the educational system in regard to economic growth, with the internal efficience of schools and with the relations between the costs of education and the methods of financing them. Mark Blaug

selected important readings on the subject in *Economics of Education* (Penguin, 1968 [vol.I], 1969 [vol.II]).

- *Health economics* is concerned with the allocation of healthcare. Kenneth Arrow's article "Uncertainty and the Welfare Economics of Medical Care" in *The American Economic Review* (53:5, December 1963, pp. 941–973) is generally considered the beginning of this discipline. Together with education economics, health economics makes up the core of the economics of human resources, a field of inquiry which silently revolutionized traditional economic research.
- *Cultural economics* studies the relationship between culture and economic behavior. Intimately connected with economic anthropology, it is interested in how norms, values, tastes, expectations, purposes and intentions are formed by a country's society, and vice versa. Economy (embedded in a social system) and culture (linked to a social system) are viewed as coupled systems. Complex interactions and feedback effects are recognized, and the dynamics are made explicit.Economists who are not interested in the cultural side of their work are narrow-minded, and their theories are lopsided, hence unreliable. Behavioral economics studies the effects of psychological, cognitive, emotional, social and cultural factors on economic decisions.[34]
- *Geographical economics* explains the distribution of economic activities across space, emphasizing that these activities do not happen on the point of a needle, as mainstream economics seems to imply. "Spatial turn" is a term often used in the literature on geographical economics; it refers to the intellectual movement in the social sciences and humanities that stresses the importance of place and space. The spatial turn followed on the cultural turn, a movement that begun in the early 1970s and made culture the focus of debates in these sciences.[35] Geographical economics should not be confused with geopolitics. [36]
- *Evolutionary economics*, which increasingly incorporates institutional economics, stresses changes over time; it emphasizes *Traditionsgebundenheit* (Gadamer) and *Strukturwandlung* (Habermas). Standing orthogonally on geographical economics, it pays attention to the interdependent processes that transform an economy through the actions and interactions of economic agents. Evolutionary economics, which goes to the heart of the ongoing discussion as to "how the West grew rich", is inspired by the ideas of Charles Darwin (1809–1882), who might have been influenced by Hegel (who historicized the revolutionary ideas of Kant). Richard Nelson is one of the leading figures, thanks to the seminal book he co-authored with Sidney Winter, *An Evolutionary Theory of Economic Change* (Cambridge, MA: Belknap, 1982).[37]
- *Demographic economics* (aka population economics) is another fast expanding field of research. It occupies itself with how variables such as fertility, mortality and migration respond to economic rules or activities and in turn affect the economic development of a society. The discipline is adjacent to population geography, which is concerned with the dynamics of the geographical distribution of a population.
- *Ecological economics* (aka bioeconomics) studies the interplay between, and coevolution of, economic and ecological systems. Nicholas Georgescu-Roegen

(1906–1994), student and protégé of Joseph Schumpeter (https://issevec.uni-jena.de), is considered its intellectual progenitor. In his masterpiece, *The Entropy Law and the Economic Process* (Harvard University Press, 1971), he argues that all natural resources are irreversibly degraded when put to use in economic activity. He saw the relevance of thermodynamics to economics and stood at the cradle of econophysics.Georgescu-Roegen's book should be read in conjunction with Vaclav Smil, *Energy and Civilization: A History* (MIT Press, 2017). The International Society for Ecological Economics is dedicated to advancing the understanding of the relationships between ecological, social and economic systems. The three nested systems of sustainability are the economy, which is contained by the society, which in turn is contained by the biophysical environment. Georgescu-Roegen was a visionary, whose thoughts were deeply rooted in those of Alfred North Whitehead (1861–1947). He died lonely at the age of 88, but his message, unwelcome to many people, is carried on by Herman Daly, Joshua Farley, Joan Martínez-Alier and Roldan Muradian. William Nordhaus is indebted to Georgescu-Roegen. See Karl-Göran Mäler et al. (eds.), *Handbook of Environmental Economics* (Elsevier, 2003ff) and Reinette Biggs et al. (eds.), *Handbook of Research Methods for Social-Ecological Systems* (Routledge, 2021).
- *Mathematical economics* uses mathematical tools and methods to analyze economic problems or to represent theories developed by economists in different fields of research.Econometrics is sited at the intersection of economics, mathematics and statistics. Nobody is equally expert in these three fast developing disciplines.
- *Computational economics* concerns the computational modeling of economic systems. It interfaces with computer science, economics and management science (the study of problem solving and decision-making processes in human organizations).[40]

A quick look at the rise and development of liberalism may drive my message home that economic thinking shall not to be dissociated from other kinds of thinking. If economic thinking is isolated and is taking hold of the mind, researchers will go astray, and man will suffer. The story of the demise of socialism has been triumphantly told many times, in particular the failure of Marxism, the voice of the humiliated and suppressed.[41] The story of the collapse and downfall of liberalism, however, is less well known. Initiated by John Locke, liberalism was celebrated by Adam Smith, enthroned by David Ricardo, redeemed by John Stuart Mill, criticized by Karl Marx, vindicated by neoclassical economists, tempered by Keynes and completely derailed by Friedrich von Hayek and his followers. The world we live in today is not what the Enlightenment prophets envisaged. John Locke (1632–1704), Baron de Montesquieu (1689–1755) and Adam Smith (1723–1790) cherished the lofty ideal of emancipating human reason from the thralldom of prejudice and superstition and of applying this emancipated reason to the cause of political, social and economic reform. Dreaming of a great society, or communal harmony, they did not serve the interests of the "*petite bourgeoisie*".

However, a great paradox seems to have escaped the attention of Locke, Montesquieu and Smith. I cannot believe that each and every individual knows his/her own interests best. Utilitarianism is a hotly debated issue (go to https://www.iep.utm.edu/util-a-r and https://plato.stanford.edu/entries/utilitarianism-history), and the "invisible hand" may lead man astray. Companies try, in an ever smarter way, to maximize the sale of their products, rather than to increase the well-being of an individual. The government of a state, which is certainly not the mere aggregate of individual beings she has been taken to be, is supposed to aim to promote and protect the *general* welfare, not the welfare of a single individual, or the wealth of a particular group of individuals. According to Pausanias (c. 110 – c. 180), "γνῶθι σεαυτόν" (know thyself) were the words inscribed in the portico of the temple of Apollo at Delphi. If nobody really knows anybody, everybody is doomed, and a communal order commensurate with the cosmic order will never be reached.

The reason why, pairwise-combined, sciences such as financial –, political –, legal –, social –, psychological –, education –, health/medical –, cultural –, geographical –, evolutionary –, demographic –, ecological –, mathematical – and computational economics are linked with each other is because they are *economy*-related; they are looking at an economic system through different disciplinary lenses.

Julius Caesar begins his famous book *Commentaries on the Gallic War* by writing: "*Gallia est omnis divisa in partes tres*". Similarly, a country's economic system is divided into three parts: the agricultural, industrial and service part. Being a system of systems, each sector is distinctive. The three sectors are the interrelated parts that belong to the economic system; they cannot be understood without the whole system being taken into account.

(1) **The Agricultural Sector**
The peasantry is an immense subject, extensively dealt with by Alexander Chayanov, Theodor Shanin, Eric Wolf, Bernard Slicher van Bath, Werner Rösener, Heinrich Niehaus, Annie Moulin, René Dumont, Norman Borlaug, Theodore Schultz, David Grigg and many other scholars. Agriculture encompasses a wide range of economic activities having one thing in common: the cultivation of land to provide crops intended for direct or indirect consumption (cereals, edible tubers, legumes, herbs and spices, coca, edible fungi, oilseeds, oil palms, soybeans, coconut, coffee and cocoa beans, tea plants, sugar beet and cane, cotton, wool, silk, kapok, jute, flax, hemp, sisal, ramie, latex, tobacco, or timber). Agriculture is dependent on soil, water (the supply of which is becoming increasingly critical), climate (light, temperature and air) and the subjects geologists are concerned with. However, man can lend nature a hand by means of lighting, heating, cooling, fertilizing, irrigation, drainage, land improvement, tillage, weeding, plant breeding, soil and crop protection or mechanization. Land can also be used to grow grass for cattle, sheep, etc. Horticulture concerns the growing of all kinds of fruit, vegetables, flowers and other cultivars (plants selected for desirable characteristics). It also includes arboriculture (the cultivation and study of individual trees, shrubs and vines) and landscape or garden design, construction and maintenance. In contrast to agriculture, horticulture does not include large-scale crop production or animal husbandry. Livestock farming, heavily relying on agriculture, concerns the breeding, feeding and management of animals of economic importance. In Western countries, the category livestock includes primarily cows, bulls, sheep, horses, pigs and goats. Animals such as donkeys, buffaloes, oxen and camels predominate in the agriculture of other countries. The commercial raising of

birds for eggs or meat (e.g., chickens, turkeys, ducks and geese) is usually treated separately. Forestry (looking after large areas of trees) is a specific method of using land to serve human needs. Its organization, like that of agriculture, can assume different forms. Like fishing, forestry straddles the agricultural and industrial sector. "Lumber industry" refers to timber cropping as well as logging (the cutting, skidding, on-site processing and loading of trees or logs onto trucks or skeleton cars). In fishing, a distinction is made between marine fishing (sea and coastal fishing) and inland fishing (in rivers, estuaries, lakes and swamps). Marine fisheries are capital-intensive (well-equipped and seaworthy fishing vessels and, on shore, processing, cooling, storing and transport facilities); they require highly skilled personnel. Inland fisheries do not necessarily call for many capital inputs; good results can be obtained by individuals operating on a small scale. Skilled management, however, is essential, and in most cases, cooling and transport facilities are needed.

(2) **The Industrial Sector**

The industrial sector can be divided into the extractive , processing and manufacturing industry. Mining (the extraction of valuable minerals or other geological material from the earth) is subdivided into six categories: coal mining, oil or gas extraction (onshore or offshore), iron ore mining, nonferrous metal ore mining (copper, lead, zinc, tin, nickel, aluminum, titanium, tungsten, magnesium, gold, silver, mercury, cadmium, uranium, etc.), gemstone mining (diamond, emerald, sapphire, jade, quartz, topaz, opal, amethyst, etc.) and quarrying (sand, clay, granite, limestone, marble, gypsum, chalk, etc.). Each category includes a variety of support operations. Processing industries use natural resources (biotic or abiotic, renewable or non-renewable) and machine them to inputs for the manufacturing industries. Examples are the coal processing, refining, food, beverage, leather, textile, chemical, glass, plastics, cement, wood, pulp and paper, tobacco, rubber, metal and packaging and labeling industry. Examples of the manufacturing industry are the fertilizer, pesticide, footwear, clothing, headgear, eyewear, furniture, ceramic, construction, elevator, tractor, earthmoving equipment, crane and hoisting equipment, infrastructure, machine and motor, hand tools, musical instruments, apparatus, kitchenware, pharmaceutical, cosmetic, jewelry, printing and graphic, automotive, tire, railway, shipbuilding, aerospace, arms, electrical, electronics and computer industry. Like the extractive and processing industries, the manufacturing industry makes use of the services of industrial engineers or production managers.

In Western countries, the development of agriculture has fostered the agricultural industry (agribusiness). Farming is only a small part of it. The modern agricultural industry has two major components: (a) the agricultural input-industry, producing inputs such as seed, fertilizers, pesticides, machinery, equipment, fuel and fodder, and (b) the agricultural output-processing industry, which takes over the farm products and transports them to factories where they are processed (slaughtered, dismembered, boned, sheared, waxed, washed, peeled, irradiated, ground, milled, pressed, minced, sawed, diced, sliced, bleached, mixed, smoked, spiced, salted, sweetened, flavored, dyed, fermented, boiled, baked, cooked, frozen, dehydrated, conserved, reconstituted, packed, bottled or labeled) and from where they are distributed to manufacturers or to wholesale or retail outlets, at home or abroad.

To reach the capillaries of a country's economic organism,[42] the reader should realize three things:

- Firstly, each of the industries enumerated has a number of suppliers. For example, a car, manufactured in the automotive industry, has about 30,000 parts, counting every part down to the smallest screw. The major parts of a car are (1) chassis and frame, (2) wheels and tires, (3) power train (engine and transmission), (4) suspension system, (5) electronic sensors and wiring systems, (6) braking system and (7) lubrication system. No manufacturer of any part is fully autonomous. It is the right configuration (pattern) of the right parts that gives what we call a car (as distinct from bicycle, ship or plane) its distinctiveness, character or essence, i.e., what makes the thing what it is: a car (no matter whether Alfa Romeo, Audi, Bentley, Bugatti, Honda, Kia, Lamborghini, Land Rover, Lotus, Maserati, Nissan, Mitsubishi, Opel, Porsche, Renault, SEAT, Škoda, or Vauxhall is the producer). To give another example, a pair of glasses, manufactured in the eyewear industry, has a relatively small number of parts (glass or hard plastic lenses and ophthalmic frames), but again: no manufacturer of any part is absolutely autonomous. The variety of eyeglasses (spectacles) is almost endless. The competition is fierce and merciless. The relentless battle for the wallet of the consumer, who is bombarded with advertisements for products he/she does not really need, pushes us to the brink of the cliff. No warning of any scientist or philosopher has been and—I am afraid—will be helping. Supply chain management involves "the movement and storage of raw materials, work-in-process and finished goods from the point of origin to the point of consumption".[43]
- Secondly, each of the industries enumerated encompasses a number of kindred but not identical industries, the products of which are made in geographically distributed factories that are run by (holding) companies or multinational corporations with different management, finance, logistics and marketing functions. For example, the list of beverages produced in the beverage industry on the basis of biotic resources and finally consumed includes alcoholic drinks (wine, beer, distilled drinks and cider) and non-alcoholic drinks (coffee, tea, chocolate and soft drinks). The manufacturers of these drinks compete with each other, using different tactics and strategies.[44]
- Thirdly, each of the industries enumerated depends on the service sector, which in turn is dependent on the agricultural and industrial sector.

(3) **The Service Sector**

The service sector delivers services rather than tangibles. Occasionally, however, the classical dichotomy breaks down (books that are only virtually available are not tangibles). Some economists see a continuum, with pure services (e.g., teaching, training) at one end, pure commodities (e.g., ballpoints and footballs) at the other end. Most products fall somewhere between these extremes. The service delivery, always involving a provider and a customer, or somebody being served, usually includes physical facilities (e.g., building, room, furniture and a parking lot) and equipment (e.g., car, plane, cash register, scissors and a computer). Services are neither cultivated (grown) nor mined (extracted), neither processed nor manufactured, but most of them are information-intensive. Services cannot be transported or stored for future use. They cannot be measured like tangibles, which is the reason why economists and statisticians have difficulty in taking services into account when compiling national accounts that measure the economic activity of a nation. Once delivered, services cannot be returned. The service sector, which is considered the third part of an economy, is often divided into the tertiary sector (commercial services) and the quaternary sector (non-commercial or governmental services).[45] The range of services is broader than people usually think.

Since "service" is a rather vague indication, I categorize service providers as follows: accountants, acolytes, actors, agents, aides, aldermen, altar boys, ambulance workers, anchormen, archivists, assistants, attendants, barkeepers, bishops, bodyguards, bookmakers, bouncers, bullfighters, butlers, camera operators, consultants, call girls, cashiers, censors, chaplains, church elders, civil servants, cleaners, clerks, clowns, columnists, comedians, commentators, concierges, conductors, copywriters, coroners, correspondents, cowboys, croupiers, detectives, dieticians, diplomats, disc jockeys, divers, dressmen, emcees, employees, entertainers, escorts, executioners, ferrymen, financial intermediaries, firefighters, fortune tellers, fundraisers, garbage collectors, gardeners, geishas, go-betweens, gondoliers, gravediggers, grooms, guards, guides, hairdressers, hangmen, hawkers, headhunters, herdsmen, hosts, hoteliers, housemaids, human billboards, HR consultants, hygienists, illusionists, impresarios, influencers, inspectors, instructors, intendants, intermediaries, internet service providers, investigating officer, journalists, judges, jurists, lackjeys, lawyers, librarians, linesmen, litter pickers, lobbyists, magicians, maids, majordomos, managers, manicurists, mannequins, masseurs, matchmakers, mayors, mentors, mercenaries, messengers, midwifes, ministers, missionaries, monks, movers, muezzins, mullahs, nannies, news anchors, notaries, nuns, nurses, occultists, operators, panelists, paperboys, paragnosts, parliamentarians, party leaders, pastors, peddlers, peons, photo models, pimps, pizza boys, podcastmakers, politicians, popes, porn stars, porters, postmen, Presbyrarians, presenters, presidents, priests, programmers, prophets, prostitutes, PR officers, psephologists, pursers, quacks, quizmasters, rabbis, receptionists, referees, reporters, representatives, rickshaw pullers, road sweepers, scalpers, secretaries, servants, shoeshine boys, showgirls, sidekicks, socialites, social service workers, sommeliers, soothsayers, sorcerers, speakers, spies, spin doctors, spokesmen, sponsors, stagehands, stand-ins, stationmasters, stevedores, stewards, stooges, strippers, studio potters, stuntmen, stylists, supervisors, supporters, surgeons, surveyors, tattooists, tax collectors, taxi drivers, teachers, televangelists, test drivers, therapists, toastmasters, touts, traders, train pushers, trainers, treasurers, ushers, valets, vassals, vendors, vergers, veterinarians, vicars, viceroys, vintners, visagists, vote counters, waiters, wardens, weathermen, webmasters, window dressers, window washers, witnesses and YouTubers.

A few remarks may be in place:

- Though sometimes meaningless work is also done in the agricultural and industrial sector, most service providers perform a "bullshit job". Statistically, the highest number of suicides can therefore be found in the service sector.[46] Would human life become meaningless if all nail studios or tattoo parlors were ordered to close down, less car brands would be marketed or less publishing houses would be active, fewer kinds of bread or beer would be for sale, and the assortment of hair sprays or body lotions would be reduced? It seems to me that a good and large-scale cleanup is what mankind needs. The COVID-19 pandemic is perhaps a blessing in disguise.
- The number and social impact of actors and actresses should not be underestimated. They appear in movies (film stars), and their talents are often deployed by the advertising industry, by professional fakers who line their pockets by creating a dream world.[47]
- A housewife is a woman who works at home, doing the cooking, cleaning, etc. Except for being a career woman, she is not employed outside the home. A housewife who has children may be called a stay-at-home mother or mom. Housewives are unpaid domestic *servants*. Many housewives have become housemates. Is it because of feminism?
- Employees are people employed by an employer. They are *Knechte,* working for, used by and being at the service of a *Herr*.Employees or laborers try to please ("suck up to") their boss, and they better do so! Over time, they are just replaced (like machine parts). If there is no, or not enough, work for them to do, they will be fired or laid off as soon as the boss thinks it is somehow possible. Legal protection is not essential to their servile relationship to an employer; power is. Power comes in many forms,and the power of those in power may be an index of freedom. Employees are economically, socially and politically bonded; they are public or civil *servants*. Equality is a myth, particularly gender equality. It doesn't rhyme or square with liberty, not to speak of fraternity. Modern society is still marred by vestiges of slavery, but the champions of freedom do not want to hear from it.
- There are different types of financial markets, different types of financial assets, different participants in the financial markets and different financial intermediaries. The last set can be divided into three subsets: depository intermediaries (commercial banks, saving banks, etc.), contractual intermediaries (insurance companies, pension funds, etc.) and investment intermediaries (investment companies, finance companies, etc.).[49]
- Kings are assumed to be serving their people, who in turn will support them. Whether they actually have been serving is a different matter. If not, they are, or should be, chased away. Kings, with royal regalia and the aureole of divinity

(which explains the close relationship between [political] power and religion), go under different designations: *Pharaoh*, *Imperator*, *Lugal*, *Malik*, *Rí*, *Czar*, *Rex*, *Roi*, *Rey*, *Principe*, *Wánax*, *Basileús*, *Autokratōr*, *Pope*, *Patriarch*, *Emperor*, *Khan*, *Caesar*, *Shah*, *Caliph*, *Sultan*, *Raja*, *Negus*, *Monarch*, *Sovereign*, *Chief*, *Fürst*, *Führer*, *Caudillo*, *Duce*, *König*, *Kaiser*, *Shōgun*, *Tennō* and *Huángdì*. Today's presidents are kings in disguise.[50]

If the reader is wondering why I give an elaborate example of the interconnectedness of the sciences, he or she should keep in mind that the agricultural, industrial and service sector, along with the parts into which they can be divided, are closely interrelated. They are the parts of a complex and dynamic economic system that has—it cannot be overstressed—different aspects: a financial, political, legal, social, psychological, educational, medical, cultural, geographical, evolutionary, demographic, ecological, mathematical and computational/technological aspect. A country's economic system (including its agricultural, industrial and service sector) is closely related to that country's interrelated financial, political, legal, social, etc., systems and shall thus not to be taken out of its context, the context that colors all the country's systems, while taking its character from these systems. That is why the economy of, say, the People's Republic of China looks different from the economy of the United States, though both economies have a divisible agricultural, industrial and service sector. It is thus crucial to investigate a country not by breaking it down, by taking one system apart and "freezing" the other parts of the whole. The country should be studied cross-disciplinarily. For a country is a living being, not a *Gesamtheit* (aggregate) of elements simply to be added up, but an evolving, constantly changing *Ganzheit* (whole) that needs to be looked at in an abstract manner, so to speak, from a distance. The zooming in and the zooming out should alternate. By zooming in, the reader will learn the importance of the parts (particular instances); by zooming out, he will realize that these parts are nothing but the parts of a whole. The crux of the matter is to see both the whole and its parts, for parts are always parts of a whole and a whole has always parts. There can be no diversity (parts) without unity (whole) and no unity without diversity. Diversity and unity are not mutually contradictory but reciprocally conditional, a subtle but crucial difference often overlooked. They are not antagonistic but complementary and correlative to each other. As Zhu Xi (twelfth century) said: "*Lǐ yī fēn shū*" (the pattern/principle is one, the parts are different).

While the problems of mankind are mounting, the struggle for political power is taking place on a global scale and whole civilizations are clashing, the big challenge is: comprehending countries by means and with the help of computers, that is, by harnessing the power of ever more advanced computers to gather, process and integrate country-related information and to organize and manage the growing corpus of country-related knowledge available in digital libraries. The new kind of country study I envision will cause an earthquake to occur at universities, research institutes and think tanks. The waiting is for the movers and shakers to pick up the idea.

The prestigious, Seattle-based, allegedly non-partisan National Bureau of Asian Research (NBR) is one of the knowledge hubs I am thinking of. The organization conducts "advanced independent research" on issues regarding the U.S. relations with Asia. Established in 1989 and "engaged in the demand-driven approach to reach out to the American Congress and to cultivate relationships on both sides of the aisle", NBR, "an institute without walls", has committed herself to "training the next generation of Asia specialists". It sounds good, but how could one person possibly be specialist on an area measuring more than 44 million km2, comprising 48 sovereign states and counting more than 4 billion people! To demonstrate the impossibility and futility of the endeavor: the Middle Eastern cuisine (style of cooking characterized by distinctive ingredients, techniques and dishes) is divided into the Bahraini, Cypriot, Egyptian, Iranian, Iraqi, Israeli, Jordanian, Kuwaiti, Lebanese, Omani, Palestinian, Qatari, Saudi Arabian, Syrian, Turkish and Yemeni cuisine. This does not really seem to concern the leadership of NBR. In June last year, the annual Asia (sic) Policy Assembly convened, a two-day conference organized by the National Bureau of Asian Research and the Institute for National Strategic Studies and bringing together "top experts from the academic and policymaking communities to address issues of strategic importance to the United States in Asia". I hope that next year the ambition is more realistic, the subject of computer-aided collaborative research regarding only one of Asia's countries being on the agenda. The focus of the scientific team could be on China. If that would be too difficult, (initially), Singapore, Vatican City, Monaco, Malta, or Bahrain could be selected as pilot projects.

I am also thinking of the much touted, Singapore-based Centre on Asia and Globalization (CAG), established in 2006 and celebrating the 100th issue of its magazine China–India Brief in September 2017. CAG "partners with pre-eminent institutions around the world to produce policy relevant knowledge on the impact of globalization on Asia and Asia's role in governing an integrated world". However, the Center has not clearly defined "Asia" and consistently plays down the formidable forces that counteract globalization. The same applies to the Asia Global Institute (AGI) at the University of Hong Kong. Inaugurated in June 2015 and with Nobel Laureate Michael Spence co-chairing its advisory board, AGI is a "multidisciplinary think tank established to assume and carry forward the Fung Global Institute's mission to generate and disseminate research and ideas on global issues from Asian perspectives". CAG and AGI are biting off more than they can chew.

The Center for China and Globalization (CCG) is a "leading Chinese independent think tank dedicated to the study of global trade and investment, global migration, global governance, global affairs and global talent policy". It boasts a "strong research team" and enjoys an "impressive record of publications and events with broad public policy impact". Founded in 2008 by Wang Huiyao (adviser to the Chinese government) and Miao Lu (secretary general of China's Global Talent Society), CCG has grown into the largest non-governmental think tank in China, with headquarters in Beijing, offices in Shanghai, Shenzhen, Guangzhou, Qingdao and Hong Kong, and representatives in major cities abroad. According to the 2018 Global Go To Think Tank Index compiled by the University of Pennsylvania, CCG ranked 94th of the top think tanks worldwide. I leave it to the reader to judge whether or not CCG is an "independent think tank", not allowing any governmental interference, but I wish to point out that, to be invited to join its advisory, academic or international council, one would better not level criticism against the line of the (leader of the) Communist Party of China.

Similar allegations could be made against (*a*) the more than 1300 members counting Japan Association for Asian Studies, established in 1953 "for the purpose of conducting and publishing results of theoretical and empirical research centering on politics and economy in Asia" and associated with the big Japan Consortium for Area Studies, and (*b*) the Asia Society, the "leading educational organization dedicated to promoting mutual understanding and strengthening partnerships among peoples, leaders, and institutions of Asia and the United States in a global context".

In their online article "Toward Immersive Asian Studies", David Leheny of Princeton University and Jin Sato of the University of Tokyo outlined their respective university's joint plan to "facilitate undergraduate exchange and education on Asia", but they made no mention of even the minimum requirement for the scientific study of a single country in that immense area. As proof of their good intention to engage in high-level international collaboration, one can read on the Web site of the Todai-Princeton partnership that "bilateral collaborative research and teaching ventures should include a balanced representation of scholars from both institutions", but a plea for high-level *interdisciplinary* collaboration is conspicuous by its absence. Criticism of the same kind could also be leveled at the Association for Asian Studies (AAS), founded in 1941 and headquartered in Ann Arbor. With approximately 7,000 members worldwide, representing all the regions and countries of Asia and all academic disciplines, AAS is "the largest organization of its kind", providing its members with a "unique and invaluable professional network". Big–bigger–biggest, rather than deep–deeper–deepest, seems to be the motto.

I am glad to learn that, under the leadership of professor Michael Szonyi, the highly prestigious Fairbank Center for Chinese Studies is pursuing "new directions in the study of China" (*Zhōngguó Yánjiū Xīn Dòngxiàng*). However, I would be delighted to know what is new about these "new directions". I sent professor Szonyi an email asking for more details. His reply was not particularly reassuring.

I am also happy to learn that professor Chris Nielsen, Executive Director of the "interdisciplinary" Harvard-China Project on Energy, Economy and Environment, is collaborating with Chinese universities and across the schools of Harvard University. I am not quite sure, however, that the project will be a resounding success because (*a*) it is "the grateful recipient of a major multiyear grant from the Harvard Global Institute of the Office of the President of Harvard University", (*b*) its major funders include the National Science Foundation, the U.S. Department of Energy, the U.S. Environmental Protection Agency, NASA, the V. Kann Rasmussen Foundation, the Henry Luce Foundation, the Ford Motor Company Foundation, the Volvo Research and Education Foundations, the Energy Foundation China, the NARI Group Corporation, the China Three Gorges Corporation and the China Council for International Cooperation on Environment and Development (The saying goes: "He who pays the piper calls the tune"), and—most importantly—(*c*) the collaboration has been assured of no fewer than twenty-seven Chinese contributors, who "have been screened for partiality or bias".

The list of "Orientalists", "Asianists" or "*Ostasienwissenschaftler*" is very long. Kishore Mahbubani, the self-promoting author of *The New Asian Hemisphere* (2008), Martin Jacques, the PRC-pampered writer of *When China Rules the World* (2012), and Evan Osnos, the self-styled China expert, are probably the most well-known among them, with the exception of "Dear Henry", the American statesman, political scientist, diplomat and geopolitical consultant who wrote (did he?) the bestseller *On China* (2011) and cannot be suspected of being prejudiced (really?). The host of CNN's *Fareed Zakaria GPS* is so smart as not to advertize himself as orientalist or Asia expert. However, in his weekly public affairs show, with his own name mentioned in super large letters but without having to pay a single dollar for the advertisement, he calls to the spectators to sign up for the newsletter "Fareed's Global Briefing" or to download and read his column in *The Washington Post*, where he offers his view of—the whole world.

Max Planck reportedly said that a new scientific truth does not triumph by convincing its opponents and making them see the light, but rather because its opponents eventually die. Whether this is an accurate description of the scientific business is debatable, but it cannot be gainsaid that many academics (*a*) fail to examine their assumptions or to do some soul searching, (*b*) cling to old beliefs which are taken for granted and (*c*) pass these beliefs down from one generation of students to another. According to Benjamin Franklin, "there are three kinds of people: those who are immovable, those who are movable, and those who move". The aim of this book is to challenge the widespread belief that area/country studies stand comparison with any scientific work in demography, geography, sociology, anthropology, economics, psychology, linguistics, political science or law. My hope is that there are people who really move, taking on the challenge presented. *Du choc des opinions jaillit la vérité*, but whoever is silent will be understood to consent.

Notes and References

1. See James Hobbs, *Fundamentals of World Regional Geography* (Cengage, 2016).
2. See Richard Lee and Richard Daly (eds.), *Cambridge Encyclopedia of Hunters and Gatherers* (CUP, 1999) and Vicki Cummings et al. (eds.), *Handbook of the Archaeology and Anthropology of Hunter-Gatherers* (OUP, 2014).
3. See Luc Ferry, *Philosophie politique* (PUF, 1984/5), Philippe Raynaud and Stéphane Rials (eds.), *Dictionnaire de philosophie politique* (PUF, 2003), George Klosko (ed.), *Handbook of the History of Political Thought* (OUP, 2011) and Robert Goodin et al. (eds.), *A Companion to Contemporary Political Philosophy* (Wiley, 2012). Useful journals are Interpretation: *A Journal of Political Philosophy* (1970ff), *The Journal of Political Philosophy* (1993ff) and *Oxford Studies in Political Philosophy* (2015ff).
4. See Kenneth Dyson, *The State Tradition in Western Europe* (Martin Robinson, 1980) and Bob Jessop, *The State: Past, Present, Future* (Polity Press, 2016).
5. Whether The People's Republic of China is a state or "a party pretending to be a country" is a point of discussion. See my article "Is Taiwan a Part of

China?" in Jean-Marie Henckaerts (ed.), *The International Status of Taiwan in the New World Order* (Kluwer, 1996, pp. 9–19), Wang Hui, *China from Empire to Nation-State*, translated by Michael Gibbs Hill (Harvard University Press, 2014) and Yangyang Cheng, "A Birthday Letter to the People's Republic", *ChinaFile*, September 28, 2019.

6. See Gerard Delanty (ed.), *Handbook of Contemporary European Social Theory* (Routledge, 2006, pp. 310–322), Émilie Tardivel, "Qu'est-ce qu'une nation?", Études, October 2016, pp. 53–64, and Eric Anceau and Henri Temple, *Qu'est-ce qu'une nation en Europe?* (SUP, 2018).
7. See Kirsty Gover, *Tribal Constitutionalism* (OUP, 2010), Lawrence Rosen, "'Tribalism' gets a bum rap", *Anthropology Today*, 32:5 (October 2016), pp. 3–4, Yoav Alon, *The Shaykh of Shaykhs* (Stanford University Press, 2016) and Stevan Hobfoll, *Tribalism: The Evolutionary Origins of Fear Politics* (Springer, 2018).
8. See Anthony Gribin, *I Am an Echo Chamber: The Basis of Tribalism* (Lulu Press: 2018).
9. See Kenichi Ohmae, *The End of the Nation State* (Free Press, 1995), William Robinson, "Beyond Nation-State Paradigm", *Sociological Forum*, 13:4 (1998), pp. 561–594, Alejandro Portes et al., "The study of transnationalism", *Ethnic and Racial Studies*, 22:2 (1999), pp. 217–237, Craig Calhoun, *Nations Matter* (Routledge, 2007), Robert Holton, *Globalization and the Nation State* (Palgrave, 2011), John Breuilly (ed.), *Handbook of the History of Nationalism* (OUP, 2013), Louis Snyder, *The New Nationalism* (Routledge, 2017), Andrea Wimmer, *Nation Building* (PUP, 2018), Jeffry Frieden et al., *World Politics: Interests, Interactions, Institutions* (W.W. Norton, 2018, chapter 14) and Yael Tamir, *Why Nationalism* (PUP, 2019). Go to http://www.nationalismproject.org, https://doi.org/10.1111/nana.12489, https://plato.stanford.edu/entries/nationalism and https://academic.oup.com/ahr (search for nationalism).
10. See Peter Burroughs and Anthony Stockwell, *Managing the Business of Empire* (Frank Cass, 1998), Susan Alcock et al. (eds.), *Empires* (CUP, 2001) and Krishan Kumar, *Visions of Empire* (PUP, 2017). For federalism, see Ann Ward (ed.), *The Ashgate Research Companion to Federalism* (Routledge, 2013).
11. For sovereignty, see Jean Bodin, *Les Six Livres de la République* (Jacques du Puys, 1576), Michael Loewe, "The concept of sovereignty", *The Cambridge History of China*, Vol. 1 (CUP, 1986, pp. 726–746), Alain Rocher, *Mythe et souveraineté au Japon* (PUF, 1997), Giorgio Agamben, *Homo Sacer* (Stanford University Press, 1998), Sylvie Mesure and Patrick Savidan (eds.), *Le dictionnaire des sciences humaines* (PUF, 2006, pp. 1119–1123), Lyndon Storey, *Humanity and Sovereignty* (Peter Lang, 2006), Jens Bartelson, "The Concept of Sovereignty Revisited", *European Journal of International Law*, 17:2 (2006), pp. 463–474, Stephen Krasner, *Power, the State, and Sovereignty* (Routledge, 2009, Part III), Lauren Benton, *A Search for Sovereignty* (CUP, 2010), David Leahy, *Beyond Sovereignty* (Davis Group, 2010), Phil Chan, *China, State Sovereignty and International Legal Order* (Brill, 2015), Lider Bal, *Le mythe de la souveraineté en droit international* (Presses Académiques

Francophones, 2016), Richard Bourke and Quentin Skinner (eds.), *Popular Sovereignty in Historical Perspective* (CUP, 2016), "Nation et souveraineté?", *Krisis*, No. 46, January 2017, Frank Chiang, *The One-China Policy* (Elsevier, 2018), Leo Klinkers, *Sovereignty, Security and Solidarity* (Lothian Foundation Press, 2019), Maria Carrai, *Sovereignty in China* (CUP, 2019) and Andrew Latham, *Sovereignty* (work-in-progress). The issue of sovereignty which is closely related to the subject of political order/system remains an open question. The Western discussion started off with the "Papal Revolution" (eleventh century) but is rooted in Augustine's work *De Civitate Dei*, a cornerstone of Western thought written in the fifth century. The key problem is whether sovereignty can be divided. Visit https://plato.stanford.edu/entries/sovereignty, https://philpapers.org/s/sovereignty and https://doc-research.org/2020/09/digital-sovereignty. See note 5 in this chapter.

12. See Mortimer Adler (ed.), *The Great Conversation: A Reader's Guide to Great Books of the Western World* (Encyclopædia Britannica, Inc., 1990) and Norman Melchert, *The Great Conversation* (OUP, 2018). In the age of globalization, comparison of the great conversations about universal order in the West, the Middle East, India and China is a *primum desideratum.* "Order", which is etymologically related to "right", is the key concept in political theory and philosophy of law. In his last book, *The Great Civilized Conversation* (Columbia University Press, 2013), William Theodore de Bary establishes links between Islamic, Indian, Chinese, Japanese, and Western classics.
13. See Ronald Arnett et al., *Communication Ethics Literacy: Dialogue and Difference* (Sage, 2009, Chap. 5) and David Bohm, *On Dialogue* (Routledge, 2013).
14. See Charles Stevenson, *Facts and Values* (Yale University Press, 1963), Hilary Putnam, *The Collapse of the Fact/Value Dichotomy and Other Essays* (Harvard University Press, 2004, Part I), Beatrice Centi and Wolfgang Huemer (eds.), *Values and Ontology* (Ontos, 2009), Giancarlo and Sarin Marchetti (eds.), *Facts and Values* (Routledge, 2017) and Michiko Kakutani, *The Death of Truth* (Tim Duggan, 2018). Go to https://doi.org/10.20935/AL131.
15. See Tony Harcup, *Journalism: Principles and Practice* (Sage, 2015) and David Randall, *The Universal Journalist* (Pluto Press, 2016).
16. See Keith Devlin, *The Joy of Sets* (Springer, 1993) and Matthew Foreman and Akihiro Kanamori (eds.), *Handbook of Set Theory* (Springer, 2010). For a critical note, see Daniel Strauss' article "The Fall and Original Sin of Set theory" (2019) at https://doi.org/10.25159/2413-3086/4983. Mathematical logic formalizes mathematical statements, including set-theoretical statements. It relates to the foundations of mathematics, meta-mathematics and theoretical computer science. Go to https://plato.stanford.edu/entries/set-theory.
17. See Tom Ritchey, "On a Morphology of Theories of Emergence", *Acta Morphologica Generalis*, 3:3 (2014), pp. 1–16, Paul Humphreys, *Emergence* (OUP, 2016), Saurabh Mittal et al. (eds.), *Emergent Behavior in Complex Systems Engineering* (Wiley, 2018, pp. 3–20), David Pines (ed.), *Emerging*

Syntheses in Science (SFI Press, 2019) and Sophie Gibb et al. (eds.), *Handbook of Emergence* (Routledge, 2019). The series *Emergence, Complexity and Computation* (2013ff) is indispensable. Also, visit https://doi.org/10.1007/s11299-020-00242-4.

18. See Tom Stonier, *Information and Meaning: An Evolutionary Perspective* (Springer, 1997), Luciano Floridi, *Information* (OUP, 2010), Paul Davies and Niels Gregersen (eds.), *Information and the Nature of Reality* (CUP, 2014) and Spyros Tzafestas, *Energy, Information, Feedback, Adaptation, and Self-organization* (Springer, 2018, pp. 39–276, 489–573). For information *theory*, which is concerned with the compression and transmission of data, see Thomas Cover and Joy Thomas, *Elements of Information Theory* (Wiley, 2006, p. 2–3), Imre Csiszár and János Körner, *Information Theory* (CUP, 2011) and Jimmy Soni and Rob Goodman, *A Mind at Play: How Claude Shannon Invented the Information Age* (Simon & Schuster, 2017). For information science, which deals with the meaning aspect of information, see Wolfgang Stock and Mechtild Stock, *Handbook of Informations Science* (Walter de Gruyter, 2013). Go to https://fis.sciforum.net.
19. See Bernardo Kastrup, "Physics Is Pointing Inexorably to Mind", *Scientific American*, March 25, 2019.
20. See David Mandel et al. (eds.), *The Psychology of Counterfactual Thinking* (Routledge, 2005) and Neal Roese (ed.), *What Might Have Been* (Psychology Press, 2014).
21. See Keith DeRose, *The Case for Contextualism* (Clarendon, 2009), Jonathan Ichikawa (ed.), *Handbook of Epistemic Contextualism* (Routledge, 2017), Hugo Mercier and Dan Sperber, *The Enigma of Reason* (Allan Lane, 2017) and Phivos Mylonas, "The Evolution of Contextual Information Processing in Informatics", February 2018, online. For context/conduct manipulation, see Erving Goffman, *Frame Analysis* (Harper & Row, 1974), Paul D'Angelo and Jim Kuypers (eds.), *Doing News Framing Analysis* (Routledge, 2010), Jörg Matthes, *Framing* (Nomos, 2014) and Vebjørn Ekroll et al., "The Other Side of Magic", *Perspectives on Psychological Science*, 12:1 (2017), pp. 91–106. For the notion of perspective, see James Elkins, *The Poetics of Perspective* (Cornell University Press, 1994), David Lindberg, *Roger Bacon and the Origins of Perspectiva in the Middle Ages* (Clarendon Press, 1996, pp. xvii-c), Brigstocke *op. cit.* pp. 553–566 and Annalisa Crannell et al., *Perspective and Projective Geometry* (PUP, 2019). For the difference between the European "convergent" and the Chinese "parallel" perspective, see Joseph Needham et al., *Science and Civilisation in China*, Vol. IV: 3 (1971), pp. 111–119.
22. See Lee Ross and Richard Nisbett, *The Person and the Situation* (McGraw-Hill, 2011) and David Matsumoto and Linda Juang, *Culture and Psychology* (Cengage, 2017). This subject is related to the rise of individualism and its attendant problems with liberalism. See Colin Morris, *The Discovery of the Individual* (Harper & Row, 1972, 158–167).

23. See Bruce Alberts et al., *Molecular Biology of the Cell* (Garland, 2014) and Francisco Ayala, *Evolution, Explanation, Ethics and Aesthetics* (Academic Press, 2016).
24. See Elsevier's open access journal *Social Sciences & Humanities* (2019ff).
25. See Blaise Pascal, *Pensées* (Guillaume Desprez, 1670, Section I) and Nikolai Berdyaev, *Freedom and Spirit* (Centenary Press, 1944, first chapter).
26. See Anne Runehov and Louis Oviedo (eds.), *Encyclopedia of Sciences and Religions* (Springer, 2013), Peter Harrison, *The Territories of Science and Religion* (University of Chicago Press, 2015), Jerry Coyne, *Faith vs. Fact* (Penguin, 2016), Gary Ferngren (ed.), *Science & Religion* (Johns Hopkins University Press, 2017) and Paul Weingartner, *Knowledge and Scientific and Religious Belief* (De Gruyter, 2018).
27. See further Nancy Cartwright, *The Dappled World* (CUP, 1999, Introduction), Katy Börner, *Atlas of Science: Visualizing What We Know* (MIT Press, 2010), James Trefil and Robert Hazen, *The Sciences: An Integrated Approach* (Wiley, 2016) and David Pines (ed.), *Emerging Syntheses in Science* (SFI Press, 2019).
28. See Steve Keen, *Debunking Economics* (Zed, 2011, pp. 37ff).
29. See *Revue d'économie politique* (1887ff), *Journal of Political Economy* (1892ff), *Scottish Journal of Political Economy* (1954ff), *History of Political Economy* (1969ff), *Review of Radical Political Economics* (1969ff), *The Review of Black Political Economy* (1970ff), *European Journal of Political Economy* (1985ff), *Review of Political Economy* (1989ff), *Review of International Political Economy* (1994ff), *New Political Economy* (1996ff), *L'Économie politique* (1999ff), *World Review of Political Economy* (2010ff) and *China Political Economy* (2018ff). Interested readers should also look at the books appearing in the series *Routledge Frontiers of Political Economy* (1995ff). For the birth of political economy, see Ernesto Screpanti and Stefano Zamagni, *An Outline of the History of Economic Thought* (OUP, 2005, pp. 19–53).
30. See Francesco Parisi (ed.), *Handbook of Law and Economics* (OUP, 2017) and Alain Marciano and Giovanni Ramello (eds.), *Encyclopedia of Law and Economics* (Springer, 2019). Law has not only an economic but also a political, social, psychological, cultural, religious, moral and literary dimension. See Dennis Patterson *op.cit.* Part III and Keith Whittington et al. (eds.), *Handbook of Law and Politics* (OUP, 2010). Visit https://doi.org/10.4337/9781782547457.
31. See Mark Granovetter & Richard Swedberg (eds.), *The Sociology of Economic Life* (Westview Press, 2011) and John Davis and Wilfred Dolfsma (eds.), *Companion to Social Economics* (Edward Elgar, 2015). Browse the journals *Economy and Society* (1972ff), *International Journal of Social Economics* (1974ff) and *Socio-Economic Review* (2003ff). For the Confucian view as to how sociology, politics, morality and economics are interrelated, see Chen Huan-chang, *The Economic Principles of Confucius and His School*, Vol. 1 (Longmans, 1911, pp. 48–118). See also Malcolm Warner, "On Keynes and

China: Keynesianism 'with Chinese Characteristics'", *Working Paper No. 2* (Cambridge Judge Business School, 2014, pp. 6–7).

32. See Peter Earl, "Economics and Psychology: A Survey", *The Economic Journal*, 100:402 (September 1990), pp. 718–755, Mattew Rabin, "Psychology and Economics", *Journal of Economic Literature*, Vol. XXXVI (March 1998, pp. 11–46, and Alan Lewis (ed.), *Handbook of Psychology and Economic Behaviour* (Cambridge University Press, 2018).
33. See Ruth Towse, *Advanced Introduction to Cultural Economics* (Edward Elgar, 2014).
34. See Alain Samson (ed.), *The Behavioral Economics Guide 2020* (online).
35. See Steven Brakman et al., *A New Introduction to Geographical Economics* (CUP, 2011).
36. See Bert Chapman, *Geopolitics: A Guide to the Issues* (Praeger, 2011).
37. See Richard Nelson et al., *Modern Evolutionary Economics: An Overview* (CUP, 2018).
38. See Kunihiro Jojima, *Ökonomie und Physik: Eine neue Dimension der interdisziplinären Reflexion* (Duncker & Humblot, 1985) and Sitabhra Sinha et al., *Econophysics: An Introduction* (Wiley, 2011).
39. See Kevin Wainwright and Alpha Chiang, *Fundamental Methods of Mathematical Economics* (McGraw-Hill, 2005).
40. See Shu-Heng Chen et al. (eds.), *Handbook of Computational Economics and Finance* (OUP, 2018, pp. 1–38.
41. See, however, Heinz Kurz (ed.), *Theory of Value and Distribution in Economics* (Routledge, 2013), Nicholas Vrousalis and Philippe van Parijs, "Analytical Marxism", *International Encyclopedia of the Social and Behavioral sciences* (Elsevier, 2015, pp. 665–667) and Christina Morina, *Die Erfindung des Marxismus* (Siedler, 2017). *Rethinking Marxism* (1988ff) is a journal produced by the Association for Economic and Social Analysis.
42. Along with the arteries, veins, coronary vessels and the heart, the capillaries belong to the cardiovascular system, which cannot function well without the good functioning of the body's other systems. See Lauralee Sherwood, *Human Physiology: From Cells to Systems* (Cengage, 2016, pp. 297–378).
43. See Hartmut Werner, *Supply Chain Management* (Springer, 2017) and Dmitry Ivanov et al., *Global Supply Chain and Operations Management* (Springer, 2019).
44. See Michael Czinkota and Ilkka Ronkainen, *International Marketing* (Cengage, 2013), Lia Zarantonello and Véronique Pauwels-Delassus, *The Handbook of Brand Management Scales* (Routledge, 2016) and David Aaker and Christine Moorman, *Strategic Market Management* (Wiley, 2017).
45. See John Bryson and Peter Daniels (eds.), *Handbook of Service Industries* (Edward Elgar, 2007). The young discipline Service Science addresses the service sector's unique problems. See Jim Spohrer et al., "Steps toward a Science of Service Systems", *Computer*, 40:1 (January 2007), pp. 71–77, Mark Daskin, *Service Science* (Wiley, 2010), Paul Maglio et al. (eds.), *Handbook of Service Science* (Springer, 2010), Robin Qiu, *Service Science* (Wiley, 2014),

Kyoichi Kijima, *Service Systems Science* (Springer, 2015) and Henriqueta Nóvoa et al. (eds.), *Exploring Service Science* (Springer, 2020).

46. The book to be read is David Graeber, *Bullshit Jobs: A Theory* (Simon & Schuster, 2018).
47. See Jerome Salinger, *The Catcher in the Rye* (Little, Brown and Company, 1951). The seminal book of Jean Baudrillard, *Simulacres et Simulation* (Galilée, 1981), shall not be overlooked.
48. See Hegel, *System der Wissenschaft*, Vol. I: *Die Phänomenologie des Geistes* (Goebhardt, 1807, IV: A) and George Ritzer (ed.), *Encyclopdia of Social Theory* (Sage, 2005, Vol. I, pp. 364–369).
49. See Stuart Greenbaum et al., *Contemporary Financial Intermediation* (Academic Press, 2016, pp. 23–86). The reader who wants to know why international crooks and their clever financial advisors walk tall should turn to Oliver Bullough, *Moneyland* (Profile, 2019).
50. See Thomas Craughwell, *5,000 Years of Royalty* (Tess Press, 2012).

Chapter 4
Cities and Countries

Abstract The similarities and dissimilarities between cities and countries are highlighted, with the emphasis being laid on the similarities. Like cities, counties are largely man-made systems. Though they are different in site and size, both cities and countries are recognizable *units*, having a unique character and being *individua* (things that cannot be broken up into parts without losing their identity or individuality). The history of a country is closely connected with the history of its different cities. For the first time, culture is defined as experience expressed, as the manifold manifestation of the ungraspable mind.

In this chapter, I introduce the subject of city-studies to clarify the case I made (the enduring confusion in regard to country studies) and to support the solution I proposed (interdisciplinary approach based on general systems science). City and country studies may seem strange bedfellows. Yet both cities and countries are human constructs, a fact not be downplayed or glossed over. Their histories are inextricably intertwined. There is no urbanization (from the Latin *urbs* [city]) without civilization (from the Latin *civis* [citizen]) and no civilization without urbanization. Citizenship (German: *Staatsbürgerschaft*; French: *citoyenneté*) is the status of a person officially recognized as belonging to a state or country; it concerns the unequivocal individual-state/country relationship. A civil society is a society of citizens, *i.e.,* a community of town or city dwellers. The city was the seat of ancient civilizations, and cities came into being with the country or culture they belong to. They begot that country and continue to be its most salient characteristic. Each and every city has a narrative to tell, making it unique. The history of French, German, Italian, Chinese, Indian and American cities tells the turbulent history of France, Germany, Italy, China, India and the USA, respectively. The deeply grooved face of these countries can be recognized, indeed their soul can be seen, by looking at the face, or gauging the soul, of, respectively, Paris, Lyon, Bordeaux, Nice, Marseilles, Toulouse, Montpellier, Clermont-Ferrand, Orléans or Chartres; Berlin, Hamburg, Stuttgart, Munich, Cologne, Dresden, Dortmund, Heidelberg, Freiburg or Wiesbaden; Rome, Florence, Venice, Milan, Turin, Genoa, Naples, Bologna, Verona or Palermo; Beijing, Hangzhou, Suzhou, Nanjing, Wuhan, Kaifeng, Guangzhou, Kunming, Harbin or

H. Kuijper, *Comprehending the Complexity of Countries*,
https://doi.org/10.1007/978-981-16-4709-3_4

Xi'an; New Delhi, Mumbai, Agra, Bangalore, Chennai, Jaipur, Lucknow, Hyderabad, Varanasi or Kolkata; or New York, Chicago, San Francisco, Philadelphia, Detroit, Baltimore, Denver, Houston, Atlanta or Cleveland. Apart from being the capital of France, Paris is a place where one unmistakably sees, hears, smells and feels not to be in Germany, Italy, China, India or the USA, but in France, a country having particular traits and famously sung of by Charles Trenet as "Douce France". Whoever visited Genoa, Florence or Naples, will agree with me by saying: "Verily, among all countries, Italy is one of a kind". A Chinese in, say, Hangzhou is not primarily proud to be a dweller of Hangzhou (with its breathtakingly beautiful West Lake); he/she is proud to be born in, and to be an inhabitant of, *zhōnggúo* (中国), the country that other countries can (and should) be compared with, but no country can be considered to be equal to. For example, it is patently obvious that China and Japan are two different universes and that China differs significantly from Europe (or any of its constituent parts), not to speak of its huge difference from the USA, a very young country (the advancement of European colonization on American territory started only in the beginning of the seventeenth century, *i.e.*, almost 1700 years after Emperor Wu of the Han and almost 850 years after Emperor Xuanzong of the Tang Dynasty!).

Sometimes, cities and states coincide. Historically, city-states included Sumerian cities such as Uruk and Ur, Egyptian cities such as Thebes and Memphis and Phoenician cities such as Tyre and Sydon. In the archaic and classical periods (c.800–325 BC), the Greek world was politically divided into some 1,500 city-states (Athens, Sparta, Thebes and Corinth being the most important ones), each consisting of an urban center with its rural hinterland.[1] The Roman Republic (followed by the Roman Empire, from the ashes of which the European Nation States would rise) grew from a city-state! Between the ninth and the fifteenth century, cities like Venice, Milan, Florence, Genoa, Piacenza, Pisa, Lucca, Cremona, Mantua, Brescia, Bergamo, Bologna, Siena, Perugia, Padua, Parma, Ancona, Verona, Reggio Emilia and Città di Castello were small independent states. Actually, up until the Italian unification in 1860, Italy was a loose conglomeration of city-states, republics and other independent entities, a little known fact that those who laugh at, or complain about, the chaotic scenes in the Italian Parliament should realize.[2] Monaco, Vatican City and Singapore, finally, are modern city-states, self-responsible and not ruled by any higher power; they are worthy to be systematically compared. Cities and countries have thus much in common, which does not mean that there are no differences between cities, between countries or between cities and countries. To explore the relationship between cities and countries, however, let us first look at the city per se. Cities have been made the subject of numerous studies, not only by taking a general position (in the way urban historians usually do) but also by focusing on a particular city (a *modus operandi* I will not be concerned with in this book). Baghdad, Athens, Jerusalem, Rome, Lisbon, Paris, London, Manchester, Edinburgh, Dublin, Berlin, Vienna, Lausanne, Amsterdam, Bruges, Salamanca, Toledo, Granada, Kiev, Moscow, Saint Petersburg, Beijing, Xi'an, Nanjing, Hong Kong, Singapore, Warsaw, Wrocław, Prague, Budapest, Sarajevo, Naples, Helsinki, Stockholm, Oslo, Grenoble, Seoul, Kyoto, Tokyo, Toronto, Vancouver, New York, Chicago, New

Orleans, Philadelphia, Phoenix, Los Angeles, Kansas City, Istanbul, Damascus, Cairo, Addis Ababa, Riyadh, Tehran, Algiers, Tunis, Bogotá, Rio de Janeiro, Buenos Aires, Ürümqi, Tashkent, Ulaanbaatar, Havana, Mexico City, Johannesburg, Lagos, Sydney, Yogyakarta, Varanasi and dozens of other cities have been described or portrayed in many ways. Visiting and exploring all the cities described or portrayed in order to be able to draw a valid general conclusion relevant to them would take more than a lifetime.

Cities are notoriously difficult to define. The criteria for distinguishing cities from towns vary between countries. The demographic definition is close to what many people intuitively conceive of as a city. In "Urbanism as a Way of Life", an article that became classic, Louis Wirth, a noted urban scholar, writes: "A city may be defined as a relatively large, dense, and permanent settlement of socially heterogeneous people".[3] But what is "relatively large"? There are many graduations between a town and a supercity. The usual ranking is as follows:

1st rank: towns having a population of 1,000–20,000 (e.g., Kufstein, Sagres, Zierikzee, Dinant and Fontainebleau);

2nd rank: large towns having a population of 20,000–100,000 (e.g., Mechelen, Delft, Avignon, Poitiers and Brindisi);

3rd rank: cities having a population of 100,000–300,000 (e.g., Leiden, Dublin, Ghent, Boulder and Rotterdam);

4th rank: large cities having a population of 300,000–1 million (e.g., Krakow, Manchester, Oslo, Miami and Bologna);

5th rank: metropolises having a population of 1–3 million (e.g., Barcelona, Rome, Paris, Agra and Dallas);

6th rank: conurbations having a population of 3–10 million (e.g., Ankara, Madrid, London, Berlin and Hong Kong);

7th rank: megalopolises having a population of more than 10 million (e.g., Cairo, Shanghai, New York, Beijing, Shenzhen, Istanbul, Moscow, Chongqing, Lagos, New Delhi, Mumbai, Dhaka, Tokyo, Mexico-City, São Paulo and Manila).

The ranking does not help very much to explain the relative strength of the interconnected forces at work in the process of urbanization. It tells nothing about the complexity and dynamics of a city. How exactly does a town develop into a large town, this into a city, this into a large city, this into a metropolis, this into a conurbation, and this into a megalopolis? Why did Volendam, at the IJsselmeer, retain its village character, and Shenzhen, at the South China Sea, explode into a supercity that has a population of nearly 13 million—substantially more than the population of Portugal or Sweden? What is the difference in quality of life between a very small town (village) and a super large city? Is it the suffocating familiarity (everybody knows everybody) or the killing anonymity (nobody knows anyone)? Is the growth of urbanization commensurate with the increase in human well-being, and does it rhyme, or can it be reconciled, with man's moral duty to care for his/her neighbor, who cries out for help, and for Mother Nature, who nurtures man in so many ways?

Cities are places where people of different age, sex or temperament and with different backgrounds, occupations or inclinations, have settled and—within the parameters laid down by the local government—(try to) live together. The functional definition of a city circumvents the problem of its size, emphasizing a city's nonlinear functions, which are dependent on other nonlinear functions, making the city a ferociously complex system.[4] Since cities can perform different functions, they are studied from different perspectives: geographically, economically, socially, anthropologically or politically.

(1) **Urban Geography**

Geography is the branch of science concerned with identifying and describing places on earth. It is a meaningful way of looking at the earth. Geography is concerned with questions as to what is in place *p*, at time *t*, due to cause *c* or reason *r*. The geographic viewpoint is based on a number of related concepts, the most important one being the concept of place. "Where" is the first thing a geographer wants to know about somebody or something. This requires the defining of its spatial relations to other persons or objects. Locating is thus essentially relating. The position of a specific place (biotope [from the Greek *βίος*, life, and *τόπος*, place]), house, street, town, city, mountain, valley, plateau, plain, desert, forest, coast, country or area can be found and indicated by means of a grid of meridians and parallels serving as coordinates. The Global Positioning System (GPS) is a satellite-based navigation system that provides geo-location information. For example, Poland is located at longitude 19.145136 and latitude 51.919438, and Singapore at longitude 103.808053 and latitude 1.351616. Knowing the coordinates of a place, however, is like having a person's address. It only tells one where to find the place, but nothing about the place itself (*e.g.*, how Poland or Singapore is organized [socially, administratively/politically, economically, or otherwise]). It tells nothing about how it is to live in one of these countries. When the place is a house, it tells nothing about how its rooms are furnished, how many people live in the house or how they live there together. Geographical insight comes only with knowledge of the place itself, that is, of both its site and situation. "Site" refers to the inner side of a place, to what is internal to it, to local forces and processes. For example, one might describe how people live in an oasis by noting and investigating the interrelations of soil, vegetation, climate, water supply, economic development, power relations, social manners and religious practices. "Situation" stands for the outer side of a place, for what is external to it, for the context in which internal affairs occur. Situation (also called relative location) shows how a place is related to other places, which impacts on the character of the place (and vice versa). Clearly, Poland (sandwiched between Russia and Germany) and Singapore (located between Malaysia and Indonesia and at the transition of the South China Sea to the Indian Ocean) have different situations and thus different sites. In terms of situation, landlocked countries like Austria, Hungary, Mongolia, Nepal and Uganda differ from seaborne countries such as Greece, Portugal, Norway, Chile and Vietnam. In this respect, the doubly landlocked country Uzbekistan (ca. 33 million inhabitants) and the island Madagascar (ca. 25 million islanders) are almost incomparable. Paying attention to the situation of a city may lead to looking at the site of the country the city is located in, the country in turn being a place not to be seen as totally isolated. Later in this book, I will argue that practitioners of international relations are inclined to be more interested in the situation than in the site of a country.

Urban geographers are geographers applying their theories to, or testing them on, cities. A small island in the Seine near the center of modern Paris marks the site of the original settlement: a defensive refuge in times of trouble and a kind of crossover in peaceful times. The most distinctive feature of Athens is a steep-sided and flat-topped hill, first chosen for its defensive qualities and later used for temples. Originally, Venice, built on 118 small islands in a shallow lagoon, was a place of refuge from marauders. London, founded by the Romans, has become one of the world's most important financial hubs and cultural cities. New York City, a collection of neighborhoods in the State New York scattered among the boroughs Manhattan, Brooklyn, the Bronx, Queens, and Staten Island, began to grow in the early seventeenth century, with the founding of a Dutch fur trading post. Though these cities have a site and living in them is definitely not the same, the key to their character and growth is not so much their site as their situation. Paris was situated at a focal point of land and waterways. It was also at the center of the emerging nation-state France. Athens grew because it became city-state number one and was, related to this, strategically located. Venice flourished as a result of its location at the head of the Adriatic Sea, the deepest penetration of the Mediterranean Sea into the European landmass. London lay astride the river Thames, about 40 miles from its estuary on the busy North Sea. New York was located at the mouth of the mighty Hudson, which feeds into a naturally sheltered harbor and then empties into the Atlantic Ocean, which connects rather than separates. As towns grew and became cities, they acquired new functions which altered their character. The Industrial Revolution, the transition to new manufacturing processes that began in Great Britain in the mid-eighteenth century and led to fundamental economic, social and political changes, was the main catalyst in this process.

The concept of place is not only important in geography.[5] It plays also a crucial role in history, the social sciences (economics, sociology and political science), mathematics, philosophy and religion. In *Les Lieux de mémoire* (Gallimard, 1984ff), an inventory of places, monuments, symbols, words, traditions and commemorations that embody the collective memory or shared pool of knowledge and information of a group of (French) people, Pierre Nora (1931) describes the group's identity.[6] This enterprise can be regarded a continuation, on grander scale, of *À la recherche du temps perdu* (in search of lost time), a novel in seven volumes written by Marcel Proust (1871–1922) and known for its convoluted sentences, which represent the working of the human mind. "*Lieux*" suggests space and "*mémoire*" time. Without time, which cannot be dissociated from space, there would be no memory, and without memory, there would be no history. The synchronic and the diachronic hang together, a remarkable fact which I will have occasion to write about further down in the book.

Any social scientist who takes time aboard has got to deal with space as well, for each change in time implies a movement in space. Conversely, taking space aboard also means dealing with time, for each movement in space takes some time. Heraclitus, who would be contradicted by Parmenides, insisted that "everything flows and nothing stays" (πάντα ῥεῖ καὶ οὐδὲν μένει), coming close to the way of thinking in ancient China, where *dào* (process) and *lǐ* (pattern) were the key concepts of thought (see preface). However, "the obscure" and "weeping philosopher" should have made two exceptions, thereby violating his "nothing": (a) change may be different (as geologists, biologists, chemists, particle physicists and sociologists can tell), but it stays (change is everywhere, all the time) and (b) the rate of change may be constantly changing. The concepts of time, change, identity, truth, thing, event, action and knowledge are interrelated. An adequate treatment of any of them requires clarification of the other concepts. While time is characteristic of any system, modern physicists are familiar with the idea that "universal constants" govern the laws of nature that extends in all directions.[7]

Space and time should be conceived of as two sides of the same coin, relation being the essence of both.[8]Geometry is concerned with objects in, and properties of, space. Topology (from the Greek τόπος, place, and λόγος, study) is concerned with the properties of space that are preserved under continuous deformation. Tracing its origin back to Gottfried Leibniz (1646–1716), who was fascinated (through his contact with Jesuits) by the ways of thought in China and preoccupied with the geometry of a situation (which he called *analysis situs*), topology developed as a field of study out of geometry. Leonhard Euler (1707–1783), August Möbius (1790–1868), Johann Listing (1808–1882), Bernhard Riemann (1826–1866), Henri Poincaré (1854–1912) and Alexander Grothendieck (1928–2014) are some of the names to be mentioned. Topological ideas are present in almost all areas of today's mathematics.

The relationship between topology and origami (the Japanese art of folding paper to represent reality) is intriguing. The word *origami* is a compound of *ori* (folding) and *kami* (paper). The art, associated with Japanese culture, can be divided into several types and has many practical applications. The history of origami is obscure, but may be traced back to the Chinese paper folding technique known as 摺紙 (*zhézhī* [bending or folding paper]). Even if the art came from China, the Japanese are the masters of it. Cutting or tearing the paper is not allowed. The book to be read is Koryo Miura et al. (eds.), *Origami*[6] (American Mathematical Society, 2015), a "unique collection of papers illustrating the connections between origami and a wide range of fields". For the 7th International Meeting of Origami Science, Mathematics and Education, visit http://osme.info/7osme.

Mereotopology combines mereological and topological concepts; it is a young and promising field of research that deals with relations among wholes, parts of wholes and parts of parts in space and time. The roots of mereotopology go back to Alfred North Whitehead (1861–1947), whose approach (explicated in *Process and Reality* [Macmillan, 1929]) is in a sense opposite to the Euclidean one. Points, lines and planes are not disregarded, but redefined and introduced in a new theory.[9]

The concept of place is not strange to philosophers either. The most famous concept in the philosophy of Nishida Kitarō (1870–1945), a student of Martin Heidegger and founder of the Kyoto-School,[10] is the logic of 場所 (*basho* [place, locus]), a non-dualistic, Zen-inspired logic meant to overcome the inadequacy of the subject-object distinction made in Western philosophy, which relies on what is called the "three principles, or laws, of thought":

(A) Whatever is, is (law of identity),
(B) Nothing can both be and not be (law of non-contradiction), and.
(C) Everything must either be or not be (law of excluded middle).

Introducing the factor time, Nishida replaced "A is A" (the first law) with "A *becomes* non-A". Whether he, in doing so, overturned the second and third law, is a matter of dispute, but it can be safely stated that becoming is not the same as being.[11] Going from A to non-A (or from non-A to A), one crosses a line, the boundary which Western philosophers have disagreed on since Anaximander (610- 546 BC), who rejected Thales' supposition that water is the ἀρχή (beginning, origin) of everything and posited the ἄπειρον (indefinite, boundless) instead. Adherents of the theory on "fuzzy sets" (Lotfi Zadeh et al.), "rough sets" (Zdzisław Pawlak et al.) and "grey systems" (Julong Deng et al.) are still discussing with the protagonists of probabilism (Andrey Kolmogorov et al.) about the existence of boundaries and the handling thereof. The whole matter could be subsumed under the heading mereotopology. See previous paragraph.

Holy places, finally, are sites which religious people consider to be of special significance; they are often visited by pilgrims. Every church, cathedral, chapel, mosque, synagogue, temple, shrine or sanctuary is a *place* of worship. Lists of religious sites and places of spiritual importance throughout the world can be found on the Internet. If "God is omnipresent" or "ubiquitous", as we are asked to believe, it is contradictory to delimit a part of space and declare it to be exclusively holy. Religious beliefs, however, are awash with cases of *non sequitur.*

> A famous story goes about two traveling Zen monks who passed by an isolated, deserted temple. One monk needed to urinate, so he urinated in the temple hall, in front of the Buddha statue. The other monk scolded him, saying: "Look, the Buddha is here. How can you urinate here?" The first monk replied: "Tell me where Buddha is not, and I will urinate there." When the scolding monk explained: "Buddha is everywhere," the urinating monk happily responded: "In that case, I can urinate everywhere."

(2) Urban Economics

Urban economists are economists in the first place. They apply economic theories to, or test them on, cities. The economic viewpoint is based on a number of related concepts, but it is different from the geographic point of view. Economics belongs to the behavioral sciences; geography only partly so (physical geography belongs to the natural sciences). Economics is the science of how people choose to employ scarce or limited resources that could have alternative uses in order to produce various commodities and to distribute them in exchange for money. Demand and supply are its basic concepts. Economists study a certain kind of human behavior: the behavior shown in a system regarding the use of different kinds of resources for the production and distribution of goods. In the next chapter, I will describe the different kinds of resources; for the time being, I am concerned with urban economics.

Urban economics is rooted in location theory, a body of theories which seeks to account for the location of economic activities. Johann von Thünen (1783–1850), a German landowner, is regarded as the father of the theory. According Fernand Braudel, he "ranks alongside Marx as the greatest German economist of the nineteenth century". Like his contemporary David Ricardo, who—like his predecessors—attempted to answer the question "How to maximize a nation's wealth?" and whose work would divide the community of economic thinkers (noticing the deadlock reached by the classical school) into a socialist and neoclassical branch quibbling about the question "What is *the source* of a nation's wealth?", Von Thünen recognized that differences in the cost of crop production resulted from utilization of land of different quality and location (i.e., distance from the market). But whereas Ricardo focused on differences in soil fertility, Von Thünen concentrated on differences in land location. For him, the problem was to devise the best (most profitable) system of land utilization. His solution was carefully worked out, but his argument was couched in a theory having rather restrictive assumptions. These would in due course be eased, but, after his death, Von Thünen's legacy became forgotten, overshadowed by the works of Alfred Marshall (1842–1924) and Léon Walras (1834–1910), who laid the neoclassical foundations of modern value theory. Though Marshall claimed to have "loved him [Von Thünen] above all my masters", he judged time to be more fundamental than space, not seeing that the two are closely related. This prejudice has been intensified ever since, the German school of location theory being an exception, as Walter Isard (1919–2010) notes in *Location and Space-Economy* (MIT Press, 1956). In this book, the author complains that classical and neoclassical economists confined themselves to the analysis of "a wonderland of no spatial dimensions" in which firms locate their plants without regard to transport costs and sell their products on markets without taking the characteristics of the sales area into account. The two theories were brought together in the Monopolistic Competition Revolution that started in the 1930s and advanced the idea of spatial product differentiation as one of the sources of a firm's monopolistic power of price-setter. However, "with the exception of a brief flicker of interest in the 1930s", Mark Blaug laments in *Economic Theory in Retrospect* (CUP, 1985, p. 629), "the neglect of spatial economics by mainstream economists largely continues to this day".

The German school of location theory prefigured the general location theory, the most fundamental theory of spatial economics.[12] Walter Christaller (1893–1969) was a towering figure in this tradition. The central place theory is his principal contribution.[13] The book in which he developed the theory has been foundational to the study of cities as systems: groups of elements organized in such a way that every element is to some degree directly or indirectly interdependent with every other element. The term urban system was introduced by Brian Berry, who applied systems analysis to the study of cities.[14] Berry is a prolific writer, having authored over 550 books and articles. This prompts the question whether such a workaholic has time to sleep, or whether he is spending some time with his wife, or partner. Berry owes a great deal to his colleague Richard Chorley (1927–2002), who was instrumental in introducing systems theory to urban studies. Christaller's central place theory laid also the foundation of geography's quantitative revolution, a paradigm shift that sought to develop a more rigorous and systematic methodology for the discipline. It was the Anglo-American response to the alleged inadequacy of continental geography to explain general spatial dynamics. The revolution, which occurred in the 1950s and 1960s, marked a shift from idiographic to nomothetic geography, from a geography tending to specify (as is typical for the humanities) to a geography tending to generalize (as is typical for the natural sciences). The revolution marked a change from *regional geography* (focusing on the interaction of different natural and cultural factors in a geographic region and practiced by Alfred Hettner, Paul Vidal de La Blache, Richard Hartshorne and Carl Sauer) to *regional science* (including location theory, spatial economics, location modeling, transportation, migration analysis, land use and urban development, interindustry analysis, environmental or ecological analysis, resource management, urban and regional policy analysis, geographical information systems and spatial data analysis) or *spatial science* (focusing on the role of space as a fundamental variable influencing both the organization of society and the behavior of its members and centering on the quantitative analysis of spatial patterns and processes in society). The next step may be the combination of the two, the integration (by scientific collaboration) of the qualitative and the quantitative approach.

Urban economics covers a number of related themes. *The study of land use* analyzes the spatial organization of economic activities within an urban area (site). It examines the intracity location choices made by firms and households and addresses such questions as to how the price of land is determined and why those prices vary across space, what the economic forces are that cause the spread of employment from the center of the city outward, by whom the use of land is controlled (difference between ownership and property) and how the urban economy is affected by these controls. *The study of market forces* also focuses on the site of the city, but addresses such questions as to how location decisions of firms and households influence the size, structure and development of the city, how market performance depends on geographical factors, and why/how the local government interferes in the market. *The study of housing* (the construction and usage of houses for sheltering) also has to do with the city's site. It addresses such questions as to why people opt for a detached single-unit, semi-detached or attached multi-unit dwelling, how the housing is financed and whether householders can count on any government support. This kind of study features in a course on housing, or more broadly, real estate economics, respectively, a sub-branch and branch of economics, which, we saw in the previous chapter, has many aspects or dimensions. *The study of urban transportation* is more concerned with the city's situation. Interaction between places implies mobility of people or products; it requires overcoming distance through transportation. Accessibility measures the degree to which a place can be approached from other places. A city can be directly or indirectly accessible by roads (car, bus and truck), railroads, waterways or air, which depends on its site and situation on the one hand and affects its size and development on the other hand. As the development of cities largely depends upon their physical, economic, social and institutional infrastructure, the importance of intra-urban transportation is paramount. Cities can be suffocating by traffic congestion. Examples are Los Angeles, New York City, Moscow, Beijing, São Paulo, Dakar, San Francisco, Bogotá, London, Atlanta, Paris, Miami, Delhi, Shenzhen, Taipei and Amsterdam. The response of the municipality of these cities to the traffic problem may be different, having different consequences in different adminstrative areas.

The relationship between cities and countryside is another concern of urban economists. "Countryside", or "rural area", refers to the geographic area outside of, or surrounding, a town or city. Divided into different districts or neighborhoods having (main) streets, side streets and alleys, cities are concentrations of people who are working and, estranged from each other, living there. A city is more than an enormous mass of stones, glass and concrete. It is a multifunctional unit lit at night (causing pollution from which the ancients who were gazing at the stars so bright did not suffer), having many buildings and amenities: offices, factories, workplaces, warehouses, houses, community centers, schools, research institutes, hospitals, laboratories, clinics, hotels, restaurants, bars, shops, clubs, malls, public spaces, supermarkets, department stores, places of worship, local government buildings, sports centers, swimming pools, public transportation, garages, public utilities, studios, theaters, music halls, cinemas, massage parlors, police stations, barracks, banks, parks, libraries, museums, etc. Cities have suburbs, areas where people live far away from, but within commuting distance of, the center of the inner city (downtown). The countryside, on the other hand, consists of vast fields or pastures interrupted by woods, crisscrossed by rivers, and dotted with farms, hamlets and villages which are not connected by turnpikes or multilane highways but by trails or country roads (immortalized by John Denver). The countryside is sparsely populated. People (who know how to read the sky) live there in harmony with nature and in tradition-bound communities.

In ancient China, "*tiānxià*" (天下, all under heaven) denoted the entire geographical world. See *Daxue* (the great learning), being one of "Four Books" in Confucianism, Alfred Forke, *The World-conception of the Chinese* (Probsthain, 1925), Joseph Levenson, "*T'ien-hsia* and *Kuo* and the Transvaluation of Values", *Far Eastern Quarterly*, 11:4 (1952), pp. 447–451, Abe Takeo, *Chūgokujin no tenka kannen* [the *Tianxia* idea of the Chinese] (Doshisha University, 1956), Kung-chuan Hsiao, *A Modern China and a New World* (University of Washington Press, 1975), Yuri Pines, "Changing views of tianxia in pre-imperial discourse", *Oriens Extremus*, 43 (2002), pp. 101-116, Tingyang Zhao, "Rethinking Empire from the Chinese concept 'All-under-Heaven'", *Social Identities*, 12:1 (January 2006), pp. 20-41, William Callahan, "*Tianxia*, Empire and the World: Soft Power and China's Foreign Policy Discourse in the 21st Century", *BICC* (British Inter-University China Centre) *Working Papers*, No. 1, March 2007, Chishen Chang, "Formation of the Key Concepts: '*Zhongguo*' and '*Tianxia*'" (Chinese), *Dongwu Zhengzhi Xuebao*, 27:3 (2009), pp. 169-257, Ban Wang (ed.), *Chinese Visions of the World Order* (Duke University Press, 2017), Tingyang Zhao, *Tianxia: Tout sous un même ciel* (Cerf, 2018), Zhaoguang Ge, *An Intellectual History of China*, Vol. II (Brill, 2018, Chap. 11) and *idem*, *What is China?: Territory, Ethnicity, Culture, and History* (Belknap, 2018, Chap. 1). The concept of *tiānxià* should be borne in mind when investigating the political status of Taiwan[15] or judging President Xi Jinping's "One Belt One Road" Initiative.[16]

"*Tiān*" (sky, or heaven) is an example of polysemy, like the Spanish word *pueblo*, which means village, or people. The Chinese literature on "*tiān*" has become very rich. See Shenzhi Li, "Reflections on the Concept of the Unity of Heaven and Man", in Karl-Heinz Pohl (ed.), *Chinese Thought in a Global Context* (Brill, 1999, pp. 115–119). Mongolians still pray to "*Munkh Khukh Tengri*" (the eternal blue sky), and Mongolia, China's neighboring country, is called "*Munkh Khukh Tengriin Oron*" (land of the eternal blue sky). See Jean-Paul Roux, "*Tängri: essai sur le ciel-dieu des peuples altaïques*", *Revue de l'histoire des religions*, 149:1 (1956), pp. 49–82; 149:2 (1956), pp. 197–230; 150:1 (1956), pp. 27–54; 150:2 (1956), pp. 173–212; and 154:1 (1958), pp. 32–66.

Readers interested in Chinese concepts of worldorder should take up Li Xing (ed.), *The Rise of China and the Capitalist World Order* (Routledge, 2010), Yunling Zhang, *Rising China and World Order* (World Scientific, 2010), Zheng Wang, *Never Forget National Humiliation* (Columbia University Press, 2012), Kenneth Lieberthal en Wang Jisi, *Addressing U.S.-China Strategic Distrust* (Brookings Institution, 2012), John Ikenberry *et al.* (eds.), *America China and the Struggle for World Power* (Palgrave Macmillan, 2015), Sulmaan Khan, *Haunted by Chaos: China's Grand Strategy from Mao Zedong to Xi Jinping* (HUP, 2018), Youngmin Kim, *A History of Chinese Political Though* (Polity Press, 2018), John Friend en Bradley Thayer, *How China Sees the World* (Palgrave Macmillan, 2018), Michael Mazarr *et al.*, *China and the International Order* (Rand Corporation, 2018), Yongjin Zhang en Teng-Chi Chang (eds.), *Constructing a Chinese School of International Relations* (Routledge, 2018), Yan Xuetong, *Leadership and the Rise of Great Powers* (PUP, 2019), Timothy Brook, *Great State: China and the World* (Profile, 2019), Huiyun Feng *et al.*, *How China Sees the World* (Palgrave Macmillan, 2019), Barbara Mittler *et al.* (eds.), *China and the World—the World and China* (OSTASIEN, 2019) and David Shambaugh (ed.), *China and the World* (OUP, 2020).

The countryside has only a few amenities, and the government, even the local one, is felt to be remote. The relationship between city and countryside is mutual; the one cannot live without the other. While the countryside directly or indirectly provides the city dwellers with all kinds of food, the city supplies the seed, fertilizers, pesticides, machinery, equipment, fuel and fodder for producing crops or breeding and feeding animals. Cities, where (or in the vicinity of which) industries are usually located, and countryside, the hinterland of cities, are complementary to one another. They are so attached to and intertwined with each other that it is impossible for a city to grow without the improvement of its hinterland and the other way around. A proven military tactic, emphasized by Mao Zedong, is to isolate a city from its countryside. If agriculture is the heart of a country, then industry must be considered its brain.[17] Cities work like magnets. They induce rural labor forces to abandon their farm work, to leave their homes (a micro-social drama) and to migrate massively to urban areas, a process once associated with the Industrial Revolution, but today resulting in an urbanization the world has never seen. Looking at it, and seeing how people behave, I cannot help wondering: who is running this show? If nobody does: where is it leading to?

It would be a mistake to view the relationship between cities and their countryside only in economic terms. The economic system of a country is closely related to, but not congruent with, the administrative, legal, environmental, demographic, ethnic, social, educational or cultural system of it. In other words, the more or less populated territory of a country is economically, politically, legally, environmentally, demographically, ethnically, socially, educationally and culturally "divided" in ways that are not the same. The rates of change of these changing systems are also different, making it absolutely necessary to study a country cross-disciplinarily. Concentration on the economic interdependence of cities and countries should thus take the importance of other dependencies into account by inviting other experts to join the discussion.

(3) Urban Sociology

Economists have a narrow view of human behavior, assuming that people rationally interrelate in terms of supply and demand only. However, *homo economicus*, the hero in mainstream economics, does not really exist. Research carried out by psychologists, sociologists, anthropologist and political scientists has made it abundantly clear that human behavior is a complex phenomenon unexplainable by focusing on nothing but economic factors. Human behavior depends on human nature, which has a cognitive, conative and affective dimension.[18] All human beings, apart from, being **Einzelwesen** and **Gemeinschafswesen** at the same time, have both a **body** and a **mind**, but they differ in the combination of these factors. Sociology, the scientific study of society, *i.e.,* of the changing patterns of social relationships and interactions, encompasses economics. It is a very broad field of research, tracing its ancestry to Henri de Saint-Simon and his secretary Auguste Comte,[19] and divided into diverse schools. These schools compete with another, but all of them are holding the view that the economy and polity of a country are only aspects/dimensions of it the former stressing the economic, the latter emphasizing the political relations. Urban sociologists apply the tools of sociologists to, or test their theories on, cities. Attitude, habitus, action, interaction, communication, position, distance, deviance, class, integration, structure, change, function, trust, control and space are some of the basic concepts used by them. "Social space" deserves special mention, because we live in a time of "time–space compression", which is said to be the essence of globalization. Both Durkheim and Simmel assumed that the importance of space would dwindle, as abstract forms of social organization (*e.g.*, monetization) became more pervasive.[21] The resurgence of interest in space (the "spatial turn"), however, has improved upon this assumption, not by replacing it but by positing that space—very much like time—is a crucial element of socialization. The question, however, is whether space should be conceived of absolutely (as natural scientists seem to do) or relatively (as social scientists are inclined to do).

There are many topics urban sociologists occupy themselves with: migration, alienation, urban structure, urban development, class formation, everyday life of (a group of) city dwellers, neighborhoods, slums, (racial) segregation, ethnic violence, ghettoization, housing conditions, urban unemployment, city vs. suburbs, commuting, gentrification, community life, leisure activities, etc. Urbanization, the process whereby large numbers of people congregate and settle in an area, has their special attention. The urban population of the world has grown from 751 million in 1950 to 4.2 billion in 2018. The Revision of World Urbanization Prospects, a publication released in May 2018 and produced by the Population Division of the UN Department of Economic and Social Affairs, notes that future increases in the size of the world's urban population are expected to be highly concentrated in three countries. Together, India, China and Nigeria will account for 35% of the projected growth of the world's urban population between 2018 and 2050. By 2050, it is forecasted that India will have added 416 million urban dwellers, China 255 million and Nigeria 189 million. Tokyo is the world's largest city with an agglomeration of 37 million inhabitants, followed by New Delhi (29 million), Shanghai (26 million) and Mexico City and São Paulo, each with around 22 million inhabitants. Today, Cairo, Mumbai, Beijing and Dhaka all have close to 20 million inhabitants. By 2020, Tokyo's population is predicted to begin to decline, while Delhi is projected to continue growing and to become the most populous city in the world around 2028. By 2030, the world is expected to have 43 megacities with more than 10 million inhabitants, most of them in developing regions. However, some of the fastest-growing urban agglomerations are cities with fewer than 1 million inhabitants, many of them located in Asia and Africa. Whereas one in eight people live in one of the 33 megacities worldwide, close to half of the world's urban dwellers reside in much smaller settlements with fewer than 500,000 inhabitants. Today, 55% of the world's population lives in urban areas, a proportion that is expected to increase to 68% by 2050. In 1950, the proportion was 30%. Projections show that urbanization could add another 2.5 billion people to urban areas by 2050, with close to 90% of this increase taking place in Asia and Africa. The decision-makers in Brussels who are worried about what they shall do with migrants from other continents are thus warned of the storm that is coming. More generally, one wonders how the world will look like in 2050 if democracy (everybody having a vote of equal weight) were the universal form of government.

Urban sociologists have different views, not only on the causes and consequences of urbanization but also on the extent to which human actors could intervene in the process. Like Georg Simmel, Ferdinand Tönnies attempted to theoretically capture the social change that was unfolding across Europe in the nineteenth century. He developed a polarity between a closely knit community and a society, between a *Gemeinschaft*, suggesting intimacy and warmth, and a *Gesellschaft*, suggesting impersonal exchanges like business transactions. These themes heavily influenced American sociologists such as Robert Park (student of Simmel), Louis Wirth, Ernest Burgess, George Herbert Mead, Howard Becker and Everett Hughes, leading exponents of the famous Chicago School.[22] In the 1970s, the School began to be challenged. Claude Fischer (1949) argued that the city was a variety of social ties and subcultures. He found the characteristic form of urban life in the neighborhood, insisting that cities are places of trust and friendship. The major critique of the Chicago School, however, came from Marxist scholars such as Henri Lefebvre (1901–1991), David Harvey (1935), Harvey Molotch (1940) and Manuel Castells (1942). More recently, the process of "suburbanization" (residential areas on the outskirts of major cities), the formation of "edge cities" (concentration of business, residence, shopping and entertainment outside a city's central business district or along major thoroughfares) and the growth of "conurbations" and "global cities" have caught the attention of urban sociologists.[23]

(4) Urban Anthropology

Urban sociologists are often criticized of paying no or not enough attention to culture. This is remarkable, because urban economic, sociological or political work makes sense if and only if culture is taken into account. In other words, without bringing the factor culture into the equation urban economists, sociologists and political scientists can be accused of dry-swimming, of not practicing the real thing. For everything they study breathes culture; everything they investigate is soaked with culture and steeped in history. Culture is exceedingly difficult to define.[24] Jacob Burckhardt (1818–1897), known as one of the major progenitors of cultural history, called culture "*die ganze Summe derjenigen Entwicklungen des Geistes, welche spontan geschehen und keine Zwangsgeltung in Anspruch nehmen*".[25]

The word culture, rarely if ever defined, refers to more than the art forms literature, painting, drawing, music, dance, architecture, sculpture, photography, printmaking, cinematography, urban designing, garden designing, landscape designing, interior designing, theatre (*art de la scène*), and gastronomy, each of which is a world apart. "Culture" refers not only to works that are primarily made for aesthetic enjoyment, spiritual contemplation or culinary pleasure; the word refers also to things of practical use, to things that show craftsmanship. A craftsman or craftswoman is someone skilled in a particular craft; he/she is an artisan. The universe of culture is thus adjacent to the universe of technology, which involves the application of science. Without mastering a technique the artist cannot produce a work of art.

Culture could be defined as the outside of the inside, as experience expressed. Experience is consciousness; it is the mind at work. For a different view, see Julie Rivkin and Michael Ryan (eds.), *Literary Theory: An Anthology* (Blackwell, 2000, pp. 1025–1027). Memory and imagination, which are inconceivable without experience, are products of the mind. So are induction and generalization (abstraction), judging and reasoning. Rationalists (as distinct from empiricists) have to acknowledge that innate ideas originate in the human mind and condition human experience. Though it tends to be forgotten, both science and culture are predicated on experience. Without mind there can be no experience, and without experience there can be neither science nor culture. Without experience, man does not differ from, and is in fact, a dead and mindless thing. Language is experience expressed in words and literature is a special kind of language, whereas arts like music, dance, painting, architecture and sculpture are forms of experience, expressed in a different way. In *Using Language* (CUP, 1997), Herbert Clark cogently argues that "language use is a form of joint action", carried out by "people who act in accordance with each other", like dancing a waltz, or sweep-rowing a boat. Speaking (or writing) and acting presuppose the process of valuing, which is a matter of the mind, because mindless things, like robots, do not value. The mind is unmeasurable. Whoever tries to grasp it is groping around in the darkness. The mind is a kind of black hole that escaped the attention (mind) of even Roger Penrose, Reinhard Genzel and Andrea Ghez, winners of the 2020 Nobel Prize in Physics for their studies of black holes. Without mind there can be no understanding, but what is understanding? Value is a hotly discussed issue.

Understanding understanding is a very difficult subject. Immanuel Kant, Edmund Husserl, Martin Heidegger, Kurt Gödel, Alan Turing, Heinz von Foerster, John Searle, Lawrence Busch, Paul Ziff, Vincent Potter, Paul Ricœur, Renate Bartsch, David Rumelhart, Richard Mason, David Chalmers, Thomas Schwandt, Wayne Waxman, Roger Penrose, Jacques Bouveresse, Richard Wurman, Gerhard Schurz, Max Tegmark, and Philip Tetlow are some of the philosophers or scientists who dealt with it. In his widely appraised book *The Road to Reality* (Alfred Knopf, 2004, pp. 17–23 and 1027–1029), for example, Roger Penrose distinguishes three domains: the physical, the mental and the Platonic (mathematical) world. He indicates and beautifully depicts the mysterious connections between them, but he does not refer to the works of Vico, Kant, Hegel, Husserl or Heidegger (not to mention the works of ancient Indian or Chinese thinkers). Understanding, or conceptualizing, is arguably a *function* of the mind. So is imagining, but the mind itself, being related to but not identical with the brain, has not been and, in my view, will never be explained, all the efforts of philosophers of mind and cognitive scientists notwithstanding. As long as matter and antimatter are not thoroughly understood, it is premature and prejudiced to claim that life (or mind) is a matter of matter and can be entirely reduced to stardust.

Cultural analysis is a wide-ranging "discipline" that collects and investigates data on cultural phenomena by using qualitative research methods. Concerned with the study of cultural practices, processes and representations, it is an intersection of cultural studies (see Chap. 1), art history, literary theory, comparative literature, sociology, psychology, anthropology, philosophy and even theology, pretending to be cross-disciplinary and to reveal how the present is shaped by the past. In 2008, Tony Bennett and John Frow edited *The Handbook of Cultural Analysis* (Sage). Ten years later, Luke Matthews and his colleagues thought they could improve on this thick volume by editing the downloadable booklet *A Manual for Cultural Analysis* (Rand). Failing to explain and describe the meaning and exact limits of culture, however, cultural analysts have, like their colleagues in cultural studies, still the habit of hiding in the mist.

"Cultural area" (*aire culturelle*) is an anthropological concept, referring to a geographic region characterized by a complex of relatively homogeneous human activities. It should not be confused with the concepts of *political* and *legal* area. Great Britain is a *political area*, a sovereign state mainly consisting of four parts: England, Scotland, Wales and Northern Ireland. Its royal coat of arms features the noticeable words *Dieu et mon Droit*. Ireland, Scotland, Whales, Cornwall, and Brittany, however, are cultural areas. The partition of Ireland (its division into two separate polities, namely Republic of Ireland and Northern Island) raised the important question of allegiance: should one be loyal to the Crown (unionists) or to one's Culture (Irish nationalists)? See Robert O'Driscoll, *The Celtic Consciousnes* (George Braziller, 1985) and Theodore Moody and Francis Martin (eds.), *The Course of Irish History* (Mercier Press, 2011, pp. 273–299). Although former journalist Boris Johnson, the current Prime Minister of the United Kingdom, "got Brexit done", one should not be surprised when, in the years to come, Scotland will regain its independency and the Irish nationalists, following suit, will celebrate their victory. Justice will then be done and "Great Britain" will be back to where it was in the beginning. "*Legal area*" refers to a legal family, tradition or system. Examples are the Romano-Germanic (civil) law system, the common law tradition, the Islamic (*sharia*) law system, the Talmudic legal tradition, the Hindu legal tradition, or the Confucian "*famille juridique*". See Patrick Glenn, *Legal Traditions of the World* (OUP, 2014) and Mathias Reimann and Reinhard Zimmermann (eds.), *Handbook of Comparative Law* (OUP, 2019). Cultural areas are subsets of civilizations (*aires de civilization* or *Hochkulturen*), examples of which are the Mesopotamian Civilization, the Egyptian Civilization, the Aegean Civilization, the Persian Empire, the Roman Empire, the Carolingian Empire, the Byzantine Empire, the Mesoamerican Civilization, The Inca Empire, the Indian Civilization, the Chinese Civilization, the Jewish Civilization, the Islamic Civilization, the Mongol Empire, the Portuguese Empire, the Spanish Empire, the Dutch Empire, the Mughal Empire, the Ottoman Empire, the European Civilization, the Negro-African Civilization, the British Empire, the German Empire, the Russian Empire, and the American Civilization. See John Haywood, *The Penguin Historical Atlas of Ancient Civilizations* (2005) and Chris Scarre (ed.), *The Human Past* (Thames & Hudson, 2018).

The main research theme of my teacher, late professor Erik Zürcher (1928-2008), was cultural interaction, *i.e.* the Chinese reception of Buddhism, from the first to the fifth century AD, and Christianity, in the seventeenth century. This was in accordance with the subject of his inaugural speech entitled "Dialoog der Misverstanden" (dialog of misunderstandings), delivered on March 1962, in Leiden. The weak points in his scholarly work, however, were (*a*) the absence of an attempt to analyze the core concept of culture and (*b*) the mixing up of "dialog" and "debate". Nicolas Standaert, in "Erik Zürcher's Study of Christianity in Seventeenth-Century China: An Intellectual Portrait" (*China Review International*, 15:4 [2008], pp. 476-502), and Jonathan Silk, in his "In memoriam Erik Zürcher" (*Journal of the International Association of Buddhist Studies*, 31:1–2 [2008], pp. 3–22), fail to notice these fundamental flaws.

I define culture as manifestation of the mind, which can never be understood, because understanding the mind presupposes a mind, the hallmark of a human being. This is the reason why I do not believe that robots (man-made machines) will ever be like humans or will ever be even more than humans, as adherents of over-hyped Artificial Intelligence want us to believe. Mathematical books, no matter how sophisticated, are manifestations of the human mind, which cannot be mathematicized. Visiting a city or country, one can see that the mind is, or has been, at work, in many ways. Few tourists are able to see this. Asked about the most fundamental criticism that I would level against current natural scientists, I would complain that most, if not all of them are—figuratively speaking—blind in one eye. They are scrutinizing (a part of) the world outside (at macro, meso or microscale), forgetting all about the world inside, without which explaining, understanding or admiring what is out there is impossible. Immanuel Kant (1724–1804) famously said: "Two things fill me with wonder: the starry sky above and the moral law within". The *portée* of this profound statement became clear in the nineteenth century, when not only semantics (the linguistic study of meaning) and semiotics (the study of signs and sign processes) came to the fore as different sciences, but also axiology, the study of values (including but not limited to ethical values). Whether meaning and value have a common ground and how they are interrelated, I do not know, but it seems to me that anything valuable has also meaning, at least for somebody. Values are criteria or standards used to form or give an opinion about someone or something. The True (science), the Good (ethics) and the Beautiful (aesthetics) are the fundamental criteria, because the worth or value of anything can be judged by referring to one of them.[26] Other criteria may be proposed as variants of the fundamental three. Norms are views on what ought to be done, or shall not be done. Normative statements are to be distinguished from assertions about what is the case (*Sachverhalt*). Aspirations or intentions imply views about what is desirable or preferable. Expectations are views on what event or action is most likely to happen. Values, norms, aspirations and expectations shall not be mixed up with each other; nor should they be confounded with rights and duties, which can be codified in a legal framework. What values, norms, aspirations and expectations have in common, however, is that the unfathomable human mind is involved. Their combination forms a system consisting of four parts that are closely knit together: a system of values, of norms, of aspirations and of expectations. The convoluted combination forms the culture of a country. Examples are the Chinese, Japanese, Korean, Mongolian, Indian, Thai, Iranian, Turkish, Russian, French, German, Spanish, Moroccan, Ethiopian, South African, Italian, Brazilian, Peruvian, Mexican and American culture. Few, if any, author has gone to the very heart of a country's culture. So let the country expert rise and speak!

There is another reason for introducing the concept of culture. Countries, and all their cities, are identified by culture.[27] In Valencia, Salamanca, Toledo and Córdoba, people eat *paella*; in Leiden, Deventer, Maastricht and Delft, *hutspot*; in Freiburg, Augsburg, Koblenz and Hannover, *Eisbein mit Sauerkraut*; in Palembang, Surabaya, Balikpapan and Makassar, *rendang*; in Sapporo, Niigata, Osaka and Fukuoka, *sushi* and in Pyongyang, Wŏnsan, Seoul and Busan, *kimchi*. The *fado* can be heard in Lisbon, Porto, Coimbra and Sagres. The *tango* is the national dance of Argentina, the *rumba* is a dance typical of Cuba, and the *samba* is danced in Rio de Janeiro and São Paulo. Italy is the place for *bel canto*, Spain for the *flamenco*, Ukraine for the *hopak* and Greece for the *sirtaki*. Hearing the *balalaika*, Russian hearts beat faster. The French are known for their *chansons*, the Germans for their *Schlagers*. To understand people from a Mongolian town or city, listen to the sound of their *limbe*. Chinese, Japanese, English, French, Dutch, Italian, Spanish, Indian and Iranian gardens look different. So does the architectural style in Germany, Sweden, Italy, China, America, Cambodia, Malaysia, Cameroon and Iceland. Dances that are performed in Dublin, Cork, Limerick or Galway are in the first place Irish dances. Famous French football players do not come from Marseille, Mâcon, Nice or Chambéry, but from France.Cities and countries imply and reveal each other. The culture of a city, like the culture of a country, is both antecedent to and consequence of human behavior, which has different dimensions and is different from natural events, because it is driven by the human mind. The mystery of life is its impossibility without, but incompatibility with, dead matter (stardust), and the mystery of mind is freedom, something very difficult to define but without which life would be deadly dull and boring.

(5) **Urban Politics**

Whereas political *science* studies the theory and practice of politics and describes and analyzes political systems and political behavior, political *theory* (aka political philosophy) is one of the fast developing parts of political science, the other parts being the study of national governments, public administration, public law, policy analysis, comparative politics and international relations. "Power" (referring to the inherently strained relationship between *Herr* [superordinate] and *Knecht* [subordinate]) is the key concept in political science. It has turned out to be difficult to define.[29] My argument that cities and countries are intimately interconnected can be strengthened by looking at urban politics, *i.e.*, politics within and related to cities. Politics is the process of making decisions concerning the πολῑτεία (citizenry). Decision-making, the process of identifying, and choosing among, alternative activities based on the values, norms, aspirations and expectations of the decision-maker(s), is a subject currently being investigated from many sides.[30] In urban politics, the decisions (to be) made concern the urban citizenry, and the decision-maker is the mayor and his/her council. City governments are not sovereign. They are subordinated to the country's central government. This means that the decisions made by a city government have to be in harmony with the decisions made at the highest level of authority of the country—"have to be", for the practice is unruly. Since the State is the chief arena of politics,[31] the term politics is usually reserved for the State. A politician is someone engaged in State Affairs, and "political" means "relating to the government of a country". Yet there is a well-established practice of referring to politics pertinent to cities as well.[32] Diversity of views, an agreed-upon procedure for reaching decisions and acceptance and enforcement of the decision reached (the common policy) are thus phenomena that are also observed at local (city) level. The power of local governments is not only restricted and derivative; the city council (often divided into political factions) is also target of outside influences such as real estate developers, environmental activists, the media, the organized small, or the medium-sized, enterprises and big companies. The decisions reached may concern matters of administrative, economic, infrastructural, financial, social, educational, medical environmental or cultural importance. The layout of a city and the architectural look of its buildings and parks betray the nature of the country, whether it is a Christian, Islamic, socialist or another kind of country. Urban politics, of which urban planning is an important part,[33] determines the socio-economic makeup of a whole city or some of its neighborhoods. Urban politics teaches us that cities are not only geographically located; they are also politically and culturally embedded in a specific country, the profile and character of which are determined by the cities (and their hinterland) that are included in and embraced by it.

The point to be stressed is that geographers, economists, sociologists, anthropologists and political scientists who have focused their mind on the city (and have had no qualms about trespassing on each other's professional territory) fail to integrate their research. In other words, genuinely interdisciplinary efforts, crucial for comprehension, are missing. For example, in *The City Reader*, a book edited by Richard LeGates and Frederic Stout and published in 2020 by Routledge (see note 4), widely different views are presented but not integrated. The book is thus a juxtaposition of views about cities rather than a composition; it is just a collection, but not an attractively arranged collection, not a bouquet.

At best, urban geographers, economists, sociologists, anthropologists and political scientists peacefully coexist, without having mobilized and pulled together their formidable resources. The accent might be on this or that aspect of the city, but grasping the whole of the city has remained an elusive object. "Urban theorists" like Christopher Alexander, Donald Appleyard, Kevin Archer, Michael Batty, Luis Bettencourt, Peter Calthorpe, Manuel Castells, Gordon Cullen, Andrés Duany, Richard Florida, Patrick Geddes, Jan Gehl, Perceval Goodman, Adam Greenfield, David Harvey, Jane Jacobs, Rob Krier, James Kunstler, Henri Lefebvre, Kevin Lynch, Donald McNeill, Faranak Miraftab, Lewis Mumford, Saverio Muratori, Elizabeth Plater-Zyberk, Juval Portugali, Denise Pumain, Céline Rozenblat, Witold Rybczynski, Saskia Sassen, Edward Soja, Elkin Velasquez, Robert Venturi, William Whyte and Frank Lloyd Wright are testimony of it.

Whereas urban geographers, economists, sociologists, anthropologists and political scientists can boast of being educated and trained in a scientific field, "urban historians", who lack a city theory of their own, can only behave promiscuously by borrowing or stealing scientific theories/models. Their presentation of city-related things is essentially nothing but putting things in chronological order. The resemblance to "country historians" is striking.

The recent, downloadable report submitted to the National Science Foundation and edited by a large group of collaborating scientists from many fields led by José Lobo, *The future of urban science: integrating the social and natural sciences*, is hope-giving. It intends to convey "the current state of the art in *urban science*" and to communicate "what makes urban science different from discipline-based examinations of cities and urbanization", but the report "does not aim at a fully comprehensive synopsis of work done under the rubric of 'urban science'". Ekistics, the science of human settlements, is perhaps the science we should be looking for. It involves every kind of human settlement and pays due attention to every aspect of it: to the geological, geographic, historic, demographic, economic, political, legal, military, psychological, social, anthropological, aesthetic, semiotic, cultural and ecological aspect. Even more importantly, it takes each of these aspects to be just that: an aspect or facet of a whole. Ekistics takes the whole (a human settlement) into consideration, but it does not underrate or discount the importance of any of its aspects/facets. The term ekistics was coined by the visionary Greek architect Constantinos Apostolou Doxiadis (1913–1975). Ekistics (the word is derived from *οἰκιστικός* [residential area]) combines chorology (the ordering of things by place) and chronology (the arrangement of things by time). Whoever left his/her home town and visited a foreign country must have realized the importance of place differences, and growing older, one becomes more aware of history as the unending context of understanding. No event or action is understandable without taking place *and* time, the conjoint dimensions of reality, into consideration. A special issue on the philosophy of the city is scheduled for publication in the Springer journal *Topoi*. In my view, to meet the challenge of urbanization, a new kind of science is needed, a city-science going beyond the *aggregation* of partial views. The mono-disciplinary approach will not do. Complexity should be handled by a team of open-minded scientists. Cities as well as countries are systems, collections of things belonging together. Cities and countries have an identity. Though Norway and Chile, like Istanbul and Tokyo, have much in common, they are unmistakably identifiable. They are *individua*, things that can not be divided up into pieces (for separate [groups of] scientists to investigate) without losing their identity. In the study of cities and in the study of countries, Humpty has to be put together again.

Notes and References

1. See Mogens Hansen (ed.), *A Comparative Study of Six City-State Cultures* (Kongelige Danske Videnskabernes Selskab, 2002), *idem* and Thomas Nielson (eds.), *Inventory of Archaic and Classical Poleis* (OUP, 2004) and Sandra Halperin, "The imperial city-state and the national state form", *Thesis* 11, 139:1 (April 2017), pp. 97–112.
2. See Tom Scott, *The City-State in Europe, 1000–1600* (Oxford University Press, 2012).
3. See *American Journal of Sociology*, 44:1 (July 1938), pp. 1–24.
4. Recent books are Patsy Healey, *Urban Complexity and Spatial Strategies* (Routledge, 2007), Sergio Albeverio et al. (eds.), *The Dynamics of Complex*

Urban Systems (Springer, 2008), Elisabeth Lichtenberger, *Die Stadt: Von dedr Polis zur Metropolis* (Primus, 2011), Tim Campbell, *Beyond Smart Cities* (Routledge, 2012), Christian Walloth et al. (eds.), Understanding Complex Urban Systems (Springer, 2014), Roger White et al., *Modeling Cities and Regions as Complex Systems* (MIT Press, 2015), Juval Portugali and Egbert Stolk (eds.), *Complexity, Cognition, Urban Planning and Design* (Springer, 2016), Octavio Miramontes et al. (eds.), *Complejidad y Urbanismo* (CopIt-arXives, 2017), Michael Smith and Thomas Bender (eds.), *City and Nation* (Routledge, 2017), Claude Rochet, *Smart Cities: Reality or Fiction* (Wiley, 2018), Michael Batty, *Inventing Future Cities* (MIT Press, 2018), Germaine Halegoua, *Smart Cities* (MIT Press, 2020), Richard LeGates and Frederic Stout (eds.), *The City Reader* (Routledge, 2020), Hermann Haken and Juval Portgali, *Synergetic Cities: Information, Steady State and Phase Transition* (Springer, 2021), Luis Bettencourt, *Introduction to Urban Science* (MIIT Press, 2021), Zachary Neal and Céline Rozenblat (eds.), *Handbook of Cities and Networks* (Edward Elgar, 2021) and Juval Portugali (ed.), *Handbook of Cities and Complexity* (EE, 2021). See also the journal *Computational Urban Science* (December 2021ff). In addition, visit https://www.ukna.asia/seannet, https://www.ubdc.ac.uk, https://doi.org/10.1162/99608f92.b3fc5cc8, https://arxiv.org (search: city science), https://philosophyofthecity.org, the website of the EAI SmartCity360° International Convention 2020, https://urbansys2021.ifisc.uib-csic.es, https://sites.google.com/view/icacs21/home and https://urbcompsys.github.io. Professor Hyejin Youn (Northwestern University) aims to develop a mathematical and computational framework to understand cities. For the transformation of European cities, see *Ethnologia Europaea*, 34:2 (2004), pp. 5–17. Google "city", "*Stadt*", or "*ville*", and a wealth of information becomes available. Many recipients of the prestigious Vautrin Lud Prize have been theorizing on the city.

5. See Harm de Blij, *The Power of Place* (OUP, 2008) and Lewis Holloway and Phil Hubbard, *People and Place* (Routledge, 2013).
6. Read Nora's preface to *Realms of Memory* (Columbia University Press, 1996, pp. xv-xxiv). In *Time Maps: Collective Memory and social Shape of the Past* (University of Chicago Press, 2003), Eviatar Zerubavel continues his endeavor to find in memory the space where time and identity meet. To celebrate the 40th anniversary of the World Heritage Convention, in 2012, UNESCO and DeAgostini Libri presented a gorgeous atlas featuring more than thousand World Heritage sites.
7. See Richard Gale (ed.), *The Philosophy of Time* (Macmillan, 1968), Stephen Hawking, *A Brief History of Time* (Bantam, 1988), Paul Davies, *About Time* (Simon & Schuster, 1996), Rosolino Buccheri et al. (eds.), *The Nature of Time* (Springer, 2003), John Barrow, T*he Constants of Nature* (Knopf Doubleday, 2009), Flavia Santoianni (ed.), *The Concept of Time in Early Twentieth-Century Philosophy* (Springer, 2016), Christophe Bouton and Philippe Huneman (eds.), *Time of Nature and the Nature of Time* (Springer, 2017) and Carlo Rovelli,

L'ordine del tempo (Adelphi, 2017). Go to https://plato.stanford.edu/entries/time.
8. See Karl Schlögel, *In Space We Read Time* (University of Chicago Press, 2016).
9. Go to https://plato.stanford.edu/entries/mereology.
10. Go to https://plato.stanford.edu/entries/kyoto-school.
11. See Uehara Mayuko, "Japanese Aspects of Nishida's Basho", in Wing Keung Lam and Ching Yuen Cheung (eds.), *Frontiers of Japanese Philosophy 4: Facing the 21st Century* (Nanzan Institute for Religion and Culture, 2009, pp. 152–164). For the relationship between being and becoming, see *Parmenides* (Plato's dialogue). 145.
12. See Masahisa Fujita, "The Evolution of Spatial Economics", *Japanese Economic Review*, 61:1 (March 2010), pp. 1–32.
13. See Walter Christaller, *Die zentralen Orte in Süddeutschland* (Jena: Fischer, 1933).
14. See Brian Berry, "Cities as systems within systems of cities", *Papers in Regional Science*, 13:1 (January 1964), pp. 147–163.
15. See Hans Kuijper, "A Proposal to Solve the 'Taiwan Problem'", *China Report*, 40:2 (2004), pp. 189–208.
16. See Su-il Jeong, *The Silk Road Encyclopedia* (Seoul Selection, 2016), Julien Chaisse and Jędrzej Górski (eds.), *The Belt and Road Initiative* (Brill, 2018), Peter Frankopan, *The New Silk Roads* (Bloomsbury, 2018), Gerald Chan, *Understanding China's New Diplomacy* (Edward Elgar, 2018), Eric de La Maisonneuve, *Les défis chinois: La révolution Xi Jinping* (Rocher, 2019), Bruno Maçães, *Belt and Road* (Hurst, 2019), Alicia García-Herrero and Jianwei Xu, *Countries' Perceptions of China's Belt and Road Initiative* (Bruegel, 2019), Daniel Drache et al., *One Road, Many Dreams* (Bloomsbury, 2019), Nazrul Islam, *Silk Road to Belt Road* (Springer, 2019), Richard Pomfret, *The Central Asian Economies in the Twenty-First Century: Paving A New Silk Road* (PUP, 2019), Chongwei Zheng et al., *21st Century Maritime Silk Road* (Springer, 2019), Liang Yang et al. (eds.), *Socio-Environmental Dynamics along the Historical Silk Road* (Springer, 2019), Li Xing (ed.), *Mapping China's "One Belt One Road" Initiative* (Palgrave Macmillan, 2019), Jeremy Garlick, *The Impact of China's Belt and Road Initiative.*

From Asia to Europe (Routledge, 2019), Hao Su and Yuanzhe Ren (eds.), *Maritime "Silk Road"* (World Scientific, 2020), Thomas Fingar and Jean Oi (eds.), *Fateful Decisions* (Stanford University Press, 2020), Alfred Gerstl and Ute Wallenböck (eds.), *China's Belt and Road Initiative* (Routledge, 2020), Deborah Brautigam, "A critical look at Chinese 'debt-trap diplomacy'", *Area Development and Policy*, 5:1 (2020), pp. 1–14, Florian Schneider (ed.), *Global Perspective on China's Belt and Road Initiative* (A'dam University Press, 2021) and Jonathan Hillman, *The Digital Silk Road: China's Quest to Wire the World and Win the Future* (Harper, 2021). Also go to https://prochetmoyen-orient.ch/routes-de-la-soie-maritimes-et-armees. To keep abreast of the latest developments, visit https://www.clingendael.org/publication/silk-road-headlines.

17. The complementarity of agriculture and industry is one of the main concerns of development economists. See Gerald Meier and James Rauch, *Leading Issues in Economic Development* (OUP, 2005, Chaps. 6 and 7), Debnarayan Sarkar, *Agriculture Industry Interactions* (Kalpaz, 2010), Dwight Perkins et al., *Economics of Development* (W.W. Norton, 2013, Chaps. 16 and 17) and Aparajita Mukherjee and Saumya Chakrabarti, *Development Economics: A Critical Perspective* (PHI Learning, 2016, Chaps. 5 and 6).
18. See Randy Larsen & David Buss, *Personality Psychology* (McGraw-Hill, 2018) and Gerald Hartung, *Philosophische Anthropologie* (Reclam, 2018).
19. Saint-Simon, Auguste Comte, Karl Marx, Herbert Spencer, Ferdinand Tönnies, Émile Durkheim, Georg Simmel, Max Weber and Karl Mannheim are generally considered to be the founding fathers of sociology.
20. The concept of time–space compression was coined in 1989 by David Harvey. In his book, *The Condition of Postmodernity* (an allusion to François Lyotard's *La Condition postmoderne* [1979]), he said that the concept refers to anything that impacts time and space. Karl Marx already wrote about the "annihilation of time and space". The literature on globalization has grown into a torrent. One could start with Or Rosenboim, *The Emergence of Globalism* (PUP, 2017) and Joseph Stiglitz, *Globalization and its Discontents* (W.W. Norton, 2017).
21. In *Philosophie des Geldes* (1900), Simmel makes the compelling case that money, as structuring agent, helps to understand the totality of life. Today, the book is more relevant than ever. The July–August 2019 number of *Esprit* devotes a special issue to *"l'argent: maître invisible"*.
22. See Martin Bulmer, *The Chicago School of Sociology* (Chicago University Press, 1984) and Lester Kurtz, *Evaluating Chicago Sociology* (Chicago University Press, 1986).
23. See Frank Eckardt (ed.), *Handbuch Stadtsoziologie* (Springer, 2012), Jan Lin and Christopher Mele (eds.), *The Urban Sociology Reader* (Routledge, 2013), Mark Gottdiener et al., *The New Urban Sociology* (Routledge, 2015) and Jean-Marc Stébé and Hervé Marchal, *La sociologie urbaine* (PUF, 2016).
24. See Friedrich Jaeger and Burkhard Liebsch (eds.), *Handbuch der Kulturwissenschaften*, Vol. I: *Grundlagen und Schlüsselbegriffe* (J.B. Metzler, 2004), Ansgar Nünning (ed.), *Grundbegriffe der Kulturtheorie und Kulturwissenschaften* (J.B. Metzler, 2005), Edward Skidelsky, *Ernst Cassirer: The Last Philosopher of Culture* (PUP, 2008), Claus Leggewie et al. (eds.), *Schlüsselwerke der Kulturwissenschaften* (Transcript Verlag, 2012), Ralf Konersmann (ed.), *Handbuch der Kulturphilosophie* (J.B. Metzler, 2012) and Doris Bachmann-Medick, *Cultural Turns* (Rowohlt, 2015). Georg Northoff, Kevin Laland, Jaan Valsinger, Michael Tomasello, David Matsumoto, Shinobu Kitayama and Shihui Han are scientists who study the subject of culture.
25. See his *Weltgeschichtliche Betrachtungen* (Alfred Kröner, 1978, p. 57).
26. See John Levi Martin, "The Birth of the True, the Good, and the Beautiful", in Harry Dahms and Eric Lybeck (eds.), *Reconstructing Social Theory, History and Practice* (Emerald, 2017, pp. 3–56).

27. The literature on urban anthropology (a subset of anthropology) is, relatively speaking, scarce. Mention should be made of Ulf Hannerz, *Exploring the City* (Columbia University Press, 1980), John Agnew et al. (eds.), *The City in Cultural Context* (Allen & Unwin, 1984), Donald Nonini (ed.), *A Companion to Urban Anthropology* (Wiley, 2014), Rivke Jaffe and Anouk de Koning, *Introducing Urban Anthropology* (Routledge, 2015) and George Gmelch and Petra Kuppinger, *Urban Life: Readings in the Anthropology of the City* (Waveland, 2017). None of these books pays due attention to the relationship between culture and mind.
28. For the distinctiveness of Chinese, Islamic and European cities, see, respectively, William Skinner (ed.), *The City in Late Imperial China* (Stanford University Press, 1977, pp. 33–100), Salma Jayyusi et al. (eds.), *The City in the Islamic World* (Brill, 2008) and Engin Isin, "Theorizing the European City", in Delanty *op.cit*, pp. 323–332. The bi(bli)ography of many cities can be found on the Internet. Books on Chinese cities are listed in Lubna Malik and Lynn White, *Contemporary China: A Book List* (PUP, 2008, pp. 36–37). Visit http://www.spatialcomplexity.info.
29. See Elias Canetti, *Masse und Macht* (Claassen, 1960) and Steven Lukes, *Power* (Palgrave Macmillan, 2005).
30. See Jonah Lehrer, How We Decide (HFH, 2009), Paul Nut and David Wilson (eds.), *Handbook of Decision Making* (Wiley, 2010), Jie Lu et al. (eds.), *Handbook on Decision Making* (Springer, 2012), Marina Sokolova & Antonio Fernández-Caballero, *Decision Making in Complex Systems* (Springer, 2012), Guangquan Zhang et al., *Multi-Level Decision Making* (Springer, 2015) and Rudolf Grünig & Richard Kühn, *Prozess zur Lösung komplexer Entscheidungsprobleme* (Springer, 2017). In *Cognition et Système: Essai d'Epistémo-Praxéologie* (L'Interdisciplinaire, 1995), Robert Vallée expounds on the loop between perception, decision-making and action. Information influences decision-making!
31. I bypass the vast and fast-growing field of international relations, but I will make some critical remarks on the subject in chapter eight. Interested readers should turn to Christian Reus-Smit & Duncan Snidal (eds.), *Handbook of International Relations* (OUP, 2010) and Frank Sauer & Carlo Masala (eds.), *Handbuch Internationale Beziehungen* (Wiesbaden: VS, 2017).
32. Decision-making is also vital for business success. The process of selecting a set of activities plays further an important role in game theory and the military (tactics, strategy). The theory and practice of decision-making often goes under the name of operations research (aka operational research), a thriving discipline that deals with the application of analytical methods to help make better decisions. If democracy is about choices made by the people (as distinct from choices on the provision of public goods!), decision-making is a serious kind of business Decision-making theory is sometimes subsumed under the category of formal sciences.

33. Architecture, civil engineering, urban ecology, geography, land development, landscape architecture, public health and real estate development are some of the disciplines related to urban planning.

Chapter 5
Scientific Feasibility of Collaborative Country Studies

Abstract In this chapter, the core of my argument, the *scientific feasibility* of collaborative country-studies, is highlighted. When taking the interdisciplinary approach, the insights of (complex) systems science can be profitably used. Systems, systems science, the roots and the branches of systems science, and cybernetics (the twin brother of systems science) are the subjects critically elaborated on. To stress the need for applying the principles of general systems theory to the study of countries, Norbert Wiener's criticism of free competition is used to express disapproval of mainstream economics.

When countries/cities are, or can be conceived of as, complex and dynamic systems of systems, the question arises whether the insights of systems scientists are relevant or pertinent to the study of these *explananda* and to what extent the tools developed in the vast and fast expanding field of computer science are applicable in such studies. In other words, once the *desirability*, even the *necessity* of integrative thinking and collaborative acting with regard to the study of countries, has been established, the questions should be answered whether countries are legitimate objects when studying systems and whether computer-supported collaborative country studies are *possible*. If the answer to these questions is affirmative, the conclusion has to be drawn that the study of a country should be a form of applied systems science, meaning that the *explanans* practitioners of country studies have been sorely missing is to be sought in systems science, a kind of science on the completion of which much hard work has to be or is being done. In this chapter, attention will be paid to the relevance of systems science. In Chap. 7, I will address the question of applicability of computer science.

Since the extensive literature on systems science is growing by the day, I select the following issues, examining them only cursorily:

(I) Systems
(II) Systems science
(III) Roots of systems science
(IV) Branches of systems science

H. Kuijper, *Comprehending the Complexity of Countries*,
https://doi.org/10.1007/978-981-16-4709-3_5

(a) Complex systems science
(b) System dynamics
(c) Network science

(V) Cybernetics

Systems

Systems (from σύστημα [constitution]) are ubiquitous. There are natural and man-made systems. Examples of the first category are the solar system, Planet Earth, a river system, the climate, an organism, an atom, a cell, a blade of grass. Examples belonging to the second category are societies, polities, political parties, companies, nuclear reactors, hospitals, health systems, universities, libraries, books, journals, archives, museums, educational systems, cognitive/mental systems, economies, industries, legal systems, military systems, houses, water supply systems, sewerages, offices, churches, pipe organs, clubs, bicycles, trains, cars, traffic systems, planes, airports, ships, seaports, cities, robots, computers, television sets, the Internet, cigarettes, Japanese tea ceremonies, baseball games, tennis matches and wristwatches.[1] Countries are huge combinations of man-made and natural systems (see Chap. 6). Man-made systems may break down, because they are not free from irrational elements.[2] It is important to bear in mind that the root meaning of the word system is constitution, an entity (whole) consisting of parts. In 1843, John Stuart Mill's massive work *A System of Logic* appeared, in London, and from 1913 to 1959, Pierre Duhem's ten-volume *Le système du monde* was published, in Paris, the former book being concerned with the realm of thoughts, the latter work with the world of things. It would be interesting to compare the monumental works of Mill and Duheim with the three volume *magnum opus* of Alexander Bogdanov (1873–1928), *Tektologiya: Vseobschaya Organizatsionnaya Nauk* (tektology: universal organization science), published from 1912 to 1917. According to the *New Oxford American Dictionary* (OUP, 2010), a constitution is "an aggregate of fundamental principles or established precedents that constitute the legal basis of a polity, organization or other type of entity, and commonly determine how that entity is to be governed". In *Weltgeschichtliche Betrachtungen* (Kröner, 1978 [1905], pp. 81–156), Jacob Burckhardt (1818–1897) explains how political, religious and cultural systems are interrelated, i.e., are to be considered as subsystems constituting a whole, their supersystem. The literature on this nexus is vast and expanding.

Countries are complex and dynamic systems consisting of complex and dynamic subsystems; they have thus related levels of systems. But what is a system and what is the difference between systems and sets? To answer these questions, I must go to the roots of mathematics.[3] Two hands, belonging to, or being parts of, one human body, usually have ten fingers. Counting on the fingers led to arithmetic, the part of mathematics that involves the adding and multiplying of numbers. Arithmetic led to algebra, the part of mathematics in which signs or letters represent numbers. Spacing, the indication of the amount (quantity) of distance, led to geometry, the area of mathematics in which space and the relationships between points, lines, curves and surfaces are studied. Locating, the indication of the exact position of something, led to topology, which is concerned with the properties of a geometric object that are preserved under continuous deformations. Descartes (1695–1650) discovered the possibilities of combining algebra and geometry, and the combination of algebra and topology resulted in differential and integral calculus, co-discovered by Newton (1642–1726) and Leibniz (1646–1715).[4] The mathematicians after them have further developed the impressive apparatus Descartes and Newton/Leibniz had created, expanding the mathematical domain in all possible directions (in height, width and depth). For example, calculus has formed the basis of "analysis, the part of mathematics concerned with sets of numbers, sets of points or sets of functions and with the processes (called limit processes) that depend on a measure of closeness between numbers, points or functions".[5]

There are other ways of looking at the result of counting on fingers (assigning numbers), the source of all mathematical thinking, as distinct from saying or writing words.[6] Balancing and weighing led to statistics, the study of numerical data. Moving led to dynamics, the study of the causes of movements or changes (calculus). Guessing led to probability theory, the study of stochastic situations and processes. Set theory, the brainchild of Georg Cantor (1845–1918), was born from the collecting of things, which led to the concept of number, the subject of number theory (according to Gauss, the "queen of mathematics"). Set theory changed the entire outlook on measure and integration[7]; it relates to group theory (see Chap. 2), because a mathematical group is a non-empty set of elements, which can be combined in a particular way and under certain conditions. The language of Cantor's theory is used for defining nearly all mathematical objects, but most of the theory dealt with "*Punktmengen*" (sets of points). Cantor defined a set as "any collection of definitely distinguished objects gathered into a whole".

Before the collecting of things, they have to be identified, but the concept of identity is far from clear.[8] The word identical, or synonyms of it, appears in sentences like "x is identical with y", "x is the same as y" and "x is equal to y", or, for the sake of brevity, $x = y$. Instead of "x is not identical with y" or "x differs from y", we write $x \neq y$. Strictly speaking, however, there is no identity, and when there is no identity, it makes no sense to speak of difference, because "identity" and "difference" are correlative (mutually involving) concepts. If a drop of water falls into the sea, there is no identity (the volume of the sea changes). With due respect to Robert Recorde (1512–1558), who invented the equality sign (=), equality is neither equal to identity nor identical with it (visit https://www.quantamagazine.org/with-category-theory-mathematics-escapes-from-equality-20191010). Yet "*Liberté, Égalité, Fraternité ou la Mort*" has been the rallying cry of revolutionary movements. Feminism is just one of them, its goal being the achievement of political, economic and social gender-equality.

In everyday language, identity is usually expressed by the use of "is", or by another form of the verb "to be" ("am", "are", "was", "were", "being", or "been"). For example: "Rome is the capital of Italy", "Shakespeare is the author of *Hamlet*" or "Oxygen is the chemical element with the symbol O". The "is" in these sentences differs, however, from the "is" in phrases like "Rome is a city", "Shakespeare is/was a playwright" or "Oxygen is a nonmetal". In the first three sentences, reversal is possible: "The capital of Italy is Rome", "The author of *Hamlet* is Shakespeare" and "The chemical element with the symbol O is Oxygen". Reversal is impossible in the last three phrases, because "is a city", "is/was a playwright" and "is a nonmetal" are predicates, giving information about, respectively, the subject Rome, Shakespeare and oxygen. We enter here the vast domain of (first- and higher-order) predicate logic, the discussion of which falls beyond the scope of this book and for which I refer the reader to the extensive and expanding literature.[9]

Once collected, things form a collection, a set. For example, the collection consisting of a book, a digital camera, a deck of cards, an guitar and a pencil, the set formed by the boys who are absent from the school in town *x*, or the assortment of vegetables in a supermarket. Classes of individuals are first-order classes.[10] Second-order classes do not consist of individuals, but of first-order classes. There are also third-order, fourth-order, etc. collections. The study of classes (their nature, notation, operations on, and relations between, them) is called set theory.

A system is a particular set. Whereas a set is a collection of things, a system is a collection of *connected* things. Their interconnectedness is to be emphasized, because, being interconnected, the parts form a unit, which has properties, a boundary and an environment. The difference between set and system is a subtle but important one. A set is just a collection of objects. For example, the men of a town form a set; what they have in common is only their being resident of the town. They are not connected. All the persons (men, women and children) of a town, however, form a system, because, living in or belonging to it, they form a community which differs from other communities. Tensions between residents do not distract from their being connected. The products for sale in a department store form a set, but the interconnected parts of a bicycle, car, train, plane or computer form a system. If the connected things belong to the physical world, it is a physical system; if they belong to the living world, it is a living or biological system, and if they belong to the human society, it is a social system. The immune system (not: set) interacts with both the nervous and the hormone system, or, to take a few examples from the social domain, a company, political party and religious community are more than sets; they are systems, operate as units and interact or communicate with their environment. External communication is often more difficult than internal communication, though the difficulties attendant on intragroup communication should not be underestimated. Further distinctions can be made. For example, social systems are divided into political, military, legal, economic, financial, manageral, linguistic, literary, cognitive, religious, cultural and technological systems. A worldview is not a collection of unrelated beliefs. It is a cognitive system, a collection of interconnected beliefs. Its most fundamental concept ("order") may be ill-conceived, but it is a system all the same. System differentiation has often led to losing sight of the nature and essence of a system *tout court*. The study of systems is called systems science and should thus be clearly distinguished from set theory.

A system is composed of a number of interrelated parts. Two aspects are involved: (a) the parts forming, composing or making up the system and (b) the way in which these parts are related to each other. Discoveries made since 1970 have shown that these relations are usually nonlinear.[11] To a degree, systems are separated from their surroundings by a boundary. Each system is, therefore, conditioned and constrained. There is nothing totally isolated. A subsystem is both a collection of connected elements (its components or constituents) and a component of a larger system (the subsystem's supersystem). A complete description of a system contains also information about its environment; the description of a system that does not contain information about its environment is thus incomplete and potentially misleading. No economic, political or social system operates in a vacuum. An overall system is a nested system; it is extremely difficult, but not impossible to describe. The change of one part of a system induces the change of another part and of the system as a whole. Over time, a series of changes forms a process that has a structure or pattern to be identified and explained. Identifying and explaining the process pattern (*Wandlungsstruktur*) of a country's history that consists of many interconnected and changing subsystems having numerous interrelated and transforming constituents is the daunting task of historians—a task I claim (in Chap. 8) they are not scientifically equipped to carry out. Historians who refuse to collaborate with, say, economists, political scientists, demographers, geographers and sociologists should apply for a job as journalist or country correspondent; universities are not the places where they belong, for they are not interested in systems, not really. Adaptation of a system depends on how well the system is adjusted to its environment. Systems often exist for the service of a person (or group of persons) having personal (or collective) interests that run counter to what is natural. The management of a system may therefore result in its failure or breakdown. As management increases in complexity (e.g., when ruling a big company or modern nation-state), the risk of system failure or breakdown will increase. This may be the reason why "*wéi wú wéi*" (letting it happen, effortless doing) is the basic tenet of Chinese thought that shines through in Chinese actions and policies. This principle is part and parcel of the Chinese worldview which I roughly outlined in the Preface, leaving an *exposé* of its details to another occasion. Suffice to say that it would be mistaken to interpret "*wéi wú wéi*" as "*laissez-faire*", the controversial principle of liberalism, a subject on which much ink has been spilled. Visit http://www.toupie.org (search for "libéralisme"), https://plato.stanford.edu/entries/liberalism and https://philpapers.org (search for "liberalism"). "*Laissez-faire*" means: Leave us alone, government; we, businessmen, know very well how to manage our own affairs ourselves. Left alone, however, these businessmen may still contravene the Chinese admonishment.

Systems can be either closed or open. A closed system does not allow transfers into or out of it. The specification of what types of transfer are excluded varies in the closed systems of physics, chemistry or engineering. For example, a system might be closed to energy, meaning that energy is neither able to enter nor to leave it. A vacuum thermos flask stops energy from leaving the system to keep the drink it contains warm. So it makes sense to treat the flask as a closed system. However, no system in the real world is perfectly and hermetically closed. An open system has external interactions, which may take the form of energy, information or material transfer into or out of it, depending on the discipline defining the system. If one pours hot drink into a mug, the heat will escape to its surroundings. Open systems are more difficult to understand than closed systems. This may be the reason why natural scientists tend to prefer working with systems they mistakenly presume to be closed. An open system may exchange material, energy, information (data), people and capital with its environment. Kostas Axelos, a Greek-French philosopher, argued that seeing the world as an inherently open system would solve many problems.[12] I do not see that/how considering the world as an inherently open system would solve all problems, but human and social scientists would be well advised to stay in touch with natural scientists and to work closely together with each other, rather than creating a monopoly whereby man, or the society/country in which he lives, is looked at through political, legal, military, economic, financial, psychological, educational, medical, linguistic, literary or cultural spectacles.

Systems Science

Systems science is the organized knowledge of systems, i.e., knowledge about systems that is methodically organized. Knowledge theory, aka epistemology (from ἐπιστήμη [knowledge] and λόγος [discourse, discussion]), is a branch of philosophy concerned with the nature, sources, possibility and justification of knowledge about reality. My concern here, however, is not epistemology but systems science. A list of epistemologists and an index of epistemology articles can be found on the Internet.

Systems science is, and this point is particularly important, one of the *formal* sciences, the other formal sciences being mathematics, logic, statistics and computer science, whereas the not formal or informal sciences (the physical, biological, psychological, social and cultural sciences) study respectively, natural, biological, psychological, social and cultural phenomena, that is to say, a concrete system, the formal sciences, which gathered momentum after the Second World War, are concerned with the structure of the informal sciences, not with their content but with their form, with the form in which they couch their theories. The formal sciences abstract away from what the of informal sciences study, attempting to aid them by providing information about the structures and models they use to describe the world. The formal sciences are thus not concerned with the validity of theories based on observations or experimentations (see Chap. 2), but with the basic properties and structures of these theories, with the characteristics of their form (hence formal sciences). Like the other formal sciences, systems science is increasingly supporting, or applied to, the informal or substantive sciences. Even more importantly, systems science is refining, and is refined by, the other formal sciences. This explains its (and their) dynamic and energizing nature. The physical, biological, psychological, social and cultural sciences are maturing by incorporating the insights of formal sciences; they are increasingly inspired by, and based on, systems science, which *idealiter* is informed by its supportive and vibrant colleagues: mathematics, logic, statistics and computer science.

The aim of systems science is to lay the foundations of, and to find the principles applicable to, all scientific disciplines, which is the reason why the founding father of systems science, the Austrian biologist Ludwig von Bertalanffy (1901–1972), is considered to be at the same level as Charles Darwin, Albert Einstein and Niels Bohr. Systems science drives the search for unity of science, an important subject closely related to the issue of interdisciplinary research (see Appendix A).[14] The themes stressed in systems science are (a) the holistic (versus the reductionist) view, (b) the complexity of systems and (c) the interactions between systems and their environment. Further down the road, I will say a few words about these points.

Systems science is related to systems *thinking* or—philosophy. Systems thinking emerged in the latter part of the twentieth century; it is reminiscent of, but not quite the same as, the functionalism of Bronisław Malinowski, Alfred Radcliffe-Brown and Talcott Parsons or the structuralism of Ferdinand de Saussure, Roman Jacobson and Claude Lévi-Strauss. According to Peter Senge, celebrated author of *The Fifth Discipline: The Art and Practice of The Learning Organization* (Currency, 1990), "systems thinking is a discipline for seeing wholes. It is a framework for seeing relationships rather than things, for seeing patterns of change rather than static snapshots. Today systems thinking is needed more than ever because we are becoming overwhelmed by complexity".[15] A special issue of the open access journal *Systems* (2013ff) will be devoted to "Systems Thinking: Concepts, Issues and Applications in "Large Complex Systems".

Systemics is an initiative taken by Mario Bunge (1919–2020). It studies systems from a holistic point of view. It attempts to develop a framework in which physical, technical, biological, cognitive and social systems can be studied. Systemics, which converges with cybernetics, network science and information science (see below), must be distinguished from *systematics*, the study of the diversification of, and relations between, past or present living forms.

Systems science should also be distinguished from systems *engineering* (SE), which is the application of systems science. Interdisciplinary in character SE is concerned with the designing and forming of systems. Engineering systems are the work of systems engineers; their products are systems of elements or parts, i.e., combinations of components working together in synergy (from *συνεργία* [collaboration]). The German physicist Hermann Haken (1927) is the founding father and undisputed leader of synergetics, the science of "*Zusammenarbeit*". For details, see the online Guide to the Systems Engineering Body of Knowledge (SEBoK) and go to https://cser2020.org.

The following list is meant to give an impression of (*a*) the still young but vibrant field of systems science and (*b*) the disciplines that keep systems scientists divided. The bibliography in the books will guide the reader further into the field.

David Easton, *The political System* (1953, third edition in 1981)

Karl Deutsch, *The Nerves of Government* (1963)

Walter Buckley, Sociology and Modern Systems Theory (1967)

Ludwig von Bertalanffy, *General System Theory* (1969)

Talcott Parsons, *The system of modern societies* (1971)

Stephen Bryen, *Application of Cybernetic Analysis to the Study of International Politics* (1971)

John Sutherland, A general systems philosophy for the social and behavioral sciences (1973)

Howard Pattee (ed.), Hierarchy Theory: The Challenge of Complex Systems (1973)

Mihajlo Mesarović and Y. Takahara, *General Systems Theory* (1975)

Talcott Parsons, *Social Systems and the Evolution of Action Theory* (1977)

Louis Le Moigne, *La théorie du système général* (1977, fifth edition in 2016)

Edgar Morin, *La Méthode* (1977–2004)

James Grier Miller, *Living systems* (1978)

Humberto Maturana and Francisco Varela, *Autopoiesis and Cognition* (1979)

Gregory Bateson, Mind and Nature: A Necessary Unity (1979)

Mario Bunge, Treatise on Basic Philosophy, Vol. 4: A World of Systems (1979)

Frederick Emery, *Open Systems thinking* (1981)

Geoffrey Vickers, Human Systems are Different (1983)

Hermann Haken, Synergetics: An Introduction (1983)

Ilya Prigogine and Isabelle Stengers, *Order out of Chaos* (1984)

Kenneth Boulding, The World as a Total System (1985)

Béla Bánáthy (ed.), Systems Inquiring: Theory, Philosophy, Methodology, Applications (1985)

Anatol Rapoport, *General System Theory* (1986)

Jean-Louis Vullierme, Le concept de système politique (1989)

Huibert Kwakernaak and Raphael Sivan, *Modern Signals and Systems* (1990)

George Klir, Facets of Systems Science (1991)

Barry Richmond, *An Introduction to Systems Thinking* (1992)

Francisco Varela and Paul Bourgine (eds.), *Toward a Practice of Autonomous Systems* (1992)

Howard Odum, *Systems Ecology* (1993)

Bernard Barber, Constructing the Social System (1993)

Murray Gell-Mann, *The Quark and the Jaguar* (1994)

Anthony Stafford Beer, Beyond Dispute: The Invention of Team Syntegrity (1994)

Robert Vallée, *Cognition et Système* (1995)

Barbara Hanson, General Systems Theory: Beginning with Wholes (1995)

Stuart Kauffman, *At Home in the Universe* (1995)

Per Bak, How Nature Works: The Science of Self-Organized Criticality (1996)

Hugh Miser (ed.), Handbook of Systems Analysis (1996)

Niklas Luhmann, Die Gesellschaft der Gesellschaft (1997)

Yaneer Bar-Yam, Dynamics of Complex Systems (1997)

Karl Åström and Björn Wittenmark, *Computer Controlled Systems* (1997)

John Holland, Emergence: From Chaos to Order (1998)

Gabriel Altmann et al. (eds.), *Systems: New Paradigms for the Human Sciences* (1998)

Walter Buckley, *Society: A Complex Adaptive System* (1998)

Robert Flood, Rethinking the Fifth Discipline (1999)

Yi Lin, General Systems Theory: A Mathematical Approach (1999)

Robert Axelrod and Michael Cohen, *Harnessing Complexity* (2000)

Günther Ossimitz, Entwicklung systemischen Denkens (2000)

Scott Camazine, Self-Organization in Biological Systems (2001)

Gerald Weinberg, *An Introduction to General Systems Thinking* (2001)
Kenneth Bausch, The Emerging Consensus in Social Systems Theory (2001)
Heinz von Foerster, *Understanding Understanding* (2002)
Radim Bělohlávek, Fuzzy Relational Systems: Foundations and Principles (2002)
Lance Gunderson and Buzz Holling (eds.), *Panarchy* (2002)
Albert-László Barabási, Linked: The New Science of Networks (2002)
Steven Strogatz, Sync: The Emerging Science of Spontaneous Order (2003)
Philip Ball, Critical Mass: How One Thing Leads to Another (2004)
Duncan Watts, Six Degrees: The Science of a Connected Age (2004)
Urie Bronfenbrenner, *Making Human Beings Human* (2005)
Lars Skyttner, *General Systems Theory* (2005)
Adolf Žižek, *Kompleksni Sistemi* (2006)
John Warfield, An Introduction to Systems Science (2006)
Seth Lloyd, Programming the Universe (2006)
Michael Luft, Systematik: Die universale Systemtheorie (2006)
Thomas Deisboeck and Yasha Kresh (eds.*), Complex Systems Science in Biomedicine* (2006)
Ned Kock, Systems Analysis & Design Fundamentals (2006)
Klaus Mainzer, *Thinking in Complexity* (2007)
Gregoire Nicolis and Catherine Nicolis, *Foundations of Complex Systems* (2007)
Hartmut Bossel, *Systems and Models* (2007)
Panos Antsaklis and Anthony Michel, *A Linear Systems Primer* (2007)
John Miller and Scott Page, *Complex Adaptive Systems* (2007)
Péter Érdi, Complexity Explained (2008)
Donella Meadows, *Thinking in Systems* (2008)
Paul Bourgine et al. (eds.), Déterminismes et complexités: du physique à l'éthique (2008)
Fredmund Malik, Strategie des Managements komplexer Systeme (2008)
Sergio Albeverio et al. (eds.), *The Dynamics of Complex Urban Systems* (2008)
Markus Schwaninger, *Intelligent Organizations* (2009)
Magnus Ramage and Karen Shipp, *Systems Thinkers* (2009)
Brian Catellani and Frederic Hafferty, *Sociology and Complexity Science* (2009)
Bert Hölldobler and Edward Wilson, *The Superorganism* (2009)
Melanie Mitchell, *Complexity: A Guided Tour* (2009)
Francisco Parra-Luna (ed.), *Systems Science and Cybernetics* (2009)
Peter Senge, The Necessary Revolution (2010)
Lance Gunderson et al. (eds.), *Foundations of Ecological Resilience* (2010)
Henk Broer et al. (eds.), *Handbook of Dynamical Systems* (2010)
Michel Loreau, From Populations to Ecosystems (2010)
Sifeng Liu and Jeffrey Yi Lin Forrest, *Grey Systems: Theory and Applications* (2010)
Johannes Zimmermann, Für eine Geschichte der Systemwissenschaft (2010)
Mario Tokoro (ed.), Open Systems *Science* (2010)

Raúl Espejo and Alfonso Reyes, *Organizational Systems* (2011)

Simon Hix and Bjørn Høyland, The Political System of the European Union (2011)

Jamshid Gharajedaghi, *Systems Thinking: Managing Chaos and Complexity* (2011)

Ton Jörg, New Thinking in Complexity for the Social Sciences and Humanities (2011)

Geert Jan Olsder et al., *Mathematical Systems Theory* (2011)

Günter Ropohl, Allgemeine Systemtheorie: Einführung in transdisziplinäres Denken (2012)

Marian Walhout et al. (eds.), *Handbook of Systems Biology* (2012)

Tamra Orr, Understanding Economic Systems (2012)

Francisco Parra-Luna (ed.), *The Performance of Social Systems* (2012)

Ika Darnhofer et al. (eds.), *Farming Systems Research into the 21st Century* (Springer, 2012)

Robert Rosen, *Anticipatory Systems* (2012)

Sifeng Liu et al., Systems Evaluation: Methods, Models, and Applications (2012)

Philippe Collard et al., *Systèmes complexes* (2013)

Jennifer Wells, Complexity and Sustainability (2013)

Helmut Wilke, *Systemtheorie* (2014)

Darrell Arnold (ed.), *Traditions of Systems Theory* (2014)

Yoshiteru Nakamori, Knowledge and Systems Science (2014)

John Mingers, *Systems Thinking* (2014)

Ricardo Hausmann et al., *Atlas of Economic Complexity* (2014)

David Byrne and Gill Callaghan, Complexity Theory and the Social Sciences (2014)

Frederick Hillier and Gerald Lieberman, *Introduction to Operations Research* (2014)

Fritjov Capra and Pier Luigi Luisi, *The Systems View of Life: A Unifying Vision* (2014)

Christian Walloth et al. (eds.), *Understanding Complex Urban Systems* (2014)

Gregorio d'Agostino and Antonio Scala (eds.), *Networks of Networks* (2014)

George Mobus and Michael Kalton, *Principles of Systems Science* (2015)

MengChu Zhou et al., Contemporary Issues in Systems Science and Engineering (2015)

Mark Denny, Ecological Mechanisms (2015)

Kyoichi Kijima (ed.), *Service Systems Science* (2015)

Hiroki Sayama, Introduction to the Modeling and Analysis of Complex Systems (2015)

Craig Kluever, Dynamic Systems: Modeling, Simulation, and Control (2015)

Claudius Gros, Complex and Adaptive Dynamical Systems (2015)

Timothy Allen and Thomas Hoekstra, *Toward a Unified Ecology* (2015)

Jean Tirole, Théorie de l'organisation industrielle (2015)

Ockie Bosch et al., Systems Thinking for Everyone (2015)

Mateus Esteves-Vasconcellos, *The New General Systems Theory* (2015)

James Lovelock et al., *The Earth and I* (2016)

Dirk Baecker (ed.), Schlüsselwerke der Systemtheorie (2016)

Andreas Tonsberg et al., *Understanding Leadership in Complex Systems* (2016)

Chih-Chun Chen and Nathan Crilly, *From Modularity to Emergence* (2016)

Axel Pelster and Günter Wunner (eds.), *Selforganization in Complex Systems* (2016)

Christian Schuldt, Systemtheorie: Theorie für die vernetzte Gesellschaft (2017)

Mary Edson et al (eds.), *A Guide to Systems Research* (2017)

Geoffrey West, *Scale* (2017)

Bruce Edmonds and Ruth Meyer (eds.), *Simulating Social Complexity* (2017)

Kelly Rainer and Brad Prince, *Introduction to Information Systems*, 2017

Stefan Thurner (ed.), *43 Visions for Complexity* (2017)

Gianfranco Minati et al. (eds.), *Second Generation General System Theory* (2017)

Yi Lin et al., Systems Science: Methodological Approaches (2017)

W.B. Lee and Farzad Sabetzadeh, Contemporary Knowledge and Systems Science (2018)

Ockie Bosch and Robert Cavana (eds.), *Systems Education for a Sustainable Planet* (2018)

Pier Luigi Gentili, Untangling Complex Systems: A Grand Challenge for Science (2018)

Jamshid Ghaboussi and Michael Insana, *Understanding Systems* (2018)

George Rzveski and Carlos Brebbia (eds.), *Complex Systems Studies* (2018)

Ricard Solé and Santiago Elena, *Viruses as Complex Adaptive Systems* (2018)

Mark Newman, *Networks* (2018)

Stefan Thurner et al., Introduction to the Theory of Complex Systems (2018)

Steven Schuster, The Art of Thinking in Systems (2018)

David Rousseau et al., *General Systemology* (2018)

Michael Jackson, Critical Systems Thinking and the Management of Complexity (2019)

Stuart Kaufmann, *A World Beyond Physics* (2019

Wassim Haddad, A Dynamical Systems Theory of Thermodynamics (2019).

Lucia Urbani Ulivi (ed.), *The Systemic Turn in Human and Natural Sciences* (Springer, 2019)

David Krakauer (eds.), *Worlds Hidden in Plain Sight* (2019)

Ted Carmichael et al. (eds.), *Complex Adaptive Systems* (2019)

David Feldman, Chaos and Dynamical Systems (2019)

Paul Davis et al. (eds.), *Social-Behavioral Modeling for Complex Systems* (2019)

Gary Metcalf et al., (eds.), *Handbook of Systems Sciences* (2021)

Albert Rutherford, The Elements of Thinking in Systems (2019)

János Kertész et al. (eds.), Computational Social Science and Complex Systems (2019)

Gurbir Bhullar and Amritbir Riar, *Long-Term Farm Systems Research* (2020)

James Ladyman and Karoline Wiesner, *What is a Complex System?* (2020)

Ray Ison and Ed Straw, *The Hidden Power of Systems Thinking* (2020).

Yoshiteru Nakamori, *Knowledge Construction Methodology* (2020)

Derek Cabrera et al. (eds.), *Routledge Handbook of Systems Thinking* (2021)

Search in the *International Encyclopedia of the Social & Behavioral Sciences* (Elsevier, 2015) for "system", "systemic" and "systems". The series *Understanding Complex systems* (2004ff), edited by Scott Kelso, focuses on the various applications of complexity; the series *Systems Science and Systems Engineering* (2012ff), edited by George Mobus, is highly acclaimed; and the series *Translational Systems Sciences* (2014ff), edited by Kyoichi Kijima and Hiroshi Deguchi, "aims at cultivating a new frontier of systems sciences". For quantitative and computational systems science, go to https://www.qcssc.uni-muenchen.de. Impressive lists of systems scientists, or engineers, and of systems science journals, or organizations, can be found on the Internet. The principal aim of *International Journal of General Systems* (1974ff) is "to promote original systems ideas (concepts, principles, methods, theoretical or experimental results etc.) that are broadly applicable to various kinds of systems. Typical subject areas covered by the journal include: uncertainty and randomness, fuzziness and imprecision, information, complexity, inductive and deductive reasoning about systems, learning, systems analysis and design, and theoretical as well as experimental knowledge regarding different types of systems". George Klir (1932–2016), who was a leading expert in mathematical systems theory, founded the journal and Radim Bělohlávek, specialist in fuzzy logic (which differs from classical, two-valued logic in that it rejects the principle of bivalence) is its current editor-in-chief.

Roots of Systems Science

The roots of systems science can be traced back to Thales, Anaximander, Anaximenes and Xenophanes, the first pre-Socratic philosophers (usually referred to as the Ionian or Milesian cosmologists), who lived on the western coast of present-day Turkey, in the period 625–475 B.C. Their inquiries spanned the workings of the natural world as well as human society. They sought explanations based on natural principles rather than on the actions of supernatural Gods. In the first volume of his massive *Kosmos*, Alexander von Humboldt (1769–1859) would deliver a general portrait of nature, suggesting that universal laws apply to the apparent chaos of the terrestrial world. However, long before Alexander the Great overthrew the Persian Empire (a truly historic action), the Ionian cosmologists were the first "Western" thinkers who introduced the notion of the world as a cosmos, an orderly arrangement of things that could be understood by rational inquiry. They replaced mythology with cosmology. Their revolutionary view was fatefully rejected by the Church Fathers, who fell back into the old (pre-pre-Socratic) habit of thinking and laid the theological foundations of Christianity.[17] The Ionian cosmologists understood the whole universe, including human society, as a system, an important fact that escaped the attention of Martin Heidegger, who focused his mind on the pre-Socratics.[18]

In *Metaphysics*, considered to be one of the greatest philosophical works of the Western world, Aristotle (384–322 B.C.) famously said:

> In the case of all things which have several parts and in which the totality is not, as it were, a mere heap, but the whole is something besides the parts, there is a cause. (Book 8, Chap. 6)

The Aristotelian worldview, i.e., the system of more or less Aristotelian beliefs, was the dominant collection of interconnected beliefs until about 1650 A.D., when it was made untenable by new discoveries and had to be replaced by a new worldview.[19]

Book III of Newton's *Mathematical Principles of Natural Philosophy* (1687), which laid the foundation of classical mechanics, is entitled *The System of the World*. At the end of it, in the *General Scholium*, the author, who caused what Thomas Kuhn (1922–1996) would call a paradigm shift,[20] writes:

> The most beautiful system of the sun, planets, and comets could only proceed from the counsel and dominion of an intelligent and powerful Being. And if the fixed stars are the centres of other like systems, these, being formed by the like wise counsel, must be all subject to the dominion of One; especially since the light of the fixed stars is of the same nature with the light of the sun, and from every system light passes into all the other systems: and lest the systems of the fixed stars should, by their gravity, fall on each other, he hath placed those systems at immense distances from one another. This Being governs all things, not as the soul of the world, but as Lord over all; and on account of his dominion he is wont to be called *Lord God παντοκράτωρ*, or *Universal Ruler*.

Sir Isaac Newton, whose body was laid to rest in Westminster Abbey and for whose epitaph Alexander Pope composed a beautiful, endlessly quoted couplet, was thus a believer, thinking that the order of things was imposed on the chaos by "somebody" outside of it. His fundamentally religious agenda still prevails, for even Albert Einstein, who allegedly replaced the Newtonian worldview,[21] could not accept the ideas of randomness or uncertainty in the universe, opining that "God does not play dice".[22] See Preface.

In and of itself, systems science began with Étienne Bonnot de Condillac's *Traité des systèmes* (1749), in which he argued for systems being studied scientifically.[23] A case can be made that Hegel (1770–1831) was a system thinker.[24] The existence, in principle, of a general theory of systems was first suggested by Alexander Bogdanov (1873–1928), a Belarusian physician, philosopher, science fiction writer and revolutionary who in his three-volume *Tektologiya: Vseobshchaya Organizatsionnaya Nauka* [tektology: the universal organizational science] envisaged a discipline that would unify all physical, biological and social sciences by considering them as sciences of systems, made up by interrelated elements. Premised on the monistic idea that the same laws work on each level of reality and all objects obey one system of laws of organization, he sought the general principles underlying every system and possibly ordering all sciences. Doing so, he was the first "modern" who developed an integrative, supradisciplinary (yet not supernatural) view and attempted to formulate the most general laws of organization. Fearful that Bogdanov wanted to replace Marxism, Lenin (1870–1924) attacked him mercilessly. Though Bogdanov has been deliberately obscured and distorted, his legacy is considerable. Talcott Parsons, Gabriel Almond, David Easton, Michael Crozier, Erhard Friedberg and many others drew on his work.[25]

In 1926, *Holism and Evolution* was published. The book, written by Jan Smut (1870–1950), the enigmatic South African statesman-philosopher who largely wrote the Preamble to the Charter of the United Nations,[26] tries to complete the evolution theory of Darwin, who gave no answer to the pressing question about evolution's direction. Smuts draws attention to its wholeness and concludes by writing that the whole interacts with the Darwinian forces, and that the immediate needs and present utilities are exceeded by the aim of the whole. The holistic vision of Smut, born out of the writings of Goethe (1749–1832) and Walt Whitman (1819–1892), was severely criticized by Karl Popper in his book *The Poverty of Historicism* (1957), but has been rehabilitated by Fritjof Capra's *Turning Point* (1982), *The Web of Life* (1996) and *The Hidden Connections* (2002). Though it threatens to be hijacked by the New Age Movement, the holistic view concurs with one of the main themes of systems scientists.[27]

In 1932, Ludwig von Bertalanffy, who probably was unfamiliar with the work of Bogdanov, made, in his groundbreaking book *Theoretische Biologie*, suggestions that are remarkably similar to the ideas of the almost forgotten Russian revolutionary.[28] Born and educated in Austria, however, Von Bertalanffy must have heard of Max Wertheimer (1880–1943), Kurt Koffka (1886–1941) and Wolfgang Köhler (1887–1967), psychologists who took a new road. Whereas William James originated the study of how the mind works, Sigmund Freud initiated psychoanalysis and John Watson founded the school of behaviorism, the *Gestalt* psychologists were primarily interested in perception. They believed that perceiving is a matter of seeing patterns. What we see is related to the background against which an object appears, the context in which a thing is perceived or the situation in which an observation is made. In other words, the whole is different from the sum of its parts, because the whole depends on the relationships between the parts. The *Gestalt* psychologists influenced social psychologists like Kurt Lewin, Fritz Heider and Solomon Asch.[29]

Discovering the general theory of systems was the grand ambition of the founders of the Society for the Advancement of General Systems Theory (1954). Three years later, the name was changed to Society for General Systems Research (SGSR), presided by Kenneth Boulding (1957–58), Charles McClelland (1959–61), William Ashby (1962–64), Anatol Rapaport (1965), Peter Caws (1966), John Milsum (1967), Milton Rubin (1968), Lawrence Slobodkin (1969), Bertram Gross (1970), Stafford Beer (1971), Margaret Mead (1972), James Miller (1973), Gordon Pask (1974), Kjell Samuelson (1975), Heinz von Foerster (1976), Geoffrey Vickers (1977), Richard Ericson (1978), Brian Gaines (1979), Robert Rosen (1980), George Klir (1981), John Warfield (1982), John Dillon (1985), Peter Checkland (1986) and Russell Ackoff (1987). In 1988, the name was changed again, to International Society for the Systems Sciences (ISSS), consecutively presided by Ilya Prigogine, West Churchman, Len Troncale, Howard Odum, Ian Mitroff, Harold Linstone, Donald de Raadt, Ervin László, Yong Pil Rhee, Gale Swanson, Béla Bánáthy, Peter Corning, Harold Nelson, Michael Jackson, Aleco Christakis, Kenneth Bailey, Enrique Herrscher, Debora Hammond, Kyochi Kijima, Gary Metcalf, Timothy Allen, Allenna Leonard, Jennifer Wilby, David Ing, Alexander László, Gerald Midgley, Ray Ison, John Kineman, Ockie Bosch and David Rousseau, each of them setting his/her mark on the organization. See David Rousseau et al., "In Search of General Systems Theory", *Systema*, 4:1 (2016), pp. 76–99. Go to https://www.systemsphilosophy.org/david-rousseau and http://www.systemology.org.

In 1969, Von Bertalanffy's masterpiece *General System Theory: Foundations, Development, Applications* was published.[30] In this book, he develops a general system theory (GST), attempting to provide an alternative to conventional models of organization. Foundational to GST are the interrelationships between the elements that form a system. It was Von Bertalanffy's reply to the madness of war (particularly the Second World War), in which each side thinks to be absolutely right and to be fighting for the just cause.[31] The dirty work of war, von Bertalanffy realized, is done by the "Flower of the Nation", young soldiers mobilized, brainwashed and sent to the front by superiors who, safely hiding behind the screen, manage to keep clear from the bloody battlefield. Though the word general does not appear in its name, ISSS's interest in the development of a GST continues through the work of its Special Integration Group (SIG), founded by Len Troncale, known for his grandiose project "unifying theories",[32] and currently chaired by David Rousseau, co-author of *General Systemology: Transdisciplinarity for Discovery, Insight and Innovation* (Springer, 2018) and "A Systematic Framework for Exploring Worldviews and Its Generalization as a Multi-Purpose Inquiry Framework", *Systems*, 6:3 (September 2018, online).

Unfortunately, systems science has become a hotchpotch, a jungle, a confusing and seemingly uncontrolled mass of concepts, methodologies and theories. There are various reasons for this deplorable situation. Some of them are:

- Insufficient awareness that systems are sets (collections of things), while sets are not necessarily systems (collections of connected things).
- Insufficient awareness that epistemic/cognitive issues are closely related to but definitely different from metaphysical/ontological subjects. These issues and subjects, in turn, are to be clearly distinguished but certainly not separated from topics linguists and philosophers of language are concerned with.
- Insufficient awareness that there are two concepts of order: a natural, spontaneous or endogenous order on the one side and a man-made, designed or engineered order on the other side.[33] These orders can be combined and, when combined, are almost inextricable (e.g., a company, or a country).
- Insufficient awareness that systems science is but one of the formal sciences, which in themselves form a system sorely neglected by systems scientists but worth to be investigated as part of the ever-growing system of sciences that consists of many branches and their applications.
- Insufficient awareness that the formal sciences (mathematics, logic, statistics, computer science, systems science) have something in common (namely the critical research of structures, or forms) and are to be distinguished from the not-formal sciences (physical, biological, human and social sciences). Systems scientists should thus be well informed about the latest developments in other formal sciences, without tacitly claiming to be an expert in any of the other fields of the formal sciences, or in any of the non-formal sciences. Writing about systems is one thing, but writing about systems as intimately related to other, connected systems and processes is quite another thing altogether.
- Insufficient awareness of the gulf between natural and cultural sciences; going from the study of physical systems to the study of biological systems is making a giant step, but going from the study of biological systems to the study of systems involving the immeasurable and unfathomable human mind is making a step that is even bigger.

The root cause of the chaos in systems science is perhaps what no formal scientist seems to realize: systems scientists, mathematicians logicians, statisticians or computer scientists are not fully informed of, or insufficiently familiar with, the work of other formal scientists. For example, a systems scientist who is really knowledgeable about, and thoroughly acquainted with, the work of his/her colleagues in mathematics, logic, statistics and computer science is hard to find, almost as hard as it is to find a concert pianist who also knows about the possibilities and limitations of playing the violin, the clarinet, the guitar and the harmonica. Like political scientists or economists who leave their home turf and trespass on the domain(s) of other scientists, systems scientists cannot resist dabbling. Talking about systems is one thing, but talking about systems *fundamentally* and in accordance with the insights of mathematicians, logicians, statisticians and computer scientists is a different cup of tea. Of course, systems scientists should primarily occupy themselves with the thing they are good at, namely the study of a particular kind of system, but they should also "take a crude look at the whole": the business of *formal* sciences.

In their interesting article, David Rousseau and his co-author Julie Billingham notice that "in the light of the importance of worldviews the study of a particular kind of systems, but they and its applications, it is surprising that systems science does not have a canonical model of the structure and dynamics of worldviews, and hence does not provide for a consistent way of working with worldviews across systems theories and methodologies". In the jungle of systems science, we find scientists who are focusing their mind on different kinds of systems: physical, chemical, biological, geographic, demographic, political, legal, military, economic, financial, management, cultural, linguistic, literary, library, medical, psychological, moral, educational, scientific, technological or environmental systems. Paying exclusively attention to one group of systems, they develop their own jargon, techniques and theories, forgetting all about abstracting away from the subject at hand. They see the trees but not the forest. Each one stays in his/her own bubble, i.e., compartment of a faculty; nobody is rising above the rest. "Worldview" means a view of the whole world (including a view of oneself), but such a comprehensive conception of reality is elusive. Therefore, it is not "surprising that systems science does not have a canonical model of the structure and dynamics of worldviews".

In order to explore the possibility of integrating worldviews, The Belgian philosopher and logician Leo Apostel (1925–1995), who advocated interdisciplinary research and bridging of the gap between natural and cultural sciences, brought together a number of researchers from various disciplines and established in 1990 the non-profit organization Worldviews Group. Deploring scientific fragmentation and the loss of overview, he dreamed about a *dialogue* among scientists from different quarters leading to a framework that would help us to cope with the complex reality and to act in a coherent manner. In 1995, the Center Leo Apostel (CLEA) was founded as an interdisciplinary research department affiliated with the Free University of Brussels (VUB). Professor Diederik Aerts, a theoretical physicist, is CLEA's present director. Few systems scientist have found their way to this unique center.[34]

Ulrich Libbrecht (1928–2017), a compatriot of Apostel, attempted to lay the foundation of comparative philosophy (aka cross-cultural philosophy), comparing the Western, Indian and Chinese way of thought. Unfortunately he paid no attention to systems science. His comparative model (based on "the paradigm-free axes of energy and information") is debatable, and his four-volume *magnum opus* appeared only in Dutch. An abbreviated English version, entitled *Within the Four Seas: Introduction to Comparative Philosophy*, was published by Peeters (Leuven, 2007). Visit https://www.iep.utm.edu/comparat and https://plato.stanford.edu/entries/comparphil-chiwes. See Joachim Lacrosse (ed.), *Philosophie comparée, Grèce, Inde, Chine* (Vrin, 2006), Robert Smid, *Methodologies of Comparative Philosophy* (SUNY Press, 2009), Stephen Angle (ed.), "Comparative Philosophy: Reviewing the State of the Art", 2016 (online) and the open access journal *Comparative Philosophy: An International Journal of Constructive Engagement of Distinct Approaches toward World Philosophy* (2010ff). The Kennesaw School of Comparative Philosophy at Indiana University (Bloomington, IN) was launched in 2011 and "aspires to avoid hegemonic thinking—where one culture is displaced by another, more dominant one—by creating a thinking space where cultures can meaningfully interact without losing their particular cultural and historical integrity". The Center for Comparative Philosophy at Duke University (Durham, NC) was inaugurated in 2014 and is dedicated to "teaching and research into the philosophies that animate different traditions". May the Han Chinese, when dealing with the Tibetans, learn something from this!

Branches of Systems Science

(a) **Complex systems science**

After decades of intensive research across the world, there is still no consensus on what makes a system complex.[35] According to the prestigious Santa Fe Institute (SFI), "complexity arises in any system in which many agents interact and adapt to one another and their environments". The nervous system, the Internet, ecosystems, economies, cities and civilizations are given as examples, but precisely how many agents are interacting and adapting is not told. A married couple walking hand in hand may not seem to be a complex system, like a rainforest, tree, cell, city, university, hospital, company, human being or country, but the walking couple certainly is, if one realizes that (*a*) countless interacting and adapting agents are involved and (b) the personal history of each partner differs.

In his monumental study, *L'origine des systèmes familiaux* (Gallimard, 2011), Emmanuel Todd describes and maps four kinds of family systems which have been decisive for the political, ideological, economic and religious developments in and far beyond Europe. No family system is, I learn from Todd, a complex system, because it is constituted by only a relatively small number of persons. Yet each family system is actually and essentially a complex system, considering not only the involvement of many human factors operative within the family system but also of the environment (*Mitwelt* and *Umwelt*), which the family had/has many exchange relationships with.[36]

Measuring the complexity of a complex system is a problem Andrey Kolmogorov, Ray Solomonoff, Gregory Chaitin, Charles Bennett, Paul Vitányi, Ming Li and many other computer scientists have attempted to solve.[37] A variable, we saw in Chap. 2, is something that varies. The degree of the variation can sometimes be measured. For example, the heights of the plants in the garden vary from 10 to 25 cm; since the end of last year, the price of the product has, in shop *x*, increased 15%; on average, the city's police arrested 300 people each day; the machine is designed to produce rods 2 cm long with a standard deviation of 0.02 cm. The measure of a complex system's complexity, however, cannot be given, because the number of the interacting agents is unknown. The number of the parts and the degree of the complexity of a whole are interrelated.

Nonlinearity is the defining, essential property of a complex system. Growth is not a matter of linear change, happening in a series of small amounts, incrementally. It is a nonlinear phenomenon. Alwyn Scott makes this abundantly clear in his book (*op.cit.* pp. 181–301). The opposite or inverse of growth is not a becoming less or decreasing in terms of what may be measured. It is decay, whereas growth is a matter of integration, decay, or senescence, is a matter of disintegration, of falling apart. Growth (e.g., economic growth) happens *non*linearly and the rate of its change cannot be measured. Surely, the increase, or decrease, of a population can be measured by comparing the number of its members at two different points in time, but neither the change nor the rate of change of a human being that belongs to the population can be measured. An individual person's feeling of well-being may not be worse but better at lesser growth of the gross national product (GNP), which is supposed to be calculable—"supposed", because the economic value of services is very hard to be calculated. The relative position of the points of a flying plane can be exactly indicated, but human beings are not planes. Treating them like such objects would violate their individuality. Many textbooks can thus be thrown into the dustbin. Our society differs from the one that our parents lived in, and their society differed from the one in still older days. Society definitely changes, but the favorite study subjects of historical sociologists cannot really be measured. I agree with Václav Smil, who in his latest book, *Growth: From Microorganisms to Megacities* (MIT Press, 2019, Preface) writes that, because of its ubiquity, growth can be studied on many levels, but I disagree with him if he means to say that growth can always and everywhere be measured. Comparative studies may be carried out, but only *partially* so, provided one realizes that (*a*) parts inextricably belong to a whole and (*b*) a society can be distinguished but shall not be separated or decoupled from a country's geography, demography, economy, polity or ecology. I am holding the same reservation in regard to the work of Martin Scheffer, an ecologist and mathematical biologist associated with Wageningen University & Research (WUR). His mind is focused on the stability, resilience and "tipping points" of complex systems ranging from ecosystems and the climate to brains and societies (https://www.sparcs-center.org). The changes of physical objects may be measurable, but measuring the growth of an organism by, e.g., its length or size would be doing deficiency to what has been really happening with it. Man can travel to the moon, but he/she is unable to comprehend the growth of a blade of grass. Go to https://doi.org/10.1038/s41567-021-01335-8.

The science of complex systems can be traced back to the late nineteenth century, when Henri Poincaré (1854–1912) struggled with the "three-body problem" in *Les méthodes nouvelles de la mécanique celeste* (1892), and the Newtonian paradigm began to break down. According to Newton's second law, the rate of change of a body's momentum is directly proportional to the force applied. Poincaré discovered that this law does not hold for the movement of three or more bodies. The modern theory of dynamical systems was born and with it the concepts of disproportionality or nonlinearity.[38] The phenomenon of nonlinearity is not restricted to the domains of physicists and chemists. In *The Nonlinear Universe: Chaos, Emergence, Life* (Springer, 2007), Alwyn Scott beautifully describes that complex systems are ubiquitous.[39]

Although the roots of the science of complex systems go back to the nineteenth century, scientific publications related to them began to proliferate in the 1970s, when the microcomputer, said to capable of coping with nonlinear equations, became popular.[40] One of the factors associated with this sudden growth was the founding of research institutes like the Center for Studies of Nonlinear Dynamics (La Jolla), the Institute for Nonlinear Science (Santa Cruz), the Center for Nonlinear Studies (Los Alamos), the Institute for Nonlinear Science (San Diego) and the Santa Fe Institute. Cybernetics, mathematical biology (Nicholas Rashevsky), general systems theory (Ludwig von Bertalanffy), non-equilibrium statistical mechanics (Ilya Prigogine), catastrophe theory (René Thom), synergetics (Hermann Haken) and the study of complex adaptive systems (Murray Gell-Mann) overlap each other considerably and can be considered precursors or components of the nonlinear-science revolution of the 1970s. SFI advocates itself as the "unique scientific enterprise attempting to uncover the mechanisms that underlie the deep simplicity in our complex world", but there can be no question that the "unique" institute (at short distance from Los Alamos National Laboratory) is developing ideas that were put forward by Bogdanov, Rashevsky, Von Bertalanffy, Prigogine and Thom already. In addition, the leadership of SFI seems to be unaware of the hotchpotch systems science (hence complex systems science) has become (see above). An article of mine clarifying this deplorable situation was flatly rejected, because "SFI doesn't accept pitches or submissions from non-affiliated authors", in my view, a perfect recipe for scientific inbreeding.

Toward the end of his widely acclaimed book, *Thinking in Complexity: The Computational Dynamics of Matter, Mind, and Mankind* (Springer, 2007), Klaus Mainzer writes:

> One of the most essential insights of this book is that the dynamics of nature and society are not only characterized by nonlinearity and chaos, but by randomness, too. Only in randomness can human free will have a real chance. In the completely deterministic and computational world of a mechanical nature, Kant had to postulate a transcendental world in order to make free will, ethical duties, and responsibility possible. In random states of nature and society, the behavior of a system is not determined in any way. … It is amazing that human free will seems to be supported just by the mathematical theory of computability. (pp. 437–438)

It would probably be asking the impossible from a German professor to question the main tenet of Kant's philosophy, but he may agree with me that the existence of "human free will" has not yet been proven. Kant only *postulated* a transcendental world; he suggested but did not prove its existence. Clearly, freedom is to be understood as a human construct. If free is defined as unbounded or unlimited, then no man can claim or said to be really free, for he is constrained by natural and man-made laws. Having a moral duty to do (or not to do) something means having a duty, being duty-*bound*. Kant's categorical imperative (“*Handle nur nach derjenigen Maxime, durch die du zugleich wollen kannst, dass sie ein allgemeines Gesetz werde*“) may not be well-founded, for it would result in awkward situations when followed by a masochist. Moral duties or obligations are not self-imposed but largely, if not completely, socially and culturally inculcated. Even Robinson Crusoe was not a free man. The predicate “is free” is said of a human being, his/her mind, or his/her will. Will cannot be equated with mind, and mind cannot be equated with human being. The ongoing discussion about freedom tends to spill over into legal and political discourses. Man may not be wholly accountable for his actions and soldiers should think twice before risking their life when ordered (by people who know how to avoid the hand-to-hand combat) to fight until death for king/queen and country.[41]

Nonlinearity explains the emergent properties of a complex system, i.e., the properties that are not trivially related to the qualities or traits of its parts. For example, the complex interactions between a man and a woman “produce” a child, the characteristics or emergent properties of which are irreducible to the character of either his/her biological father or mother. A child is not a replica. Humans are not clones, each one is different. Countries are complex systems and *ipso facto* unpredictable. Their characteristics or emergent properties shall and can be investigated, neither by an individual researcher nor by a group of researchers working independently from each other, but by collaborating specialists from different disciplines. The geology, geography, demography, economy, polity, society and ecology of a country are interconnected domains. Separate studies, carried out by scientists who hardly know each other, will be useless exercises, a squandering of talent and a waste of money. The complexity or architecture (unifying and coherent form) of complex systems such as human beings, rainforests, trees, cities, families, companies and countries is what Paul Valéry (1871–1945) defined as “*une intelligible imprévisibilité essentielle*”. Complexities are both organized and, recursively, organizing.

Though complexity scientists are mainly concerned with the behavior of nonlinear systems consisting of physical, chemical, biological and ecological elements, they are increasingly, dealing with mental issues as well. This makes their research highly relevant to the rapidly expanding field of cognitive science.[42] The book *Intercorporeality: Emerging Socialities in Interaction* (OUP, 2017) offers a “new perspective, conceiving of the human body as interacting with other bodies and as open to, and engaging with, both the natural and cultural world”. Its authors, Christian Meyer, Jürgen Streeck and Scott Jordan, present an “integrating framework for research on multimodal interaction in the context of natural and cultural developments”. In *Social-Behavioral Modeling for Complex Systems* (Wiley, 2019), a book intended to be “a milestone document that sketches a research agenda for the next decade”, its editors (Paul Davis et al.) outline “the frontiers of social-behavioral modeling, identifying the various challenges to be met and drawing on the work of physicists, neuroscientists, psychologists, sociologists, economists, political scientists, anthropologists, health professionals, communication specialists, computer scientists and complexity scientists”.

Potentially, cognitive semiotics is a fruitful field of research. The biennial conference series, organized, since 2014, by the International Association for Cognitive Semiotics (IACS), seeks to bring together "experts in fields such as cognitive science, semiotics, anthropology, linguistics, psychology, communication science, media studies, content analysis and visual analysis". The fourth conference, in Aachen, was scheduled for July 2020 (postponed to July 2022) and its theme is going to be: Semiotic Complexities. IACS4 aims "to account for the complexities of semiosis, that is, for the intricacies of semiotic processes in a broad range of semiotic systems, with a focus on either cognitive, mono-modal, cross-modal or multi-modal processes of meaning construal". It is intended "to encourage contributions showing how theoretical frameworks and methods of analysis may mutually enhance each other and thus offer truly new insights into the merits of semiotic inquiry and advance the cross-disciplinary enterprise of cognitive semiotics". By "theoretical frameworks" is meant the inclusion of influential semiotic theories (as advanced by Umberto Eco, Algirdas Greimas, Roman Jakobson, Yuri Lothman, Charles Peirce, Ferdinand de Saussure, Thomas Sebeok, John Deely, Sandra Moriarty, Marcel Danesi and Göran Sonesson), as well as theoretical concepts and principles that are central to the interconnected cognitive sciences linguistics., psychology, anthropology, artificial intelligence, neuroscience and philosophy. By "methods of analysis" is meant the encompassment of (*a*) qualitative, quantitative and experimental methods and tools, (*b*) mixed-methods approaches developed to derive emerging patterns and regularities and (*c*) unique instances of meaning-making in both static and dynamic semiotic contexts.[43]

The *Encyclopedia of Complexity and Systems Science*, chief-edited by Robert Meyers and published by Springer, in 2009, may be the best starting point for studying complex systems. Meant to be read by a wide audience, the *Fundgrube* "provides an authoritative single source for understanding and applying the concepts of complexity theory together with the tools and measures for analyzing complex systems in all fields of science and engineering". This monumental work, the structural similarities of which can be exploited to transfer analytical methods and understanding from one field to another, is divided into 35 topical sections, each of which is edited by a world-leading expert in the field.

Complex systems scientists approach, from a unifying point of view, a large body of objects belonging to domains which traditionally fell within the purview of different scientists. They are thus at the crossroad of different disciplines. For many years, the idea prevailed that complexity is tantamount to incomplete information because of the multitude of variables to be dealt with. It is now realized that the study of complexity is essentially interdisciplinary, drawing heavily upon the knowledge gained in various disciplines and stimulating the cross-fertilization of theories and experience. The uniqueness of complex systems science is its potential to forge bridges between sciences. This is an absolute requirement for comprehending the complexity of countries, arguably the most complex and dynamic systems that exist, even more so than cities (see Chap. 4). However, modeling complex systems, scientists usually work with a group of "domain experts"; they are not trying to capture all knowledge, but only that group's view of the domain. The reason for this is that those experts are the ones with detailed hypotheses and research questions. Complex systems scientists deliberately avoid trying to add to the knowledge of experts, so that they understand exactly what the model and simulation include. In that way, their results can be directly compared with the experimental or theoretical evidence used by the "domain experts".

In "Complexity according to Peirce", Jaime Nubiola, professor of philosophy of language and methodology at the University of Navarra (Spain), explains that complexity is the core of Charles Sanders Peirce's philosophy. The main lesson to be learned from the man whom Daniel Everett called "the American Aristotle" is, according to Nubiola, "to listen to each other, in spite of different backgrounds and specialties".[44]

(b) System dynamics

System dynamics (SD) is an approach to understanding the nonlinearity of complex systems over time, using stocks, flows and feedback loops (see below, under cybernetics). It reveals the underlying dynamics of organizational, societal and global systems through computer modeling. Jay Forrester (1918–2016) is founder of the System Dynamics Society and author of *World Dynamics* (Wright-Allen, 1971), the book that served as basis for the World3 model used by Donella Meadows and Dennis Meadows in their popular 1972 book *The Limits to Growth*. Forrester, who had invented the "coincident core memory" enabling 3D storage of information, summed up his understanding of SD in the following elevator pitch (statement short enough to be made in an elevator ride):

> System Dynamics deals with how things change through time, which includes most of what most people find important. It uses computer simulation to absorb the knowledge we already have about details in the world around us and to show why our social and physical systems behave the way they do. System dynamics demonstrates how most of our own decision-making policies are the cause of the problems that we usually blame on others, and how to identify policies we can follow to improve our situation.

A system dynamic approach typically proceeds in five steps:

(1) *Formulating questions*: The questions that would define the scope of a project are formulated. It is of strategic importance to know what is and what is not to be modeled.
(2) *Understanding the business*: The understanding of the business' concrete situation is deepened and all the stakeholders that will be affected by the model are identified. A sketch is made of the dynamics of the business, using a causal loop diagram. This may require the refinement of stage one.
(3) *Simulating the business*: A detailed stock and flow model is built and scenarios to be investigated in order to answer the strategic questions are described. Based on these scenarios, the data needed to test the model are identified and gathered.
(4) *Evaluating scenarios*: Once the model is considered powerful enough to answer the strategic questions, detailed investigation of the future scenarios is started.

(5) *Making recommendations*: Findings are used and recommendations concerning future development of the business are made.

System dynamics[46] is to be distinguished from dynamical systems theory, a kind of mathematics for describing the behavior of dynamical systems by means of differential or difference equations. In dynamical systems theory, the variables are dated. When differential equations are used, the theory is about continuous dynamical systems. When difference equations are used, the theory concerns discrete dynamical systems. The words continuous and discrete mean without and with pause or interruption respectively, that is to say, in the former case, the variables change at each point of time, and in the latter case, the variables change only once within a period of time. The continuous case can be regarded as the limit of the discrete case. Dynamical systems modeling is thus a kind of modeling used to represent the dynamics of a system, the process which a system goes through, the diasynchronicity of a system. It focuses on *how* the system changes over time, on the course of a *Strukturwandlung*. "Dynamical" means that time is incorporated as an essential feature of the model; "system" indicates that research questions are conceptualized as involving the parts or members of a whole; and "modeling" suggests that the dynamical system is represented by mathematical equations.[47] *Strukturwandlung* must be distinguished, but shall not be separated, from *Wandlungsstruktur*.

The discussion may be limited to a model with respect to one differential (or difference) equation only. Sometimes, however, the dynamical quality of a system requires simultaneous determination of the time path of all the variables involved. This would imply simultaneous differential (or difference) equations, which may be first-order or higher-order equations. All these equations can be linear or nonlinear. The assumption of linearity means losing or neglecting a great deal of realism. Furthermore, the use of constant coefficients for the sake of mathematical manageability will "freeze" the environment of the problem to be investigated. The endogenous adjustment of the model will then be studied in a sort of vacuum. Loss of realism will be the result, for realistic systems are open and function within a constantly changing environment. Insistence on the use of constant coefficients would thus mean that really existing connections are cut through. The study of nonlinear *stochastic* systems has led to better understanding of the behavior of complex, dynamical systems. Chaos theory focuses on the behavior of dynamical systems that are sensitive to initial conditions. Henri Poincaré, Edward Lorenz, Stephen Smale, David Ruelle, Mitchell Feigenbaum, Doyne Farmer, James Crutchfield, Alfred Hübler and Steven Strogatz made important contributions to the field.[48]

(c) Network science

Network is one of the metaphors we live by,[49] and the concept of community implies nothing but connectivity. Confucius (551–479 BC) was a sociologist or social network thinker *avant la lettre*. He emphasized the importance of *wŭ lún*, the "five human relationships": between father and son, between ruler and minister, between husband and wife, between elder and younger brother and between friends. Confucius regarded the establishment and maintenance of correct human relationships as paramount to keeping social and political order. Asked about the core concept of his thinking, he replied: *shù* (reciprocity). Though the teachings of the Master have been the subject of numerous studies[50] and are the cause of widespread misunderstandings, *guānxìxué* (the study of connections) is today a flowering branch of research within and outside of China.[51] In *Ways of Thinking of Eastern People* (The University Press of Hawaii, 1964, pp. 407–530), Hajime Nakamura gives a detailed analysis of the principal characteristic of the Japanese way of life, of the high esteem of a particular social nexus influenced by Confucius.

Anthropologists like Franz Boas, Marcel Mauss, Alfred Radcliff-Brown, Bronisław Malinowski, Edward Evans-Pritchard, Claude Lévi-Strauss, Edmund Leach, Louis Dumont, Mary Douglas, Eric Wolf, Clifford Geertz, Marvin Harris and Marshall Sahlins studied how people in different cultures lived together. Social psychologists deal largely with intra- or intergroup structures and processes. *Kulturwissenschaftler* are concerned with culture, which consists of individuals carried by a collective.[52] Western cognitive scientists, finally, are discovering that man is deeply embedded in, and closely related to, his *Mitwelt* and *Umwelt*,[53] that is to say, in/to the environment which systems scientists are particularly interested in. Indeed Western cognitive scientists are discovering that the individualistic thinking on which liberalism has been predicated is fundamentally flawed.[54]

Network *science* started with the research and publications of Steven Strogatz (an American mathematician), Albert-László Barabási (a Romanian-born Hungarian-American physicist), Duncan Watts (an American sociologist) and Mark Newman (a British physicist). It needs to be stressed that this "gang of four" became active *after* Ludwig von Bertalanffy (an Austrian biologist) had advanced his revolutionary ideas. Network science has exploded into a big industry. The number of publications on the subject runs into tens of thousands and there are dozens of journals about it. In a paper entitled "The network takeover", Barabási, who is also a good businessman, once said:

> Data-based mathematical models of complex systems are offering a fresh perspective, rapidly developing into a new discipline: network science. ... It is making a real impact from science to industry. .... Therefore, if we are ever to have a theory of complexity, it will sit on the shoulders of network theory. (*Nature Physics*, 8 [January 2012], pp. 14–16)[55]

A bold statement indeed, for network science is still in its infancy. It may work well in the field of natural sciences (visit https://doi.org/10.1098/rstb.2019.0314), but in the social and human sciences domain it has still a long way to go. Stephen Wolfram, a British-American computer scientist, physicist and, like Barabási, a keen businessman, suggests he has found a path to the fundamental theory of physics (visit https://arxiv.org/abs/2004.08210 and https://www.wolframphysics.org), but I wonder whether that path also leads to the fundamental theory of the human mind, for example, the mind of Stephen Wolfram himself.

In a letter addressed to him, I raised this question. Dr. Wolfram's Physics Project Team responded on his behalf by referring me to some paragraphs of his (controversial) book, *A New Kind of Science* (2002). Unsatisfied, I thereupon asked whether, according to the team, there are things that can *not* be counted, quantified or computed, and, if they thought there are such things, how they could possibly know or model them. The question has never been answered satisfactorily, for Dr. Wolfram's modest reply, passed on by his team, was: "My theory is intended to be an underlying theory of the whole universe, in perfect detail. If one could run the model long enough, then it is intended to reproduce everything about the universe, including the writing of this answer. The amount of computation required to do this would be immense—and the phenomenon of computational irreducibility implies that there cannot in general be shortcuts".

However, how long is "long enough"? A theory "intended" to be a theory of the "whole universe, in perfect detail" is not yet a theory of the whole universe, in perfect detail, and a model "intended to reproduce everything about the universe" is not yet a model reproducing everything about the universe. It seems to me that whoever claims to be able to identify, delineate or model the human mind forgets about his/her own mind, which is also part of reality (the universe). Perhaps the solution to this maddening problem can be found in the *Upanishads*. Whether man will ever be able to construct a system that can successfully emulate human thinking remains to be seen. Man is essentially more than a machine. Each computer, no matter how sophisticated, needs somebody to write a program for it, even somebody to switch the computer on. When no computer is capable of tying my shoelaces, or folding a necktie (not to speak of growing grass, bearing fruits, or creating a piece of art), Dr. Wolfram and his team might be a bit too optimistic.

Compared to natural events, individual *actions* are particularly complex and more difficult to understand. *Social* actions are particularly complex and even more difficult to understand. When more than two individuals are involved, understanding social actions by means of the tools used in the natural sciences becomes quickly impossible. In comparison, the notorious three-body problem pales. "The real goal of knowledge is to become knowledge of a community, not the knowledge of one or two persons", Jacob Lurie, winner of the 2015 Breakthrough Prize in Mathematics, said, apparently ignorant of Plato's problem, as unfolded in his dialogue *Parmenides* (see Sandra Peterson, "The greatest difficulty for Plato's Theory of Forms", *Archiv für Geschichte der Philosophie*, 63:1, January 1981, pp. 1–16). I do not know whether this points towards the insight gained by the authors of the *Upanishads*, but I hope the problem identified by Lurie will be tackled at the 9th International Conference on Complex Networks and Applications, scheduled for December 2020 and to be organized in Madrid (https://complexnetworks.org). Go also to https://arxiv.org/pdf/2009.13620v2.pdf.

Practitioners of econometrics, a composite discipline virtually founded by Jan Tinbergen and Ragnar Frisch, make a fundamental mistake in assuming that natural laws are ruling economies, a mistake overlooked by Roy Epstein in his book *A History of Econometrics* (North Holland, 1987). The same goes for econophysicists, who apply theories and methods developed by physicists in order to solve economic problems. Whether Erwin Dekker in his forthcoming biography of Tinbergen will notice the flaw remains to be seen. The basic problem is: Can the knowledge of physicists be applied to the human and social domain or should they learn something from the human and social scientists?

Scientists in the field of "multiagent systems", "agent-based modeling" or "multi-agent artificial intelligence systems" seemingly base their impressive theories on the premise that the human mind is reducible to the human brain (matter). I am not so sure, as the mind cannot be counted, weighed or quantified, precisely the activities natural scientists rely on. Albert Einstein reputedly said: "Not everything that counts can be counted". I take the human mind, which I think cannot be perceived, to be something that really counts, but cannot be counted. Natural scientists could learn a thing or two from cultural scientists. Read John Pavlus's online article "Machines beat humans on reading test, but do they understand?" (*Quantamagazine*, October 17, 2019).

Human beings are not like ping pong balls in a lottery mill or swirling molecules in a vat. Humans are unique, not only relative to other living things (plants, other animals, etc.) but also relative to each other. They communicate with each other in a great variety of ways. Everybody has two inseparable faces, a physiological and a psychological one. Though we have many things in common, nobody is the same as, or a clone of, somebody else. The differences between people are not restricted to their fingerprint or teeth mark. Depending on his/her personality (which is formed by countless factors) and unpredictable situation, everyone takes a particular route. Not only the length of this way varies (people die at different ages under different circumstances), the where, the when and the how of a person's living vary as well. From A to B, a man or a woman can be accompanying, ambling, ascending, bouncing, bumping, catwalking, changing, chugging, climbing, crawling, creeping, crossing, cruising, cycling, dangling, descending, deviating, diving, dribbling, drifting, driving, entering, exiting, falling, flashing, floating, flying, following, gliding, going, hiking, hitchhiking, hobbling, hopping, hovering, jay-walking, jogging, jumping, leaping, limping, marching, motoring, moving, navigating, pacing, paddling, parading, passing, plunging, plying, pouncing, promenading, racing, riding, rising, rolling, ranging, rocking, rotating, rowing, running, rushing, sailing, sauntering, scampering, scooting, scuba diving, scurrying, shifting, shooting, shoving up, shuffling, sinking, skating, skidding, skimming, skipping, sleepwalking, sliding, slipping, slogging, sneaking, soaring, spinning, springing, sprinting, staggering, stepping, street-walking, striding, strolling, stumbling, surfing, swaying, swerving, swimming, swinging, swirling, switching, swooping, teetering, thundering, tottering, tramping, transiting, travelling, tripping, trotting, trudging, tumbling, turning, twirling, veering, waddling, wading, wagging, walking, wavering, whipping, whirling, wobbling, wriggling, zigzagging or zooming. He/she will be doing so in a certain period of time, in a certain space/country and under conditions made up by actors or factors too numerous to list. As movements of humans are almost never in a straight line and at a constant speed, I am not quite sure whether network scientists are, or will soon be, able to map out the different kinds of human movement and to come forward with a satisfactory, tested theory applicable to any human society, that is to say, with a theory also taking into account the different modes of transportation and the unbelievable intricacies of communication.

In "The hidden universality of movements" (https://arxiv.org/pdf/2002.06070.pdf), Markus Schläpfer, Michael Szell, Hadrien Salat, Carlo Ratti and Geoffrey West develop a general theory that "predicts that the spatio-temporal structure of movement in cities follows remarkably simple universal rules that are well supported by data from across the globe". Well supported by data across the globe? In contrast to what Thomas Hobbes, Julien de la Mettrie, Ludwig Büchner, Paul and Patricia Churchland, and Dick Swaab want us to believe, man, whose life is inextricably "*vernetzt und verknotet*" with his *Mitwelt* and *Umwelt*, is not simply matter-in-motion. Man is said to be "teetering on a knife's edge"; his/her behavior is rarely if ever predictable. Complexity and network scientists should take a leaf out of Sandra Mitchell, *Unsimple Truths: Science, Complexity, and Policy* (University of Chicago Press, 2009).

Studying (a) society is the daunting task of social scientists and philosophers. Suffice it to browse the table of contents of the *International Encyclopedia of the Social & Behavioral Sciences* (Elsevier, 2015), to visit http://encyclo-philo.fr/ontologie-sociale-gp and to explore the Web site of the American Sociological Association (ASA) and the Cambridge Social Ontology Group (CSOG). The study of societies, or countries, becomes particularly complex when man's *Umwelt* (environment) is taken into account (https://plato.stanford.edu/entries/ecology).

Long ago, as freshman at Leiden University, I had the rare opportunity to listen to a world-famous expert in Zen Buddhism who gave a semi-public lecture on *satori* (悟り), which is—according to Daitetsu Suzuki—the Japanese Buddhist term for comprehending reality. At Q&A time, I was bold enough to ask the speaker what possibly could remain of my capability to comprehend reality when I achieved the *satori* stage of becoming one with the One, like a ripple disappearing in the ocean. The question left the speaker (and the attending professors) dumbfounded but I would never get a satisfactory answer from anybody. Many years later, I happened to find the English translation of the *Upanishads* by Patrick Olivelle, a famous Sri Lankan Sanskritist. Having read this wonderful book from cover to cover, I couldn't but conclude that the thinkers in ancient India (like their Chinese counterparts) conceived of mind as superior to matter. In contrast to it, reading *I Am a Strange Loop*, the extolled sequel to Douglas Hofstadter's bestseller *Gödel, Escher and Bach: An Eternal Golden Braid* (Basic Books, 1979), I had to conclude that the American author conceived of matter as superior to mind.[56] However, I found the ambiguity of the first word in the title of the book a serious problem. If "I" equals the French "je", Hofstadter would be making only a subjective statement, not allowing him to speak for anybody or everybody else. However, if "I" equals the French "*moi*", he would be stating the obvious, because natural scientists (in particular the neuroscientists) have long shown the brain to be an extremely complicated *object*. In short, Hofstadter seems to be mixing up the first- and third-person perspective. Unfortunately, my efforts to receive some clarification from him electronically have been futile.

The great mystery, with which the European symbolists (among others) wrestled, is that I am a part of nature and apart from it. Without body/matter my mind would not exist (I presume), but—and here I disagree with the materialists or physicalists—my mind cannot be completely reduced to my body. The great variety of human beings cannot be explained by only pointing at the anatomy or physiology of their body/brain. Mind cannot be equated with matter; mind implies something more than matter. See Paul Valéry, *Monsieur Teste* (Gallimard, 1926). Ergo, a kind of dualism (Descartes?) is unescapable. In the words of Arthur Rimbaud (1854–1891), who attempted to eliminate his senses in order to reach the absolute: "*Je est un autre*" (I is another). Visit https://plato.stanford.edu/entries/self-reference, https://plato.stanford.edu/entries/self-consciousness-phenomenological, http://encyclo-philo.fr/subjectivite, http://encyclo-philo.fr/philosophie-de-lesprit-a and https://doi.org/10.1080/05568641.2018.1450160. The key question is: what precisely is the nature of, and difference between, matter and mind?

Charles Rathkopf, a philosopher at Iona College, in New Rochelle (NY), claims that network science exploits the very properties that make a system complex. Rather than stripping them away, it would bring these properties to the fore, using them to supply new forms of explanation. Network science would be approaching complexity head-on. In *Temporal Network Theory* (Springer, 2019), its co-editors, Petter Holme and Jari Sarimäki, maintain that "network theory has, over the past 20 years, proven to be one of the most powerful tools for the study and analysis of complex systems" and in "Nonlinearity + Networks: A 2020 Vision", a recent article uploaded by Mason Porter, it is claimed that "through its exciting combination of graph theory, dynamical systems, statistical mechanics, probability, linear algebra, scientific computation, data analysis and many other subjects, and through a comparable diversity of applications across the sciences, engineering, and the humanities, the mathematics and science of networks has plenty to offer researchers for many years".[57] Whether these wild claims can be substantiated remains to be seen. I am rather under the impression that network scientists have not been able to solve the riddle of self-reference and to deal successfully with the phenomenon of nonlinearity, so prevalent among people. The number of connections of a node with other nodes is called its degree, and a scale-free network is a network the degree distribution of which satisfies a power law, whereby quantity y varies as a power of quantity x, thus $y = ax^k$. Mathematicians profoundly disagree on whether scale-free networks exist universally. Visit https://arxiv.org/pdf/1811.02071v1.pdf.

In a recent collection "bringing together selected research, comments and review articles on how networks are structured, how networks can describe healthy and disordered systems, how dynamics unfold on networks, and about community structures and resilience in networks", the editors of the prestigious journal *Nature* boldly claim that "network science is now a mature research field, giving mathematical descriptions of systems containing many interacting components, including power grids, neuronal networks and ecosystems" (December 19, 2019). I venture to say that network science is *not yet* a mature research field; its practitioners have still a long way to go.

At the end of her brilliant book, *Complexity: A Guided Tour* (OUP, 2009), Melanie Mitchell, student of Douglas Hofstadter, muses about the future of the study of complexity, writing that the waiting is for another Carnot, who originated some of the key concepts of thermodynamics. "Can we invent the calculus of complexity, the mathematical language that captures the origins and dynamics of self-organization, emergent properties, and adaptation in complex systems", she wonders, mentioning the names of scientists who are *attempting* to accomplish this heroic task. Remarkably, the name of no network scientist appears on her short list. "What's needed is the ability to see the deep relationships between information, computation, dynamics and evolution", Mitchell laments, concluding her reflections with the translation of a beautiful phrase taken from André Gide's novel *Les Faux-monnayeurs* (1925):

"*On ne découvre pas de terre nouvelle sans consentir à perdre de vue, d'abord et longtemps, tout rivage*". She could also have recalled the wisdom of thinkers in Ancient China who, neglected by natural scientists in the West, had written about 道 (*dào* [way, process]) and 理 (*lǐ* [structure, pattern]).

Network scientists study complex networks such as physical, chemical, neural, biological, ecological, economic, commercial, financial, political, legal, military, medical, educational, social and urban networks. They are also concerned with computer, telecommunication, transportation and traffic networks, drawing heavily on statistical mechanics, graph theory and data mining. Statistical mechanics is one of the pillars of modern physics. It is foundational to econophysics and its history goes back to the eighteenth century (Daniel Bernoulli). Barabási, leading expert in network science, earned his Ph.D. under the direction of Gene Stanley, who made important contributions to statistical mechanics and in 1995 coined the word econophysics. However, the approach of statistical mechanics is based on statistical methods. They are strong in averaging out, which is a rather dangerous thing to do when dealing with human beings, who are more than bunches of molecules. Head in the freezer, feet in the oven, and on average the body is on the right temperature, right? The length, age or income of two people can be averaged out, but not their character, personality or experience. The standard deviation of a group of people, being the number that shows the amount by which members of the group are different from the mean value for the group, may be calculated and have interesting mathematical properties, but what is value? Economists strongly disagree and there is no consensus in axiology (a branch of philosophy concerned with the nature of value and valuation, and with the kinds of things that are valuable).[59] In addition, what is the mean value of a group of *people who are different*? If the king is unique, so am I.

Leaving the study of society to social physicists or the study of economy to econophysics, as experts in statistical mechanics seem to advocate,[60] would be a terrible impoverishment. Graph theory is a sub-branch of mathematics studying primarily the *connections* between things, but not the things themselves. The point is important, particularly when the dots or vertices stand for human beings, the study of whom still lacks a grand, unifying theory[61] and about whom the great works of literature and the great movies narrate. Graph theory is thus mainly concerned with the study of relations, not with the *relata*. Pioneered by Leonhard Euler (1707–1783), it grew out of a fusion of algebra and topology (algebraic topology), the combination of which had also resulted in differential and integral calculus. Network science is just one of the application fields of graph theory, which is interrelated with group theory (see Chap. 2). Whether network science can be—lock, stock and barrel—transposed from the natural to the social sciences, is highly questionable.[62] Great caution has to be exercised. Human relationships differ significantly from biological relations, which in turn differ widely from physical relations. These discrepancies shall not be muddled up. The step from studying physical relations/networks to analyzing social relations/networks is a giant one. Human beings are more than living bodies, let alone inanimate objects, because the mind is involved, the mysterious mind which thus far no physicist, chemists or neuroscientist has been able to grasp or measure. Actions like writing, smoking, driving and playing an instrument in an orchestra cannot be equated with events such as sneezing, falling, dying and raining, whereas events have causes, actions (made possible by, but not completely reducible to physical factors) have reasons. Whoever ignores this plain fact makes a serious mistake. Indiscriminately treating physical, biological and human relations as just connections between things is, in my view, the main error of physicists in general and of network scientists in particular. Data mining is the process of finding patterns in sets of data and big data analytics is the examination of large datasets (big data) to uncover hidden patterns, unknown correlations or trends that can help to make decisions. "Data mining" is a thus a misleading term, because the goal is not mining and scrutinizing data but finding a pattern between them.[64] Data can be extracted, collected, stored and transmitted. Binary data are pulses of light, but human beings are more than just data. Light can be measured, but humans, being matter *and* mind, can't.

The United States National Research Council defines network science as "the study of network representations of physical, biological, and social phenomena leading to predictive models of these phenomena". Interested in the development of network science (network-centric warfare), the U.S. military has formed an alliance between its Command, Control, Computers, Communications, Cyber, Intelligence, Surveillance and Reconnaissance Center, in Aberdeen (MD), and a consortium of dozens of American industries and universities, its goal being the development of deep understanding of the commonalities among networks and improvement of the ability to analyze, predict, design and influence complex systems.[65] Properly connected with the science of complex systems, I would suggest, network science could be used in understanding *countries*. The study of countries would then require identification of the nodes (dots) within, and investigation of the connections between, them. Such studies would not be meant to wage wars more efficiently, but to comprehend countries in order to *prevent* warfare and *promote* welfare.

Announcing the fifth edition of its program, in May 2019, the Lake Como School of Advanced Studies said:

> Many real systems can be modeled as networks, where the elements of the system are nodes and interactions between elements are edges. Networks represent the backbone of many complex systems, and their theoretical and computational analysis makes it possible to gain insights into numerous applications. Networks permeate almost every conceivable discipline—including sociology, transportation, economics and finance, biology, and myriad others —and the study of "network science" has thus become a crucial component of modern scientific education.[66]

Abstracting their article "Multilayer Networks in a Nutshell" (April 2018), Alberto Aleta and Yamir Moreno were writing:

> Complex systems are characterized by many interacting units that give rise to emergent behavior. A particularly advantageous way to study these systems is through the analysis of the networks that encode the interactions among the system's constituents. During the last two decades, network science has provided many insights in natural, social, biological and technological systems. However, real systems are more often than not interconnected, with many interdependencies that are not properly captured by single layer networks. To account for this source of complexity, a more general framework, in which different networks evolve or interact with each other, is needed. These are known as *multilayer* networks. Here we

> provide an overview of the basic methodology used to describe multilayer systems as well as of some representative dynamical processes that take place on top of them. We round off the review with a summary of several applications in diverse fields of science.[67]

This sounds good and hope-giving, provided Aleta and Moreno realize that societal networks are dynamic, encoding the extremely complex interactions among human beings, who cannot be equalized with molecules in a vat.

When a network shows multiple connections (edges) between any pair of nodes, it is defined as *multidimensional*. A general theory about multidimensional networks is still missing, but the number of studies on them is growing.[68] A set of objects (indicated as nodes, or vertices) and the relations between them (indicated as edges, or links) is represented by a graph. Whereas a multigraph (in contrast to a simple graph) is a graph in which two vertices may be connected by more than one edge, a hypergraph is a graph in which an edge can connect any number of vertices. If the joining is nonlinear, the question is whether a general graph model under uncertainty can be built. With such a model, we would have the principle of all ordering, i.e., a powerful tool to approach the complexity of a country, provided (*a*) the realization that human beings are more than gas molecules the trajectories of which might be statistically or computationally calculated and (*b*) the right dataset with respect to the country is available. In Chap. 7, it will be shown that we can be hopeful about the latter proviso, but I am not sure that, *rebus sic stantibus*, network theory will have the future.

In "networks beyond pairwise interactions: structures and dynamics" (https://arxiv.org/abs/2006.01764), Federico Battiston and seven other scientists present an overview of the emerging research-field of networks beyond pairwise interactions, claiming that "taking the higher-order structure of real complex systems into account, the field can greatly enhance our modeling capacities". However, the authors of this recent article overlook the fundamental difference between events and actions. They have not built, or suggested how to build, a model that could possibly help us comprehend a country's complexity.

As networks of interconnected nodes (or systems of interrelated parts) play a key role in the cognitive sciences, network science (or systems science) and cognitive science are converging. Indeed, a new network science is developing, providing insight into system properties in contexts such as metabolic reactions (chemical processes occurring within a living cell or organism). This kind of network science could be helpful to study the complexity of countries. The confluence of network science and cognitive science may radically change the way in which political, military, economic, financial, social, demographic, geographic, environmental and other situations and processes are understood.[69]

Network scientists may be interested in knowing that in Indian philosophy *Indrajāla* (Indra's net) is a metaphor that can be traced back to the Vedic period (c. 1500–c. 500 BC) and resurfaced in the Buddist concept of *Pratītyasamutpāda* (dependent origination). It has been developed by the *Huáyán* (flower adornment) School, a tradition of Mahayana Buddhist philosophy that flourished in China during the Tang dynasty (618 to 907 AD) and strongly influenced *Chán* (Zen) Buddhism. Visit https://

plato.stanford.edu/entries/buddhism-huayan. Whereas the *Tiāntái* School (another Buddhist tradition, having its main seat on Mount Tiantai, in the province Zhejiang) concentrated on the relationship between the noumenon (what is thought) and the phenomenon (what is perceived), the *Huáyán* School focuses its attention on the intricate pattern of the particulars, on the convoluted network or web of relations between the phenomena, maintaining that these particulars/phenomena are in harmony with one another, because they are manifestations of the one and immututable noumenon. Notably, while scientist of complex systems and networks keep an eye on the *environment* of a system/network, taking care that environment and system/network are not mixed up with each other, *Huáyán* Buddhists do not impose such restriction on themselves. "Flower Adornment" suggests the crowning glory of Buddha's profound understanding of *dharmadhatu* (the ultimate reality). The image of Indra's net is used to describe the interconnectedness of everything in the universe. In *Hua-Yen Buddhism: The Jewel Net of Indra* (Pennsylvania State University Press, 1977), Francis Cook beautifully describes Indra's net as follows:

> Far away in the heavenly abode of the great god Indra, there is a wonderful net which has been hung by some cunning artificer in such a manner that it stretches out infinitely in all directions. In accordance with the extravagant tastes of deities, the artificer has hung a single glittering jewel in each eye of the net, and since the net itself is infinite in dimension, the jewels are infinite in number. There hang the jewels, glittering like stars in the first magnitude, a wonderful sight to behold. If we now arbitrarily select one of these jewels for inspection and look closely at it, we will discover that in its polished surface there are reflected all the other jewels in the net, infinite in number. Not only that, but each of the jewels reflected in this one jewel is also reflecting all the other jewels, so that there is an infinite reflecting process occurring. (p. 2)

Network scientists who are willing to study the way of thinking in ancient China will realize that they still have a long way to go for another reason. There is no road without a beginning or end and there is no beginning or end without a road. Equilibrium, the balance between two weights, means motionlessness. Paradoxically, this stillness or oneness implies, and cannot exist without, duality. Emphasizing the importance of one weight only leads to narrow-mindedness, bias, predisposition and separatism. Furthermore, it would be difficult to ascertain where a line ("edge") in a constantly changing pattern ("graph") begins or ends, i.e., to pinpoint the "vertices" of a dynamic social network, given the fact (acknowledged by all sociologists) that "no man is an island entire on itself" (John Donne). The Chinese were (and probably are) thinking holistically rather than reductionistically, inclusively rather than exclusively. To them, there is no community without members of, or participants in, that community, any more than there would be participants without a community they belong to, or participate in. Parts and the whole are inseparably interrelated. The wholeness or unity of a pattern implies a multiplicity, a variety of items, which are only conceivable when the system they constitute is taken into account. In fact, there are no parts, because parts are always related to a whole, which is part of a bigger whole, ad infinitum. Conceiving of something as being a part is already making a mistake. And yet, it is impossible to conceive of a whole without parts. Reality is not an aggregate; all things hang together and nothing stands alone. The relationship between the one and the many (monism and pluralism) is a maddening problem with which philosophers in ancient Greece, India and China already occupied themselves. Werner Heisenberg wrote a marvelous book on the subject (*Der Teil und das Ganze: Gespräche im Umkreis der Atomphysik* [Piper, 1969]) and mereology, a thriving field of research, is the study of parts and the wholes they form (see Andrzej Pietruszczak, *Foundations of the Theory of Parthood: A Study of Mereology* [Springer, 2020] and below, Chap. 9).

Cybernetics

Cybernetics is not a branch but the twin brother of systems science. Co-evolving with (complex) systems science and network science and studying how humans, animals and machines control and communicate, it is the brainchild of Norbert Wiener (1894–1964), a Jewish-American mathematical genius. "Cybernetics" stems from *κυβερνήτης* (steersman, helmsman, captain, pilot, leader, governor). The word was first used by Plato to signify the governance of people. In his classification system of human knowledge, André-Marie Ampère (1775–1836) used the word *cybernétique* to denote the sciences of government. Entangled with information theory, cybernetics deserves special attention, because it takes system *control* (information processing) into account. A bicycle is a system (a set of connected things), but someone riding a bicycle is, from a cybernetic point of view, an interesting subject for investigation. The cyclist who wants to ride on a road and perceives a deviation from the norm he has established will correct the deviation by turning the handlebar. The cyclist controls thus the direction he/she takes by receiving, and reacting to, information. This may look a simple cybernetic process, but numerous physical, biological, neurological, psychological, sociological and even anthropological factors are involved. Perceiving, for example, is a highly complex process the number of publications on which is increasing rapidly. Politics (the art of governing), management and organization are arguably nothing but exercises in cybernetics. Over the last sixty years, cybernetics has advanced enormously. Sociocybernetics, which is based on general systems theory and cybernetics, is currently a thriving field of research. Strikingly (and regrettably), sociocyberneticists pay much attention to the mechanism and techniques of "governing", but little to the character and moral behavior of the "governor".[72]

Cybernetics, complexity science and informatics are tightly interrelated. The science of complex systems and cybernetics are based on the same principles; complexity science and informatics impact every field of research where communication plays a central role; and cybernetics is the source discipline of informatics. See John Motloch, "Big History Understanding of Complexity, Informatics and Cybernetics", *Systemics, Cybernetics and Informatics,* 15:6 (2017), pp. 54–60. On March 11–13, 2020, the 11th International Multi-Conference on Complexity, informatics and Cybernetics (IMCIC) was held in Orlando, FL. The 13th is scheduled for March 2022.

Cybernetics was one of the main subjects discussed at the famous Macy Conferences, in New York, organized by the Josiah Macy, Jr. Foundation and held from

1941 to 1960. The proclaimed purpose of these conferences was to restore the unity of science by forging bridges between the disciplines. The isolation of medicine was viewed as particularly problematic. Two innovations were designed to facilitate and encourage interdisciplinary exchanges: the Macy conferences and their transactions. At the conferences, participants presented ongoing research, their papers having been prepared ahead of time and circulated. The transactions, which were transcriptions rather than formal proceedings or selected and revised versions of the conference papers, were sent to a wide and international group of potentially interested people, in particular to those who had not been able to attend the conferences. They were meant to invite as many scientists or scholars as possible to join the discussion. The transactions were also valuable to the participants, because they enabled them to better remember what had been said during the conferences and to read the comments that perhaps had been missed. Participants in at least one of the cybernetics conferences were Harold Abramson, Nathan Ackerman, Vahe Amassian, William Ross Ashby, Yehoshua Bar-Hillel, Gregory Bateson, Alex Bavelas, Julian Bigelow, Herbert Birch, John Bowman, Henry Brosin, Yuen Ren Chao, Jan Droogleever-Fortuyn, Malin Ericsson, Frederick Fitch, Lawrence Frank, Ralph Gerard, William Walter, Molly Harrower, Evelyn Hutchinson, Heinrich Klüver, Lawrence Kubie, Paul Lazarsfeld, Kurt Lewin, Joseph Licklider, Howard Liddell, Donald Lindsley, William Livingston, David Lloyd, Rafael de Nó, Duncan Luce, Donald MacKay, Donald Marquis, Warren McCulloch, Turner McLardy, Margaret Mead, Frederick Mettier, Marcel Monnier, Oskar Morgenstern, Filmer Northrop, Walter Pitts, Henry Quastler, Antoine Remond, Ivor Richards, David Rioch, Arturo Rosenblueth, Leonard Savage, Theodore Schneirla, Claude Shannon, John Stroud, Hans-Lukas Teuber, Mottram Torre, Gerhardt von Bonin, Heinz von Foerster, John von Neumann, Heinz Werner, Norbert Wiener, Jerome Wiesner and John Young.[73] I mention the conferences, because, given the basic problem with area or country studies, the Macy conferences could possibly provide the format for an international meeting on comprehending countries.

Basically, there are two kinds of control (see Neculai Andrei, "Modern Control Theory: A Historical Perspective", February 10, 2005 [online] and Cihat Karaali, *Grundlagen der Steuerungstechnik* [Springer, 2018]). In open-loop (feedforward) control, the control is independent of the process output. For example, an irrigation sprinkler system that is programmed to turn on at set times. Even if it is raining cats and dogs, the sprinkler system is activated on schedule. Open-loop control is only useful when the relationship between input and output can be mathematically modeled. In closed-loop (feedback) control, the control is dependent on the process output (see Karl Åström and Richard Murray, *Feedback Systems* [PUP, 2021]). According to the British Standards Institution, a closed-loop control systems is a system "possessing monitoring feedback, the deviation signal formed as a result of this feedback being used to control the action of a final control element in such a way as to tend to reduce the deviation to zero". Feedback, which thus occurs when the outputs of a system are routed back as inputs, plays an important role in cybernetics. There are two types: positive and negative feedback. Both types play a prominent role in system dynamics (see above). Positive feedback occurs when the disturbance of a system brings about an increase in the magnitude of the perturbation. For example, alarm in a herd of people can spread by positive feedback to cause a stampede. A bank run occurs when customers of the bank withdraw their deposits due to concern about its solvency; as more people withdraw their funds, the probability of default increases, thereby prompting more people to withdraw their deposits. Agriculture and human population can be considered to be in a positive feedback mode: one drives the other with increasing intensity. This process may end with a catastrophe, as modern agriculture is using up the available phosphate and resorting to monocultures which are susceptible to systemic risk. The positive feedback loop between technology and human population may explain the hyperbolic growth of the population (and the increasing pressure on agriculture); the growth rate accelerates because of a second-order positive feedback. However, the question remains how to measure growth, which is a nonlinear phenomenon (see above, under "complex systems science"). Negative feedback tends to promote the finding of a balance, because it reduces the effects of perturbations. It is widely used in mechanical and electronic engineering, the thermostat that runs a central heating or cooling system being the usual example. Negative feedback also occurs in living organisms, where feedback enables body temperature, blood sugar level, blood oxygen content, arterial blood pressure or volume of body water to be maintained within a desired range by homeostatic processes. Satisfying thirst is an important homeostatic process.[74] "Homeostasis" comes from *ὅμοϊος* (similar) and *στάσις* (standing). Many diseases, of which there are many types, are the result of a homeostatic failure. Self-organization or self-ordering is the capability of certain systems to organize their own behavior or structure. It would thus be interesting to investigate the relationship between self-organization and anarchism, which holds that the state (who is supposed to impose order on what otherwise would be chaos!) is undesirable, unnecessary and harmful. Mainstream economists assert that the market price mechanism operates to match supply and demand, because mismatches between them feedback into the decision-making of suppliers and demanders of goods, altering prices and thereby reducing any discrepancy. Norbert Wiener, however, wrote in 1948:

> There is a belief, current in many countries and elevated to the rank of an official article of faith in the United States, that free competition is itself a homeostatic process. Unfortunately, the evidence is against this simple-minded theory. (*Cybernetics: Or Control and Communication in the Animal and the Machine*, pp. 158–159)

The notion of an economic equilibrium that is driven to, or maintained by, market forces is currently questioned by "heterodox economics", an umbrella term covering various schools of economic thought or theories.[75] What they have in common, however, is their critique of mainstream or orthodox economists, who are accused of overemphasizing *homo economicus* and taking things out of their context. Economic man, heterodox economists claim, is a human construct, invented to please some people. A critique of this construct is the speech delivered by Nick Hanauer at MIT, in September 2018: "How to destroy Neoliberalism: Kill '*Homo Economicus*'" (online). In addition, see also Richard Wolff, *The Sickness is the System* (Democracy at Work, 2020) and visit https://evonomics.com/why-behavioral-economics-cant-fix-a-broken-discipline.

> Economics is the study of how decisions are made with regard to the following fundamental and interdependent questions which every society has to answer:
> (I) How many/much of which goods shall be produced?
> (II) By what resources and techniques and by whom shall the goods be produced?
> (III) For whom shall the output be?

Heterodox economists stress that orthodox economics is the study of just one kind of human behavior, namely the behavior shown in a subsystem that is part of a bigger system, the context of the subsystem. As this subject strikes at the heart of the problem I attempt to solve in this book, it may be useful to dwell a bit longer on the issue of human behavior.

> Human capital is an important, if not the most important, factor of production; it is a collection of traits: the knowledge, talents, skills, abilities, experience, intelligence, training, judgment and wisdom possessed by individuals. Theodore Schultz (1902–1998), Jacob Mincer (1922–2006) and Gary Becker (1930–2014) pioneered the study of this production factor. Land (including natural resources), labor, capital and leadership (management) are usually taken to be the four factors of production. Without peasant or miners, however, land would not provide what it could produce and natural resources would not be mined. Capital goods such as pitchforks, pump, computers, oilrigs, battleships, tractors, office buildings, batteries, cameras and kitchen utensils incorporate both labor and one or more than one kind of natural resources. Leadership, often regarded as the production factor without which land, labor and capital would be unproductive, is only a form of labor. There is thus no valid reason for remunerating laborers and leaders fundamentally differently. Admittedly, laborers need to be managed and organized, but without workers there would be little left for leaders to manage or organize. The production of commodities would come to a standstill. Sooner or later, this will have grave consequences for the leaders and owners of the company. That is why going on strike is a powerful weapon. Trade unions are necessary to prevent laborers from being exploited. Economic transactions take always place with a tug-of-war between the parties involved. We must not deceive ourselves: as in the jungle, the strongest wins; civilization is nothing but a thin veneer.
>
> The subject of human resources (as distinct from natural and financial resources) can be treated quantitatively or qualitatively, and it involves not only economic but also demographic, geographical, sociological, anthropological, linguistic, psychological, educational, political, legal and even religious factors. For example, the effects of the size, composition and dynamics of the population of a country on the economic qualities of its labor force make it necessary to acquire knowledge of the country's demography and its government policies towards it. What are the socioeconomic correlates of declining fertility and how might government policies affect these correlates? If public policies can reduce mortality rates, can they also diminish fertility rates? While fertility, health and education affect one another, each of them in turn influences the incomes, hence the performance of an economy.
>
> Physiologically, a human being is a fine-tuned system of systems made up of cells organized according to specialization to maintain homeostasis. The body is an intricate and complex system consisting of a muscular, skeletal, integumentary, immune, endocrine, nervous, respiratory, circulatory, reproductive, digestive and urinary system. The complexity is not only physiological or biological, for a human being is more than a fleshy machine, more than a physical-biological system. Man, the "featherless biped", the *homo symbolicus*, is both a part of and apart from nature. He has experiences and a *Lebenswelt* (Edmund Husserl, Alfred Schütz, Jürgen Habermas, Björn Kraus), gathers and processes information, is able to learn,

has emotions, and is endowed with the mental states that may trigger actions: moving one's tongue and lips in a certain way while speaking, growing crops, making medicines, buying or selling a house, driving a car, writing a letter, reading a book, performing a ritual, playing the violin, rescuing (or killing) somebody etc. There are many types of human behavior (aggressive, irritating, charming, rude, impulsive, diplomatic etc.); they can be exhibited under different circumstances (e.g. in an economic situation which may affect, or be affected by, social, psychological, cultural, political or other factors). Man is a being driven by interests and full of fears, hopes and biases. Man is an open system, hence interacting with, and influenced by, his/her environment. Framers of people's mind such as religious leaders, commercial advertisers and political demagogues know this full well. The reader may think of political rallies, pop festivals, or sporting events. The scenes at Mecca during the Hajj, at Varanasi (Benares) when a crowd of Hindus gathers on the banks of the Ganges, and at St. Peter's Square in Rome when the Pope gives his blessing *Urbi et Orbi* may also spring to mind. The mass hysteria in Thailand after the death of King Bhumibol the Great, in England after the death of Princess Diana (the "British nation in shock"), or in North Korea after the death of Kim Il-sung (and Kim Jong-il) are other example of public pressure on individual behavior. "The mob is man voluntarily descending to the nature of a beast," Ralph Waldo Emerson said. Talking about the wavering multitude, Shakespeare put it more pithily: "the blind monster with uncounted heads".[76]

According to the European Association of Social Psychology, "Social psychology is a scientific discipline that is aimed at understanding how the behavior, motivations, thoughts, and feelings of people are influenced by other people, real or imagined. What is special about social psychology is its strong focus on the study of the underlying, psychological mechanisms that intervene in social phenomena".[77] Due to political constraints on field research, a reliable and comprehensive book on social psychology in China is sorely missed, adding to the misunderstanding of the country prevalent among "China experts". At the Institute of Psychology of the Chinese Academy of Sciences, in Beijing, some socio-psychological research has been carried out, but the institute is directed and closely watched by the government (read: communist party).

For mass hysteria (aka collective obsessive behavior), see Elias Canetti, *Masse und Macht* (Claassen, 1960), John McClelland, *The Crowd and the Mob* (Routledge, 1989), Adrian Kear and Deborah Steinberg (eds.), *Mourning Diana* (Routledge, 1999), Lisa Blackman and Valerie Walkerdine, *Mass Hysteria: Critical Psychology and Media Studies* (Palgrave, 2001), Thomas Brudermann, *Massenpsychologie* (Springer, 2010), Jeff Fleischer, *The Latest Craze: A Short History of Mass Hysterias* (Fall River Press, 2011), David Miller, *Introduction to Collective Behavior and Collective Action* (Waveland, 2014, Chap. 5) and Fabrizio Fogliato, *Paolo Cavara, Gli occhi che raccontano il mondo* (Il Foglio letterario, 2014). Though these books describe and analyze the irrational behavior of "*homo sapiens*", a satisfactory explanation for the behavior of *kamikaze* pilots, the followers of Charles Manson, Muslim suicide bombers, or football hooligans is yet to be given.[78]

Human beings are gregarious; they form groups varying in size and nature. *Intra-group* dynamics are the underlying processes that give rise to a set of norms, roles, relations and common goals that characterize a particular social group. "*Inter-group* dynamics" refers to relationships between two or more groups. What this noun phrase refers to includes perception of, opinion about, and attitude and behavior towards another group.[79]

In economics, the standard way of grouping is dividing the participants in the economic process into households, firms, and government. Households vary greatly in purchasing power, social status, political orientation and cultural background. Aristotle compares "the art of acquiring money" (*χρηματιστική*) with "the art of housekeeping" (*οἰκονομικά*).[80] This important distinction is now blurred. A firm (aka business, company, or enterprise) is an organizational and legal entity; it is an association of people having an economic purpose: the profitable (but not always ethically responsible) provision of goods or services to customers. Firms differ enormously in size, ownership, management, production and marketing. The *theory* of the firm is a cluster of theories that explain and predict the nature of a firm, including its existence, behavior, structure, and relationship to the market, where demand and supply of goods meet each other. Business administration (to be distinguished from public administration) concerns the organization and management of a firm. Public and business administration researches are subsumed under the broad category of organization studies, which are akin to the study of systems or networks (see next chapter). Policy analysis, having its roots in systems science and cybernetics, is usually deployed in the public sector, but can also be applied to other kinds of organization. Government[81] is the authoritative expression of the State. Its formal functions, broadly defined, correspond to legislative and administrative functions. Governments can be centralized or decentralized. Households, firms and governments have different objectives, but all of them are engaged in decision-making, the process of identifying and choosing alternatives based on the culturally determined norms, values, preferences and beliefs of the decision-maker. The decision theory (aka theory of choice) is very rich; because of its wide range of applications, it is counted among the formal sciences.

Institutions are recurring patterns of behavior and systems/networks of established social rules. Even though the structures or mechanisms of social order develop over time, institutions govern the behavior of a group of individuals and are the key factors of social life. Their nature, role and development are subjects of research in several academic disciplines, including (but not limited to) economics. Slavery, private property, the competitive market, wage labor, lending and borrowing, and taxation are economic institutions with non-economic aspects. Patterns of behavior become stabilized through repetition, reflecting their advantageous character to at least some members of the community. They become routine, their stability being reinforced by social approval. Belief systems make historic institutions appear as "natural", the royal family and the Roman Catholic Church being cases in point. The ultimate reinforcement of social behavior takes place through legal codes and sanctions that are formed by powerful and privileged people. Destabilization of institutions is usually initiated by community members who are not happy with them, as witnessed by the French and Russian Revolution. Institutional Economics (Thorstein Veblen, John Commons, Wesley Mitchell, Clarence Ayres and John Maurice Clark) studied the institutions that shaped economic behavior and focused on the role of evolutionary processes. *New* Institutional Economics (not to be confused with Neo-institutional Economics [Kenneth Galbraith, Allan Gruchy, Geoffrey Hodgson, Malcolm Rutherford and Warren Samuels]) extends economics by focusing on the social and legal norms and rules (institutions) that underlie economic activity and on the effects economic activity may have on social or legal institutions. Ronald Coase, Armen Alchian, Herbert Simon, Douglass North, Eirik Furubotn, Harold Demsetz, Oliver Williamson, Elinor Ostrom, Masahiko Aoki, Richard Posner, Thráinn Eggertsson and Ha-Joon Chang are notable representatives of this heterodox school. In *The Civil Sphere* (OUP, 2006, pp. 69–192), sociologist Jeffrey Alexander differentiates between "regulative institutions" (voting, parties, bureaus and offices of the State) and "communicative institutions" (public opinion, mass media, polls, associations).[82]

Rituals, a difficult to rationalize feature of all known human societies, are activities involving words, actions (not: events!), gestures and objects; they are performed in a set sequence. Activities are ritualized and codified, which make them repeatable performances.[83] Rituals regulate and stabilize social institutions and are characterized by formalism, traditionalism, invariance, rule-governance, and symbolism. Rituals, reinforcing the status quo and often accompanied by music, include not only worship rites, ceremonies, sacraments or other religious practices, but also rites of passage and initiation, calendrical and commemorative rites, oaths of allegiance, dedication ceremonies, coronations and presidential inaugurations, wedding ceremonies and funerals, fastings and festivals, school rushings and graduations, club meetings, sporting events, processions and military parades, rites of exchange, and political and legal rituals. Even common actions like hand-shaking and saying "hello" may be termed rituals. See Andréa Bellinger and David Krieger, *Ritualtheorien* (Springer, 2013, pp. 7–34), Richard Schechner, *Performance Studies* (Routledge, 2013) and the series *Oxford Ritual Studies*. Comprehending a country is tantamount to understanding its institutions and rituals, both of which are structured as well as dynamic. They must be studied interdisciplinarily

There is a rich variety of economic schools, but an economic theory taking all the relevant factors into account is missing. Following Alfred Marshall, one could maintain that economics boils down to questions of demand for, and supply of, goods. However, the number of, and relations between, the factors relevant to demand and supply are far from clear. Many schools interconnect with, or spill over into, disciplines different from economics, indicating that economists should cast a wider net, not by trespassing on territories that belong to other disciplines, but by engaging in genuine scientific collaboration (see Chap. 3). For example, *behavioral economics* and its subfield behavioral finance study the effects of psychological and social factors on decisions made by individuals or institutions in regard to, e.g., resource allocation and market prices. Behavioral economics and finance overlap with new institutional, evolutionary, and social economics.[84] They cannot be disconnected from (*a*) *constitutional economics* (which explains the choice of alternative sets of legal-constitutional-institutional rules that constrain the choices and activities of economic and political agents), (*b*) *legal economics* (which applies [micro-]economic theory to the analysis of law), (*c*) *regulatory economics* (which applies law to economic issues such as remedying market failure, protecting the environment and benefiting politicians) and (*d*) *political economics* (which draws upon economics, sociology, law and political science in explaining how economic, social, legal and political institutions influence each other).[85]

This clearly points in the direction of contextual economics, which has more relevance to real-world concerns than the dominant, mainstream economic paradigm.[86] Ever more scientists are realizing that isolation of economics, i.e., removing economic processes from their political, legal, military, geological, geographical, demographic, social, psychological, cultural, religious, moral and ecological context, runs the danger of missing important aspects of the economic reality. In *Cents and Sensibility: What Economics Can Learn from the Humanities* (PUP, 2018), Gary Morson and Morton Schapiro make the strong case that "the humanities offer economists ways to make their models more realistic". They trace the connection between Adam Smith's classic, *The Wealth of Nations* (1776), and his less celebrated book, *The Theory of Moral Sentiments* (1759), and allude to Jane Austen's novel, *Sense and Sensibility* (1811).[87] In *Licence to be Bad: How Economics Corrupted Us* (Alan Lane, 2019), according to Ha-Joon Chang [author of *23 Things They Don't Tell You About Capitalism*] a "must read", Jonathan Aldred cogently argues that mainstream economics is based on the utterly unrealistic assumption that men are rational beings who, aiming at utility maximization, have all the information necessary for making a sound business transaction at their disposal. His final verdict on the economic courses still offered at highly prestigious universities is, therefore, merciless: "junk in, junk out".

Economics is a social science. This central idea, guiding the *Journal of Contextual EconomicsSchmollers Jahrbuch* since its founding in 1871, has lost none of its relevance. In philosophy, the humanities and the social sciences, contextualism has grown in importance over the last several years. The core postulate is that a fuller understanding of human thought and action requires a grasp of the context in which they are embedded. In line with its original intention, the main objective of the journal has been to offer a research forum where economic life is set in relation to its social and physical environment, particularly a forum where the interface between the economic system and other systems is studied. Institutional approaches appear to be relevant to this purpose. Quantitative and qualitative methods are equally welcome. The editorship wishes to solicit "contributions that devote themselves to questions of long-term human development and societal change, including those from the perspective of economic history and the history of economic thought". Consideration of social and normative questions has been an important thematic focus in the long tradition of the journal.

Over the last few decades, in the social sciences and humanities, contextualism has grown in importance. Contextualism emphasizes the embeddedness and deeply ingrained character of utterances and actions. The key idea is that fuller understanding of the human condition requires a grasp of the context in which it is embedded. This understanding can only be attained interdisciplinarily (as distinct from multidisciplinarily), by truly collaborative research. Yet, mainstream economists tend to take the easy way out by invoking the *ceteris paribus* clause. Examining the effect(s) of one variable in their model, they take the values of all related variables to be constant, that is to say, they wrongly *presume* them to be constant. Their reasoning is thus counterfactual, for change is all-pervasive. Whoever ignores this makes a big mistake. The *ceteris paribus* assumption lies behind the orthodox analysis of supply and demand in a single market, the caveat being that correlation does not imply causation. At higher levels of analysis, this assumption is said to be dropped for the more realistic proposition that all things are interrelated. However, this more general analysis, rather than being more realistic, is in fact a model of how the whole economy can reach, and stay in, equilibrium (Kenneth Arrow, Gerard Debreu, Frank Hahn, Herbert Scarf, Andreu Mas-Colell). *The non-economic aspects are not taken into account* and, out of respect for the requirements of labor division, left to colleagues at other faculties ("*divide et impera*"!). General equilibrium theory, which *conditionally* explains the behavior of a whole economy consisting of many parts (interacting markets), is both the crowning achievement of orthodox economics and its greatest failure.[88] General equilibrium theory contrasts to the theory of partial equilibrium, which only analyzes single markets. This point may need some clarification.

An economic system is a whole the parts of which are connected with, and react to, each other. An increase in the income of the producers of commodity A will affect the demand for commodities B, C, etc. and the income of the producers hereof. The reactions of these producers will affect the demand for commodity A. The Russian-American Nobel laureate Wassily Leontief (1906–1999) is credited with developing an input–output model, a quantitative economic technique that represents the interdependencies between different branches of a national economy or different regional economies. Influenced by François Quesnay (1694–1774), who had developed a cruder version of this technique, and Léon Walras (1834–1910), father of the general equilibrium theory, Leontief was the first to use a matrix representation of a national or regional economy. Input–Output (I-O) Analysis[89] and Industrial Organization[90] were developed at Harvard University in the 1930s. The two fields, having led an almost entirely separate existence, deal with industrial structure and performance (the main subjects of meso-economic research) and often use a similar vocabulary. I-O Analysis is subset of a set of related methods which show how a change of one part of a system affects the other parts. Specifically, I-O Analysis shows how industries are interlinked, how an economy, taken out of its context, hangs together. For example, coal is required to produce steel and steel in the form of tools is required to mine coal. Without coal no steel, without steel no coal. An economy is full of such circular dependencies, which are direct or indirect in character. I-O Analysis may be regarded as a simple form of linear programming, which is a special case of mathematical programming, which in turn is a special case of equilibrium analysis. I-O Analysis, which arose to deal with the problem of connectedness between industries or industrial branches, provides thus the simplest form of the general equilibrium theory. General equilibrium is a state of balance, but when things are of equal weight or force in all directions, it is impossible to explain change. Economists appealing to the theory of general economic equilibrium cannot explain the economic growth of a country without taking non-economic factors into consideration.

There are many sorts of systems: physical, biological, psychological, social and cultural systems. All these systems are–like their subsystems (and the sub-subsystems thereof)–closely connected with each other in ways increasingly made clear by the scientific community that resorts to computers. As explained in Chap. 3, a country's economic system can be divided into three interconnected sectors (agriculture, industry and services), each sector being a system of systems by itself. The economic system of a country is inseparably linked with its ecological, geological, geographical, financial, political, legal, ideological, demographic, educational, medical and religious system, each of which has been (and will be) evolving. The interconnected physical, biological, psychological, social and cultural systems to which these interwoven subsystems belong constitute the all-including, overarching, comprehensive suprasystem, viz. the reality ancient philosophers were pondering about, but, given the ever-expanding system of sciences, no single human being can dream of firmly grasping or fully comprehending one of its parts. The physical, biological, psychological, social and cultural systems are investigated by, respectively, the physical, biological, psychological, social and cultural sciences, meaning that these sciences are forms of applied systems science. Though systems biology, which is the computational and mathematical modeling of complex biological systems, and systems ecology, which takes a holistic approach to the study of ecological systems, are probably the most advanced among them,[91] the systemic (vs. analytic) approach is increasingly appropriated by all scientists.[92] This is a welcome development, but Ludwig von Bertalanffy's aim to lay the foundations of, and to find the principles applicable to, all scientific disciplines, threatens to be forgotten, as many cooks spoil the broth. Von Bertalanffy reacted against reductionism. Regarding it as disastrous, he attempted to revive the unity of science. The crusade against scientific fragmentation he initiated still continues but his lofty idea has not yet been realized.

Though the *International Journal of General Systems* (1974ff) focuses on general systems, there is no journal of general system *theory*.[93] My hope is, therefore, that this book, pleading for country studies uplifted onto a higher level (i.e., being a form of computer-supported applied systems science), will inspire future generations of students who see the crucial difference between ὅλον (*Ganzheit*) and πᾶν (*Gesamtheit*) and are seriously interested in a foreign country, not only in and of itself but also in relationship to other countries (its foreign relations). The list of systems scientists available on the Internet may sufficiently indicate the fertility of the field of research opened by von Bertalanffy. Despite the infelicitous name International Society for the Systems Sciences (infelicitous because the word general is left out), the search for Archimedes' point, for a truly general system theory, i.e., for a theory applicable to all systems, must go on as long as it takes.

The search for a general system theory resembles Leibniz's as-of-yet unfulfilled dream of a *characteristica universalis*, a recurrent concept in his writings. Gottfried Leibniz (1646–1716) imagined a universal and formal language to express mathematical, scientific and metaphysical ideas, i.e., a language that could be used within the framework of a universal logical calculation or *calculus ratiocinator*. Writing in French, he sometimes employed the phrase *spécieuse générale* to the same effect. Leibniz might have been intending his *characteristica universalis* to serve as an architectonic structure of human knowledge. He was inspired by Descartes (1596–1650), whose work (based on the logical fallacy of *petitio principii* "I think, therefore I am"!) aroused so much hope that the second half of the seventeenth century was animated by the vision of a future science termed *mathesis universalis*, a vision that seems to have resurfaced in the masterpiece of Alexander Bogdanov in the beginning of the 20th century. The hypothetically universal science would be modeled on mathematics.[94] However, Philip Davis and Reuben Hersh's *Descartes' Dream* (Dover, 2005), a book that "passionately pleads against the use of formal mathematical reasoning for solving mankind's problems", is "an antidote to the Cartesian view that mathematical and scientific knowledge will suffice to solve the central problems of human existence" (*The New York Times*). "Leibniz's dream of a *characteristica universalis* is partially realized", Marciszewski and Murawski write on p. 230 of their book, and the dream of a universal logic is carried on by Jean-Yves Béziau in (*a*) the book he edited in 2005, *Logica Universalis: Towards a General Theory of Logic* (Birkhäuser, 2007), (*b*) the world congresses he has been organizing since 2005 and (*c*) the journal he has edited since 2007, *Logica Universalis*. However, whether Béziau is aware of it that logic is a product of the unfathomable human mind, which cannot be mathematically modeled, is a moot question. From 2013 to 2017, the project "*Bolzanos und Husserls Weiterentwicklung von Leibnizens Ideen zur 'Mathesis Universalis*'" was funded by the *Deutsche Forschungsgemeinschaf*.[95]

Logicians like Nuel Belnap, Gerald Sacks, Dag Prawitz, Per Martin-Löf, Dov Gabay and Johan van Benthem are concerned with the structure of language (a subject also of concern to linguists, the number of which is amazing), the general laws governing concepts (the primary and basic issue of philosophers), and the correctness or soundness of reasoning. However, their lectures are of little value, because, requested to elaborate on, and justify, their statements, these logicians shy away from clarifying their epistemological and ontological stance. The soundness of reasoning they are talking about is not firmly grounded in everyday life. Logic, epistemology and ontology are tightly interconnected and, to understand the fundamental structure of human thought, language and action (the three are interrelated), they should not be separated, as Peter Strawson (1919–2006) cogently argues in his little but profound book *Analysis and Metaphysics* (OUP, 1992, pp. 51–58). The richly varied human actions, which are intimately related to thougt and language, have not only a logico-epistemological but also an ethico-political side, a side that Western (particularly Anglo-Saxon) logicians have neglected to deal with or tend to underexpose. In the empirical-analytical tradition initiated by the English philosophers Francis Bacon and John Locke, Western logicians largely prefer to remain silent on axiological subjects, thereby clipping man's wings and ignoring or deliberately forgetting the wisdom of Socrates who pointed out that both truth (*Wahrheit*) and value (*Wert*), i.e., both *Sein* and *Sollen*, are the two focal points around which man's *Leben und Reden* elliptically orbit. In other words, *to be* is not congruent and often conflicts with *ought to be*. If they were the same, the circle would have been a better metaphor. The interface between logic (to the extent that it is concerned with the forms of valid inference) and ethics (insofar as it is concerned with questions as to what man should, or should not, be doing) tends to be forgotten by logicians who are inclined to that misguided movement in Western philosophy called logical positivism.

Descartes, Spinoza and Leibniz were the main protagonists of classical rationalism, the view that reason is the chief source and test of knowledge. "Reason" stems from the Latin word *ratio* (relation between numbers or quantities). The question as to what "reason" exactly means has been a bone of contention in Western philosophy since the Pythagoreans discovered the irrational numbers, which are not ratios, fractions, or quotients of integers (a discovery that caused the first crisis in mathematics). Leopold Kronecker (1823–1891) reportedly said: "*Die ganzen Zahlen hat der liebe Gott gemacht, alles andere ist Menschenwerk*", but—leaving aside the question whether *Gott* exists and if so, whether He/She is *lieb*—reality, which is not limited to what natural scientists are concerned with, cannot be modeled by a set of Diophantine equations. Even the world that is untouched by man and is not *Menschenwerk* (that is to say, the purely physical world) cannot be captured by means of integers only. The laws of thermodynamics, which are not formulated in exclusively rational numbers, govern the transfer of energy in and among all systems and the direction of spontaneous processes. The wonderful order and harmony that nature shows (superbly documented by David Attenborough) cannot be understood by the use of nothing but "whole numbers". Other types of numbers are needed.

Yet the concept of proportion, or analogy, plays an important role not only in logic, biology, psychology, law, linguistics and literary science but also in art. Among many other artists, Vitruvius, Alberti, Dürer, Michelangelo, Palladio, Kandinsky, Mondrian, Mies van der Rohe, Le Corbusier and Dom Hans van der Laan discussed a work's proportions (equalities of ratios), claiming that mathematics is somehow related to art.[96] In mathematics, two quantities are in the golden ratio if their ratio is the same as the ratio of their sum to the larger of the two quantities,[97] but the reader will remember that (I) not everything is quantifiable and (II) a system is not the sum of its parts.

Percentages, being fractions of 100, are ratios. In the modern world, one cannot do without them. Percentages are used in almost every field of activity. In a shopping center, one is bound to see advertisements with discounts of a certain percentage. This book is not the place to go into the different opinions expressed on the issue of ratio, or into the arguments used to support them. Suffice it to say that empiricists like John Locke, George Berkeley and David Hume disagreed with Descartes and his followers and that Immanuel Kant (1724–1804) attempted to reconcile rationalism (traceable to Plato) and empiricism (traceable to Aristotle) in three groundbreaking, interrelated books: *Kritik der reinen Vernunft* (1781), *Kritik der praktischen Vernunft* (1788) and *Kritik der Urteilskraft* (1790). Whether he succeeded in doing so is still a matter of dispute, but in the late 1700s, the very concept of reason, in the name of which the medieval way of thought had been criticized, began to show cracks. Romanticism, the artistic, literary and intellectual movement that would reach its peak in the period from 1800 to 1850, had begun. The Enlightenment, predated by the pre-Socratic philosophers, more than 2000 years ago ("Ionian Enlightenment"), and being the offspring of the Scientific Revolution that was heralded by Francis Bacon (1561–1626) and Descartes, enshrined rationality as the supreme value, but the movement that dominated the world of ideas in Europe in the eighteenth century and is still eulogized in many quarters has been suffering from inner tensions which are not yet relieved. While criticism seemed to end in skepticism, naturalism appeared to result in materialism, the fateful philosophy which reduces reality to matter. The books to be read are Frederick Beiser, *The Fate of Reason* (HUP, 1987), Terry Pinkard, *German Philosophy 1760–1860* (CUP, 2002) and the volumes 3 (2005), 6 (2008) and 14 (2016) of *Internationales Jahrbuch des Deutsche Idealismus*, dealing with "*Deutscher Idealismus und die gegenwärtige analytische Philosophie*", "*Romantik*" and "*Der deutsche Idealismus und die Rationalisten*", respectively. For the history of irrationality, which prevails in human life, see Justin Smith, *Irrationality: A History of the Dark Side of Reason* (PUP, 2019).

On the first page of his excellent book, *An Introduction to Complex Systems* (Springer, 2017), Paul Fieguth writes:

> For a vey good reason of pedagogy, the vast majority of system concepts to which undergraduate students are exposed are (*a*) lineair (since there exist elegant analytical and algorithmic solutions to allow linear problems to be easily solved), (*b*) Gaussian (since Gaussian statistics, resulting from the Central Limit Theorem, are well behaved and lead to linear problems) and (*c*) small (since high-dimensional problems are impractical to illustrate or to solve on the blackboard). Unfortunately, nearly all major environmental, ecological, and social problems facing humankind are non-linear, non-Gaussian, and large-scale. To be sure, significant research has been undertaken in the fields of non-linear, non-Gaussian and large-scale problems, so a great deal is in fact known. However, the analysis of such systems is really very challenging.

I would like to add: the combining of systems is even more challenging. The annual International Symposium on Frontiers of Combining Systems, which aims "to develop techniques and methods for the integration of formal systems" and the proceedings of which are published in the famous Springer series *Lecture Notes in*

Artificial Intelligence (LNAI), is propitious, hopeful and heartening, [98] if we can be reassured that due attention will be paid to the incomprehensible human mind.

Notes and References

1. The astronomia tourbillon is the ultimate wristwatch.
2. See Chris Clearfield and András Tilcsik, *Meltdown: Why Our Systems Fail and What We Can do About It* (Allen Lane, 2018).
3. See Ivor Grattan-Guinness, *The Rainbow of Mathematics* (W.W. Norton, 2000, pp. 18–103) and Mickaël Launau, *Le grand roman des maths* (Flammarion, 2016, pp. 11–40). I bypass the controversial work of Florentin Smarandache, editor-in-chief of the journal *Neutrosophic Sets and Systems* (2013ff). Neutrosophy seems to be a generalization of Hegel's dialectics.
4. Grattan-Guinness *op. cit.*, pp. 234–301.
5. See Fritz Reinhardt and Heinrich Soeder, *Atlas Mathematik*, Vol. 2 (DTV, 1998) and Timothy Gowers et al. (eds.), *The Princeton Companion to Mathematics* (PUP, 2008, VI.11ff). The quote is from The Encyclopædia Britannica, Vol. 13, 1990, p. 441.
6. See Karl Menniger, *Zahlwort und Ziffer* (Vandenhoeck & Ruprecht 1957/58), James Hurford, *Language and Number* (Blackwell, 1987, Chap. 4), Tobias Dantzig and Joseph Mazur, *Number: The Language of Science* (Penguin, 2007), Stanislas Dehaene, *The Number Sense* (OUP, 2011) and James Hurford, *The Origins of Language* (OUP, 2014). According to Leonard Bloomfield (1887–1949), "Language enables us to count".
7. See Grattan-Guinness *op. cit.*, pp. 493–530 and José Ferreirós, *Labyrinth of Thought* (Birkhäuser, 1999). For the intriguing relationship between set theory and category theory, read Giovanni Sommaruga (ed.), *Foundational Theories of Classical and Constructive Mathematics* (Springer, 2011, pp. 113–125) and visit https://plato.stanford.edu/entries/category-theory.
8. Visit https://plato.stanford.edu/entries/identity. In addition, see Seth Schwartz et al. (eds.), *Handbook of Identity Theory and Research* (Springer, 2011) and Peter Brooks, *Enigmas of Identity* (PUP, 2013).
9. See Willard Van Orman Quine, *Methods of Logic* (Harvard University Press, 1982).
10. See Peter Strawson, *Individuals* (Methuen, 1959, first three chapters).
11. See Note 8 in Chap. 1 and visit https://cnls.lanl.gov.
12. See Kostas Axelos, *Systématique ouverte* (Éditions de Minuit, 1984). For an interview with Axelos, see *Radical Philosophy*, 130:1 (March/April 2005), pp. 25–28.
13. See Ludwig von Bertalanffy, "An Outline of General System Theory", *British Journal for the Philosophy of Science*, 1:2 (August 1, 1950), pp. 134–165, and David Pouvreau, *Une histoire de la systémologie générale de Ludwig von Bertalanffy* (EHESS, 2013). Visit https://www.isss.org and http://www.bcsss.org. Daniël Strauss is critical of Von Bertalanffy in "The Scope and Limitations of Von Bertalanffy's Systems Theory", *South African Journal of Philosophy*, 21:3 (2002), pp. 163–179.

14. See Brian McGuinness (ed.), *Unified Science* (Springer, 1987), Gerald Midgley, "Rethinking the Unity of Science", *International Journal of General systems*, 30:3 (2001), pp. 379–409, and Gerhard Schurz, *Philosophy of Science: A Unified Approach* (Routledge, 2014). Visit https://plato.stanford.edu/entries/scientific-unity. The series *Logic, Epistemology, and the Unity of Science* (2004ff) also covers the subject. Őrsan Şenalp, a researcher affiliated with the Centre for Systems Studies (University of Hull), currently focuses on Aleksandr Bogdanov's groundbreaking but marginalized work *Tektologiya* and on the topics of the unity of science and the feasibility of a universal system paradigm. See Note 3 in Introduction and Note 25 of this chapter.
15. See further Frederick Emery, *Systems Thinking: Selected Readings* (Penguin, 1969), Magnus Ramage and Karen Shipp, *Systems Thinkers* (Springer, 2009), Jamshid Gharajedaghi, *Systems Thinking* (Morgan Kaufmann, 2011), Xing Pan et al., "Systems Thinking", *Procedia Computer Science*, 16 (2013), pp. 1027–1035 (online), John Mingers, *Systems Thinking* (Routledge, 2014), Michael Jackson, *Systems Thinking* (CreativeSpace, 2017) and Steven Schuster, *The Art of Thinking in Systems* (CreativeSpace, 2018). The 10th Relating Systems Thinking & Design Symposium (RSD10) will be hosted by Delft University of Technology (NL), in November 2021). Go to https://www.edge.org/conversation/mary-catherine-bateson, https://thesystemsthinker.com and https://stream.syscoi.com (the Systems Community of Inquiry is a worldwide network of systems thinkers, scientists and practitioners). Also notice the Springer series *Contemporary Systems Thinking* (1990ff). For the mathematical theory of systems, visit https://mtns2020.eng.cam.ac.uk. The publication of my article 'The Foundation of Chinese Systems Thinking' is forthcoming, in *Systems Research & Behavioral Science.*
16. See Mario Bunge, *A world of Systems* (Reidel, 1979), Gianfranco Minati and Eliano Pessa, *From Collective Beings to Quasi-Systems* (Springer, 2018, Chaps. 1, 7 and 8) and Michael Matthews (ed.), *Mario Bunge: A Centenary Festschrift* (Springer, 2019). Visit https://doi.org/10.1080/01969728208927689 and https://a-g-i-l.de/obituary-mario-bunge-physicist-philosopher-champion-of-science-and-citizen-of-the-world-1919-2020. See also Charles François, "Systemics and Cybernetics in a Historical Perspective", *Systems Research and Behavioral Science*, 16:3 (1999), pp. 203–2019.
17. An informative website is https://www.churchfathers.org.
18. See Anthony Long (ed.), *The Cambridge Companion to Early Greek Philosophy* (CUP, 1999, pp. 45–65), Daniel Graham, *Explaining the Cosmos* (PUP, 2006) and Carlo Rovelli, *The First Scientist: Anaximander and his Legacy* (Westholme, 2011). See also Stephen Hawking, *A Brief History of Time* (Bantam, 1988), Steven Weinberg, *The First Three Minutes* (Basic, 1993), Jim Peebles, *Cosmology's Century* (PUP. 2020) and Ghada Chehade, 'Not If, but when: Cosmology in Crisis & The Coming Paradigm Shift", part I, March 2021, and part II, April 2021 (online). A timeline of cosmological theories can be found on the Intetnet.
19. See Richard DeWitt, *Worldviews* (Wiley, 2018).
20. Visit https://plato.stanford.edu/entries/scientific-revolutions.

21. See Christoph Lehner et al. (eds.), *Einstein and the Changing Worldviews of Physics* (Springer, 2012).
22. See Thomas Ryckman, *Einstein* (Routledge, 2017). The Einstein archives, "including scientific writings and correspondence, non-scientific writings and correspondence, family letters and travel diaries", are online. See Note 1 in Preface and Note 21 of this chapter.
23. See Ellen Hine, *A Critical Study of Condillac's Traité des systèmes* (Nijhoff, 1979).
24. See Vyacheslav Korotkikh, "The Idea of the System of Philosophy in Hegel's Works", *Indian Journal of Science and Technology*, 8: Special Issue 10 (December 2015), pp. 1–12.
25. See Ilmari Susiluoto, *The origins and development of systems thinking in the Soviet Union* (Suomalainen Tiedeakatemia, 1982), John Biggart et al., *Alexander Bogdanov and the Origins of Systems Thinking in Russia.* (Ashgate, 1998), Giulia Rispoli, "*La tectologia di Bogdanov come nuovo paradigma fra la cibernetica e la teoria generale dei sistemi*", *Epistemologia*, 36:2 (2013), pp. 315–330, James White, *Red Hamlet* (Haymarket, 2019) and Maja Soboleva, "*Vom Empiriokritizismus zum Empiriomonismus: Aleksander Bogdanovs Rezeption der Epistemologie von Ernst Mach*", in Stadler *op.cit.*, pp. 87–97. Visit https://bogdanovlibrary.org, https://brill.com/view/serial/HM (volumes 111 [2015] and 197 [2020] and https://monoskop.org/Alexander_Bogdanov. See also Notes 14 and 16 of this chapter.
26. See Richard Steyn, *Smuts: Unafraid of Greatness* (Jonathon Ball, 2015) and *idem*, *Churchill's Confidant* (Robinson, 2018).
27. See Fritjof Capra and Pier Luigi Luisi, *The Systems View of Life: A Unifying Vision* (CUP, 2014, Part II). Capra's early ideas have been chastised by Jeremy Bernstein, Leon Lederman, Peter Woit and Eric Scerri.
28. See Manfred Drack, "Ludwig von Bertalanffy's Organismic View on the Theory of Evolution", *Journal of Experimental Zoology, Part B: Molecular and Developmental Evolution*, 324:2 (March 2015), pp. 77–90. Biology is the study of life and living organisms, including their structure, function, growth, origin, evolution, distribution, and taxonomy. Whereas the second half of the twentieth century was dominated by biochemistry, molecular biology, genetics and cell biology, the development of new technologies to probe the complexity of living systems has soared since the year 2000.
29. See Susan Nolen-Hoeksema et al., *Introduction to Psychology* (Cengage, 2014, Chap. 5) and Elliot Aronson et al., *Social Psychology* (Pearson, 2018, Chap. 4).
30. In June 2015, a reprint of the book, with a foreword by Wolfgang Hofkirchner and David Rousseau, was published in New York. In the meantime, four important books had been published: Gerald Weinberg, *An Introduction to General Systems Thinking* (Dorset House, 2001), Lars Skyttner, *General Systems Theory* (World Scientific, 2005), Michael Luft, *Systematik: Die universale Systemtheorie* (Pro Business, 2006) and Günter Ropohl, *Allgemeine Systemtheorie* (Sigma, 2012). See Note 13 of this chapter and https://doi.org/10.1007/BF02169018.

31. For the concept of a just war, see John Mattox, *Saint Augustine and the theory of Just War* (Continuum, 2006) and Thom Brooks (ed.), *Just War Theory* (Brill, 2013). Polemology (aka war studies) is the interdisciplinary study of war; is is a subset of the study of international relations. A timeline of wars and an interactive map of the battles fought around the world in the last 4,000 years are available on the Internet.
32. See Len Troncale, "Towards a Science of Systems", *Systems Research and Behavioral Science*, 23:3 (May, 2006), pp. 301–321. Explore the websites https://lentroncale.com, http://pespmc1.vub.ac.be/CSBOOKS.html and https://systemsinnovation.io. "The Science of Systems and the System of Sciences" could be the catchy title of an article about the desirability and feasibility of coherent, computer-aided country studies.
33. See Friedrich Hayek, *Law, Legislation and Liberty* (Routledge, 1982, Vol.1, pp. 35–54). The distinction between two types of order (to be treated in the following chapter) is important, because "order/system" is a core concept, playing a central role in building scientific models.
34. Visit http://isss.org and www.newciv.org/ISSS_Primer/seminrya.html. See Notes 30 and 32 of this chapter.
35. The literature on the study of complex systems is very large. See the Springer book series *Understanding Complex Systems* (2004ff), *Proceedings in Complexity* (2013ff) and *Emergence, Complexity and Computation* (2013ff), and the ISCE book series *Managing the Complex* (2005ff). The Journal of *Systems Science and Complexity* (2006ff) is dedicated to "publishing high quality papers on mathematical theories, methodologies, and applications of systems science and complexity science". The online open-access journal *Northeast Journal of Complex Systems*, run by a community of complex systems researchers, was recently launched. Informative websites are http://www.art-sciencefactory.com/complexity-map_feb09.html, https://www.santafe.edu, http://necsi.edu, http://csh.ac.at, https://warwick.ac.uk/fac/cross_fac/complexity, http://www.intelligence-complexite.org, https://www.paralimes.org, https://www.nico.northwestern.edu, https://www.isc.cnr.it, http://complex.upf.edu, https://cssociety.org, http://yrcss.cssociety.org, http://coco.binghamton.edu, http://complex.um.si, http://www.comdig.org, https://ccs2020.org, http://www.scholarpedia.org/article/complex_systems, https://www.springer.com/gp/physics/complexity, https://www.complexthinking.org and https://doi.org/10.1177/030913250603000317. For a list of complexity organizations, visit http://evodevouniverse.com. In April 2019, the 4th World Conference on Complex Systems was held, in Quarzazate, Morocco, and the 10th International Conference on Complex Systems was organized in July 2020, in Boston, USA. There is a growing community of mathematicians who regard the equal sign (=) as a veneer hiding important complexities that could unlock solutions to a great number of problems. Go to https://www.quantamagazine.org/with-category-theory-mathematics-escapes-from-equality-20191010. "The greatest challenge today, not just in cell biology and ecology but in all sciences, is the

accurate and complete description of complex systems" (Edward Wilson). It needs to be realized that complex systems are a special kind of systems (see Notes 13–16 of this chapter).

36. See Ritzer *op.cit.*, pp.735–740, Aronson *op.cit.*, Chaps. 8–12, and Paul Van Lange et al. (eds.), *Handbook of Theories of Social Psychology* (Sage, 2012, Parts IV and V).
37. See Melanie Mitchell, *Complexity: A Guided Tour* (OUP, 2009, pp. 13–14, 94–111) and go to https://arxiv.org/abs/1909.13243v1 and https://doi.org/10.3389/fphy.2020.00035.
38. See Klaus Mainzer, *Thinking in Complexity* (Springer, 2007, pp. 1–16). The central thesis of Manuel de Landa, *A Thousand Years of Nonlinear History* (Swerve, 2000) is that all structures surrounding us "are the products of specific historical processes" (p. 11).
39. Leading journals are *Journal of Nonlinear Science* (1991ff), *Advanced Nonlinear Studies* (2001ff), *Journal of Nonlinear Sciences and Applications* (2008ff) and *International Journal of Applied Nonlinear Science* (2013ff).
40. See Toney Hey and Gyuri Pápay, *The Computing Universe* (CUP, 2015, pp. 141–173).
41. See Brian O'Shaughnessy, *The Will* (CUP, 2008), John Baer et al. (eds.), *Are We Free?* (OUP, 2008), Galen Strawson, *Freedom and Belief* (OUP, 2010) and Sven Walter, *Illusion freier Wille?* (J.B. Metzler, 2016). Go to https://www.iep.utm.edu/freewill, https://plato.stanford.edu/entries/freewill and https://philpapers.org/browse/free-will. The word freedom sums up the thoughts of Zhuangzi (369–286 BC).
42. See Scott Jordan et al., "The Role of Complex Systems Theory in Cognitive Science", *Cognitive Processing*, 16:4 (November 2015), pp. 315–317.
43. Go to https://iacs4.signges.de. See Jordan Zlatev, "Cognitive Semiotics: An emerging field for the transdisciplinary study of meaning", *The Public Journal of Semiotics*, 4:1 (October 2012), pp. 2–24. For a possible relationship between semiotics and category theory, go to https://arxiv.org/abs/1311.4376.
44. Go to http://www.digitalpeirce.fee.unicamp.br/complex and read Daniel Everett, "The American Aristotle", *AEON*, August 15, 2019 (online).
45. See Donella Meadows et al., *Limits to Growth: The 30-year Update* (Chelsea Green, 2004) and Graham Turner and Cathy Alexander, "Limits to Growth was right. New research shows we're nearing collapse", *The Guardian,* September 2, 2014.
46. Visit https://www.systemdynamics.org. See Bilash Bala et al., *System Dynamics* (Springer, 2017) and John Sterman, *System Dynamics at Sixty: The Path Forward* (September, 2018, online).
47. Visit https://doi.org/10.1002/9781118901731.iecrm0074. See Boris Hasselblatt and Anatole Katok (eds.), *Handbook of Dynamical Systems* (Elsevier, 2002).
48. See David Ruelle, *Chance and Chaos* (PUP, 1993). This book is to be supplemented by David Feldman, *Chaos and Dynamic Systems* (PUP, 2019). The Elsevier Series *Handbook of Dynamical Systems* is indispensable. For

quick reference, go to http://www.scholarpedia.org/article/Encyclopedia:Dynamical_systems. Also visit http://www.math.rug.nl/dsmp. In addition, see the *Journal of Dynamical and Control Systems* (1995ff), *Far East Journal of Dynamical Systems* (1999ff), *Dynamical Systems* (2001ff), *Journal of Applied Dynamical Systems* (2002ff) and *Global Journal of Dynamical Systems and Applications* (2011ff). For application of dynamical systems theory to economies, see Wainwright and Chiang *op.cit.*, part 5.

49. See Ralf Konersmann (ed.), *Wörterbuch der philosophischen Metaphern* (WBG, 2014, pp.252–264).
50. See Yiji Tang (†) and Zhonghua Li (eds.), *Zhōnggúo Rú xué shǐ* [The History of Chinese Confucianism], 9 vols. (Beijing University Press, 2010–2011). This monumental work should be critically compared with *The History of Christianity*, 9 vols. (CUP, 2005–2009).
51. See Mei-hui Yang, Gifts, *Favors and Banquets* (Cornell University Press, 1994), Kwang-Kuo Hwang, *Foundations of Chinese Psychology* (Springer, 2012, pp. 297–326), Ailei Xie, *Family Strategies, Guanxi, and School Succes in Rural China* (Routledge, 2016) and Yanjie Bian, *Guanxi: How China Works* (Polity Press, 2019). Many articles dealing with the subject can be found in *The China Quarterly*.
52. See Jesse Prinz, *Beyond Human Nature* (W.W. Norton, 2012), David Matsumoto and Linda Juang, *Culture and Psychology* (Cengage, 2017) and Larsen and Buss *op.cit.*, part V.
53. See Jenann Ismael, *The Situated Self* (OUP, 2007) and Philip Robbins and Murat Aydede (eds.), *Handbook of Situated Cognition* (CUP, 2008). Browse the pages of *Journal of Cognition and Culture* (2001ff).
54. See Georges Gusdorf, *La découverte de soi* (PUF, 1948), John O'Neil (ed.), *Modes of Individualism and Collectivism* (Heinemann, 1973), Louis Dumont, *Essais sur l'individualisme* (Seuil, 1983), Philippe Ariès and Georges Duby (eds.), *A History of Private Life* (HUP, 1987), Alain Renaut, *L'Ère de l'individu* (Gallimard, 1989), Charles Taylor, *Sources of the Self* (CUP, 1989), Enno Rudolph, *Odyssee des Individuums* (J.B. Metzler, 1991), Aaron Gurevich, *The Origins of European Individualism* (Blackwell, 1995), Steven Lukes, *Individualism* (ECPR, 2006), Jeffrey Alexander, *The Civil Sphere* (OUP, 2006, pp. 31–32, 178–184), Alan Macfarlane, *Individualism, Capitalism and the Modern World* (Create Space, 2013), Larry Siedentop, *Inventing the Individual* (Penguin, 2015), Henry Rosemont Jr., *Against Individualism* (Lexington, 2015) and Peter Callero, *The Myth of Individualism* (R&L, 2017).
55. Barabási is the co-author of *Network Science* (CUP, 2016). See also Guido Caldarelli and Michele Catanzaro, *Networks: A Very Short Introduction* (OUP, 2012), Ulrich Weinberg, *Network Thinking: Was kommt nach dem Brockhaus/-Denken?* (Murmann, 2015), Ernesto Estrada, *The Structure of Complex Networks* (OUP, 2016), César Hidalgo, "Disconnected, fragmented, or United? A trans-disciplinary review of network science", *Applied Network Science*, 1:6 (2016), pp. 1–19, Peter Groenewegen et al. (eds.), *Structure, Content and Meaning of Organizational Networks: Extending Network Thinking* (Emerald,

2017) and Filippo Menczer et al., *A First Course in Network Science* (CUP, 2020). Goran Putnik and Maria Cruz-Cunha (eds.), *Encyclopedia of Networked and Virtual Organizations* (IGI Global, 2008) shall not be overlooked. Professor Cruz-Cunha is co-editor of the *International Journal of Web Portals* (2009ff). See Naoki Masuda et al. (eds.), *Proceedings of NetSci-X2020* (Springer) and visit https://netscisociety.net, https://iuni.iu.edu/networks2021, http://isidcn.dieei.unict.it, https://appliednetsci.springeropen.com and https://www.cnn.group.cam.ac.uk.

56. See Douglas Hofstadter, *I Am a Strange Loop* (Basic Books, 2007) and Pam Weintraub, "The Biology of Love", *AEON*, February 13, 2020. Unfortunately, Hofstadter has nothing to say about the way of thought in ancient India, or China, nor about the much debated nature of, and relationship between, time and space, in which we are all living.
57. See Charles Rathkopf, "Network Representation & Complex Systems", *Synthese*, 195:1 (January 2018), pp. 55–78. Compare this article with Melanie Mitchell, "Complex systems: Network Thinking", *Artificial Intelligence*, 170:18 (2006), pp. 1194–1212. See further Petter Holme and Jari Saramäki (eds.), *Temporal Network Theory* (Springer, 2019). Visit https://arxiv.org/abs/1911.03805, https://www.complexnetworks.org and https://www.frontiersin.org/articles/10.3389/frobt.2020.00057/full. See Note 55 of this chapter.
58. See Jeremy Butterfield and John Earman (eds.), *Philosophy of Physics* (Elsevier, 2007, pp. 923–1074) and Luca Peliti, *Statistical Mechanics in a Nutshell* (PUP, 2011).
59. Visit https://plato.stanford.edu/entries/value-theory and https://philpapers.org/browse/value.
60. See Luciano Pietronero, "Complexity Ideas from Condensed Matter and Statistical Physics", *Europhysics News*, 39:6 (November–December 2008), pp. 26–29. For social physics, go to https://arxiv.org/abs/2110.01866v1.
61. See Larsen and Buss *op.cit.* pp. 2–21.
62. See Serge Galam, *Sociophysics* (Springer, 2012), Parongama Sen and Bikas Chakrabarti, *Sociophysics: An Introduction* (OUP, 2013), Alex Pentland, *Social Physics* (Penguin, 2015), Nilanjan Dey et al. (eds.), *Social Networks Science* (Springer, 2018) and Stephen Borgatti et al. (eds.), *Analyzing Social Networks* (Sage, 2018).

    The research interest of Joshua Epstein, director of the Johns Hopkins University Center for Advanced Modeling in the Social, Behavioral, and Health Sciences (CAM), has been the modeling of complex social dynamics by using mathematical and computational methods. Whether his models include the ways of the socio-culturally embedded human mind, however, remains to be seen, but such a model is a *sine qua non* for overcoming the covid-19 pandemic. See Giuliano di Bernardo, *La conoscenza umana* (Marsilio, 2010, Chap. 7) and go to https://doi.org/10.1177/1368431021991046.
63. Even chemists struggle with the problem of relationship. Visit https://doi.org/10.1002/ange.200800480.
64. See Daniel Larose et al., *Discovering Knowledge in Data* (Wiley, 2014).

65. Visit https://www.army.mil/article/217033/combat_capabilities_development_command_overview and http://www.ns-cta.org.
66. Visit http://ntme.lakecomoschool.org.
67. Visit https://arxiv.org/abs/1804.03488 and https://arxiv.org/pdf/2011.04623v1.pdf. In addition, see Mark Dickison et al., *Multilayer Social Networks* (CUP, 2016) and Ginestra Bianconi, *Multilayer Networks* (OUP, 2018). Multidimensional networks are a kind of multilayer networks. Ever more sophisticated attempts are made to model real-world systems as multidimensional networks.
68. See Michele Berlingerio et al., "Foundations of Multidimensional Network Analysis", 2011 (online).
69. Visit https://arxiv.org/ftp/arxiv/papers/1304/1304.6736.pdf. The 17th International Symposium on Neural Networks was held in Cairo, on December 4–6, 2020. May the participants have realized that measuring the brain's operations is not he same as understanding the human mind! See the journal *Network Neuroscience* (2017ff).
70. See Leone Montagnini, *Harmonies of Disorder: Norbert Wiener: A Mathematician-Philosopher of Our Time* (Springer, 2017). In *God and Golem, Inc.* (MIT Press, 1964), Wiener concerns himself with major points in cybernetics which may be relevant to religious issues. His *Collected Papers* are located at the MIT library. Also go to https://norbertwiener.org. For the history of cybernetics, on the principles of which spaceflight is based, see Slava Gerovich, *From Newspeak to Cyberspeak* (MIT Press, 2004), Thomas Rid, *Rise of the Machines: A Cybernetic History* (W.W. Norton, 2016) and Dmitry Novikov, *Cybernetics: From Past to Future* (Springer, 2016), a book published in the series (edited by Janusz Kacprzyk) *Studies in Systems: Decision and Control* (2014ff). Also visit http://www.rossashby.info. Aksel Berg, Sergei Sobolev, Alexey Lyapunov, Leonid Kraizmer, Anatoly Kitov, Victor Glushkov and Yuri Gastev were notable Soviet cyberneticists. Charles François, editor of the *International Encylopedia of Systems and Cybernetics*, Harold Lawson, pioneer of work in hardware, software and real-time system technologies, and Albert Müller, president of the Heinz von Foerster Society, attempted to bridge the fields of systems science (see Notes 13, 15 and 16 of this chapter) and cybernetics. See further *Cybernetics and Systems Analysis* (1965ff), *Cybernetics and Systems* (1971ff), Leonid and Anton Grinin, *The Cybernetic Revolution and the Forthcoming Epoch of Self-Regulating Systems* (Uchitel, 2016) and Sergio Barile et al. (eds.), *Cybernetics and Systems* (Routledge, 2018). Interesting is also *Cybernetics & Human Knowing* (1992ff), *Cybernetics and Information Technologies* (2001ff) and *Journal of Systemics, Cybernetics and Informatics* (2003ff). For *third*-order cybernetics, go to https://chaturvedimayank.wordpress.com/tag/third-order-cybernetics.
71. See *International Journal on cybernetics & Informatics* (2012ff).
72. See the special issue of *Kybernetika*, 44:3 (2008), pp. 297–335, on "Sixty Years of Cybernetics", Albert Müller and Karl Müller (eds.), *The Beginning of Heaven and Earth Has No Name* (Fordham University Press,

2014) and Alexander Riegler et al. (eds.), *New Horizons for Second-Order Cybernetics* (World Scientific, 2018). Interesting is also Klaus Krippendorff, *Content Analysis* (Sage, 2018). Currently, some theorists invoke a third, or even higher, order cybernetics. Go to http://cybsoc.org, http://www.ieeesmc.org, http://pespmc1.vub.ac.be, https://gws-kybernetik.org, http://pcp.vub.ac.be/Papers/Cybernetics-EPST.pdf, www.wissenskulturen.de (> Soziokybernetik), https://www.isa-sociology.org/en/research-networks/research-committees/rc51-sociocybernetics and http://wosc.co. Volume 67, Issue 4 (July 2019) of *Current Sociology* is dedicated to sociocybernetics. The online *Journal of Sociocybernetics* (2000ff) and *Brill Research Perspectives in Sociocybernetics and Complexity* (2020ff) are also notable. Sociocybernetics should be compared but not confused with sociophysics, for which visit https://doi.org/10.1038/s41598-019-53300-4. See Note 62 of this chapter and *Int. Journal of Systems and Society* (2014ff).

73. See Steve Heims, *The Cybernetics Group* (MIT Press, 1991) and *Constructing a Social Science for Postwar America* (MIT Press, 1993).
74. See Susan Nolen-Hoeksema et al., *Introduction to Psychology* (Cengage, 2014, pp. 346–347).
75. See Robert Kuttner, *Everything for Sale* (Knopf, 1997), Felicity Lawrence, *Not on the Label* (Penguin, 2004), Tony Lawson, "The Nature of Heterodox Economics", *Cambridge Journal of Economics*, 30:4 (July 2006), pp. 483–505, Marc Perelman, *Le sport barbare: Critique d'un fléau mondial* (Michalon, 2011), Joanna Blythman, *Swallow This* (Fourth Estate, 2015), Maja Göpel, *The Great Mindshift* (Springer, 2016), Simon Springer et al. (eds.), *Handbook of Neoliberalism* (Routledge, 2016), Kate Raworth, *Doughnut Economics* (Cornerstone, 2018), Paul Collier, *The Future of Capitalism* (Harper, 2018), Damien Cahill et al. (eds.), *Handbook of Neoliberalism* (Sage, 2018), Adrian Pabst, *Liberal World Order and Its Critics* (Routledge, 2018), Edmund Fawcett, *Liberalism: The Life of an Idea* (PUP, 2018), John Quiggin, *Economics in Two Lessons* (PUP, 2019), Abhijit Banerjee and Esther Duflo, *Good Economics for Hard Times* (PublicAffairs, 2019), Jonathan Aldred, *Licence to be Bad* (Allen Lane, 2019), Fabio Ciconte and Stefano Liberti, *Il grande carrello* (Laterza, 2019), Waldemar Zeiler, *Unfuck the Economy* (Goldmann, 2020), Blair Fix, "Why Free Market Ideology is a Double Lie", *Evonomics*, September 26, 2020 (online), Henri Sterdyniak and Stéphanie Villers, *Crise économique 2020* (Economica, 2020) and Mariana Mazzucato, *Mission Economy: A Moonshot Guide to Changing Capitalism* (Harper, forthcoming). *Routledge Advances in Heterodox Economics* (2008ff) publishes "innovative books in heterodox economic theory, policy, philosophy, intellectual history, institutional history, and pedagogy". Visit http://hetecon.net, http://www.paecon.net, http://fixcapitalism.com, https://www.worldeconomicsassociation.org, https://www.ineteconomics.org and http://www.atterres.org.
76. See Elliot Aronson et al., *Social Psychology* (Pearson, 2018, pp. 246–288).

77. Leading journals are *Kölner Zeitschrift für Soziologie und Sozialpsychologie* (1921ff), *Journal of Social Psychology* (1929ff), *British Journal of Social Psychology* (1962ff), *Journal of Personality and Social Psychology* (1965ff), *Journal of Experimental Social Psychology* (1965ff), *Journal of Applied Social Psychology* (1971ff), *European Journal of Social Psychology* (1971ff), *Japanese Journal of Experimental Social Psychology* (1971ff), *Personality and Social Psychology Bulletin* (1974ff), *Les Cahiers Internationaux de Psychologie Sociale* (1978ff), *Basic and Applied Social Psychology* (1980ff), *European Review of Social Psychology* (1990ff), *Journal of Community & Applied Social Psychology* (1991ff), *Revue Internationale de Psychosociologie* (1994ff), *Personality and Social Psychology Review* (1997), *Asian Journal of Social Psychology* (1998ff), *Psychologie et Société* (1999ff), *Nouvelle Revue de Psychologie Sociale* (2006ff), *Social Psychology* (2008ff) and *Journal of Theoretical Social Psychology* (2017ff). The website https://www.socialpsychology.org is an inexhaustible source of information. Network scientists should more cooperate with social psychologists.
78. A list of mass hysteria cases is provided on the Internet. For "swarm intelligence", go to http://www.scholarpedia.org/article/Swarm_intelligence.
79. For the important differences between intrapersonal, interpersonal, intragroup, and intergroup processes, see Michael Hogg and Joel Cooper (eds.), *Handbook of Social Psychology* (Sage, 2007, pp. xv–xviii). For intergroup contact and collective action, see the special issue of *Journal of Theoretical Social Psychology*, 3:1 (2019).
80. See Spencer Pack, *Aristotle, Adam Smith and Karl Marx* (Edward Elgar, 2010, pp. x–xv), Ricardo Crespo, *A Re-Assessment of Aristotle's Economic Thought* (Routledge, 2014, pp. 9–49) and Robert Gallagher, *Aristotle's Critique of Political Economy* (Routledge, 2018).
81. Like "cybernetics", the word government stems from κυβερνάω (to steer, to navigate, to rule, to govern). Few governors or heads of government (helmsmen) are acquainted with the latest in cybernetics, to the detriment of those being governed. When the herd goes astray, the herdsman is to blame.
82. The literature on institutions, institutionalization and institutional change is vast and growing. See Douglass North, *Institutions, Institutional Change and Economic Performance* (CUP, 1990), Robert Boyer and Yves Saillard (eds.), *Théorie de la régulation: l'état des savoirs* (La Découverte, 2002, Chaps. 7–9 and 49), Karl-Ernst Schenk, *Economic Institutions and Complexity* (Edward Elgar, 2003), Thráinn Eggertsson, *Imperfect Institutions* (University of Michigan Press, 2005), Elinor Ostrom, *Understanding Institutional Diversity* (PUP, 2005), Ha-Joon Chang (ed.), *Institutional Change and Economic Development* (Anthem, 2007), Bernard Chavance, *L'économie institutionelle* (La Découverte, 2012), Rudolf Richter, *Essays in New Institutional Economics* (Springer, 2015) and Francesco Parisi (ed.), *Handbook of Law and Economics* (OUP, Volume 3, 2017). Go to http://winir.org, https://www.sioe.org, http://coase.org/niereadinglist.htm and www.vcasi.org/en. A special issue of *Social Anthropology* is devoted to "rethinking institutions" (19:4 [November 2011],

pp. 363–494). For social institutions, go to https://plato.stanford.edu/entries/social-institutions.

83. For performances, see the works of Erving Goffman and John Searle. A "speech act" not only provides a piece of information but also performs an action. According to the American philosopher Kent Bach, "almost any speech act is the performance of several acts at once." For research on rituals in ancient India, see Axel Michaels (ed.), *Ritual Dynamics and the Science of Ritual* (Harrassowitz, 2010). Research on rituals in ancient China, as recorded in the Classics Zhōulǐ, Yílǐ and Lǐjì, is still a *desideratum*.
84. The scientific activities of the International Sociological Association (ISA) are decentralized in many interrelated research committees, each dealing with a specialty in sociology. The research committee Economy and Society is only one of them.
85. For the relationship between law and religion, go to http://www.iclars.org. For the relationship between religion and social order, visit https://www.sociologyofreligion.com. For a list of the (interrelated) research committees of the International Political Science Association (IPSA), visit https://www.ipsa.org (> research > list of research committees). IPSA and ISA (see previous note) are members of the International Social Science Council (ISSC), which recently merged with the International Council for Science (ICSU).
86. See Neva Goodwin et al., *Principles of Economics in Context* (Routledge, 2015).
87. In addition, visit https://plato.stanford.edu/entries/ethics-business and http://isbee.org, and read Robert Frederick (ed.), *A Companion to Business Ethics* (Blackwell, 2002), Jan Peil and Irene van Staveren (eds.), *Handbook of Economics and Ethics* (Edward Elgar, 2009), Patrick O'Sullivan et al. (eds.), *Business Ethics* (Routledge, 2012), John Tomasi, *Free Market Fairness* (PUP, 2013), Adam Arvidsson and Nicolai Peitersen, *The Ethical Economy* (Columbia University Press, 2013), Christoph Lütge (ed.), *Handbook of the Philosophical Foundations of Business Ethics* (Springer, 2013), George DeMartino and Deirdre McCloskey (eds.), *Professional Economic Ethics* (OUP, 2016) and Manuel Velasquez, *Business Ethics: Concepts and Cases* (Pearson, 2017).
88. See Steve Keen *op.cit.* pp. 175–202), John Cassidy, *How Markets Fail: the logic of Economic Calamities* (FSG, 2009) and Michael Jacobs and Mariana Mazzucato (eds.), *Rethinking Economics* (Wiley, 2016). See Note 75 of this chapter.
89. See Ronald Miller and Peter Blair, *Input–Output Analysis: Foundations and Extensions* (CUP, 2009) and Thijs ten Raa (ed.), *Handbook of Input–Output Analysis* (Edward Elgar, 2017). The Nobel laureates Paul Samuelson, Robert Solow, Vernon Smith and Thomas Schelling were doctoral students of Wassily Leontief, who passed away in 1999. Leontief was influenced by Léon Walras and influenced George Dantzig and Jan Tinbergen.
90. See Jean Tirole, *Théorie de l'organisation industrielle* (Economica, 2015), *International Journal of Industrial Organization* (1983ff) and *Review of Industrial Organization* (1984ff).

91. See Werner Dubitzky et al. (eds.), *Encyclopedia of Systems Biology* (Springer, 2013); Edda Klipp et al., *Systems Biology: A Textbook* (Wiley, 2016); Uri Alon, *An Introduction to Systems Biology* (CRC Press, 2019) and Sven Jørgensen, *Introduction to Systems Ecology* (CRC Press, 2012). Visit https://www.springer.com/gp (fill in: systems biology). Systems ecology can be seen as an application of general systems theory to ecology. Visit https://www.sciencedirect.com/referencework/9780124095489/earth-systems-and-environmental-sciences (search systems ecology).
92. In the physical sciences, the systemic approach is virtually common place. For the systemic (complexity) turn in the social sciences and humanities, see Eric Beinhocker, *The Origin of Wealth* (Harvard Business School Press, 2006); Brian Castellani and Frederic Hafferty, *Sociology and Complexity Science* (Springer, 2009, pp. 33–65); Ton Jörg, *New Thinking in Complexity for the Social Sciences and Humanities* (Springer, 2011); Ricardo Hausmann et al., *The Atlas of Economic Complexity* (Puritan Press, 2011); Andrzej Nowak et al. (eds.), *Complex Human Dynamics* (Springer, 2013); Paul Youngman and Mirsad Hadzikadic, *Complexity and the Human Experience* (Routledge, 2014); W. Brian Arthur, *Complexity and the Economy* (OUP, 2015); Wolfram Elsner et al., *The Microeconomics of Complex Economies* (Elsevier, 2015) and the Routledge series *Complexity in Social Science* (2017ff). See note 35.
93. See Ludwig von Bertalanffy, "An Outline of General System Theory", *The British Journal for the Philosophy of Science*, 1:2 (August 1, 1950), pp. 134–165.
94. See Paul Weingartner, "The Ideal of Mathematization of All Sciences and of 'More Geometrico' in Descartes and Leibniz", in William Shea (ed.), *Nature Mathematized* (Reidel, 1983, pp. 151–195), and Witold Marciszewski and Roman Murawski, *Mechanization of Reasoning in Historical Perspective* (Rodopi, 1995, pp. 77–112).
95. Readers who want to know how far man has progressed on the road toward automated reasoning should look for Alan Robinson and Andrei Voronkov (eds.), *Handbook of Automated Reasoning* (Elsevier, 2001), John Harrison, *Handbook of Practical Logic and Automated Reasoning* (CUP, 2009) and Stefania Centrone et al. (eds.), *Mathesis Universalis, Computability and Proof* (Springer, 2019, pp. 1–6). They should also visit https://www.ontology.co/mathesis-universalis.htm and https://ijcar2020.org. Whereas Western logic, adhering to the principle of bivalence and upholding contradictions rather than contraries, has been excluding, the logic of the ancient Chinese was including, because they were thinking holistically (a way of thought modern scientists in the West are becoming increasingly attracted to).
96. See Brigstocke *op. cit.* pp. 606–609, Lynn Gamwell, *Mathematics + Art: A Cultural History* (PUP, 2016, Chap. 2) and Kristóf Fenyvesi and Tuuli Lähdesmäki (eds.), *Aesthetics of Interdisciplinarity: Arts and Mathematics* (Birkhäuser, 2017). Visit http://bridgesmathart.org.
97. See Mario Livio, *Golden Ratio* (Broadway Books, 2002) and go to https://goldennumber.net.
98. Visit http://frocos.cs.uiowa.edu.

Chapter 6
Countries Consist of Both Spontaneous and Man-Made Systems

Abstract The important distinction is made between self-organizing (spontaneous) and not self-organizing (man-made) systems. The universe and living organisms are examples of the former; companies and political parties typify the latter. Countries are combinations of spontaneous and artificial systems. The thriving study of organizations is shown to have close affinity with the science of systems.

In the previous chapter, I stated that countries are huge combinations of man-made and natural systems. This calls for an explanation, for this important difference is often overlooked. Before further exploring the possibility of collaborative country studies, I attempt to provide the exploration in this chapter, by way of *intermezzo*.

The discussion about the distinction and relationship between nature and convention, custom or tradition, i.e., between the natural and the artificial, can be traced back to ancient Greece. Socrates, the first moral philosopher in the West, and Plato, his main pupil, who became Aristotle's teacher, profoundly disagreed with Sophists like Callicles and Trasymachus over the question as to whether laws were a matter of *physis* (given by nature) or *nomos* (made by man). The debate was not about how to distinguish facts from values, but about the *source* of value and the *justification* of morality. The key question was: IS MIGHT RIGHT? Centuries later, this question would be rephrased in: "Can the waging of war [struggle for might] be justified?" or "Which price is the *iustum pretium*?" The answer of the Sophists to the first question was: Yes. Socrates and Plato forcefully argued against the idea of their opponents. The books to be read are Gregory Vlastos, *Socrates* (CUP, 1991, pp. 179–232), Pierris Apostolos, *Physis and Nomos* (Patras, 2007), Marc Garcet and Serge Dalla Piazza, *Entre Physis et Nomos* (L'Harmattan, 2016) and Thanos Zartaloudis, *The Birth of Nomos* (Edinburgh University Press, 2020; publication of the companion volume *The Birth of Physis* is forthcoming). The Web site to be visited is https://plato.stanford.edu/entries/callicles-thrasymachus.

The question is still relevant. Whereas Thucydides, Machiavelli, Hobbes, Nietzsche, George Kennan, Hans Morgenthau, Raymond Aron, Kenneth Waltz, Zbigniew Brzezinski, Robert Gilpin, Stephen Krasner, Paul Kennedy, John Mearsheimer, Robert Kaplan, Charles Glaser, Stephen Walt, Robert Kagan and Randall Schweller emphasized or are emphasizing the ubiquity of power (might) and Elias Canetti

H. Kuijper, *Comprehending the Complexity of Countries*,
https://doi.org/10.1007/978-981-16-4709-3_6

(1904–1994) explores the power of the crowd in his masterpiece *Masse und Macht* (Claassen,1960), Jeffrey Alexander advances, in his widely reviewed and hugely informative book *The Civil Society* (OUP, 2006), the thesis that justice, which depends on a feeling of "being part of something larger than ourselves" (on what Adam Smith designated as fellow-feeling), is possible, ending his massive book by writing that civil society is a project, "inspiring hope for a democratic life".

The idea of justice (rightfulness) became central to the revival of political theory unleashed by the publication of John Rawls' controversial book *A Theory of Justice*, published by Oxford University Pres, in 1972. Whether justice is to be taken as fairness or equity, is a question for others to answer, but I find it remarkable that Rawls (*a*) relegates Plato's *Republic* to two short footnotes on pp. 454 and 521 and (*b*) seems to be unaware of the etymological relationship between "order" and "right". See Chandran Kukathas and Philip Pettit, *Rawls:* A Theory of Justice *and Its Critics* (Polity Press, 1990). Though the discussion continues unabated (visit https://www.iep.utm.edu/dist-jus, https://www.iep.utm.edu/justwest, https://plato.stanford.edu/entries/justice and https://philpapers.org/s/justice), Yushun Huang, a prominent philosopher currently engaged in the creative development of Confucian philosophy in China, countered with *Voice from the East: The Chinese Theory of Justice* (Path International, 2016).

Countries are more than sets; as explained in Chap. 5, they are systems. If the political side is emphasized, countries are called "states"; if the cultural side is stressed, they are called "nations". Countries have a population, a collection of individuals forming a community. Since individuals are nonlinearly connected, their country is a complex system, and because they interact with, and adapt to, each other, their country is a complex, dynamic system. Finally, the complexity and dynamics of a country are reinforced by the mutual relations individuals have with their constantly changing natural environment (*Umwelt*). For example, when it starts raining (a natural event), people will, in different ways, act differently from when the sun breaks through. A tsunami, earthquake or volcanic eruption (natural events) will cause all kinds of human actions, which differ from the actions caused by a bombardment (certainly not a natural event).

There are different kinds of community: local, national, academic, business, financial, rural, urban, political, ethnic, religious, gay, health-related, football, online, old age or protest communities. The following sentences make this clear: "He is well known in the local community"; "There is a large black community living in this area"; "Her speech caused outrage among the gay community"; "There is a real sense of community in that neighborhood"; "His ideas have attracted a lot of attention in the scientific community"; "The tragedy has shocked the close-knit rural community"; "The club is dedicated to the business community"; "The restaurant is popular with the old age community"; and "He insulted the Chinese community as a whole".

"Community" refers to a group of people living in an area or being considered as a unit because of their commonality. With countries there is a fusion of area and commonality. Countries are thus a particular kind of community. As we saw in Chap. 3, countries are bordered stretches of land with a politically or otherwise organized people who, to some extent, share a history and have a sense of belonging. Countries are locally defined and historically shaped organizations. Both locality and organization are important for characterizing a country. Think of Iceland versus Nepal, Japan versus Indonesia, Bolivia versus Afghanistan or the United States of America (USA) versus the People's Republic of China (PRC).[1] Since there are natural (grown, spontaneous or endogenous) and non-natural (constructed, man-made or exogenous) organizations, it may be useful to say a few words about "organization", a concept I conceive of as synonymous with "system". The natural vs. artificial organizations distinction, highlighted by Montesquieu (1689–1755), is important. It seems to have escaped the attention of, most notably, Béla Bánáthy (1919–2003) and George Klir (1932–2016), who were at some time co-editors of the newsletter of the International Federation of Systems Research. One may even wonder whether Ludwig von Bertalanffy himself was aware of the fundamental difference between the natural (spontaneous) and non-natural (constructed) order of things.

In the course of time, there can be a change from relying on the natural to relying on the non-natural order of things. For example, whereas Christopher Columbus, Vasco da Gama, Pedro Cabral, Ferdinand Magellan and James Cook were dependent on the system of trade winds, the captain of a modern seagoing vessel relies on the chief engineer in charge of the engine room, the engine being a system of parts put together by man. The book to be read on this epoch-making change in the maritime history is Lincoln Paine, *The Sea and Civilization* (Knopf, 2013). It opens with the catchy sentence: "I want to change the way you see the world".

"Organization" may refer either to a process (e.g., "The organization of the conference was time-consuming") or to the result of it (e.g., "The organization has been changed"). As the result of a process of organizing, an organization has a structure, a framework within which organizational changes occur or are brought about. There is a difference in meaning between the sentences "The organization has changed" and "The organization has been changed"; in the former case, the verb "to change" is used intransitively, in the latter case, transitively. Bogdanov's *tektology* (from the Latin *tectum* [roof, the completion of a dwelling] and the Greek λόγος [discourse]) did not stress the importance of differentiating between natural and man-made organizations. However, in a country where physical thinking was officially promulgated, he prudently diverted to organic thinking and from there to human organization.

There is yet another distinction to be made. Organization is what gives a variety of parts a unity; it is a holistic concept. The parts of an airplane per se do not fly; the *organization* of the parts flies. Similarly, chemicals per se do not make life; their *organization* causes the life capacity of an organism. Biologically, life is a matter of natural organization; sociologically, life is a matter of living together in a way determined by man. "Order" means simply configuration of elements. For example, crystals show a configuration of atoms; books, having been ordened by somebody, show an order in a bookcase; in a game of football, the players at both sides show a dynamic configuration, which football analysts use to explain critically. It should be kept in mind that a flying plane consists of connected parts. Life without connected chemicals is biologically impossible, as much as life without one or more than one companion is sociologically inconceivable. Whereas "order" emphasizes the parts of a whole, "organization" stresses the whole of parts. "Order" and "organization" are interchanged, however.

Natural Order

A natural order, organization or system is an order that is not ordered by people and needs no human activity to be an order; it is an *inner* order, an order which is self-organized.[2] For example, there are many kinds of trees, e.g., beeches, oaks, elms, birches, poplars, ashes, planes, willows, lime trees, apple trees, pear trees, peach trees, plum trees, cherry trees, almonds and walnut trees. Trees are wonderful, natural systems. Nobody is involved in their growth, which may only be *facilitated* by man. The roots of a tree "know" how to suck up the nutrients-containing-water in the soil, and how much of it to suck up. The up-flowing sap "knows" where to change direction in order to feed the spreading branches and their leaves or fruits. Adapting to the vagaries of the weather, trees "know" when to blossom, whether to bear a certain kind and number of fruits and when to drop their leaves. Silently, trees show a recurrent pattern in space and time, without somebody ordering them to follow the pattern. Trees are indescribably beautiful, providing shelter, coolness and shade to anybody, without distinction of persons ("*Ombra mai fu*"). Trees dance, without somebody setting or making the rhythm. Only when they crash onto the ground and die, at man's brutal hand, they make a sound.

"Nature" (from Latin *natus* [born]) is a word with different but related meanings.

- It may refer to everything that is distinct from culture, which is inherently a human affair. Man defines nature as contrasted with culture and culture as "*das Andere des Anderen*".[3] The questions may then arise: how are nature and culture related with each other and is one of them reducible to the other? This would open the door to the controversy between naturalists and idealists, an argument in which I do not want to get bogged down.[4] However, reconciling (the sciences of) nature with (the sciences of) culture or mind is a *desideratum* no university can afford not to deal with.[5]
- "Nature" may also refer to what is considered to be essential of something. For example: "It is in the nature of young men to be foolhardy and impetuous" and "People are by nature inclined to be gregarious". The essential properties of things, individuals or relations are collectively called their essence. They are the properties, attributes, qualities, features or characteristics that things, individuals or relations *must* have.[6] By contrast, the accidental properties of a thing are the properties that it *may* have.[7] Properties are attributed to, and can be predicated of, things, individuals and relations. Readers schooled in analytic philosophy,[8] may consider the phrases that express essential properties to be analytic propositions, statements in which "young men" and "people" already contain what is being said of them, statements that are true by virtue of the meaning of the terms involved (as opposed to synthetic sentences like "bachelors are unhappy" and "ophthalmologists are rich"). Though essential properties indicate *what* a thing is, they make not clear *that* a thing is or exists. If the existence of something is not

convincingly evidenced, however, there is no point in talking about its properties. Essence is thus to be sharply distinguished from existence (*esse*), a distinction theologians tend to overlook. Existence is not a property of anything or anybody. Before the collecting of things, we saw, they have to be identified. The identification of things, however, is a matter of ascertaining their essential properties. So without being sure that a thing exists one cannot discover its essence; without being sure of the essential qualities of things there can be no identification; and without identification there can be no collecting.

Objects belonging to the same natural kind have something in common. For example, a bracelet, brooch, chain, choker, medallion, necklace, pair of cufflinks, pair of earrings, pendant, tiara and wedding ring are made of gold when they are composed of atoms having atomic number 79. The objects made of gold belong to the same natural kind. A fundamental question is whether there is a unique way of grouping things into natural kinds. Differently put: is there a general theory of classifications? Whereas Aristotle stated that kinds must be grouped hierarchically, modern chemists classify natural kinds according to the periodic table drawn up by Dmitri Mendeleev (1834–1907).

- "Nature" may finally refer to the cosmos, to the grand order of all things, which must have filled the ancients who, at night, gazed at the starry sky with deep awe. They did not suffer from light pollution as we do, and they had no telescope at their disposal. Although nature's manifestations are manifold, nature/cosmos implies the unity in diversity which the first cosmologists, who lived in Miletus (a Greek colony on the western coast of Anatolia in present-day Turkey), found particularly beautiful.[9] "*Kósmos*", which can be taken as referring to the system of all systems, also meant "the result of activities that make something look more beautiful by putting things on it or around it (decoration)". Compare "cosmos" with "cosmetics". Whoever looks up to heaven and cannot believe any myth related to it, must wonder when and how the unfathomable cosmos came into being; by what elements it is basically constituted; by what kind of rules it is ruled; and whether these rules can be conceived of as being operative without a ruler or lawmaker, i.e., without falling into the trap of anthropocentrism.[10] Can these ultimate questions be answered by the universe itself; is there no other way but to appeal to revelations written down in "Holy Scriptures" by men claiming, or claimed, to be reliable; or is resignation the third option, the attitude of accepting that the questions are unanswerable and that there is no possibility for mortals to go beyond the ultimate or absolute?

In Chap. 5, I stated that the Ionian cosmologists understood the whole universe, *including human society*, as a system, replacing mythology with cosmology,[11] Their revolutionary view was fatefully rejected by the Church Fathers, who fell back into the old (pre-pre-Socratic) habit of thinking, developed the human construct of a God who created the world (and sent his virgin birth son to rescue humanity) and laid the theological foundations of Christianity. These foundations began to be shaken in the nineteenth century, when biblical criticism reached full flower and natural history largely gave way to more specialized scientific disciplines. Since about 1650 A.D., natural scientists have built on, and refined, the ideas that had been put forward by some of the early Greek Philosophers, leaving the explanation of human activities to colleagues in the faculties of social and human sciences. However, the scientific community has never been able to get rid of the God idea completely.[12] The concept of a creator, who imposed order on the chaos, has been clogging up the thinking of almost all Western scientists. Since Charles Darwin (1809–1882), who made a great contribution to the theory of evolution, there has been a fierce debate between evolutionists and creationists,[13] but Darwin was not an atheist. He eventually settled into a kind of agnosticism and never railed against Christianity.[14] Readers who are doubtful as to whether scientists could do without some idea of God should read Jacques Gernet, *La raison des choses* (Gallimard, 2005).

The Ionian Cosmologists, who made keen observations and boldly speculated without recourse to revelation or supernatural experience, were the first natural philosophers,[15] but they never tested their "theories". Furthermore, they tried to explain the entire universe, whereas modern scientists, who are baffled by the complexity of nature, have been dividing their labor. Physicists are concerned with the nature of, and relations between, matter and energy; chemists with the composition of matter; astronomers with celestial objects; geologists with Planet Earth; and biologists with living things. The sciences of physics, chemistry and astronomy are increasingly subdivided and have been the object of historical or analytical studies. Consideration of the Earth's properties presupposes physical, chemical and astronomic knowledge. On the other hand, biology is based upon knowledge of Planet Earth. Similarly, the social and human sciences, which (from different points of view!) are concerned with human activities, are grounded in, but not reducible to, biology. The individual sciences as well as the formal sciences, which deal with them structurally and formally, are thus closely connected parts of Science as a whole. In 2018, the International Science Council (ISC) was inaugurated. Being a merger of the International Council for Science (founded in 1931) and the International Social Science Council (founded in 1952), the ISC may be expected to serve and support the interests of *all* scientists. Whether this is really the case is something I leave to the reader to find out.

This does not mean that natural philosophy has been replaced by the natural sciences. In 2008, Miguel Espinoza, a philosopher associated with the University of Strasbourg, founded the *Cercle de Philosophie de la Nature*. Its researchers address problems such as the transformation of metaphysics to science, the relationships between mathematics and the sensory world, naturalism, causality and liberty, and the links between the physical, biological, social and psychological levels. In 2009–2012, the proceedings of the conferences of the members of the Circle were published in *Eikasia: Revista de Filosofía*. Since 2014, they have been published in *Scripta Philosophiae Naturalis*. In 2017, Espinoza's widely acclaimed book *La matière éternelle et ses harmonies éphémères* was published by L'Harmattan, in Paris.

The universe is not (man-)made. The belief that the universe was or has been created and made by "God" cannot really be justified, meaning that a good reason for such belief cannot be given. It is an unwarranted, annoying belief. Indeed, it is just a case of make-believe. In physics, chemistry, astronomy, geology and biology, a *part* of the universe is considered to be a system and selected for analysis. Everything outside their system is regarded as environment, which is either left to "outsiders" to deal with or completely ignored. No self-respecting physicist, chemist, astronomer, geologist or biologist will dare to step outside his/her purview. A physical, chemical, astronomical, geological and biological system is a collection of, respectively, physically, chemically, astronomically, geologically and biologically relevant entities belonging together. Received wisdom has it that labor division increases productivity, but the division has also a downside: Labor division (scientific fragmentation) is bound to lead to social alienation, to a situation in which nobody knows where the compartmentalized ship of science is going to. Many a scientist does not understand what the guy or girl next door is doing. The Dean, who is responsible for the organization and management of his/her department (faculty), or the Chancellor, who at best worries about the international ranking of his/her university, has no idea where science is navigating or is navigated to. At higher education, nobody has insight and overview (small wonder, for scientists tend to shun philosophy). The "governor" is nowhere to be seen and the ship of science is floating around. So it remains to be seen whether the benefits of labor division outweigh its disadvantages. Universities (places where both the one and the many should be honored) have degenerated into a number of kingdoms, that is to say, into *multi*versities, where departments are disconnected and operate largely autonomously. The scientist who digs deeply without losing sight of the big picture is hard to find. The *uomo universale* or Renaissance polymath is and will remain a thing of the past. In the age of exploding information, we should pin our hope on interdisciplinary or cross-disciplinary research, which differs fundamentally from multidisciplinary research, for a symphony is performed by an orchestra that consists of professional musicians playing different kinds of instruments under the *bâton* of a conductor. The opposite of symphony is cacophony, a big mess, chaos, madness, "*ein großes Durcheinander*", something like the winning "song" of the Eurovision Song Contest of 2021. See Appendix A and B.

Like other animals, humans are not made or manufactured; we are born, at some place and at some time; we are organisms naturally endowed with a mind. Nature works spontaneously. Nature is not created or constructed; it regulates itself. Nature follows its own dictates and obeys its own laws. Nature is mysteriously self-organized, without being ordered from the outside and without its order or fine-tuned organization being imposed by somebody or something outside of it. Countries are different; they are constituted of both natural and man-made systems. People have to be organized and ruled, with the firm hand of a helmsman. A country left to its own devices is uncontrolled and will turn into chaos. A ship without helmsman goes adrift and a plain without captain will crash.

Organized entities (natural or man-made) come up, appear, can be identified and, slowly or rapidly, disappear. A company that has gone bankrupt does not exist anymore; it has ceased to exist. When I die (at some place, at some time and under certain circumstances), I will return, sink in, immerse and disappear, like a wave or ripple that disappears in the ocean. A cycle has then ended: rising–shining–declining–disappearing, for ever. There will be other waves and ripples, but not the ripple I was or the wave I pretended to be. Human life lasts only for a very short period of time, between two eternities. The belief that we will either go to Paradise or to Hell is a myth, concocted by parasitic clerics, professional swindlers, peddlers of pieces of baked air for people unable or unwilling to face the truth.

"Emergence" (not: "emergency") is a notoriously difficult concept.Philosopher and literary critic George Henry Lewes (1817–1878), known for his relationship with George Eliot, coined the term in 1875 by writing:

> Every resultant is either a sum or a difference of co-operant forces; their sum, when their directions are the same—their difference, when their directions are contrary. Further, every resultant is clearly traceable in its components, because these are homogeneous and commensurable. It is otherwise with emergents, when, instead of adding measurable motion to measurable motion, or things of one kind to other individuals of their kind, there is a co-operation of things of unlike kinds. The emergent is unlike its components insofar as these are incommensurable, and it cannot be reduced to their sum or their difference.

Jens Greve and Annette Schnabel end their marvelous co-edited book *Emergenz* (Suhrkamp, 2017) by writing:

> *So bleibt also ein wichtiger Anspruch auf Emergenz auch in der Welt der modernen Biowissenschaften bestehen, da sich die Frage nach den Eigenschaften des Ganzen noch immer nicht aus einer blossen Addition seiner Teile erklären lässt.*

The emergent properties of an entity or system are thus *new* properties, properties that are irreducible to the properties or qualities of any of its constituent parts. The properties of water cannot be reduced to the properties of hydrogen or oxygen, and the properties of the wood cannot be explained by the properties of the different trees in the wood. Generally, the properties of a whole cannot be explained by those of its parts. Reduction means impoverishment. Knowing more of less ultimately results in knowing everything about nothing. As the British pop singer William Emmanuel Bevan (alias Burial) declaimed with ample voice:

> We persevere in looking at small questions instead of large ones and our view of the forest is forever obscured by the trees. Yet specialized knowledge derives its meaning from the context of larger perspectives and questions. When it loses touch with that larger context, it loses its coherence.

Emergent properties are not to be confused with *essential* properties, which, as we saw, are the properties, attributes, qualities, features or characteristics that things, individuals or relations must have. Essential properties indicate *what* a thing is; they make not clear *that* it is, or exists. If the existence of something is not convincingly evidenced (a condition hard to fulfill), there is no point in talking about whatever essential properties it may have. Properties (essential or emergent) are attributes to something and should not be confounded with the thing itself. Endo- or exogenously organized entities such as cells, plants, animals, political parties, companies and hospitals emerge, exist and can be identified. Like essential properties, emergent properties are properties of something that exists. It would be meaningless to talk about the essential or emergent properties of something that does not exist at all, or is not present in a particular place at a particular time. Whereas essential properties are necessary, emergent properties are new. Both endogenously and exogenously organized entities have come to existence ("*sind da*"); their defining difference is that organizations/systems which belong to the former group are spontaneous, self-organized, and those belonging to the latter group are man-made, constructed. If the organizers of organizations violate the laws of nature, they are reaching for trouble. Nature cannot be dictated; a wound cannot be ordered to heal; love, care and friendship cannot be forced. Violation of the laws of nature is what we see all the time and everywhere. Daoists urge us not to be so stupid.

Man-Made Order

Whereas natural systems like the solar system and Planet Earth can perfectly do without man, man, or the varied systems of his making, cannot do without nature. Man is dependent on nature (in more ways than he cares to acknowledge), but nature is not dependent on man. Ebb and flow will obey the laws of nature, irrespective of the rules of man. The alternation of day and night disregards any order of man. Sooner or later, death follows on birth. No man can change the order of nature or the course of things. What he can and often does change, however, is the result of man's organizing activities. Politicians change the ruling of a country; religious leaders steer the herd of believers in a certain direction; and businessmen induce the flock of consumers to buy their products. Politicians, religious leader and businessmen are (re)organizing something that can be changed; they issue orders, which can be either obeyed or disobeyed. The Protestant Reformation challenged the authority of the Pope (*Pontifex Maximus*); the French Revolution changed France from a monarchy to a republic; and the (neo) liberal world order is a human construct. So let us look at the phenomenon of organization in man-made orderings.

Artificial, man-made organizations are arrangements which are exogenously made. They differ fundamentally from natural, spontaneous and endogenously ordered or self-regulatory organizations, like snowflakes and organisms.[17] For example, economic inequality is a human construct; it is enduring and tenacious, but certainly not ordained from Heaven (nature). Though Karl Marx, Max Weber, Douglass North, John Rawls, Immanuel Wallerstein, Amartya Sen, Robert Nozick, Anthony Giddens, Joseph Stiglitz, Angus Deaton and many others had done or are doing extensive and in-depth investigations into this issue and Plato had already raised the problem of justice in *The Republic* (see Gerasimos Santas, *Understanding Plato's Republic* (Wiley-Blackwell, 2010, Chaps. 2–5) and visit https://www.isjr.org, https://www.iep.utm.edu/justwest and https://plato.stanford.edu/entries/justice), rockstar economist Thomas Piketty needed (and obtained) more than 2000 pages to cover the subject in his books *Le Capital au xxie siècle* (Seuil, 2013) and *Capital et idéologie* (Seuil, 2019), the much-publicized publication of which he is following up with international trips (not paid for by himself) to promote their sale.

Another example of a human construct is the business organization. Being part of a division of a country's economy, a company is man-made; it is an artifact. An external factor is needed, namely the organizer, who not only directs and controls the organization but also divides and coordinates the labor to be performed. While the Moon orbits the Earth and the Earth orbits the Sun aimlessly (or so it seems), the leader or manager of an organization has, indeed must have a purpose. He steers the organization in a particular direction, just like the cyclist I earlier spoke about, like the driver of a car, the captain in charge of a ship or the pilot controlling an aircraft. Study of the leadership of a company is a form of applied cybernetics. When the organization under his leadership grows, the leader will employ assistants or delegate some of his responsibilities. His core task, however, will remain the same: making strategic decisions. These decisions involve employment and investment decisions, which means that he should be informed of the situation at, and developments in, the labor market and money or capital market respectively. Exploiting the potential of these markets, he continuously tries, possibly by means of his research and development department, to improve the quality of his company's products and looks, by means of his sales and marketing division, for opportunities to sell his products on the local, national or international market. With the R&D division, the sales and marketing department and other parts of the organization nicely in place (set up according to his directives), the leader shows that he knows how to divide, coordinate and control the jobs or tasks to be carried out, that is, how to establish effective relationships between employees, means, tools and activities in order to achieve an objective, the relationships being defined quantitatively, qualitatively, spatially, temporally or in terms of power.

Countries are certainly not companies, but countries and companies are largely human constructs. Leadership (the directing of people) is their essence. Another word for leadership is "management", which implies organization of the company, division of the tasks to be performed and control of activities carried out.[18] When it comes to performing tasks or controlling activities, ruling a country may show a considerable degree of latitude; it implies organization and government (management) nonetheless. This indicates that study of the leadership of both a company and a country is essentially a form of applied cybernetics (see Chap. 5). Whereas natural, endogenous organizations or systems could be represented by a self-contained sphere (containing everything that is needed), man-made, constructed organizations or systems could be symbolized by a vector (having size and direction).

Organization Studies

Organization studies are rooted in the writings of the founding fathers of sociology, who attempted to cope with the transformations wrought by the Industrial and French Revolution. Saint-Simon (1760–1825) "was probably the first to note the rise of modern organizational patterns, to identify some of their distinctive features, to insist on their significance for the emerging society and to predict that planned organization would play a new role in the world".[19] He believed that the logic of organization dominates modern society. In *Télécommunications et philosophie des réseaux* (PUF, 1998), Pierre Musso notes that in his book *Du système industriel* (1821) Saint-Simon used the word network (*réseau*) to characterize the future industrial system and the social transition to it. But while his argument showed a fine "*esprit de système*", Saint-Simon did not spell out in detail the nature of the problem he signaled, leaving this to later generations of social scientists.

Frederick Taylor (1856–1915), an American mechanical engineer who sought to improve industrial efficiency, followed in the footsteps of Saint-Simon and introduced, in 1911, the principles of scientific management. According to Taylor, there were three reasons for inefficiency: (a) the fallacy among workmen who were thinking that increase in their output would result in throwing them out of work, (b) the "defective management system", making it necessary for laborers to work slowly in order to protect their interests and (c) the ingrained idea that workmen operate inefficiently. Scientific management (aka Taylorism) analyzes and synthesizes workflows to improve labor productivity.[20] In *Administration industrielle et générale* (1916), one of the first comprehensive statements of a general theory of management, Henri Fayol (1841–1925), a French mining engineer, came up with five primary management functions: planning, organizing, staffing, directing and controlling. Probably due to their profession, both Taylor and Fayol were inclined to consider man as a machine, taking industrial organizations to be closed systems.

Elton Mayo (1880–1949), took a different view and saw industrial organizations as social systems, observing the failure of the rationalistic organization theory to address the problem of integration and realizing that authority is only effective with cooperation. In *The Human Problems of an Industrialized Civilization* (Harvard University Press, 1933), he stressed that the informal organization-structure of a company is at least as important as the formal one, the former being spontaneous and unplanned, the latter being man-made and imposed. The book laid the foundation of the Human Relations Movement. Chester Barnard (1886–1961), who was for many years President of Bell Telephone Company, analyzed the functions of the executive, starting from the premise that an organization functions because her members are willing to make a contribution. He asked himself what managers have to do in order for this contribution to continue.

Over the last few decades, the literature on corporate finance, marketing and business logistics has grown by leaps and bounds, and every businessman or marketeer is now familiar with the idea of "customer journey", the trajectory between producer (supply) and customer (demand). See Asbjørn Følstad and Knut Kvale, "Customer journeys: a systematic literature review" (*Journal of Service Theory and Practice*, 28:2 [2018], pp. 196–227). The HR-variant, the "employee journey" (the path between employee and employer), has been equally paid attention to, and the human relations theory of Elton Mayo (father of human resource management [HRM] and the famous Hawthorne Studies) has been developing in many directions, the labor market and vocational training being just two of them. Suffice it to browse through journals like *Human Resource Management* (Wiley, 1961ff), *Human Resource Management Journal* (Wiley, 1990ff), *The International Journal of Human Resource Management* (Taylor & Francis, 1990ff) and *Human Resource Management Review* (Elsevier, 1991ff). Business administration (aka business management) includes all aspects of running any type of company (a list of business theorists, among whom many are systems thinkers, can be found on the Internet). Being twin sister of the theory of the firm (see Chap. 5), it is not limited to the financial, marketing, logistic and human relations aspect, though the whole of what the Germans call *Betriebswirtschaftslehre* comes down to the management of people.

By the late 1940s and early 1950s, the conception of industrial organizations as social systems converged with the ideas which Ludwig von Bertalanffy had laid out in his trailblazing book *Theoretische Biologie* (1932). However, it was structural functionalism (Talcott Parsons, Kingsley Davis, Robert Merton, and others) that would dominate theoretical developments and empirical investigations until social-conflict theories became popular (Gerhard Lenski, Ralf Dahrendorf, Randall Collins, among others) and systems science took off in the late 1960s. Structural functionalism looks at society as a whole in terms of the functions of her constituent elements; it sees society as a system whose parts work together to promote solidarity and stability. Inspired by Karl Marx (and Mao Zedong) and fueling the May 1968 events in France, social-conflict theorists were also macrosociologists, but they emphasized society's conflicts rather than its consensus.[21]

The organization approach of systems scientists is primarily holistic. This quality distinguishes (but by no means separates) them from (*a*) cyberneticists (or cyberneticians), who are mainly interested in the controlling of organizational processes and (*b*) experts in the fast expanding field of information science, which further develops the idea of cybernetic control. In contrast to researchers like Taylor and Fayol, systems scientists take industrial organizations to be open systems, and in line with Mayo and his followers, they compare industrial organizations with living organisms, which interact with their environment for survival.

Organization studies and systems science have now largely merged into one discipline; "largely", because for systems scientists the word system has wider connotations.[22] The problems which physicists, chemists, astronomers, geologists, geographers, demographers, biologists and social scientists are coping with are restated and reformulated by systems scientists in terms of the connected parts of a whole that cannot be disconnected from its constantly changing environment. Systems scientists are formal scientists; they have a keen eye for the evolving connectedness of, or relationships between, things, without losing sight of the nature and track of the things related (*relata*). It would thus be off the mark to accuse them of ignoring or neglecting the importance of the constituents of a system. For systems scientists, both the relations and the related things (the *relata*) are important. The two are coexisting and interdependent. Societies (communities) and their members (individuals) are investigated as complex, dynamic systems belonging (as subsystems) to a supersystem and consisting of intimately connected and continuously adapting parts.

The literature on organization studies is overwhelming[23] and the worldwide community of researchers in this field keeps on expanding. Leading journals are (in alphabetical order): *Academy of Management Journal*, *Academy of Management Review*, *Administrative Science Quarterly*, *American Journal of Management*, *American Journal of Management Studies*, *European Management Journal*, *Harvard Business Review*, *International Journal of Interdisciplinary Organizational Studies*, *Journal of Japanese Management*, *Journal of Leadership and Organizational Studies*, *Journal of Management*, *Journal of Management Development*, *Journal of Management & Organization*, *Journal of Management Studies*, *Journal of Organizational Behavior Management*, *Management and Organization Review*, *Organization*, *Organization Management Journal*, *Organization Science*, *Organization*

Studies, *Organizational Research Methods*, *Scandinavian Journal of Management* and *Strategic Management Journal*.

Special attention is to be paid to Aleksandr Bogdanov's trailblazing but marginalized (by Lenin) masterpiece *Tektologiya* (general organization science), a massive, three-volume treatise which was published between 1912 and 1928. In 1996, the Centre for Systems Studies, in Hull (UK), partly published an English translation (by Vadim Sadovsky and Vladimir Kelle), edited and introduced by Peter Dudley. A new and integral translation is planned by Brill and will be incorporated in its *Alexander Bogdanov Library* (https://bogdanovlibrary.org). In *Essays in Tektology* (1921), Bogdanov (1873–1928) condensed his invaluable *magnum opus*. The result is a book that has been translated into English by George Gorelik and, in 1980, published by Intersystems, in the series *Systems Inquiry*. In 1984, the second edition came out.

There are four books I wish to draw the reader's attention to, because they stress the importance of approaching company-related problems holistically, i.e., by making use of the advances in *complex* systems science:

- Hans-Jörg Bullinger et al. (eds.), *Handbuch Unternehmensorganisation* (Springer, 2009);
- Peter Allen et al. (eds.), *Handbook of Complexity and Management* (Sage, 2011);
- Timothy Ludwig and Ramona Houmanfar (eds.), *Understanding Complexity in Organizations* (Routledge, 2012);
- Harold Koontz and Heinz Weihrich, *Essentials of Management* (McGraw-Hill, 2015).

The Routledge series *Complexity and Emergence in Organizations* (2001ff) "explains how the application of complex systems science to today's organization could have radical implications for management practice". This being said, I cannot but hope that one day a comparative study of successful American, European, Chinese and Japanese companies will be published, that is, a study solidly based on theoretical developments and empirical investigations in the rapidly expanding field of systems science. The step from company to country studies is a short one, for running a company is essentially the same as ruling a country and the word administration is not only related to a business but also to a country: it is a matter of cybernetics, of steering a ship, of governing a whole. While the subject of complexity is often discussed from the business point of view (the Center for Systems Studies, in Hull, being a case in point), it may be useful to apply (with the support of computers and ICT) the burgeoning science of complex systems to comprehending a country, an idea that may seem absurd with regard to big countries such as Russia, Canada, the USA, China, Brazil, Australia and India but is certainly not outlandish in relation to relatively small countries like Vatican City, Monaco, Malta, Singapore, Bahrein, Luxembourg, Brunei, Cyprus, Lebanon, Qatar, Kuwait, Kroatia, Israel, Djibouti, Haiti, Albania, Switzerland, Belgium, The Netherlands, Estonia, Slovakia, Slovenia, Sri Lanka, Iceland, Serbia, Austria, Portugal, Hungary and Cuba. What first applies to the study of small countries may finally also be applicable to the study of bigger countries.

Dr. Naomi Stanford, who is leading authority in the field of organizational design and is currently writing the third edition of her book *A Guide to Organization Design*, is scheduled to launch, later his year, the series *Talking Systems—Organisational Design in a Systems Context*. Her website can be found at http://www.naomistanford.com. Also visit https://derroth.com and https://systemic-design.org. Hakan Gulliksson's *Pervasive Design* (fourth edition in 2015) should also be mentioned.

Perhaps the best way to see the discrepancy between natural and man-made order is to realize the fundamental difference between *actions* and*events*. Whereas events are caused, actions have reasons, that is to say, the human mind is always involved in actions. Causes may be difficult to identify, because the chain of causes is often long and their interdependence is hard to ascertain. Even more difficult is it to gauge the human mind and to fathom out what humans are moved by. "*Le coeur a des raisons que la raison ne connaît point*", the French mathematician and physicist Blaise Pascal (1623–1662) said. The habits of the heart are firmly grounded in one's culture. *Trans*actions are actions of more than one person. A transaction is a deal or agreement (often after lengthy negotion), mostly between a seller and a buyer. It is legally, ethically or socially binding and it is usually associated with paperwork (or electronic data exchange) and financial transfer. Marriage contracts are prenuptial agreements, transactions between two human beings, solemn promises that must be obeyed. Economic transactions, involving the transfer of property rights, have been the subject of intensive research by Armen Alchian, Ronald Coase, Harold Demsetz, Donald McCloskey, Rchard Posner, Joseph Stiglitz, Gordon Tullock, Douglass North, Oliver Williamson, Louis Putterman and many other economists. *Inter*actions are also a kind of actions. They occur when two, or more than two, people communicate with or react to each other. A meeting or reception may illustrate this. Symbolic interactionism, "the study of how individuals shape society and are shaped by society through meaning that arises in interactions" (Herbert Blumer, *Symbolic Interactionism: Perspective and Method* [Prentice-Hall, 1969, pp. 1–6]), is one of the main theoretical traditions in sociology. Against this backdrop, the renaissance in Hegel appreciation that has bloomed in the past few decades and has been led by Robert Pippin, Terry Pinkard, Kenneth Westphal and Robert Brandon may be understood. See Willem deVries, "Hegel Today", AEON, 26 October 2021.

When people interact, they communicate and exchange information. Communication study, the origin of which can be traced back to antiquity, became an academic discipline in the early twentieth century. Closely connected with the development of *information* theory in the 1920s, it is now an immense field of research adjacent to semiotics (the science of signs) and encompassed by social psychology ("the scientific study of the way in which people's thoughts, feelings and behaviors are influenced by the real or imagined presence of other people" [Elliot Aronson]). Communication can be at short distance, for example, when lovers are whispering to each other. At long distance it becomes *tele*communication, which the International Telecommunication Union (ITU), in Geneva, defines as "any transmission, emission or reception of signs, signals, writings, images and sounds or information of any nature by wiring, radio, optical or other electromagnetic systems". Shifting my attention to telecommunication, however, I enter the "virtual"/online world and its blurring boundaries with the "real"/offline world. In other words, I enter the infosphere, populated by what Luciano Floridi in his book *The Fourth Revolution* (OUP, 2014) aptly calls "inforgs" (informational entities). The infosphere, dominated by Apple, Microsoft, Amazon, Google, and Facebook (in the West), will be dealt with in the next chapter.

Notes and References

1. See Chap. 4 (under urban geography).
2. See Norman Barry, "The Tradition of Spontaneous Order", *Literature of Liberty*, 5:2 (1982), pp. 7–78, and Brian Albrecht, "The Breakdown of Spontaneous Order: Smith and Hayek Diverge", *New York University Journal of Law and Liberty*, 11:1 (2017), pp. 346–370. The Chinese characters for the word natural are 自然 (*zìrán*), self so, so of its own, so of itself. Their metaphor for naturalness, the quality of not being artificial, is 樸 (*pǔ*), unworked wood. See Qingjie Wang, ""It-self-so-ing" and "other-ing" in Lao Zi's concept of *Zi Ran*", January 25, 2003, online.
3. See Jaeger and Liebsch *op.cit.* pp. 60–75 and Jan Schmidt, *Das Andere der Natur: Neue Wege der Naturphilosophie* (Hirzel, 2015).
4. Go to https://plato.stanford.edu/entries/naturalism and https://plato.stanford.edu/entries/idealism.
5. Bridging the gap between (the sciences of) nature and (the sciences of) culture is a subject Jean-Jacques Rousseau (1712–1778) already wrestled with; see Dale Goldhaber, *The Nature-Nurture Debates* (CUP, 2012, Chap. 2). See also C.P. Snow, *The Two Cultures and the Scientific Revolution* (OUP, 1959); Stephen Jay Gould, *The Hedgehog, the Fox, and the Magister's Pox: Mending the gap between Science and the Humanities* (Harmony Books, 2003); Jerome Kagan, *The Three Cultures: Natural Sciences, Social Sciences, and the Humanities in the 21st Century* (CUP, 2009); Martha Nussbaum, *Not for Profit: why democracy needs the humanities* (PUP, 2010); Edward Slingerland and Mark Collard (eds.), *Creating Consilience: Integrating The Sciences and the Humanities* (OUP, 2012) and Christian Madsbjerg, *Sensemaking* (Hachette, 2017).
6. Philosophical anthropologists are dealing with questions of man's nature, an issue in which Indian and Chinese philosophers have been particularly interested.
7. For the perennial problem of the nature of, and relation between, necessity and contingency, see Jules Vuillemin, *Necessity or Contingency* (CSLI, 1996).
8. Visit https://www.iep.utm.edu/analytic and https://plato.stanford.edu/entries/analysis. See Erich Reck (ed.), *The Historical Turn in Analytic Philosophy* (Palgrave Macmillan, 2013); Aaron Preston (ed.), *Analytic Philosophy* (Routledge, 2017) and *Journal for the History of Analytical Philosophy* (2011ff).
9. See Walther Kranz, *Kosmos und Mensch in der Vorstellung frühen Griechentums* (Vandenhoeck & Ruprecht, 1938) and Maria Sassi, *The Beginnings of Philosophy in Greece* (PUP, 2018).

10. See Panayot Butchvarov, *Anthropocentrism in Philosophy* (De Gruyter, 2015).
11. See Martin Meyer, "Die Bedeutungsgenese der Begiffe 'Mythos' und 'Logos' in der Griechischen Antike", *Archiv für Begriffsgeschichte*, Vol. 41 (1999), pp. 35–63 and Ken Dowdenand Niall Livingstone (eds.), *A Companion to Greek Mythology* (Wiley-Blackwell, 2011, Part I).
12. The January 2013 number of *Esprit* devotes a special issue to "*Dieu, un absent si présent*".
13. For the creation–evolution controversy, I refer to the Internet and to Jim Stump (ed.), *Four Views on Creation, Evolution, and Intelligent Design* (Zondervan, 2017).
14. See John Haught, *Making Sense of Evolution* (John Knox Press, 2010, pp. XI–XVII) and Alister McGrath, *Why God Won't Go Away: Engaging with the New Atheism* (SPCK, 2011).
15. See Pierre Hadot, *Le voile d'Isis: Essai sur l'histoire de l'idée de Nature* (Foliot, 2008) and Thomas Kirchhoff et al. (eds.), *Naturphilosophie* (UTB, 2017).
16. Visit https://www.iep.utm.edu/emergenc, https://plato.stanford.edu/entries/properties-emergent, http://www.scholarpedia.org/article/self-organization, https://philpapers.org/browse/emergence, https://www.d-iep.org/diep and https://doi.org/10.3390/e22101163.
17. For the difference between natural order and market order, see Friedrich Hayek, *Law, Legislation and Liberty* (Routledge & Kegan Paul, 1982, Volume I, pp. 35–54, Volume II, pp. 107–132).
18. See George Goethals et al. (eds.), *Encyclopedia of Leadership* (Sage, 2004) and Eric Kessler (ed.), *Encyclopedia of Management Theory* (Sage, 2013).
19. See Robert Merton et al. (eds.), *Sociology Today* (Harper & Row, 1965, pp. 400–428).
20. See Daniel Nelson (ed.), *A Mental Revolution: Scientific Management since Taylor* (Ohio State University Press, 1992) and Sonia Taneja et al., "Frederick W. Taylor's Scientific Management Principles: Relevance and Validity", *Journal of Applied Management and Entrepreneurship*, 16:3 (July 2011), pp. 60–78.
21. See Bryan Turner (ed.), *The New Blackwell Companion to Social Theory* (Wiley, 2008, pp. 106–123) and George Ritzer and Jeffery Stepnisky, *Sociological Theory* (Sage, 2018, pp. 234–273). In chapter 10, I will go into the matter of structure, structuralism and formalism.
22. See Cornell Chikere and Jude Nwoka, "The Systems Theory of Management in Modern Day Organizations", *International Journal of Scientific and Research Publications*, 5:9 (September 2015), pp. 1–7.
23. See Stewart Clegg and James Bailey (eds.), *International Encyclopedia of Organization Studies* (Sage, 2008) and Mairi Maclean et al. (eds.), *Historical Organization Studies: Theory and Applications* (Routledge, 2020).

Chapter 7
Technical Feasibility of Collaborative Country Studies

Abstract The exploration of the possibility of studying countries in a cross-disciplinary and international manner is continued. I will investigate whether or to what extent computers could be deployed to uplift this kind of study, but first a few words will be devoted to that strange and yet so familiar thing called computer. Although computers are ever more sophisticated instruments (man-made systems), they are and will always remain just instruments, mindless objects to be instructed and activated by human beings, whose mind is immeasurable and unfathomable. The potential of digitized and electronically connected libraries and archives is shown to be underestimated. Computer science and systems science could stimulate each other and the deployment of the outcome of this mutual stimulation may revolutionize the study of countries.

Though rarely noticed, it is important to realize that the words complexity and computer have the prefix *com-* (together) in common. Complexity is the state of being complex, of having parts the relationships of wich are difficult to understand; it is a whole consisting of nonlinearly interconnected parts. Complexity is thus an ontological category. The word computer, however, suggests a human activity. On the one hand, coming from the Latin *putare*, it means "to estimate". On the other hand, *computare* means "to calculate (on the fingers)". John Atanasoff (1903–1995), who is credited with co-constructing (together with Clifford Berry) in the late 1930s the first electronic digital computer, noticed that his machine took eight hours to solve a certain set of equations. He also noticed that the mechanical parts of a computer make use of a decimal and binary system. See Jane Smiley, *The Man Who Invented the Computer* (DoubleDay, 2010).

Computers are information processing devices and supercomputers (as distinct from personal computers and microcomputers) are computers with a high level of performance. The TOP 500 project, which started in 1993, regularly ranks the 500 most powerful computer systems in the world. Currently (February, 2020), China dominates the list with 229 supercomputers, followed by the USA, counting 121 of them. Ranking is based on the LINPACK benchmark, a measure of a system's floating point computing power. "Floating point" refers to the fact that a number's decimal point can "float". In contrast to integers, floating point numbers are numbers

H. Kuijper, *Comprehending the Complexity of Countries*,
https://doi.org/10.1007/978-981-16-4709-3_7

that contain floating decimal points. A petaflop is a computer's capability to perform one quadrillion floating point operations per second (flops). When IBM's Roadrunner broke the petaflop barrier, in May 2008, Thomas D'Agostino proudly said: "If all six billion people on Earth used hand calculators and performed calculations 24 h a day and seven days a week, it would take 46 years to do what Roadrunner can do in one day". In June 2012, Roadrunner was replaced by Sequoia, reaching 16.3 petaflops. A year later, China revealed a supercomputer (Milky Way2) having a speed of 33.9 petaflops. Since June 2018, the American Summit is the world's most powerful supercomputer, reaching 143.5 petaflops on the LINPACK benchmark (peaking at 200 petaflops). This awesome IBM machine, covering the space of two basketball courts and having 250 petabytes storage capacity, can be utilized in fields such as cosmology, physics, chemistry, material science, earth science, biology, and computer science. POWER9 is Summit's dual-core processor, which is a CPU with two linked processors (in 2020, POWER10 has been introduced). Visit and explore the websites https://www.olcf.ornl.gov/summit, https://www.nextplatform.com/2020/03/04/lawrence-livermore-to-surpass-2-exaflops-with-amd-compute and http://aurora.alcf.anl.gov. Scientists and engineers are closing in on quantum computing, which uses quantum mechanical phenomena to perform computation. Visit https://www.nist.gov/history-and-future-quantum-information and https://www.nextplatform.com; subscribe to *Quanta Magazine* (https://www.quantamagazine.org).

The microprocessor is called the heart of a computer, hence the name central processing unit (CPU). The word heart is somewhat misleading, because the CPU is only comparable, and certainly not identical, to the human heart—that wonderful organ in our chest conceived of (in some cultures) as the seat of the human mind. The CPU is to be distinguished from a computer's peripherals, *e.g.* keyboard, mouse, graphics tablet, microphone, digital camera, webcam, projector, printer, speaker, modem and storage devices. The microprocessor, which is fabricated on a tiny piece (or chip) of semiconductor material (usually silicon), is the electronic circuitry that executes the instructions making up the program/software of the computer. The number of instructions executed at computer run-time is in the order of millions, even billions of instructions per second (IPS) and the performance of a CPU is measured by it. The transistor (from "transfer" and "resistor") is an electronic component relying on the properties of semiconductor material and used to amplify or switch electronic signals. Replacing the vacuum tubes of the first generation computers, transistors became the identifying feature of the second generation of computers. Integrated circuits (IC's) were the next technological breakthrough. Integrated circuits are sets of electronic circuits. "Integrated" suggests the idea of man-made ordering (see Chap. 6). Further research and development made the production of microcomputers possible (laptops, smartphones etc.). Miniaturization was the defining characteristic of the fourth generation of computers. There is a great variety of microprocessors and a list of them can be found on the Internet. Until 2017, Intel, an American company headquartered in Santa Clara (CA), was the dominant chip manufacturer in the world. In that year, Intel was surpassed (in monetary terms) by Samsung Electronics, headquartered in Daewu (South Korea). ASML, headquarterd in Veldhoven

(NL), is the main manufacturer of machines for the semiconductor industry. With dozens of subcontractors across the world and Intel and Samsung as customers, the company (an offspring of Philips) is "the world's only manufacturer of lithography machines that use extreme ultraviolet light", that is to say, light with a wavelength of just 13.5 nm (nearly X-ray level).

The often drawn analogy between the human brain/mind and a computer is deceitful. The computer, a combination of hard- and software, is a mindless thing waiting to be activated from the outside—by a human being. The mind is not computable and the human heart is a mystery. The source and the course of a river and its tributaries can be mapped, but I fail to see how such mapping can ever be done of the source and course of the human mind, whose vageries are beyond anybody's comprehension, because of its uncountable and untraceable "tributaries". The human mind reaches out to explore the macro- or microworld outside, not only struggling to understand the world out there but also the world inside. Go to http://dx.doi.org/10.2139/ssrn.3516946; see Raymond Tallis, *Seeing Ourselves* (Agenda, 2020); and remember what is said in the *Vedas*: "*Ātman* [the self] cannot see, hear or know *Ātman*, because it is *Brahman* [the reality, the binding unity behind diversity]".

Data or information *processing* is to be distinguished from data or information *storage* and *transmission*. In telecommunication, transmission is the process of sending and propagating information signals over a transmission medium. In the West, the telecommunication business (aka telecom or telco industry) is becoming increasingly dominated by the "Big Five": Google, Amazon, Facebook, Apple and Microsoft (GAFAM). These giants are fiercely competing with each other (and with Chinese companies like Baidu, Alibaba, Tencent and Xiaomi), have many subsidiaries (a list of their mergers and acquisition can be found on the Internet) and are heavily criticized on account of their employment policy, tax evasion, cartelization/monopolization and privacy infringement. See Sheera Frenkel and Cecilia Kang, *An Ugly Truth: Inside Facebook's Battle for Domination* (Harper, 2021). Metaphorically, the Big Five form the thickening trunk of a tree rooted in computer hardware and devices, precious metals, data center facilities and GPS satellites, and ramifying in different but increasingly interconnected areas such as the admistrative, legal, military, economic, financial, social, cultural, medical, educational, scientific and technological sector. Clearly, the telecom industry could possibly play an important role in collaborative country-studies. Bob Merritt, *The Digital Revolution* (Morgan & Claypool, 2016) focuses on the potential impact of technologies in the field of telecommunication and José van Dijck et al., *The Platform Society* (OUP, 2018) analyze the digitally connected world we live in, but the article to be read first and foremost is Andreas Hein et al., "Digital Platform Ecosystems", uploaded in November 2019 (https://doi.org/10.1007/s12525-019-00377-4). Katy Börner and David Polley, *Visual Insights* (MIT Press, 2014) is "a practical guide to making sense of data". Meanwhile, China is the rising star in information and communication technology (ICT). See Kai-Fu Lee, *AI Superpowers: China, Silicon Valley, and The New World Order* (HMH, 2018). Dr. Lee, who had obtained a degree in computer sience from Columbia University, earned his PhD in computer science at Carnegie Mellon University and played a key role in establishing the research division of Microsoft in Beijing, was (from 2005 to 2009) President of Google China. In 2009, he started Sinovation Ventures, a leading Chinese technology venture capital company (http://www.sinovationventures.com). Dr. Lee, who doesn't shy away from the camera and pretends to know the future, predicts that "AI will reshape teaching". It needs to be borne in mind that Cecilia Kang, who covered technology and regulatory policy fot the *New York Times* since 2015 and managed to look inside Facebook, is a *Chinese* woman who might have a secret line with the Chinese authorities, who would love to hear about the bankruptcy of that company.

Computers are information processing devices. The word hardware refers to their interconnected physical and tangible parts, and "software" to the instructions that are executed by the hardware, which is designed and manufactured for it. Software (aka program) is the sequence of external orders given to the hardware to carry out a specific task. Hardware and software are interdependent. One can either go down into the hardware layers or go up into the software levels. According to Daniel Hillis, a well-known American computer scientist, "This hierarchical structure of abstraction is our most important tool in understanding complex systems." (*The Pattern on the Stone* [Basic Books, 1998, p. 18]).

It is not easy to give a simple, precise definition of the word information. The idea is mysteriously related to energy, but "information" is not interchangeable with "knowledge". "Information" is a computational, "knowledge" an epistemological concept. We have received information when *what we know* has changed, See Richard Gregory, *Companion to the Mind* (OUP, 2004) and Ihor Lubashevsky, *Physics of the Human Mind* (Springer, 2017). The organization of knowlege is known as science or *Wissenschaft*. Information can be exchanged and translated. Unless one understands the language in which it is couched, the information given is not informative. Computers are capable of receiving, remembering, processing and outputting information, provided the input is in a language they "understand". Many languages have been developed to facilitate the communication between humans and computers.[1]

Computers are mindless, even lifeless objects. They are non-natural, man-made systems (see previous chapter). No matter how technologically advanced they are, computers are, and will probably always be, lagging behind human beings, who are uniquely able to *act*.[2] Computers have to be exactly told and instructed what to do, step-by-step. The process of achieving the required level of detail is called problem analysis, which can be divided into six more or less distinct parts:

- Identifying and ascertaining the problem, *e.g.* the problem of how to cope with the interrelationship between, or with the system of, sciences, or the problem of comprehending the internal coherence and external (foreign) relations of a country, to which these interrelated sciences are applicable;
- Choosing a solution method, *e.g.* the method of complex systems science (theory formation, empirical research and—*der Dritte im Bunde*—computer simulation);

- Describing, step by step, the solution method, *i.e.* formulating a solution algorithm, a sequence of instructions;
- Converting or restating the algorithm into a program, which is a sequence of statements the computer can understand by means of a translator, or compiler, available in its memory;
- Testing, and removing bugs from, the program; and
- Validating the algorithm. A list of algorithms can be found on the Internet. The question of how to find the fastest algorithm to solve a particular problem has become of prime importance in the computer science community, where the hierarchy of problems has been delegated to the second place. A problem is said to be intractable, when only exponential time algorithms can be found. This awkward situation may change in the future, as science and technology are steadily advancing. Quantum computing may be the technological breakthrough we are waiting for (go to https://plato.stanford.edu/entries/qt-quantcomp).

Man is more than a machine, and I consider the making of computers as capable as, or even more capable than, human beings to be a pipe dream. A computer can do certain things faster than man and has a better memory, but a computer will never have man's power of imagination. According to Gregory Chaitin, it's not all in the numbers. See David and Ricardo Nirenberg, *Uncountable* (The University of Chicago, 2021) and read page XII of the Preface. Perhaps one day, biophysicists, biochemists, molecular biologists, geneticists, neuroscientists and computer scientists will triumphantly declare to have established beyond reasonable doubt that reality consists of nothing but atoms and bits. For the time being, however, I venture to doubt whether man will ever be able to build a computer or robot which is alive and conscious, has compassion and a conscience, can philosophize, criticize, laugh, cry, love, mourn, or dream, can commit a crime, rise in revolt, create a piece of art, have an encounter (as understood by phenomenologists), take an oath of loyalty, show solidarity, perform a ritual, or dies, "*multis bonis flebilis*". A robot cannot even dress itself yet. Verily, man is more than a set or aggregate of things that can be counted or measured; he is not a construct, but a composition of matter and mind; man is a wonderful system, both a part of nature and apart from it. See Melanie Mitchell, *Artificial Intelligence* (FSG, 2019).

Computer *science* is the science concerned with computers. It is a vast, ramifying and fast developing field of research that belongs to the formal sciences.[3] Nobody has comprehensive and in-debt knowledge of the whole field. Even in small print, a book containing all the computer-related information that is available would be a very large and thick volume. Whereas the physical, biological, psychological, social and cultural sciences study natural, biological, psychological, social and cultural systems respectively, the formal sciences are concerned with their structure; they are aiding the non-formal sciences by providing information about the structures or models they use to describe the world. The formal sciences, which comprise mathematics, logic, statistics, computer science and systems science, are thus not concerned with the validity of theories based on observations or experimentations, but with the properties of the structures of these theories, that is to say, with the characteristics of their form (hence "formal sciences"). Like systems science, mathematics, logic and statistics, computer science is increasingly supporting, or applied to, non-formal sciences and—most interestingly—to the other formal sciences. This explains the dynamic nature of the formal and non-formal sciences. So it would be no exaggeration to say that the formal sciences are the driving force of science and technology (see Chap. 5, systems science).[4]

While the fundamental relationship between the formal sciences is of utmost importance, it is hard to find a book about the subject. In 1999, Benedikt Löwe initiated a series of interdisciplinary conferences on the Foundations of the Formal Sciences. The last one was held in September 2013, in Cambridge. The conferences dealt with mathematics, logic, statistics, philosophy, sociology of science, linguistics, and computer science, but systems science and network science were conspicuous by their absence, philosophy is not a science, and whether sociology of science and linguistics are formal sciences is debatable. With the foundations of the formal sciences still being thoroughly investigated, however, the literature on the relationship between logic and mathematics is large, and the field of statistics, interfacing with mathematics and logic, is immense. Whereas mathematics is a fundamental tool in computing, computing is increasingly used as a key component in mathematical problem-solving. The Logic in Computer Science symposium, organized by the Institute of Electrical and Electronics Engineers ("the world's largest technical and professional organization dedicated to advancing technology for the benefit of humanity") is an annual international forum on theoretical and practical topics in computer science that relate to logic. As systems science is closely connected with the other formal sciences, it seems to me that "order", "structure", or "form" is the underlying, overarching or unifying concept of the five *formal* sciences, to which network science could be added, for each network has a structure.

Plato took εἶδος (idea, form) to be transcendent (see David Sedley, "An Introduction to Plato's Theory of Forms", *Royal Institute of Philosophy Supplements*, 78 [July 2016], pp. 3–22). Aristotle, his main student, disagreed and considered it immanent. There is a fierce, still ongoing debate over this issue, and the relationship between epistemology (the study of knowing things) and ontology (the study of what it means to exist) is far from clear. What is "immanent" should perhaps not be opposed to what is "transcendent", for without relationship there can be no order, structure, or form, and vice versa. It should also be kept in mind that the Greek *eidos* and the Sanskrit *Veda* come from the Proto-Indo-European root **u̯eidu̯eid*-, meaning "to see" or "to know"—to see/know reality, that is.

Science and Technology

Whereas science is organized knowledge about a piece of reality, technology (from τέχνη [art, craft, cunning of hand] and λόγος [discourse]) is the skillful *application* of science in a way that may be useful or practical. What we get is thus theory versus practice, knowing-that versus knowing-how. In Chap. 2, it was explained that there is a relationship between theory and observation/experiment. In some sciences, it may appear that experimentation (the process of testing ideas or methods) is unnecessary and only mathematics is needed to find the solution of a set of equations. However, this view overlooks the importance of observation and experimentation in raising or setting up a problem. It has led, particularly in the twentieth century, to the assumption that theorizing and experimentation (from "experience") are distinct, if not separated activities, rarely performed by the same person. Nevertheless, the creative theorist must be familiar with the way things are working, in nature or society. By the same token, the inventive experimenter cannot make progress without being well informed by theory. Theory and practice (must) go hand in hand. Engineering and technology are greatly indebted to science, and the other way around. Modern science could not have soared without telescope, camera, microscope, spacecraft, power plant, sensor, autoclave, chromatograph, robot, collider, computer, television, microphone, telecommunications network and many other artifacts constructed and engineered by craftsmen, by people with special knowledge and a particular skill.

Technology involves the application of organized knowledge and, in turn, has an important and noticeable effect on science. It has also played a major role in man's social, economic, political, legal, military, and educational activities. It has further had a conditional effect on the development of arts (literature, theatre, motion pictures, music, dance, architecture, garden landscape, design, sculpture, painting, printmaking, photography, and decorative art). To a degree, technology has even affected developments in man's religious life. In short, technology is a major dimension in the history of mankind and its branches have become the subject of historical and analytical studies.[5] Military –, civil –, mechanical –, chemical –, petroleum –, biological –, nuclear –, aerospace –, industrial—and electrical and electronic engineering are the major fields of technology. Interestingly, they are all cut through by systems engineering, which is concerned with the way of designing and managing complex systems (see Chap. 5, systems science).

Electrical and electronic engineering deals with the practical application of electricity in all its forms. The distinction between electrical and electronic engineering is based on the strength of the electric current used; the former deals with "heavy current" (electric light and power systems), the latter with "light current" (wire and radio communication, computers, radar, and automatic control systems). With recent technological progress, the distinction has become blurred. Electronic engineering (EE) utilizes integrated circuits to design and produce Very Large Scale Integration (VLSI) devices and their systems. VLSI is the process by which millions of transistors are combined. EE, in turn, has many divided subfields, *e.g.* signal processing, electromagnetics and telecommunications –, control –, instrumentation—and computer engineering. It is rooted in computer science and is closely related to Silicon Valley, home of many high-tech corporations and thousands of start-ups. EE deals with the design of computers or computer systems. Details can be found in Neil Storey, *Electronics: A Systems Approach* (Pearson, 2017). Computer engineers may work on a computer system's hardware or software. However, the research, development and design of computer software are usually the domain of software engineering, which is considered a fast expanding different discipline.

Computer science is the scientific and practical approach to computation and its applications. Somewhat technically, it is the study of algorithmic processes that scale (reduce or increase in size according to a scale). Its mutually related and reinforcing subfields include: mathematical foundations, algorithms and data structures, artificial intelligence, communication and security, computer architecture, computer graphics, parallel, concurrent and distributed systems, databases, programming languages and compilers, scientific computing, software engineering and theory of computation. Each of these subfields is a vast and fast expanding field of research. A timeline of computer history, a list of pioneers in computer science, a list of important publications in computer science and a list of best paper awards in computer science can be found on the Internet (see also Note 3 of this chapter). For the man who is considered the father of theoretical computer science and artificial intelligence but, during his lifetime, due to his homosexuality was "not fully recognized" in his home country (impact of culture on science and technology!), see Jack Copeland et al., *The Turing Guide* (OUP, 2017). Virtually all universities have now a Department of Computer Science. *Lecture Notes in Computer Science* (LNCS) is a renowned series of books published by Springer. *ACM Computing Surveys*, published by the New York-based Association for Computing Machinery, presents "comprehensive, readable tutorials and survey papers that give guided tours through the vast literature and explain topics to those who seek to learn the basics of areas outside their specialties".[6]

In a famous paper entitled "Mind, Brains, and Programs" (*Behavioral and Brain Sciences*, 3 [1980], pp. 417–424), John Searle, professor emeritus of philosophy of mind and language at the University of California (Berkeley), challenged the computer science community by advancing his "Chinese Room" argument.[7] Cognitive scientist Stevan Harnad, founder and editor of the journal, was one of those who leveled criticism against Searle, but he was also skeptical about the claims and pretensions of computer scientists. Mathematical physicist Roger Penrose put forward other objections to the view that mental states are computational states, appealing to Alan Turing who had written: "There are certain things that a digital computer cannot do." (*Mind*, 59:236 [October 1950], p. 444).

Yet with the study of information and knowledge processing deepening, human–computer interaction getting more and more sophisticated, machine learning making giant strides, computational linguistics and natural language processing making major breakthroughs, big data analytics rapidly developing, AI expected to be able to tackle global complex problem (visit https://www.frontiersin.org/articles/10.3389/frobt.2020.00050/full and https://eccai.org) and soft computing soaring, it should (soon?) be possible to mine, collect, process and integrate information from different sources and to near the ideal of interdisciplinary and international country research that benefits from the result of worldwide investigations in distinctive disciplines, formal as wel as non-formal. Data can already be stored on a single atom and quantum computing, predicted to dwarf conventional computing, is said to be on the horizon.[8] We only need the will and determination to apply available technology to collaborative country studies. It may cause a sea change in and far beyond academia. In a provocative article entitled "The big idea: one change that could transform universities" (*Times Higher Education*, November 29, 2018), the challenging question was raised as to what single change of university practice would have the most positive and far-reaching consequences. Only eight academics cared to give their views (or succeeded in getting their ideas published in THE), but none of them suggested that country studies should be cross-disciplinary and inter-national exercises firmly based on systems science and using the latest in computer science and ICT. Just imagine that the scientists and engineers in Silicon Valley or in one of the world's science/research parks, pool all their resources, or are mobilized, to uplift the current study of countries! It's easy if the reader tries. He may say I am a dreamer, but "I'm not the only one" (John Lenon).

Computer science could stimulate systems science. Concerned with electronic computation and its applications, it could be profitably deployed to investigate and understand the composition of, and relationships between, systems, in particular the system of systems called country. Conversely, systems science could stimulate computer science. Concerned with the nature and functioning of systems and aiming to lay the foundations of all scientific disciplines (including the discipline of computer science), it could be profitably deployed to analyze and understand the problems to be solved by computers (for problem analysis, see above), particularly the problem of comprehending a country. Systems science is pre-eminently suitable to map the systems of a country and to chart the relationships prevailing among them. If both computer science and systems science are further developing, the possibility of interaction between them should be explored. Instead of heavily leaning on either the computer science or systems science side, a more balanced approach would take advantage of the strength at both sides. An international conference on this issue is urgently needed—a conference that picks up the thread where it was left off by Benedikt Löwe in 2013 (see above). Go to https://www.quantamagazine.org (fill in: maths and computer science).

The question may be raised (a) how to digest the huge chunk of information collected in a conventional country library, (*b*) how to find the main avenues between the scientific disciplines that have produced some but by no means all of the volumes added to the enormous book collection, or (*c*) which books to select and read in what order to become a country expert (see Preface). Having pointed out that systems and computer science, particularly when stimulating each other, could be helpful in studying and comprehending the complexity of a country holistically rather than piecemeal, I think there is a reason to be cautiously optimistic about the applicability of computer science to such studies. To substantiate my view, I would like to draw the reader's attention to the growing potential of digitized libraries, for they may help to cope with the complexity of a country in a way undreamed of by scholars in the past and their current colleagues.

Digital Libraries

A digital library is a collection of digital objects (possibly including text and visual, audio or video material) stored electronically along with means for retrieving, organizing and managing the information contained in the objects. The digital library idea can be traced back to the innovative work of Vannevar Bush (1890–1947) and Joseph Licklider (1915–1990). Bush invented and developed a device for mechanically storing books capable of being consulted with exceeding speed and flexibility. Licklider realized that computers were getting to be powerful enough to support the type of automated library system that Bush had described. In1965, he wrote a book (*Libraries of the Future*) about how a computer could provide people at different places with simultaneous and remote access to the objects in a digitized library.[9]

Digital libraries, the concern of Library and Information Sciences (LIS), vary in scope, can be maintained by individuals or organizations, can be affiliated with physical libraries or academic institutions, and can be integrated, which is essentially a case of knowledge integration.[10] The content of digital libraries may be accessed remotely or via a network of computers. There is a center for Digital Scholarship in Ann Arbor, Atlanta, Notre Dame, Providence, Oxford, Leiden and many other places. There is even a *Guide to Digital Scholarship.*[11] Digital geology, geography, biology, demography, linguistics, literacy, anthropology, sociology, psychology, education, medicine, politics, law, economics, finance, ecology, paleontology, archaeology and musicology are hot topics. For digital Buddhology, search the Web.

The humanities are currently undergoing a revolution similar to what happened in the life sciences twenty years ago with the collection and centralization of genomic data. The marriage between computer science and humanities has resulted in the thriving field of digital humanities. Very large databases, new computational approaches and digital collaborative technologies are reshaping the humanities, opening up promising avenues for research and education in general and for country studies in particular. The discipline of digital humanities explores the potential of these new methods and the scholarly practices fostered by digital technologies (see Note 9 in Chap. 1). *Digital Scholarship in the Humanities* (1986ff) is probably the oldest journal in the field. The series *Advances in Human and Social Aspects of Technology* (2010ff) "seeks to explore how social scientists and scholars in the humanities have been affected by technology and aims to publish cutting-edge research on the ways in which the digital age is changing society". Digital Humanities is the special theme of *ERCIM News*, number 111 (October 2017), and *Frontiers in Digital Humanities* (2017ff), an open-access journal, "publishes rigorously reviewed articles on discoveries made at the intersection of computer science and humanities, aiming to bring all digital humanities areas together on a single platform".[12]

Myriads of books and scholarly journals are stacked in libraries throughout the world. Some of these libraries are digitized and their number is rapidly growing. The advantages are obvious:

- The fast growing potential of "human–computer interaction" can be deployed.

- Digitized libraries can be quickly, remotely, simultaneously and systematically searched.
- Double/multiple counting can be avoided.
- Books and journals (articles) on a particular country can be selected.
- Books and journals (articles) on a particular country can be arranged by year/month of publication.
- Books and journals (articles) on a particular country can be arranged by publisher.
- Entries can be alphabetized.
- Entries can be systematically cataloged by (sub) discipline or topic.
- Entries can be listed in order of some kind of frequency.
- Entries can be listed by the time a subject is first addressed, or further advanced.
- Indexes (alphabetical lists of names or subjects, with their location) can be drawn up and integrated.
- Concordances (alphabetical lists of the words used in a book or article, with their immediate context) can be compiled and integrated.
- Contents mentioned in tables of contents can be systematically arranged and synoptically outlined.
- Digitalization facilitates analysis and cross-checking.
- Semantic fields can be explored.
- Schemes of concepts or keywords can be mapped.
- Plagiarism can be detected.
- Technical terms and expressions can be clarified by relating and organizing them.
- The tools, techniques and methods of text analysis (aka text mining) and topic modeling can be used. With text analysis, computational linguistics and computer-aided translation sufficiently developed, it may even be possible to discover and sort out re- or paraphrasing.
- The tools, techniques and methods of information and knowledge integration can be used.
- The tools, techniques and methods of information fusion can be used.
- The tools, techniques and methods of natural language processing (NLP) can be used. The Natural Language Processing Group at Stanford University may be appealed to.
- Quantitative data can be collected and subjected to statistical analysis. This kind of analysis is particular important, as the volume of available quantitative data is growing and the statistical apparatus is becoming increasingly sophisticated. Quantitative linguistics is booming. Go to http://www.iqla.org.
- The desirability and feasibility of performing a meta-analysis can be explored. A meta-analysis combines the results of multiple scientific studies.[13]
- National data on authors, contributors or translators can be sifted and orderly mapped.
- Personal data on authors, contributors or translators can be sifted and orderly mapped (education, appointment[s], institution[s] associated with, prize[s], award[s], publication[s] etc.).
- A master bibliography can be alphabetically or thematically compiled, kept up-to-date and explored.

- IT-connections can be established with publishers of books or journals on a country.
- IT-connections can be made with databases which are partly or wholly related to a country.
- The last two points will enable the aspirant author to acquaint him/herself with, or stay abreast of, recent information on the subject he/she is interested in or wants to write about.

A good example of the usefulness of electronically accessible databases is the digitized administration of the toll which the king of Denmark levied on shipping through the strait between Sweden and Denmark from 1497 to 1857, the so-called Sound Toll Registers (STR). The registers are an "important source for research into production, trade, transport and consumption in Europe". Visit http://www.soundtoll.nl/index.php/en.

Computer-Supported Cooperative Work (CSCW) on, Computer-Supported Collaborative Learning (CSCL) about and Collaborative Information Seeking (CIS) in regard to a country (see next chapter) can be facilitated by the use of digital libraries. In other words, the cooperation and collaboration between country-focused scientists from different countries and different disciplines, and their information seeking can be stimulated, and the coherence and integration of their work can be enhanced. By computationally collating data from diverse sources a well-planned project geared to international and interdisciplinary country-research will come to fruition and novel insights into how a country functions and develops will be provided. The wiring of the country, being an intricate network of networks, will be discovered and its overall pattern or deep structure will show up.

Computationally visualized relationships, interconnections, processes, developments, trends, interactions or interdependences may suggest further research, possibly by means of graph theory made applicable to societies. The importance of computationally creating an image of both the weft and the warp of a country's carpet, that is to say, of the country's order of things crosswise (*querschnittlich*) and lengthwise (*längsschnittlich*), is to be emphasized, because many scholars and scientists, bewildered about the intimate relationship between the spatial (simultaneous) and the temporal (consecutive), are unfamiliar with, or inclined to underrate, the possibilities of information visualization.[14]

In December 2017, the colloquium "Modeling and Visualizing Science and Technology Developments" was held in Irvine, CA. It brought together "researchers and practitioners from multiple disciplines to present, discuss, and advance computational models and visualizations of science and technology". Realizing that (*a*) large datasets can now be utilized to simulate the structure and evolution of science and technology, (*b*) advances in computational power have created the possibility of implementing scalableand empirically validated computational models and (*c*) both the data and the models tend to exceed human comprehension, because the databases are massive and multidimensional, the participants in the colloquium explored the possibility of advances in data visualization. This technique was "effectively employed to communicate the data, the models and the model results to diverse stakeholder groups". Videos of the colloquium talks are available on the Sackler YouTube Channel. Explore https://www.pnas.org/page/about/colloquia.

In her book *Die Bildpädagogik Otto Neuraths* (Springer, 2015), Angélique Gross describes the beginning and development of "*eine Bildsprache für die Wissensvermittlung*", designed by Otto Neurath (1882–1945), an Austrian sociologist, political economist and philosopher of science who co-authored (together with Rudolf Carnap and Hans Hahn) the famous 1929 manifest entitled "*Wissenschaftliche Weltauffassung der Wiener Kreis*".[15] The design of Neurath, champion of the unity of science movement, became known as ISOTYPE (international System of Typographic Picture Education).[16]

In his book *Diagrammatology* (Springer, 2007), Frederik Stjernfelt attributes the term diagrammatology to art historian W.J.T. Mitchell, who in 1981 wrote of the need for "something like a diagrammatology, a systematic study of the way relationships among the elements of a system are represented and interpreted by graphic constructions". The term is also used by Olga Pombo and Alexander Gerner in the title of the book they co-authored, *Studies in Diagrammatology and Diagram Praxis* (College Publications, 2010). This volume is a collection of the papers given at an interdisciplinary workshop organized by the Center for Philosophy of Science of the University of Lisbon and aiming to analyze the place of images and diagrammatic thinking in different epistemological and semiotic contexts.

Network diagrams are visual representations of a network, but they are not iconic models such as photographs and scale models (see Chap. 2). A network diagram is the mapping of the structure or architecture of a network, showing the nodes or components that make it up and how they are interconnected. Depending on its scope and purpose, the diagram may contain many details or just provide a broad overview. A network diagram can either be physical or logical. A physical one shows the physical arrangement of the elements that constitute the network or system. The diagram gives a bird's eye view of the network in its physical space, like a ground plan. A logical network diagram describes the way information flows through the network. A network diagram may include hundreds of meaning conveying symbols. Some of these symbols stand for physical entities, while others represent relations between these entities. Network diagrams are ideal for sharing the layout of a network or for planning and tracking a project. Visual representations of a network make it easier for users to understand how items are interconnected and to decide on the actions to be taken or activities to be pursued. A network diagram could thus be useful for planning and executing a country study on project basis. "Could", but not can, because I am doubtful whether present-day network theory is capable of mapping a country adequately (see Chap. 5). Well-known complements to network diagrams include the PERT and Gannt charts. PERT (program evaluation and review technique) is a method of analyzing the tasks involved in completing a given project. The critical path method (aka critical path analysis) is kind of recipe for project activities. It is commonly used in conjunction with PERT. A Gantt chart (named after its inventor Henry Gannt [1861–1919]) is a bar chart that illustrates a project's schedule. It lists the tasks to be performed on the vertical axis, and time intervals on the horizontal axis. The width of the horizontal bars shows the duration of each activity. Modern Gantt charts also show the relationships between activities and current schedule status. The International Symposium on Graph Drawing is an annual conference (since 1992) on graph theory, graph drawing and network diagrams.[17] For network science, see Chap. 5. For the collaborative study of countries carried out on project basis, see Chap. 8.

Multi-volume dictionaries and thesauri can be systematically searched in a minute, a modern atlas can display detailed information about a geographical area's different aspects, and the streets and side streets of any city can be almost instantly mapped and visualized, but the potential of digitized and electronically connected libraries and archives is at present heavily underestimated. Digital libraries are huge databases, large collections of data which are electronically accessible to big data analysts using cutting-edge software that runs on state-of-the-art supercomputers. Completely new and hitherto unknown structures of, and heretofore unimagined patterns between, countries could be discovered. As integrated and interactive computer systems, such libraries could make computationally visible that any country-topic chosen for research is only part of a complex and dynamic system, in other words, is nothing but the node of a poly-dimensional and multi-layered network. All the relevant dots can and shall be connected. The deployment of digital libraries could give an enormous boost to collaborative country research. It could improve the understanding of countries needed by foreign affairs ministers and world leaders.

The electronically accessible *Encyclopedia of Life Support Systems*[18] is "one of the largest e-books repositories on the Web", an "integrated compendium of twenty-one encyclopedias", and "a virtual dynamic library of about 600 volumes". The EOLSS body of knowledge "attempts to forge pathways between disciplines in order to show their interdependence". It "deals in detail with interdisciplinary and transdisciplinary subjects, but it is also disciplinary, as each major core subject is covered in great depth by world experts". EOLSS is testimony of it, that partial studies which are not nicely dovetailed or firmly interlocked with each other present nothing but a *spectacle coupé*, a Humpty Dumpty broken into bits. The parceling up of a whole neglects relations that matter and the compartmentalization of a complex system into "more manageable" subsystems easily results in losing sight of the context. Conducting partial studies makes thus only sense if they are conceived of as, and demonstrated to be, the constituents of a coherent whole. If the repository *Encyclopedia of Life Support Systems* is available online and serves the interests of many scientists, it should be possible to create a similar storehouse of information related to a country. The electronic hookup between country-related depots, which network scientists call the nodes (or vertices) of a network, would make comparative research possible and bring the international context or foreign relations of a country to light.

The latest edition of the *Manual of Online Search Strategies* will be an invaluable tool to start using digital libraries optimally. The third edition contains references to hundreds of databases and electronic information resources. It "sets itself the task of offering to the searcher sound advice on database selection, search service selection and search strategy compilation in order to maximize the chances of finding the best information available for a given task". Volume I deals with chemistry, the biosciences, agriculture, and the earth sciences; volume II with business and law;

and volume III with the social and behavioral sciences (25 pages), the humanities (88 pages) and education (122 pages), together with an introductory chapter on search strategies (22 pages).

In 2005, James Billington, late chief librarian of the Library of Congress, suggested to UNESCO director-general Kōichirō Matsuura that a World Digital Library (WDL) might be created which would "have the salutary effect of bringing people together by celebrating the depth and uniqueness of different cultures in a single global undertaking". Dr Matsuura enthusiastically welcomed the proposal as a "great initiative that will help to bridge the knowledge divide and promote mutual understanding". The WDL was launched in April 2009 and developed by a team of the Library of Congress, with technical assistance of the Bibliotheca Alexandrina, in Alexandria. Institutions contributing to the WDL include national libraries and cultural and educational institutions in Brazil, China, Egypt, France, Iraq, Israel, Japan, Mali, Mexico, Morocco, the Netherlands, Qatar, the Russian Federation, Saudi Arabia, Serbia, Slovakia, South Africa, Sweden, Uganda, the United Kingdom and the United States.[19] The WDL is an impressive undertaking, but the potential of digitized and electronically connected libraries and archives is still far from being fully exploited.

Mention should finally be made of the *list of lists of lists* available on the Internet, particularly of the *lists of country-related topics*. The references are hyperlinked; visiting the websites will start a snowballing effect. Country information available on the Internet would tremendously increase, if conventional book publishers, realizing that making scientific research freely available to the public would support a greater global exchange of knowledge, were willing to loosen their open-access conditions. Their current business model blocks scientific progress. Generally, publishing companies are only commercially interested in the output of scientists and scholars. This hampers the progress of science, because making profit can not be harmonized with searching for the truth. Visit https://glasstree.com/shop.

Notes and References

1. Alphabetical, categorical, chronological and generational lists of program ming languages can be found on the Internet. For human–computer interaction, see Julie Jacko (ed.), *The Human–Computer Interaction Handbook* (CRC Press, 2012), the Springer *Human–Computer Interaction Series* (2002ff) and Constantine Stephanidis (ed.), *HCI International 2019-Posters* (Springer, 2019). In addition, go to www.hci.international, http://tochi.acm.org, https://hcipioneers.wordpress.com and http://hcibib.org.
2. Human action has been the subject of extensive research. See Samuel Gutenplan, *Mind's Landscape* (Blackwell, 2000), a thought-provoking and enjoyable book.
3. For a first encounter with computer science, I recommend Edward Blum and Alfred Aho (eds.), *Computer Science* (Springer, 2011), George Dyson, *Turing's Cathedral* (Pantheon, 2012), Eric Schmidt and Jared Cohen, *The New*

Digital Age (Vintage, 2014), Hey and Pápay *op.cit.*, Peter Denning and Craig Martell, *Great Principles of Computing* (MIT Press, 2015), Glenn Brookshear and Dennis Brylow, *Computer Science: An Overview* (Pearson, 2015), Bob Merritt, *The Digital Revolution* (Morgan & Claypool, 2016), Ken Steiglitz, *The Discrete Charm of the Machine* (PUP, 2019), or Hal Abelson et al., *Blown to Bits* (Addison-Wesley, 2020). Visit https://www.springer.com/gp/computer-science. For computer science societies, go to https://www.references.net/societies/compsci_soc.html. Lists of computer scientists, computer science journals, computer science conferences and unsolved problems in computer science can be found on the Internet.

4. See Sal Restivo (ed.), *Science, Technology, and Society* (OUP, 2005); Steve Fuller, *New Frontiers in Science and Technology Studies* (Polity Press, 2007) and Ulrike Felt et al. (eds.), *The Handbook of Science and Technology Studies* (MIT Press, 2017). Visit https://easst.net, http://www.aascit.org, https://www.jst.go.jp/EN, and https://erc.europa.eu. *Science and Civilisation in China* (SCC) is a book series initiated by Joseph Needham (1900–1995) and published by Cambridge University Press (1954–2015). The project is lopsided towards chemical and biological technology. Except in the volumes III and VII (not written by a mathematician and logician respectively!), SCC pays no attention to *all* the formal sciences and none of the social or human sciences is covered. The project needs to be refurbished and revamped, with the connected concepts of science, civilization and technology being thoroughly discussed. Read Raymond W.K. Lau, "Undoing Joseph Needham? A critique of G.E.R. Lloyd and Nathan Sivin on comparing China and Greece" (June 2021, online) and visit http://www.nri.cam.ac.uk and https://isheastm.org/institutes.
5. For the history of technology (to be read in conjunction with the history of science), see Wolfgang König (ed.), *Propyläen-Technikgeschichte* (Propyläen, 2000). For the philosophy of technology, the twin brother of the philosophy of science, go to https://www.iep.utm.edu/technolo and https://plato.stanford.edu/entries/technology. For the foundations of technology, not to be dissociated from the foundations of science, see Tomas Wright et al., *Foundations of Engineering and Technology* (Goodheart & Willcox, 2017). See previous note.
6. For a list of computer science links, go to https://www.dagstuhl.de/en/library/weiterfuehrende-links, and for computer science bibliographies, visit https://liinwww.ira.uka.de/bibliography. Also visit https://www.nextplatform.com.
7. See John Preston and Mark Bishop (eds.), *Views into the Chinese Room* (Clarendon Press, 2002, pp. 51–69).
8. See Ethan Siegel, "Has Google Actually Achieved 'Quantum Supremacy'?", *Forbes*, September 27, 2019.
9. See Hey and Pápay *op.cit.* pp. 161–163, 192–242). For a thoroughly researched biography of Licklider, see Mitchell Waldrop, *The Dream Machine* (Stripe Press, 2018).

10. Lists of digital library projects, open-access journals, online encyclopedias, online databases, and academic databases and research engines can be found on the Internet. The *International Journal on Digital Libraries* (1997ff) examines the theory and practice of acquisition, definition, organization, management and dissemination of digital information via global networking. See further *International Journal of Library Science* (2010ff). Visit http://www.wdl.org, http://dl.ifip.org, https://dl.acm.org/dl.cfm, http://ecdlconference.isti.cnr.it, http://www.tpdl.eu, http://www.dlorg.eu, https://icadl2018.org, https://www.clir.org, http://www.jcdl.org, https://doi.org/10.1109/JCDL.2014.6970177, https://fox.cs.vt.edu, https://ircdl2019.isti.cnr.it, http://lhldigital.lindahall.org, https://www.ifla.org and http://www.globalcc.org. For digital archives, go to http://www.wileydigitalarchives.com. For knowledge integration, see Mitsuru Kodama, *Knowledge Integration Dynamics* (World Scientific, 2011). See also W. Boyd Rayward, "A history of Computer Applications in Libraries: Prolegomena", *IEEE Annals of the History of Computing*, 24:2 (2002), pp. 4–15, Ravindra Sharma (ed.), *Libraries in the Early 21st Century: An International Perspective* (De Gruyter, 2012), Frederik Tell et al. *Managing Knowledge Integration Across Boundaries* (OUP, 2016), Christine Borgman, *Big Data, Little Data, No Data: Scholarship in the Networked World* (MIT Press, 2017) and John McDonald and Michael Levine-Clark, *Encyclopedia of Library and Information Sciences* (CRC Press, 2020). Ikujiro Nonaka, Hirotaka Takeuchi and Yoshiteru Nakamori are Japanese authorities on knowledge management and integration.
11. Go to https://libguides.tulane.edu/dh.
12. See also Marylin Deegan and Willard McCarty (eds.), *Collaborative Research in the Digital Humanities* (Routledge, 2012), Bryan Carter, *Digital Humanities* (Emerald, 2013), Jerome McGann, *A New Republic of Letters* (Harvard University Press, 2014), Steven Jones, *The Emergence of the Digital Humanities* (Routledge, 2014), Eileen Gardiner and Ronald Musto, *The Digital Humanities* (CUP, 2015), Julie Thompson Klein, *Interdisciplinary Digital Humanities* (University of Michigan Press, 2015), Robert Simanowski (ed.), *Digital Humanities and Digital Media* (OHP, 2016), Susan Schreibman et al. (eds.), *A New Companion to Digital Humanities* (Wiley-Blackwell, 2016), Bojan Bozic et al. (eds.), *Computational History and Data-Driven Humanities* (Springer, 2016), Patrik Svensson, *Big Digital Humanities* (University of Michigan Press, 2016), Matthew Gold and Lauren Klein (eds.), *Debates in the Digital Humanities* (University of Minnesota Press, 2016), Richard Lane, *The Big Humanities* (Routledge, 2017), David Berry and Anders Fagerjord, *Digital Humanities* (Wiley, 2017), Olivier Le Deuff, *Digital Humanities* (Wiley, 2018), Lewis Levenberg et al. (eds.), *Research Methods for the Digital Humanities* (Springer, 2018), Agiatis Benardou et al. (eds.), *Cultural Heritage Infrastructures in Digital Humanities* (Routledge, 2018), Shun-han Rebekah

Wong et al., *Digital Humanities and Scholarly Research Trends in the Asia–Pacific* (IGI Global, 2019) and Hannes Werthner et al. (eds.), *Perspectives on Digital Humanities* (Springer, 2021). Useful websites are http://www.adho.org, http://dhcenternet.org, http://shc.stanford.edu/digital-humanities, http://dhawards.org, https://sr.ithaka.org (search for digital humanities), http://dev.zfdg.de, http://eadh.org, http://journalofdigitalhumanities.org, https://www.kcl.ac.uk/ddh, http://ceur-ws.org/Vol-2461, https://networks.h-net.org/h-digital-history, https://jdmdh.episciences.org/page/histoinformatics, https://www.digihuman.eu and https://www.parmenides-foundation.org/parmenides-center-for-digital-humanism. See also Note 9 in Chap. 1.

13. See Michael Borenstein et al., *Introduction to Meta-Analysis* (Wiley, 2009) and Nicola Döring and Jürgen Bortz, *Forschungsmethoden und Evaluation in den Sozial- und Humanwissenschaften* (Springer, 2016).
14. See Jason Lankow et al., *Infographics* (Wiley,2012), Colin Ware, *Information Visualization* (Elsevier, 2013), Nathan Yau, *Data Points: Visualization that Means Something* (Wiley, 2013), Chaomei Chen, *Mapping Scientific Frontiers* (Springer, 2013), Robert Spence, *Information Visualization* (Springer, 2014), Weidong Huang (ed.), *Handbook of Human Centric Visualization* (Springer, 2014), Matthew Ward et al., *Interactive Data Visualization* (CRC Press, 2015), Tamara Munzner, *Visualization Analytics & Design* (CRC Press, 2015), Alexandru Telea, *Data Visualization* (CRC Press, 2015), Andy Kirk, *Data Visualization* (Sage, 2016), Lindy Ryan, *The Visual Imperative* (Morgan Kaufmann, 2016), Kieran Healey, *Data Visualization* (PUP, 2018), Jean Lopez et al., *Infographie de la Seconde Guerre mondiale* (Perrin, 2018), *Computer Graphics Forum* (1982ff) and *Information Visualization* (2002ff). Visit http://ieeevis.org, https://www.see-conference.org and http://www.visigrapp.org.
15. See Friedrich Stadler, The Vienna Circle (Springer, 2015).
16. Visit https://plato.stanford.edu/entries/neurath.
17. See Arthur Benjamin et al., *The Fascinating World of Graph Theory* (PUP, 2015). For planning, the process of thinking about the activities required to achieve a desired goal, see Ernest Alexander, *Approaches to Planning* (Routledge, 1992), Christoph Schneeweiß, *Planung* (Springer, 1992), Patsy Healey, *Collaborative Planning* (Palgrave, 2005) and Stanley Abraham, *Strategic Planning* (Emerald, 2012).
18. Visit https://www.eolss.net.
19. Visit https://www.wdl.org/en.

Chapter 8
Implications for Higher Education: Connect the Dots!

Abstract In this chapter, I explain why the lifting of country studies onto a new, scientific level could possibly have profound implication for higher education. The problems universities have to cope with are reviewed; integration of research methods and integrity of mentality are proposed to solve these problems. The need for planning and executing cross-disciplinary and international country research on project basis will be emphasized and the meaning, necessity and possibility of interdisciplinary research will be outlined. Historical or era studies turn out to suffer from the same shortcomings as area or country studies. Up to this point, we were like ascending a mountain. I have written what I wanted to write. From now on, we will descend, but along the imaginary road some surprises will be awaiting the practical-minded reader.

People interested in the state of affairs in countries such as China, Russia, Iran, Israel, South Africa, the UK and the US are overloaded with information from various sources which they struggle to filter down to what is essential, reliable and relevant, setting aside what is–for the time being–of minor or secondary importance. It would be no exaggeration to say that these people are facing *une mer à boire*. To use another metaphor, whoever engages in reading area/country-books threatens to get stuck in a quicksand of information. I believe that he/she can be rescued by trying to separate the chaff from the grain, by putting away the books that are written by authors who habitually veer off to areas outside their expertise. The undeniable fact that all things hang together does not imply that everybody can speak or write authoritatively about everything, and yet this is what I daily hear talking-heads doing on radio and television. What I advocate is to see both the whole and its parts (the general and the particular), but to leave the parts, which should always be put or seen in context, to specialists to expound on. "Connect the dots!" is the battle cry that should resound from this book. Depending on the accentuation, the call can be understood in two different ways: "*connect* the dots", or "connect *the dots*". I call for *connecting* the dots, but a few words on dots might be useful. Readers who are not interested in the concept of dot may skip the following digression.

H. Kuijper, *Comprehending the Complexity of Countries*,
https://doi.org/10.1007/978-981-16-4709-3_8

Like holes, dots are difficult to define. "Tiny things" will not do. Atoms are tiny things, but they are in fact natural systems consisting of tinier things (particles). However, what is a particle? Visit https://www.quantamagazine.org/what-is-a-particle-20201112. At night, the sky is dotted with stars, not particularly "tiny things". Stars are also natural systems. "Dot" and "point" are interchangeable. Euclid opens *Elements*, one of the most influential works in the history of mathematics, with the following definitions:

> A point is that which has no parts", "A line is length without breadth" and "A surface is that which has length and breadth only.

Interestingly, the first two definitions are negative. Euclid did not say what a point and a line actually are. Perhaps he was not quite sure of them, but it deserves to be remarked that Euclid's illustrous book is entitled *Στοιχεῖα*, the constituent parts of a whole, whether as minutes, letters, numbers, parts of speech or physical components. The constituents form a *σύστημα* (system, the constituted). The word *stoichos*, from which the word *stoicheion* (plural: *stoicheia*) stems, means a line or row of people or things standing or being put next to each other. The suffix *-eios* is the same as the Latin *-eius* and expresses what belongs to, or has the quality of, something. Therefore, as *stoichos* means a row, *stoicheion* means what belongs to, has the quality of, or constitutes, a row. A hand or foot cannot exist in isolation from the body as a whole, for a detached part is not a part properly so called. Bodily parts are functionally *identified by their relation* not only to other parts but also to the functioning of the system into which they fit. Physically isolated from that context, they are at best potential parts and at worst dead, *i.e.* former parts. In modern parlance, the word element, having lost the subtle meaning of *part of a whole*, refers to something that is quite the opposite of "system". Modern scientists are inclined, even keen, to reduce a system to its parts, forgetting that the parts are the parts of a whole. Analysis is the act of isolating and examining something in detail, in order to discover more about it. To understand something, however, the thing (part) has to be seen in context, in relation to the system (whole) it belongs to. In modern science, the fine line between distinguishing and separating is blurred, which, in my view, is a basic flaw. The fatefulness of scientific fragmentation can be attributed to it.

In Chap. 10 of his great booklet *The Greeks* (Penguin, 1951), the notable classical scholar Humphrey Kitto (1897–1982), surveying the character of the Greek Mind and some of its achievements, writes:

> A sense of the wholeness of things is perhaps the most typical feature of the Greek mind. The modern mind divides and specializes, thinking in categories; the Greek instinct was the opposite, taking the widest view and seeing things as an organic whole. (p. 169)

I further refer the reader to the pages 170–194, where professor Kitto illustrates this wholeness, beginning with "that very Greek thing, the Greek language". What intrigues me is the similarity between the ways of thought in Ancient Greece and Ancient China. Their similarity (organic worldview) is striking, but the subsequent divergence between the Western and Chinese ways of thought is even more striking. Was it because of the overwhelming power of Christianity that culminated in what Colin Morris beautifully and inimitably describes in his landmark book *The Papal*

Monarchy: The Western Church from 1050 to 1250 (Clarendon Press, 1989)? Was it because of the impact of the Reformation, the 16th-century religious, political, intellectual and cultural upheaval which splintered Catholic Europe, setting in place the structures and beliefs that would define the continent in the modern era? Or was the Enlightenment responsible for the divergence? I do not know, but I refer the interested reader to Arthur Cooper (edited by Imre Galambos), *The Other Greek* (Brill, 2018) and to the work of the French Hellenist and Sinologist (!) François Jullien.

Nicomachus of Gerasa (c. 60–c. 120 AD), less known than Euclid but celebrated for his *Ἀριθμητικὴ εἰσαγωγή* (introduction to arithmetic), improved on his illustrious predecessor by saying:

> A point is the beginning of a dimension, but not the dimension itself, and likewise the beginning of a line, but not the line itself. A line is the beginning of a surface, but not the surface itself, and likewise the beginning of a two-dimensional, but not the extending in two directions. Naturally, too, a surface is the beginning of a body, but not the body itself, and likewise the beginning of a three-dimensional, but not the extending in three directions. Exactly the same can be said with regard to numbers. Unity is the beginning of all number that advances unit by unit in one direction. Linear number is the beginning of plane number, which spreads out like a plane in one more dimension, and plane number is the beginning of solid number, which possesses a depth in the third dimension, besides the original ones. (Book II, Chap. vii, paragraph 3).

Though the use of "unity" (*μονάς*) is understandable (because the Greek had no symbol for zero [*μηδέν*]), it is worrisome, for unity implies the unification of at least two parts and can thus not be "the beginning of all number that advances unit by unit in one direction **(ἀρχὴπαντὸς ἀριθμοῦ ἐφ' ἓν διάστημα κατὰ μονάδαπροβιβαζομένου)**".

Set theory, an important branch of mathematics initiated by Georg Cantor (1845–1918), does not use unity as "the beginning of…", but relies on there being (existing) something called empty set, symbolized by Ø, like a memberless group of people. The empty set is the set (*Menge*) containing no elements, 1 being the set that contains one set, namely the empty set (symbolized by {Ø}), 2 being the set that contains two sets, namely the empty set and the set containing the empty set (symbolized by {Ø, {Ø}}), 3 being the set that contains three sets, namely the empty set, the set containing the empty set, and the set containing the empty set and the set containing the empty set (symbolized by {Ø, {Ø}, {Ø, {Ø}}}) etc. The principle believed to be true is that everything is based on nothing. This axiom (!) flatly contradicts the thesis of Parmenides (fifth century BC), who argued that "nothing comes from nothing", but it befits the theological doctrine that Cantor supported (*creation ex nihilo*). This doctrine is incompatible with the Christian key truth: "God created the world". If he/she (?) created the world, the origin of the world cannot be said to be nothing, because saying that God is nothing would be tantamount to saying that God doesn't exist. In the next chapter, I will return to this fundamental issue. In the meantime, the reader may see Alister McGrath, *Christian Theology: An Introduction* (Blackwell, 2001, pp. 296–307) and Thomas Oord (ed.), *Theologies of Creation: Creatio* Ex

Nihilo *and its New Rivals* (Routledge, 2015). For Parmenides, he may explore https://www.parmenides-foundation.org.

The generalization of arithmetic into algebra resulted from the extension of symbolism, but it would last fifteen hundred years until Descartes (1596–1650), who is often dubbed the father of modern philosophy and mathematics, discovered that algebra can be used to describe geometry, a branch of mathematics concerned with the shape, size, and relative position of things. He showed that by the use of coordinates any algebraic formula could be plotted as a curve in space and, vice versa, any curve in space could be converted into algebraic terms. Analytical geometry (aka Cartesian, or coordinate, geometry) was born, the study of geometry using a coordinate system, whereby the position of points is uniquely determined by numerical coordinates. The study would turn out to be revolutionary, because it could be used in the graphical analysis of motion (change of position) and process (series of changes), a subject that captured the mind of Newton and Leibniz, who, independently from each other, developed modern calculus (the mathematical study of continuous change). However, "modern physics" (Max Planck, Albert Einstein, Niels Bohr and many other physicists) has shown that Descartes had not the last word on the matter.

The exhortation "*Connect* the dots!" (in the winged words of E.M. Forster: "Live in fragments no longer, only connect") should be taken as a call to see how things belong together and form a system, to realize that a river is more than an aggregation of droplets of water, to throw the stovepipe away, come out of one's box or cubicle and look at the bigger picture. In short, the exhortation is a call to be truly intelligent. Connecting the dots, or showing unity in diversity, is beautifully illustrated by the complex numbers formula of Leonhard Euler (1707–1783), who unified trigonometric with exponential functions and connected the five most important constants in analysis: 0, 1, e, π and i. Euler's complex numbers formula is to be distinguished from his equally famous polyhedra formula, which relates the number of a polyhedron's vertices, edges, and faces.[1] I do not know whether Euler's formulas are or can be connected, but it strikes me that relationship is what the formalas have in common. Professional mathematicians assured me that the two formulas have nothing to do with each other.

In *Tales of Impossibility* (PUP, 2019),[2] David Richeson argues that the complex numbers formula was "one of the key ingredients in proving that *pi* is a transcendental number and that it is impossible to square the circle". If this is the case, it seems to me that going from linearity to nonlinearity, as scientists of complex systems are doing, is like jumping up to the transcendental number, which does not solve a nonzero polynomial equation with integer coefficients—a "leap of faith" (Kierkegaard) that cannot be explained in ordinary ways. A transcendental number is a number beyond normal human understanding, but not necessarily beyond any understanding. If the two formulas have nothing to do with each other and the circle cannot be squared, it would mean that we have one Euler concerned with linearity (polyhedral formula) and another Euler concerned with nonlinearity (complex numbers formula). If the two formulas cannot be bridged, the stunning conclusion has to be drawn that linearity and nonlinearity are irreconcilable. However, the formulas were created not by a split or schizophrenic but by an integrative mind. Euler *must* have thought whether or how, on a deeper level, the two formulas were connected. My hunch is that Euler's formulas are somehow related to each other. Whether it is a conjecture that can be added to the list of mathematical conjectures (available on the Internet), I leave to mathematicians such as Ron Calinger, Lokenath Debnath, David Richeson, Hanspeter Kraft, John Baez, Nick Gurski and Tom Leinster to decide.

In August 2019, the University of Amsterdam launched a major recruitment campaign called Connecting Science in the Areas of Physics, Chemistry, Mathematics, and Computer science. Starting a campaign called Connecting the Sciences would have been more ambitious and more responsive to the needs of modern society. It would have been taking on the challenge of reconciling the natural and cultural sciences. Connecting the dots and establishing relationships between seemingly diverse objects is what I advocate. Rethinking country studies, I attempt to redesign, renew and uplift them by stressing the need for collaborative, *round*-table research and by pleading for use of the latest insights of complex systems and computer scientists. Connecting connectivies, ordering orders, or systematizing systems, is ultimately needed to comprehend a country. The garment of current area/country studies, which have a long pedigree,[3] is old-fashioned and worn-out. Patching and darning will not be sufficient anymore. *A complete renewal is needed: genuinely interdisciplinary and truly international collaboration based on general systems science and aided by computer science.* Countries such as China and Japan, where collective behavior has been favored and the individual is relegated to a subsidiary position, seem to be more apt to scientific collaboration than countries such as the United Kingdom and the United States, where individual actions have been glorified. Privacy protection has been more an issue in the U.S. or the U.K. than in China or Japan. See the online article of the German-American anthropologist and historical geographer Berthold Laufer (1874–1934), 'East and West', *The Open Court*, 47:927 (December 1933), pp. 473–478.

Since the late twentieth century, the word *network* has appeared regularly in the media and the term network society is on nearly everybody's lips. Examples of networks (discrete structures consisting of dots and their connections) are the Internet, social media, transport networks, neural networks, organizations (natural or human-made), financial systems and ecosystems. In Chap. 5, I dealt critically with network *science*. Nevertheless—and remarkably—the current education system is still based on a predominantly reductionist mindset. Teaching takes place on a subject-by-subject or module-by-module basis, and little, if any, attention is paid to the complexity of things or situations, under the pretext that youngsters are not interested in, or cannot cope with, complexity. At higher education, the interconnection between the subjects covered by lecturers or professors from different faculties tends to be neglected. In other words, our Information Age, highlighted by Manuel Castells, Daniel Bell, Alvin Toffler, John Naisbitt, Thomas Friedman and many other scholars, suffers from complexity illiteracy. Though the relationship between network and complexity science would be an intriguing and riveting subject to study (Euler), universities, originally meant to *educate* rather than to train, demonstrate a shocking lack of the *systemic* approach. Graduates have learned a special trick or to give a paw on demand instead of being really critical, adopting an active and integrative lifestyle, making a real contribution to society, and leaving the world better than they found it.[4]

Practitioners of area or country studies have been investigating a wide range of topics: education, sexual behavior, ancestor worship, family life, caste system, migration, criminality, conflict resolution, decision-making, pollution, government, bureaucracy, corruption, art(s), religion, folklore, literacy, gardening, business, negotiating style, sport(s), urbanization and many other research areas. They have been doing so in basically two different ways:

- By choosing a research method they have not invented or developed themselves (*e.g.* literature review, archival research, textual analysis, discourse analysis, conversation analysis, interview, focus group method, survey, observation, comparison, photography, audio or video recording, statistical analysis).
- By using concepts from, or drawing and relying on theories developed in, disciplines ranging from geography to demography, from economics to politics, from psychology to sociology, and from linguistics to literary science, that is to say, in an array of disciplines they cannot possibly be graduated in, the result being that these concepts or theories have been like a razor blade in the hands of a baby. What can we expect from a treatise on the Russian government written by a "Russia expert" with no schooling in political science! What can we expect from a discourse on the population problems of India delivered by an "India expert" who is not a demographer! What can we expect from a book about China's economy written by a "China expert" having no knowledge of the principles of economics!

Practitioners of area or country studies fail to put Humpty Dumpty together, to organize their pieces of knowledge scientifically, to build a house that features a distinctive architecture, to weave a tapestry bearing the stamp of a genuinely scientific discipline. A patchwork, a collage of pictures (of poor quality at that) is all they can produce. The eloquent lecturer gets away with it and the docile students write dutifully down every word from his/her mouth.

Intellectual walls have to be torn down, academic fences must be removed, and mental silos are to be taken away—without compromising somebody's expertise. Cross-disciplinary teams should be formed and empowered. It is time for decompartmentalization and open-mindedness, time for carefully looking at the parts without losing sight of the whole they belong to, time for mastering or developing an academic discipline without forgetting its modest but legitimate place in the ever more sophisticated system of sciences. It is time for learning and practicing the art of flower arrangement, where the beauty of each individual flower is optimally reflected by prioritizing the whole over the parts. It is time, not for thinking socially, politically, legally, economically, geographically, demographically, or militarily, but for synthesis, for the grand design,[5] for computationally collecting, processing and combining the data acquired from different sources. It is time for thinking systemically, for putting things into perspective, for taking "a crude look at the whole" (Murray Gell-Mann), for applying the principles of systems science. It is time for a new kind of country study that (*a*) takes into account that a country's polity, judiciary, military, economy, society, demography, ecology, geography, history, technology and philosophy are intimately interconnected, (*b*) recognizes that each and every country has a unique set of emergent properties, none of which can be the property, quality or characteristic of any constituent part of the country, (*c*) draws on the insight that systemics, cybernetics and informatics are closely interlinked[6] and (*d*) makes full use of the potential provided by digital libraries.

The kind of country study I have in mind is *radically* new, having profound implications for both the social and the human sciences, hence for higher education. When the science of systems is seriously applied to the study of countries, it will transform the system of sciences. The new kind of country studies advocated in this book may be the cause for many things; it may:

(1) Revolutionize country studies by widening and deepening them. Able to generalize the new method of studying a single country to the way of studying *any* country (small, medium-sized, or big), practitioners of country studies may organize themselves into an international discussion forum and launch a rigorously edited journal devoted to the history, theory, practice and principles of their research and standing head and shoulders above the dozens of philosophically and scientifically unsound area or country journals currently published.

(2) Adjust the work of Western social and human scientists by prompting them to test their theories out on countries that have been lying outside of their traditional purview. When, on the basis of computational systems science, the developments in non-Western countries are thoroughly investigated and systematically compared with those in Western countries, the multifaceted and polydimensional project of modernizing the world and the postmodern reactions to it (a huge project with which Western thought has been bound up) may be redesigned and overhauled.[7]

> The social sciences, under which area studies are usually but not consistently subsumed, largely encompass the human sciences, because the human mind, the various aspects of which are studied by the human sciences, is firmly embedded in, and inextricably linked with, its *Mitwelt* and *Umwelt*, as Ana Abrantes, Enrique Bernárdez, Leon De Bruin, René Dirven, Guy Dove, Andreas Elpidorou, Roslyn Frank, Shaun Gallagher, Peter Hanenberg, John Haugeland, Shinobu Kitayama, Matthew Lieberman, Robert Logan, David Matsumoto, Nicolas Payette, Jesse Prinz, Norbert Ross, Geoffrey Samuel, Arnold Sameroff, Daniel Schmicking, Jaan Valsiner, Lev Vygotsky, Anne Watson, Ning Yu, Dan Zahavi, Tom Ziemke, Jörg Zinken and many other scientists have been making abundantly clear in their work. Social psychologists unanimously subscribe to the view of John Donne (1572–1631): "No man is an island; every man is a piece of the continent, a part of the main".

One may wonder whether social psychologists really know how mean and nasty humans can be. See how man behaves and you will know what kind of a person he is: a kicker downwards, a licker upwards, or a sincere and honest man, able and willing to stand on his own feet. In *Social Organization: A Study of the Larger Mind* (1909),

Charles Cooley (1864–1929), who influenced Herbert Blumer (one of the protagonists of symbolic interactionism) and must be regarded as a pioneer of communitarianism (Amitai Etzioni, Charles Taylor, Robert Putnam, Michael Sandel), formulated the key role of family, playgroups and community of elders as the sources of man's ideals and morality. He defended the thesis that "the mind is social". On the *Reichsparteitag* in 1934, during the *Totenehrung*, in Nuremberg, who dared *not* to extend the right arm from the neck into the air with a straightened hand and shout "*Heil Hitler*"! When the king is about to pass by, everybody pushes and is eagerly waiting. Whoever reads Elias Canetti, *Masse und Macht* (1960) is forever cured from his belief in the existence of personal autonomy or individual independence (visit https://www.iep.utm.edu/autonomy and https://www.iep.utm.edu/autonomy). The human mind (also the concern of *philosophers* of mind) is the invisible side of the human body, studied by biologists and medical scientists. However, human beings are complex, adaptive systems influenced by and influencing their environment; they are deeply grounded in their *Lebenswelt* (Edmund Husserl) and this lifeworld is undeniably located in a particular part of a country in a certain period of time.[8]

The social and human sciences must be rethought by taking the diversity of countries into account. Readers should take a good look at *Le Déjeuner des canotiers*, a painting wrought by Pierre-Auguste Renoir, who lived from 1841 to 1919. What they see is not (first look) fourteen different characters, but (second look) the different relations between them, *in the summer of 1881*. It is not the juxta- but the typical *com*position, the order making *French* impressionist painting so wonderful. See International Science Council, *Conversations on Rethinking Human Development* (2020, online).

(3) Make an improvement on the work of demographers by urging them to make their ideas more concrete and to put their analyses into a comparative country-perspective. Many demographers are inclined to thinking and writing abstractly, not paying (enough) attention to the factors that formed, or to the processes that produced, the demographic situation in a particular country.

(4) Realign the work of sociologists by asking them to be also concerned with the nature of the *location or territory* where people, living together, are involved in persistent social interaction. Many sociologists, who take a prominent position among social scientists, can be accused of parochialism, of not considering whether or to what extent their theories are applicable to situations in non-Western countries. This tendency to show interest in only a geographically or culturally narrow range of matters, especially in affairs that somehow affect people in Western countries, is intolerable, because it goes against the grain of true science (see Chap. 2).

(5) Invigorate the work of anthropologists by asking them to be also concerned with the characteristics of the *location or territory* where people, living in community, have a different mindset. Many anthropologists, whose work is to be distinguished (but by no means separated) from that of sociologists, can be accused of ethnocentrism, of subconsciously grafting their own system of ideas and beliefs onto the connected concepts and assumptions prevalent in another

culture, of judging another culture on the basis of values, norms, aspirations and expectations in their own culture.

> Ideas have a history (see *Journal of the History of Ideas* [1940ff], *Archiv für Begriffsgeschichte* [1955] and Maryanne Horowitz [ed.], *New Dictionary of the History of Ideas* [Thomas Gale, 2005]); they are not "*freischwebend*" (Max Weber), but bound to changing places. Concepts, Bertrand Russell once said, constitute the building blocks of man's thinking and galvanize him into action; they form, subtly interconnected, the fabric of man's life. As long as some notions and their cognates remain vague, others must share this defect. The requirement not to be vague about ideas that have been most potent and persistent is a country is thus paramount. Epistemic relativism, the view that the truth of knowledge-claims is relative to the standard a country uses in evaluating such claims, is an incoherent view.[9] If it is right, the very notion of rightness is undermined: the rightness of epistemic relativism. On the other hand, the non-relativists also face a tall order: how to develop a view on rationality that "is not dogmatic and rejects any notion of a privileged framework in which knowledge-claims must be couched" (Michael Krausz). Universalists, claiming that ideas such as democracy, liberty, equality, property and individuality are universally applicable, tend to be arrogant, as exemplified by the Western dominance in—notice the designation—"less developed countries". The attitude and behavior of the Spaniards, Portuguese, British, French, Dutch and Americans in overseas territories were/are downright appalling. They never treated the indigenous people as equals, enjoying the same rights as the colonists. Missionaries and pastors, having followed the gunners, considered the natives heathens to be rescued. Not long ago, cultural anthropologists were speaking of "savages". The Chinese will never forget the sign "Dogs and Chinese forbidden" that was put up at the entrance of China's Foreign Concessions. The Dutch government has never apologized sincerely for the crimes committed at her behest in the Dutch East Indies.[10] I cannot agree with the relativist, who maintains that culture-bound disciplines block our ability to understand another country, but neither can I agree with the non-relativist, who posits the opposite view. The emic-etic debate among cultural anthropologists revolves around the question as to whether an account of actions should be given in terms that are meaningful to the actors living in or being native to the country under study, or in terms applicable to actions in other countries as well. Whereas the emic perspective focuses on intrinsic distinctions, only meaningful to the members of a given country, the etic view relies heavily upon the extrinsic concepts and categories of foreign scientific observers. This opposition is unnecessary and the disagreement may result in a deadlock. The points of view can be reconciled. A sensible combination of the emic and etic lens will yield a binocular vision, making depth-perception possible.[11] The fact that the great bulk of knowledge of social and human scientists is only based on the investigation of Western things does not imply the impossibility of cross-cultural research.[12]

Cross-cultural research involves a process in which the parties gradually learn to understand each other. The Okanagan Indians call this *en'owkin*, solving a conflict through respectful dialogue, being a gentle process of clarification and integration of ideas, through a discussion not meant to draw distinctions that drive apart, but to find a common ground or denominator and to reach a consensus each of the contestants can peacefully live with.[13] Extremism is antonym to compromise, which, depending on the context, may be associated with tolerance. To compromise is to reach an agreement in which each party accepts less than what she wanted at first. *En'owkin* goes further than this. It does not imply stepping back or giving something up in order to reach a calculated agreement, but transcending one's own position and having due regard for the one taken by the other. It means not tolerance (whereby each goes his or her own way, allowing others to believe what they want to or to pursue their own interest without being criticized) but respect.[14] Careful comparing may result in a change of mind, a mental leap, a conceptual reconfiguration. It may bring about true enlightenment. The interdisciplinary and international or intercultural country research I envisage is not based on a debate over who is right and who is wrong, but on a dialogue, on a sincere and honest exchange of views. Such a dialogue is only possible between scientists who respect, and actively listen to, each other, between scientists who feel they are stepping on holy ground when visiting the country of their foreign colleagues.

(6) Complement the work of geographers, who primarily and essentially *spatialize*, whatever the degree of their differentiating.

(7) Complement the work of historians, who primarily and essentially *temporalize*, whatever the degree of their differentiating. Some historians opt for the thematic treatment of their material, but, in the absence of an historical *explanans* (see below), they cover the chosen subject or idea in a manner that is primarily and essentially chronological. Only the sequence of time keeps the things together and no attention is systematically paid to space, the twin brother of time. Here, three related issues should be noticed:

    (A) Whereas geography may be conceived of as history in space, history may be conceived of as geography in time. Geography and history are like conjoined twins. For example, the ancient "Silk Road" is to be viewed as spreading over space and time. Whoever ventures to describe that network of trade routes[15] should take both its geography and its history into account. Neglecting one side of the coin would distort the picture of the other side. In the mid-1960s, inspired by the space-time model in physics, Torsten Hägerstrand (1916–2004) created time-geography.[16] Paraphrasing his main message, one could say that each person's life-course is unique and comparable to a barcode, for the path he/she has taken in space as well as in time nobody else has taken or will be taking. How different individuals are able to form a community is the central problem of sociology and anthropology. In *Sein und Zeit* (1927), Martin Heidegger (1889–1976) investigates what he claims to be the "long-neglected question of being", but whereas the subject *Dasein* und *Zeitlichkeit* (existence and temporality) is extensively treated (pp. 231–437), the subject of *Dasein* und *Räumlichkeit* (existence and spatiality) receives scant attention (pp. 101–113). Heidegger seems to have forgotten that the two basic dimensions of being are intimately interrelated. The book, "*in Verehrung und Freundschaft dedicated to Edmund Husserl*" (his Jewish colleague whose great cause Nazi-sympathizer Heidegger would not champion [the friendly dedication was missing in a later edition!]), suffers not only from arcane language-use but also from a fundamental imbalance. Its author should have written a crystal-clear book entitled *Sein · Zeit · Raum*. No philosopher after Heidegger has accepted the challenge.[17] However, in his *magnum opus*, posthumously edited by Kajsa Ellegård and Uno Svedin (with a long bibliography by Bo Lenntorp), Hägerstrand rose to the challenge. In 1992, the highly prestigious Vautrin Lud Prize was awarded to him. See also Georges Benko and Ulf Strohmayer (eds.), *Geography, History and Social Sciences* (Kluwer, 1995), Vladimir Batagelj et al., *Understanding Large Temporal Networks and Spatial Networks* (Wiley, 2014) and Tony Taylor et al., *Place & Time* (Pearson, 2018). Currently, geography, history and the social sciences are increasingly integrated.

(B) The close relationship between people, space and time can be illustrated by looking at globalization, the process in which the world becomes smaller. Opinions about globalization differ, but there is agreement on the core meaning of it: change in man's spatial and temporal experience. Time is the essence of everybody and everything. "Time, like an ever-rolling stream, bears all its sons away; they fly forgotten, as a dream dies at the opening day", the poet said. Geographical distance is typically measured in time. As the time necessary to connect distinct geographical locations shortens (due to technological breakthroughs), space also contracts. Time–space compression is the result. The human experiences of space and time are thus interconnected. As the French geographer Élisée Reclus (1830–1905) already said: "*Vue de haut, dans ses rapports avec l'homme, la géographie n'est autre chose que l'histoire dans l'espace, de même que l'histoire est la géographie dans le temps*" (*L'Homme et la Terre*, Vol. I, Chap. 1, page 50).[18] In the words of Clifford Kinvig (1934–2017): "It must never be forgotten that geography is not something fixed and unalterable, but something which is essentially dynamic." Theorists of globalization tend to agree that the change in man's experience of space and time undermines the importance of local and even national boundaries. Globalization is, therefore, a subject closely related to modernization. Intensively studied in the field of International Relations and having far-reaching implications for virtually every facet of human life, globalization necessitates the rethinking of key questions in (geo) political, legal, economic, financial, psychological, social, educational, environmental, demographic and cultural theories. The importance of *anti*globalization movements should not be neglected, however.[19]

(C) My critique of **area** studies might as well apply to **era** studies. Historians also have an *explanandum* (an era to be explained), but they do not have a distinctive (historian) *explanans* (a theory of their own that does the explaining). Whereas journalists, who are habitually short-sighted, report on things that are somewhere happening (at present) or have been happening in the near past,historians give a subjective description of what happened long ago or has been happening for a long time. If geographic, demographic, ecological, political, legal, polemological, economic, financial, psychological, sociological, educational, religious, artistic and anthropological ideas or theories (the staple of geographers, demographers, ecologists, political scientists, jurists, polemologists, economists, financial experts, psychologists, sociologists, educationists, experts in religious matters, aestheticists and anthropologists respectively) were erased from history (as account), little would be left that is scientifically interesting. Like so-called area or country experts, historians can do nothing but to show the feathers of other people and to flaunt what is not theirs. Consequently, what the reader of a book on the history of, say, Iran or Turkey written by an "Iran/Turkey expert" gets is easy to guess: certainly not the *explanans* (theory) of a historian. Generally, books entitled "*The history of*..." or "*The story of*..." are thus *mit viel Vorsicht zu genießen*. Classical studies encompass all the studies related to the classical era and medieval studies include all the studies related to the medieval period, but no classical scholar or medievalist disposes of a distinctive theory about the classical era and the medieval period respectively. How the two periods should be theorized about separately is far from clear, not to speak of the pre-modern (or the modern) era. A full-blown (not one-sided and thus misleading) account of the European Renaissance, Protestant Reformation, Age of Enlightenment, Industrial Revolution, America, French or Russian Revolution, First, or Second, World War, Cold War or History of the European Union can never be the work of a single person, of a shoemaker who sticks to his last (assuming he has one). The writing of an "integral history", as aimed at by the "*marxistisch angehauchter*" scholar and Huizinga-student Jan Romein (1893–1962), should be interdisciplinary in character. Juxtaposing a number of analyses only checked on grammatical or typographical errors in an edited book will not be enough, because the architecture, the overarching or underlying idea and the theoretical justification of putting the analyses together are missing. The book will be a hodgepodge, not a composition. The present is nothing but the tiny tip of a huge iceberg above the waterline. The Second Word War, an enormous entanglement of actions and events (notice the difference!), cannot be comprehended without thorough knowledge of the First World War, which cannot be comprehended without comprehensive knowledge of the previous period in the world history etc. etc., that is to say, without the all-inclusive and all-embracing knowledge no historian, or practitioner of Big/World History (see below), will ever have. Historians are curious to know how things *have come to be* as they are. However, this presupposes that they know already how these things *actually were* in the way they were—a highly questionable assumption. Historians are missing a scientifically-rooted principle of historiography and have no distinctive *explanans*. They are nothing but narrators and storytellers ventilating their personal opinion. Definitely, they are not scientists properly so called.

"Historiography" refers to one or more than one of the three aspects of history-writing:

(A) *Descriptive* historiography accepts what historians usually do and describes the standard methods and procedures.
(B) *Historical* historiography traces the ways in which history has been written throughout the ages. For a long time, nationalism was the framework in which history was written. Nationalism was the driving force behind Europe's historiography. According to the American medievalist Patrick Geary:

> The modern history was born in the nineteenth century, conceived and developed as an instrument of European nationalism. As a tool of nationalist ideology, the history of Europe's nations was a great success, but it has turned our understanding of the past into a toxic waste dump, filled with the poison of ethnic nationalism, and the poison has seeped deep into popular consciousness (*The Myth of Nations* [PUP, p. 2015])

Enoug has been written about Eurocentrism, a term coined by Samir Amin. It refers to a world-view biased towards Western civilization. See Peter Gran, *Beyond Eurocentrism: New View of Modern World History* (Syracuse University Press, 1996), Dipesh Chakrabarty, *Provincializing Europe: Postcolonial Though and Historical Differences* (PUP, 2008), Rajiv Malhotra, *Being Different: An Indian Challenge to Western Universalism* (HarperCollins, 2011), Nick Hostettler, *Eurocentrism: A Marxian and Critical Realist Critique* (Routledge, 2012) and Ilia Xypolia, *Eurocentrism and Orientalism*, in Sangeeta Ray et al. (eds.), *Encyclopedia of Postcolonial Studies* (Wiley-Blackwell, 2016).

Like history itself, historical historiography is divided into periods, areas/countries, or issues. It also encompasses studies of individual historians. Archaeology, epigraphy, sigillography, palaeography, prosopography, genealogy, heraldry, phaleristics, numismatics, onomastics, diplomatics, codicology, philately, toponymy and vexillology are some of the auxiliary studies relied upon by "era experts" who—like "area experts"—cannot stand on their feet, need crutches and have miserably failed to come up with a distinctive *explanans*, a theory that does the explaining.

(C) *Analytical* historiography (aka critical historiography) discusses the concepts and philosophical problems arising from the writing of history. It overlaps with analytical or critical philosophy of history, the main difference being that the approach of analytical historiography is historical and the approach of analytical philosophy of history is philosophical. There are two sorts of philosophy of history: whereas history-as-account belongs to analytical or critical philosophy of history, history-as-happening belongs to speculative philosophy of history.[21]

A diasynchronic (complex and dynamic) approach to the *explanandum* may be the solution of the fundamental problem of both era and area studies: thus a *längsschnittliche Betrachtungsweise* in combination with a *querschnittliche Betrachtungsweise*, the former emphasizing the diachronical without neglecting the synchronical, the latter laying stress on the synchronical without being indifferent to the diachronical.

Dirk Vollenhoven (1892–1978) struggled many years with the problem of the relationship between dia- and synchronicity, but his view was restricted to the history of philosophy and hampered by his religious belief. In *Working: Researching, Interviewing, Writing* (Knopf, 2019), Robert Caro, celebrated biographer of Lyndon B. Johnson and said to be the nestor of American investigative journalism, cogently argues that both the time and the space in which a person lived are important for writing his/her biography. Structure and process may be distinguished but they are inseparable and complementary. They are even mutually revealing dimensions, as anthropologists such as Claude Lévi-Strauss, Edmund Leach and Marshall Sahlins emphatically taught.Evolution, or history (as a series of events and actions), does not take place along a razor-thin line but over the entire width and full breadth of life. There is unity in diversity. That is what is meant by the "great interdependence of all differences", the central message of the *Yìjīng*.

However, there is a caveat. Whereas determination of a structure leads to a scalar-pattern (things seen as beings or entities), determination of a process results in a vector-pattern (things considered to be entities *in statu nascendi*). Notice that in both cases a pattern emerges. It is very difficult, if not impossible, to view existent and nascent entities at the same time. According to quantum mechanics, a scalar field cannot be described in terms of a vector field, and vice versa. Interestingly, whereas Laozi (6th century BC) opted for becoming, Parmenides (late 6th century BC) went for being. The Chinese (and the Japanese) *décompose* reality and carve it up differently, which has been hampering a fruitful conversation between East and West. All this leaves unimpeded that practitioners of area and era studies, being aware of the mutual relationship of their work, should be open-minded about the possibilities of computer-aided collaborative research based on the science of complex and dynamic systems. Chengxin Pan, professor of International Relations at Deakin University (Australia), champions a new relational ontology in global politics, advocating a paradigm shift from the Cartesian-Newtonian ontology to an "holographic relational ontology". Go to https://doi.org/10.1093/irap/lcy010 and https://doi.org/10.1177/1354066120938844. My hunch is that viewing the compositionality, or dynamic complexity, of a country is basically the same as considering the country to be a holographic relational entity. For compositionality, read Wolfram Hinzen et al. (eds.) *Handbook of Compositionality* (OUP, 2012) or visit https://plato.stanford.edu/entries/compositionality.

If geography (space) and history (time), being two sides of the same coin, are inseparable, it may be necessary to rethink topology in general and mereotopology in particular. Indeed, mathematics may have to be placed on a new footing. A list of topologists can be found on the Internet. Spatio-temporal pattern formation is a subject thus far only studied in the natural sciences and the technique of multidimensional data analysis (research of many data from several angles) is still in its infancy.

(8) Countervail the work of globalists by urging them to pay attention to both the national (local) and the international (global) and by pressing them to synthesize or reconcile the thesis of homogenization (defended by Herbert

Schiller, Armand Mattelart, George Ritzer, and Todd Gitlin) and the antithesis of heterogenization (defended by Ulf Hannerz, Gayatri Spivak, Homi Bhabha, and Arjun Appadurai).

> "Global Studies" and "World/Big History" are in. Leften Stavrianos, David Christian, Merry E. Wiesner-Hanks, Craig Benjamin, Jürgen Osterhammel, Andrey Korotayev, Leonid Grinin, David Baker, Fred Spier, John Mears, Eric Chaisson, Ken Baskin, Dmitri Bondarenko, Esther Quaedackers, Cynthia Brown, Ken Gilbert, Barry Rodrigue, Tracy Sullivan, William McNeill, Robert McNeill, Carroll Quigley, Jerry Bentley, Peter Stearns, Patrick Manning, Dominic Sachsenmaier, Peter Frankopan, Robert Tignor, Felipe Fernández-Armesto, Noel Cohen, Tonio Andrade, Manfred Mai, Frank Welsh, Laurent Binet, Yuval Harari, Sari Hanafi and Richard Haass are some of those who were/are active in this field. They were/are prolific writers, whose products are selling like hot cakes, much to the delight of their publishers. However, how somebody can be expert on everything and can attribute meaningful significance to it is beyond me. If scholars like the ones I just mentioned did/do *not* claim to be such an expert, how then could/can they–following in the footsteps of Hegel, Spengler, or Toynbee–call themselves, "world historians"! Whereas specialists tend to know everything about nothing, generalists (the predicate of "world historians") tend to know nothing about everything. There is no tested theory of everything, no all-encompassing, coherent theoretical framework that connects and explains all phenomena, all events and all actions. Those who claim, or believe, that such a theory exists are fooling not only themselves but also their gullible readers. Thucydides, Herodotus and Polybius might have been able to write a panoramic history single-handedly; aiming to do so in the 20th or 21st century is sheer madness. Many authors, with different views, contributed to *Saeculum Weltgeschichte*, *Propyläen Weltgeschichte* and UNESCO's seven-volume *History of Humanity*. Even *The Cambridge Ancient History* has been compiled by dozens of scholars who are considered to be professional in their field. I am ready to believe that the general public wants to be provided with a broad overview and are not interested in the details, but historians or sociologists who are, for the sake of publicity or pecunia, eager to meet these needs should know one thing: "*Man greift nicht nach den Sternen*".

Recently, Sari Hanafi, Professor of sociology at the American University of Beirut and President of the International Sociological Association, wrote an online article entitled "How could we rethink our conceptual understanding of human development?" I think he might have been a bit too ambitious in doing so. It is certainly important to rethink our conceptual understanding of human development, but we should first try to rethink our conceptual understanding of individual countries as diverse as Libanon, Israel, Iran, Russia, China and the United States by seriously carrying out interdisciplinary and international research. According to Franz Boas (1858–1942), "father of American antropology", ethnologists ultimately aim to comprehend *a*) individuals influenced by their society and *b*) societies influenced by the actions of the individuals constituting them.

What to think of Richard Haass, author of *The World: A Brief Introduction* (Penguin, 2020)? The title of the book contains already a contradiction, for the world is an extremely complex theater with a very long history; it cannot be *briefly introduced*. Whoever thinks he can write a short introduction to the whole world shows dangerous symptoms of self-overestimation. Fernando Reimers, professor at the Harvard Graduate School of Education, writes:

> This is a great book that every high school and college student, as well as teachers and parents, should read. It is an indispensable resource to educate global citizens

Madeleine Albright, who served as Secretary of State under President Bill Clinton, writes:

> This is the book that explains how the world really works, how it is changing, and why it matters. Just what every citizen and student needs to read;

And Fareed Zakaria, in whose self-promoting TV program Haass is a welcome quest, writes:

> This book is a superb introduction to the world and global issues. Richard Haass has been able to write something that is brief, readable and yet comprehensive.

Not involved in political, academic or media affairs, I would rather think that Mr. Haass, who was close advisor to the late Secretary of State Colin Powell in the Bush administration and is president of the American Council on Foreign Relations (an "independent, nonpartisan membership organization"), has written a book the reliability and relevance of which is debatable since the worlwide outbreak of the coronavirus.

One should consider the history of the world lengthwise and breathwise. The vision of it will be dizzying. Any attempt to fully embrace and understand what happened in the past and has been happening will inflict upon the mind a "sense of profound mystery" (John Henry Newman). The practitioners of big history may fancy that they have found the direction that all happenings (events and actions) are taking, but as long as they do not know towards which point these happenings are finally drifting, *i.e.* where history will end and how the cosmic attractor looks like, they do not differ from high prieststs who fob people off with follies, or from "those who claim to know the mind of God and who are prepared to use force, if necessary, to make others conform" (Barbara Taylor). Peter Turchin, Harvey Whitehouse, Pieter François and Kevin Feeney may want us to believe that historical patterns can be detected by big data analytics, as long as they cannot explain the workings of the human mind (which have not yet been shown to be completely reducible to the workings of the human brain) I do not buy their stories. Making statements about spatio-temporal patterns on macro scale, one quickly gets into murky waters.

(9) Shed new light on world-system research by warning for sweeping statements and emphasizing the importance of the parts of a system, *i.e.* the importance of the individuality of the countries making up the world system.

> World-system research or analysis (aka world-system theory) is a macro-scale approach to world history, emphasizing the world-system (instead of the nation-state) as primary unit of social analysis. "Social" indicates that non-social aspects of the world-system are relegated, which is a guarantee that the analyses are lopsided, for there is no scientific reason to prioritize the social side above the political, legal, military, economic, financial, ecological, educational, religious or cultural side of the world-system. This type of research emerged in the 1970s and is rooted in sociology. World-systems thinkers like Immanuel Wallerstein Samir Amin, Giovanni Arrighi, Andre Gunder Frank, Christopher Chase-Dunn, Jackie Smith, Beverly Silver, Volker

Bornschier, Janet Abu Lughod, Terence Hopkins, Thomas Hall, Patrick Manning, Andrej Grubačić, Theotônio dos Santos, Dale Tomich, Michael Burawoy, Fernando Cardoso, Jason Moore, Ravi Palat, Arno Tausch and many others aim to replace modernization theory, which Wallerstein (https://doi.org/10.5195/JWSR.2020.995) criticized for three reasons: (*a*) its focus on the nation-state, (*b*) its assumption that there is only one path of development, and (*c*) its disregard of transnational structures. The main influences on world-systems research were: (I) the *Annales* school (founded by Lucien Febvre [1878–1956], Henri Hauser [1866–1946] and Marc Bloch [1886–1944] but represented most notably by Fernand Braudel [1902–1985]), (II) the Marxist tradition (which stresses social conflict and capital accumulation) and (III) the dependency theory (a neo-Marxist explanation of development processes). By focusing on the system of the *world* instead of attending to the individuality of the nation-states that constitute the world-system, Wallerstein and his followers avoided Scylla, but fell into Charybdis.

With the demise of Marxism and the downfall of socialism, the torch carried by world-system researchers has been taken over by globalists, who fall into the trap of historicism (see previous entry). Leonid Grinin and Andrey Korotayev are only two examples of world-system researchers who knew when and how to change course and where to move the beacons to. Grinin, senior research professor at the Institute for Oriental Studies of the Russian Academy of Sciences, in Moscow, and author of more than 440 "scholarly works", published in *Journal of Big History* (1:2 [2018], pp. 15–26) an article entitled "Evolution of the Early Solar System in Terms of Big History and Universal Evolution", but he is not an astronomist or cosmologist by profession. Korotayev is —hold your breath— a "Russian anthropologist, economic historian, comparative political scientist, demographer and sociologist, with major contributions to world-systems theory, cross-cultural studies, Near Eastern history, Big History, and mathematical modelling of social and economic macrodynamics". He is also co-editor of the journals *Social Evolution & History* and *Journal of Globalization Studies*. It sounds very impressive, but how can one person be expert in many fields of research! "Being expert in more than one field of reseach" feels like contradictory. Thomas Hall (ed.), *Comparing Globalizations: Historical and World-Systems Approaches* (Springer, 2018) explores the debate on globalization and world-system analysis, acknowledging that each of these approaches requires interdisciplinary research.

(10) Affect the study of International Relations (IR) by urging scholars in this fast expanding field of research to pay due attention to both the individual countries (as distinct from the *relations* between them) and their context, *i.e.* to both the diverse subsystems which the countries consist of and the extremely complex world system which the countries belong to. In *China-US Focus* (December 7, 2018), renowned Harvard scholar Joseph Nye ventures his opinion that "the rise of China does not pose an existential threat to us" and "we should think of the US-China relationship as a cooperative rivalry rather than a revival of the Cold War". However, emeritus professor Nye, who served under President Clinton, may be an expert on international relations; he is certainly not an expert on China, that is to say, on its historically

evolved and closely connected political, legal, military, geological, demographic, geographic, economic, financial, social, educational, cultural and ecological systems. Conversely, experts on China's internal situation and development would probably fall short of being experts on its external (international) relations. See also my remarks on globalization, "Global Studies" and world-system research.

Students of international relations are concerned with governmental, legal, demographic, ecological, economic, financial, cultural, humanitarian, or military relations between sovereign or semi-sovereign states. However, a relationshould not be confused with the things being related (*relata*). A relation consists of three sets: (1) a set A of elements or members, (2) a set B of elements or members, and (3) a set R of sentences saying something about A *and* B. If "A" and "B" stand for a country, the relations between them (expressed by a host of different sentences) can be political, legal, military, economic, financial, environmental, scientific, technological, social, educational, religious, or cultural in character. Any of these types of relations is bound up with hard or soft power, with dominance on the one side and weakness on the other side. There can also be one or more than one relation between the different kinds of international relations. For example, a change (for one reason or another) in the political relations between Russia and the USA may be of influence to their military, economic or financial relations, and a change (for whatever reason) in the economic relations between China the United States may affect their political, cultural or technological relations. There is a wide variety of possibilities here, making the study of international relations exceedingly complex. As we saw in Chap. 5, the change of one part of a system induces not only change of the other parts but also change of the system as a whole. International relations may be driven by different ideologies or theoretical perspectives, but any account or *exposé* of these relations (the relevant literature is abundant and still growing) is incomplete without taking the *relata* into account, *i.e.* without paying due attention to the size, site and situation of the countries involved (for the difference between a country's site and situation, see Chap. 4, at "urban geography"). IR studies would greatly benefit from making a clear distinction, but by no means a separation, between countries (*relata*) and their international relations. Writing about the latter without taking the former into account would be a senseless exercise. For example, it would be pointless to write something about USA obligations under *international* law when America's *constitutional* law is not taken into full account. Reading a book about the rise and decline of the British Empire, one would like to be also informed of the multi-faceted situation in, and dynamic development of, all the countries and territories formerly ruled or administered by the United Kingdom, as "British Empire" refers to a tangled web of relationships, *i.e.* to an enormous human organization the parts of which belonged to a whole and the headquarters of which were located in London.

The importance of making a distinction betwee relations and *relata* may be further illustrated by the age-old, convoluted relations between Judaism, Christianity and Islam. It makes no sense, would be misleading and could even be dangerous to speak or write about the former without in-depth knowledge of the latter. There is no dearth of description and discussion of the hostilities between Christians and Muslims. They are elaborated on and expanded by the media, but the religions themselves are rarely or never thoroughly investigated.[23]

The main thrust of the highly acclaimed work of Erik Zürcher (1928–2008) is the Chinese reception of Buddhism and Christianity, but—*amicus Plato, sed magis amica veritas*—all his scholarly work in the field of intercultural interaction was groundless without any real engagement in studying the *relata*, that is, without a thorough investigation of the historically and culturally embedded basic tenets of Buddhism or Christianity on the one side and Daoism and Confucianism on the other side. Therefore, the critical remarks made by Leonardo Sacco in "Matteo Ricci and the Metaphor of the Bridge between Civilizations" in *Storia, Antropologia e Scienze del Linguaggio, Anno XXXIV-Fasc.1* (Domograf, January–April 2019, pp. 41–116) are not critical enough.

Similarly, comparative studies of the Chinese reception of Marxism should go hand in hand with deep understanding of the fundamental ideas of Marxists and Confucianists. The problem of understanding international relations cannot be solved without comprehending the complexity of countries. A real intercultural dialog requires the active and careful listening to the other party and the clear seeing of the point of view of those at the other side of the table. If not, it will be a "*dialogue de sourds*".

Even if foreign/international relations experts associated with Western universities, research institutes or think tanks would acknowledge that they have unduly emphasized the relations between countries and too much neglected their size, site, situation, nature, history and dynamic structure, they should not overlook the following points:

(A) Much more interesting things could be written on the political situation in Russia, if the authors were well up in the political ideas not only of Machiavelli, Grotius, Hobbes, Milton, Locke, Montesquieu, Hume, Rousseau, Kant, the Federalists, Paine, Burke, Tocqueville, Marx, Gabriel Almond, Karl Deutsch, Robert Dahl, David Easton, William Riker, Michael Oakeshott, John Rawls, Giovanni Sartori, John Polcock, John Dunn, Quentin Skinner, Robert Axelrod, David Miller, Michael Walzer, Theda Skocpol, and Étienne Balibar, but also of *Russian* political scientists and political philosophers. Similarly, books or articles about the Chinese economy would considerably gain in importance, if the writers thereof were acquainted with the economic ideas not only of Adam Smith, David Ricardo, John Maynard Keynes, Joseph Schumpeter, Friedrich Hayek, Jacob Viner, Jan Tinbergen, John Hicks, Ronald Coase, Kenneth Boulding, Armen Alchian, James Buchanan, Hyman Minsky, Hannah Arendt, Douglass North, Thomas Schelling, Kenneth Arrow, William Baumol, Edmond Malinvaud, Gary Becker, Milton Friedman, Pierangelo Garegnani, Irma Adelman, Amartya Sen, Robert Solow, Herman Daly, Joseph Stiglitz, Hal Varian, Vernon Smith, Jean Tirole, Olivier Blanchard, and Paul Krugman, but also of *Chinese* economists such as Sun Yefang, Ma Yinchu, Xue Muqiao, Zhou Shulian, Yu Guangyuan, Wang Mengkui, Li Yining, Dong Fureng, Chen Yizhi, Jiang Yiwei, Liu Guoguang, Zhou Xiaochuan, Wu Jinglian, Zhang Weiying, Mao Yushi, Zhang Xiaoqiang, Hu Angang, Zhang Yichun, Xu Xiaonian, Wan Guanghua, Pan Gongsheng, Fan Gang, Zhang Jun, Xiang Songzuo, Su Dongshui, Zhu Min, Cai Fang, and Song Ligang. If Western economists who want to write about the Chinese economy are unable to read Chinese, they may appeal to "China experts" to translate He Wei and Wei Jie (eds.), *Zhōnggúo*

Zhùmíng Jīngjìxuéjiā Lùn Gǎigé (famous Chinese economists discuss the reforms). This invaluable book, published by Beijing Press, in 1992, contains the extensive and reasoned opinion of 40 prominent Chinese economists about the problems of socialist market economy and is the first volume of a series called *Zhōnggúo Jīngjì Gǎigé Wénkù* (depot of works about China's economic reforms), a treasure trove for those seriously interested in the well and woe of China's economy. Journals like *China Economic Review* (1998ff), *Frontiers of Economics in China* (2006ff) and *China Economic Journal* (2008ff) come not even close to it.

(B) The study of, and writing about, the political situation in Russia, or the economy in China, shall not be exclusively carried out by, respectively, political scientists and economists educated at a Western university. Testing their Western ideas/theories out in Russia or China, they should closely cooperate with their Russian and Chinese counterparts respectively. The best solution, however, would be a combination of international and interdisciplinary research.

(C) I cannot agree with the "new approach" suggested by Thomas Rawski and Lillian Li in the book they co-edited, *Chinese History in Economic Perspective* (University of California Press, 1992, pp. 1–32). The *modus operandi* they advocate would imply a kind of academic imperialism, which I reject. To get a "fully rounded" picture of it, Chinese history, being the change of a multidimensional and multifaceted country, should not be put into economic perspective only and it should not be studied from a single point of view. Indeed, (the history of) no country can be fully understood without scientific collaboration, without working together, internationally as well as interdisciplinarily.

The polity, military, economy, society, geology, geography, ecology and demography of a country are interconnected, forming the parts, dimensions or aspects of a historically molded whole. The studies related to them can thus not be dissociated from each other. If they are, a distorted and misleading picture will be presented and any foreign policy based on it is to be rejected out of hand.

Consequently, the study of International Relations properly so called is (in addition to being a study of the relations between countries) both interdisciplinary and international in regard to the countries themselves (the *relata*). The making of distinctions may be, but the making of separations is rarely or never useful. Like a city, a country is a dynamic and complex system of dynamic and complex subsystems or component systems; it cannot be comprehended by isolating its parts, by putting its history into one or another perspective. One-sidedness means lopsidedness, which is wasteful and dangerous. Only describing or analyzing relations between countries will not be sufficient. The things that are related (*les termes des relations*) should be given as much attention. Otherwise, something essential is missing in the equation; there is abstraction without real understanding. IR studies are thus due for refurbishment.

(D) Perhaps the best example of seeing the importance of putting (the history of) a country into different perspectives and of carefully comparing them is provided by development economics, a well developed branch of economics concerned with the economic aspects of the development process in low income countries. It took development economists many years to realize that the process of lifting a country out of poverty involves more than economic factors.[24] Positively (+) or negatively (−), political, legal, military, geographic, demographic, social, educational, technological, psychological, religious and cultural (mental) developments contribute to economic developments, which in turn affect them. In other words, an interdisciplinary and international approach is needed to figure out how economic and noneconomic forces interact. Gunnar Myrdal (1898–1987), Albert Hirschman (1915–2012) and Irma Adelman (1930–2017) were among the first to acknowledge this, but there are very few development economist who have drawn the inevitable conclusion that the economic development of a country can only be understood and the right policy to be pursued towards it can only be formulated and implemented by taking a systemic, cross-disciplinary view, by taking *all* relevant (and not always measurable) factors into account. Even Nobel laureate Jan Tinbergen (1903–1994) did not seem to be imbued with this idea.[25]

(11) Give an impetus to European studies by pointing out the mutual relationship between Europe (the whole) and the States of Europe (its parts), *i.e.* by explaining that the whole and the parts, being interlinked, balance each other and should not be pitted against one another.

> In 1954, the European Cultural Foundation (ECF) was inaugurated. Robert Schuman, the famous architect of the European Union, Denis de Rougemont, a Swiss philosopher, and HRH Prince Bernhard of the Netherlands stood at its cradle, passionately believing that culture would be of vital importance to Europe's post-war rebuilding and healing. However, there were two problems: Europe was not clearly defined and the concept of culture remained vague. In 2019, the Foundation celebated its 65th anniversary, but the problems have not yet been resolved (https://www.culturalfoundation.eu/our-history).

In 2000, referring to the diversity of its member states, the European Union assumed the motto "United in Diversity" (from the Latin saying "*In Varietate Concordia*"), promulgating that "it [European Union] signifies how Europeans have come together to work for peace and prosperity, while at the same time being enriched by the continent's many different cultures, traditions and languages." For the sake of convenience, the "European Union" was taken to be identical with the European continent, the westernmost part of that enormous landmass called Eurasia. The sensitive difference was smartly hidden in the quotation. The fundamental problem with the European "Union" is that it is not a union at all; its people have no sense of belonging to it; they do not constitute a community of destiny and share no clearly defined values. None of the *EU leaders* (three President: of the Commission, of the Parliament, and of the Council, that is to say, three captains on the ship!) has been able to instill that sense into them and to cement them into a family, into a United States of Europe that knows where it is heading. The "European Union" is not a *discordant concord*, the fight about jurisdiction over the European Central Bank between the German Constitutional Court, in Karlsruhe, and the European Court of Justice, in Luxembourg, being the latest testimony of it. This fundamental weakness is adroitly exploited by states having their acts together, the PRC in the first place. Depending on the emphasis being laid, the motto "Unity in Diversity" can be interpreted differently.

Whoever overemphasizes the importance of the European Union and the process of European integration demonstrates to have no eye for, or to downgrade, the considerable differences between North, South, Eastern, Western and Central or Middle Europe, between the philosophical tradition of France (Descartes), Great Britain (Locke) and Germany (Kant), between states such as Italy and Spain, Poland and Portugal, Ireland and Hungary, Greece and Germany, Finland and France, Bulgaria and Austria, Cyprus and Denmark, the United Kingdom and Romania, and Belgium and the Netherlands (read Inés María Gómez-Chacón [ed.], *European Identity* [University of Deusto, 2003]). On the other hand, anti-Europeanists, who tend to overemphasize the significance of national heritage and of the preservation of national identity, are overlooking what has held "Europe" together over a surprisingly long period of time, a cohesion that may be captured by three keywords: *LOGOS* (reason), *FIDES* (revelation) and *LEX* (rule). Anti-Europeanists have no eye for the foundations,the pillars of Western civilization, for the uniqueness of "Europe". They should read *the Cambridge History of Europe*, said to be an "innovative new textbook series covering the whole of European history from c. 600 to the present day, combining chronological and thematic approaches to the past, surveying Europe in its entirety, and situating European developments within a global context".Taking a shortcut, they should read the special issue "Europe's Culture(s)" in *Journal of CrossCultural Psychology*, 49:6 (july 2018). Europeanists and anti-Europeanists are short-sighted, because they do not realize that nationalism and supranationalism are correlative concepts. If only they could see the close relationship between parts and whole! The *European Green Deal*, darling pet of Frans Timmermans (First Vice President [*sic*] of the European Commission), is said "to have the potential to become the defining mission of the EU", but it stands or falls with the realization that parts and whole are interdependent, that is, they are not antagonistic towards but complementary to each other.

The developments of the parts (countries), constituting the whole of Europe, are like variations on a musical theme featuring two caesurae. The shift from mythological to (proto-) scientific thinking was the first revolution, beautifully described by Bruno Snell (1896–1986) in his masterpiece *Die Entdeckung des Geistes: Studien zur Entstehung des europäischen Denkens bei den Griechen* (1946) and clearly expounded by Martin Meyer in "*Die Bedeutungsgenese der Begriff 'Mythos' und 'Logos' in der griechischen Antike*" (*Archiv für Begriffsgeschichte*, 41 (1999), pp. 35-63). The rise and development of Christianity, claiming that the "City of God" was more important than the "pagan world", and that the Christian way to happiness was to be preferred to the one canvassed by the philosophers in ancient times, was essentially a relapse into mythological thinking, a regression to old habits of thinking, because nobody knows the *Oberwelt* (as distinct from the *Mitwelt* and *Umwelt*). *Fides* was pitted against, even deemed to be more valuable than, *logos*. However, the attempt of Thomas Aquinas (1225–1274) to reconcile faith and reason failed. Duns Scotus and William of Ockham, harbingers of modernity, found his way of thougt unacceptable.[26]

In the 17th century, the age when science became "modern", Descartes, Spinoza and Leibniz picked up where Plato had left off, but the rationalism they defended was heavily criticized by Locke, Berkeley and Hume, the major figures of British empiricism who maintained that the origin of all knowledge is sense experience. Kant attempted to resolve the dispute by arguing that knowledge arises from the application of innate concepts to sensory experience. By the end of Chap. 5, I expressed my doubt as to whether the genius from Königsberg succeeded in doing so.[27] The second *Umbruch* in Europe's development took place in the 17th century, when a rift began to appear in the concept of law (*lex*), a crack that has so much deepened and widened that by now the gulf seems to be unbridgeable. The "conversation" between scientists of nature and scientists of culture has become a *dialogue de sourds*. The Scientific Revolution, about which myriads of books have been written, had a downside. It gave birth to the fatal alienation between defenders of the *descriptive* laws, or the laws of nature (in the sense of natural sciences), and those who were more interested in the *normative* laws, or the laws of man (in the juristic sense), emphasis on the former triggering the stormy development of science and technology, stress on the latter nurturing the tempestuous development of political doctrines such as anarchism, liberalism, conservatism, socialism and feminism. Natural scientists (physicists, chemists, astronomers, geologists, and biologists) began to consider themselves cleared of the obligation to engage in ethical and other axiological matters, which are not measurable. In her highly acclaimed book *Beyond Justice* (Basil Blackwell, 1987, pp. 48–115), the Hungarian philosopher Ágnes Heller (1929–2019) explores this breakdown, taking her cue from Hegel's penetrating analysis of modernity. There is profound disagreement over the nature, scope and ramifications of both laws of nature and laws of man. For example, the defenders of normative laws have been strongly disagreeing on the question as to whether such laws are given or made (natural law vs. positive law). Thomas Hobbes (1588–1679), who set out to create the science of the state in his landmark book *Leviathan* (1651), is considered to be the watershed in the development.

The interesting thing here is that the two kinds of law, namely the laws of nature and the laws of man, have a common root (see Joseph Needham [ed.], *Science and Civilisation in China* [CUP, Volume 2, 1956, pp. 518–583]). Articulated and distinguished in Europe, *LOGOS*, *FIDES* and LEX are interconnected. There, on that small and deeply indented "*cap de l'Eurasie*", sandwiched between the United States of America and the Russian-Chinese *entente*, they are even distinctively and exclusively interconnected.

In the Age of Globalization, we are experiencing a third *Umbruch*. Living through the third Axial Age, the difference with the first two *Umbrüche* is that we have reached the anthropocene, the geological epoch in which human beings have got a profound and enduring impact on their natural environment. See Erle Ellis, *Anthropocene* (OUP, 2018), Jason Kelly et al. (eds.), *Rivers of the Anthropocene* (University of California Press, 2018), Eva Horn and Hannes Bergthaller, *The Anthropocene: Key Issues for the Humanities* (Routledge, 2019), Katharine Zywert and Stephen Quilley (eds.), *Health in the Anthropocene: Living Well on a Finite Planet* (University of Toronto Press, 2020) and Lucas Stephens et al., "The Deep Anthropocene", *AEON*, 1 oktober 2020. The tragedy is that most people do not realize that they are living at this critical juncture; while the titanic is sinking, they keep on dancing and making fun.

It is thus very important that anti-Europeanists (including the Brexiteers) not only realize what has made Europe unique (particularly *vis-à-vis* China and India) but also how the countries, the national identity of which they are so keen to preserve, are just parts of a whole which is but a small part of a turbulent world in big trouble. A comparative study of the major legal systems in today's world, a difficult task once bravely shouldered by René David (1906–1990) and carried forward by scholars like Konrad Zweigert, Heinz Kötz, Pastick Glenn, Uwe Kischel, Raymond Legeais, Stefan Goltzberg, Mathias Reimann and Reinhard Zimmermann, is of utmost importance. Being a challenge to be taken on by scientists who are willing to collaborate internationally and interdisciplinarily, it would bring important things to light.

The European Union is said to be a *Rechtsstaat*, where the rule of law is supposed to prevail. It would be good, if (*a*) the historical roots of the *Rechtsstaat* were explored and scrutinized, (*b*) the historical interconnections between *LOGOS, LEX and FIDES* were thoroughly investigated and (*c*) irrefutable reasons were given why the civil law system should be preferred to the common law tradition, the Islamic law system (*sharia*), the Talmudic legal tradition, the Hindu legal tradition, or the Confucian "*famille juridique*". Until then, it will be premature to uphold Europe as a model to be universally emulated.

After the so-called Warring States period (476–221 BC), with which the current state of competing, if not warring, nations could (and should) be compared, Qin Shi Huang (259–210) founded the Qin dynasty and became the first emperor of a unified China. The Imperial China period, divisible into three sub-periods, ended in 1912 with the proclamation of the Republic of China (ROC). This would turn out to be a hollow phrase, because the ROC became the major stake of a power struggle, first among warlords and then between Chiang Kai-shek and Mao Zedong. In 1949, with Chiang Kai-shek (1887–1975) having fled to Taiwan, Mao Zedong (1893–1976) proclaimed the People's Republic of China (PRC), apparently unaware of the pleonasm "people's republic", but making sure that the Chinese people would have no say in governmental affairs. Mao was consecutively succeeded by Deng Xiaoping, Jiang Zemin, Hu Jintao and Xi Jinping, the *de facto* yellow emperors (in taylor-made suit) of "modern" China. The contrast with the United States of America and the European Union cannot be more striking. Whereas China is a unitary state, the US is a failing federal state and the EU is not a political union at all.

(12) Promote understanding of the close but often overlooked relationship between the mutual dynamics of nature and culture at country (local) on the one side and at world (mondial) level on the other side, *i.e.* bolster the understanding of planet Earth (human predicament) in relationship with the environmental situation in individual countries. This subject, which is a matter of analysis at different though related levels, must appeal to:

   (a) The Club of Rome, describing itself as "an organisation of individuals who share a common concern for the future of humanity and strive to make a difference" and the mission of which is "to promote understanding of the global challenges facing humanity and to propose solutions through scientific analysis, communication and advocacy";

(b) The Intergovernmental Panel on Climate Change, "dedicated to providing the world with an objective, scientific view of climate change and its political and economic impacts";

(c) The Global Network of Science Academies, "committed to making the voice of science heard on issues of crucial importance to the future of mankind";

(d) The Cambridge-based Centre for the Study of Existential Risk, "dedicated to the study and mitigation of risks that could lead to human extinction or civilizational collapse";

(e) The Oxford-based Future of Humanity Institute, exploring "what we can do now to ensure a long flourishing future";

(f) The International Institute for Sustainable Development, "championing sustainable solutions to twenty-first century problems";

(g) The International Institute for Applied Systems Analysis, "conducting research into the critical issues of global environmental and the economic, technological and social change that we face in the twenty-first century";

(h) The World Future Council, claiming to be "the voice of future generations";

(i) The World Economic Forum, engaging "the foremost political, business and other leaders of society to shape global, regional and industry agendas";

(j) Columbia University's Earth Institute, "blending research in the physical and social sciences to help guide the world onto a path toward sustainability";

(k) The High-level Political Forum on Sustainable Development, "promoting system-wide coherence and coordination of sustainable development policies"; and

(l) The team supporting the FuturICT project, the goal of which is "to understand and manage complex, global, and socially interactive systems, with a focus on sustainability and resilience".

In scientific circles across the world, the necessity of developing the new discipline of sustainability science is keenly felt.[28] However, the sustainable development idea, which was first advanced in the 1987 Brundtland Report, is strongly linked to political agendas. Politicians who are only interested in their own career hijack the environmental protection issue, the most glaring example being the tug of war over policies to halt the progressive deterioration of the climate. To really understand (locally and globally) the dynamic relationship between nature and culture, and to know how mankind can get out of the dire situation she has brought herself into, serious collaborative country studies should start being conducted as soon as possible. All hands on deck! If we do not hang together now, we will sooner or later all be hanged, one by one.

In a touching online article published in *The New York Times* of August 11, 2018, Erle Ellis, professor of geography and environmental systems at the University of Maryland, in Baltimore (MD), and celebrated author of *Anthropocene* (OUP, 2018), writes:

> The real question is how we better negotiate among ourselves, across all our many diverse peoples and cultures, so that we can navigate together toward the better futures we wish for, in our different ways.

I venture to say that redefining the way countries are studied is needed to negotiate across cultural or national boundaries successfully.

(13) Bridge the gap between naturalistic positivism and anti-naturalistic hermeneutics by clarifying the connection between notions, which, while being culture-shaped, are also culture-shaping. Concepts (notions) have to be contextualized and put into historical perspective. It should be further borne in mind that both positivists and hermeneutists gravitate to experience; the former look at the world outside (the object), the latter at the world inside (the subject). Though positivists and hermeneutists are often talking back-to-back, they have thus something of crucial importance in common: EXPERIENCE.[29]

(14) Stimulate intercultural conversation and improve mutual understanding among the members of the "United Nations". The chairman of the Okanagan Nation Alliance (https://www.syilx.org) should be invited to explain the meaning of the concept of *en'owkin* at the the United Nations General Assembly. Dialogue goes far beyong debate. See Note 13 of this chapter.

The Syilx or Okanagan people live in what is currently south-central British Columbia. One of their leading scholars and cultural ambassadors is Jeannette Armstrong, who founded the En'owkin Centre (http://www.enowkincentre.ca). More information about her can be found online. The sites provided will lead the reader to her works, which are very informative. Her article "Enowkin: what it means to a sustainable community" was written for the Center for Ecoliteracy (https://www.ecoliteracy.org), in 2009. It could have been written by a Daoist.

(15) Be a corrective on cultural studies by urging their practitioners to be more specific about what they mean by "cultural".

Like "art", "beauty", "being", "experience", "fate", "goodness", "happiness", "honor", "justice", "knowledge", "liberty", "life", "love", "meaning", "relation", "time", "truth", or "value", "culture" is an evasive concept. Culture is the manifold manifestation of mind, which is mysteriously related to, but cannot be identified with, matter. It cannot be explained, or grasped by the mind, a hard fact which physicists, whose work is based on measuring and counting, stubbornly refuse to acknowledge. "Culture" is the central concept in cultural anthropology, the study of phenomena that are, by means of social learning, transmitted in human societies. However, culture, being found in all human societies, is the *pièce de résistance* of all human and social sciences; its study is not limited to, or the prerogative of, cultural anthropology. Each and every country is soaked with culture.[30] Culture is a multifaceted, polydimensional phenomenon, to be distinguished but not separated from nature, which is the subject of physical and biological studies. There are myriads of books and articles on culture in general and on cultures in particular. Experience is essential to culture, and language is expressive of it (*Erfahrung die zur*

Sprache kommt), language not being restricted to the subject with which linguists are concerned. Meaning is probably the bridge between nature (matter) and culture (mind). For seeing the order of things (the way in which they are both *querschnittlich* and *längsschnittlich* related and connected) is nothing but seeing their true nature and awakening to their resemblance, which is expressed by means of the tightly-knit network of signs spreading out over the world. Comprehending a country is nothing but seeing its wiring and understanding what the country in its innermost core holds together, as a unique entity or system. To understand the meaning of something is to be aware of it as part of a whole and as a whole consisting of parts. To understand something is to be able to put it into perspective and to see it in context, that is, to take the systemic approach, which is typical of the truly intelligent person.

(16) Make a valuable contribution to the current debate about higher education.

Higher education is in the midst of a major crisis, and the media, driven by money-making, are aggravating the problem. In his swansong, *Die Krisis der europäischen Wissenschaften* (1936), written just before the emigration of European intellectuals to the United States gained momentum, Edmund Husserl (1859–1938) masterly diagnosed the illness of modern science. In his scholarly and widely acclaimed book *From Dawn to Decadence* (Perennial, 2001), Jacques Barzun (1907–2012) even writes about "a culture in decay". In the words of computer scientist David Wheeler, "a cancerous, consumer-driven capitalism has weakened higher education". Knowledge, sport, art, healthcare, education and man's intimate life have been poisoned by the world of commerce and industry. What counts at universities is not *Bildung* (*zum Leben*), the ideal of Wilhelm and Alexander von Humboldt,[31] but *Ausbildung* (*zum Beruf*). Students, downgrading themselves (without being aware of it), learn how to be useful, not how to be humane. Higher education has been perverted into higher instruction, marked by gender and race fault lines. Drill instructors, who are brainwashed themselves, have replaced the sages, who used to take young pupils under their wings. The lofty project of civil society that was meant to inspire hope for a democratic life has gone awry. The search for the useful has become more important than the pursuit of the good.

In a companion essay to the work of Abraham Flexner (1866–1959), an American educator known for his role in the reform of higher education and founding director of Princeton's Institute for Advanced Study (IAS), Robbert Dijkgraaf, the institute's current caretaker, makes the case that society can achieve deeper understanding and practical progress only by valuing and funding the pursuit of useless knowledge. However, the author had a trick on his sleeve: by "usefulness of useless knowledge" he meant usefulness in the long run of knowledge that might seem to be useless in the short run. In the words of Eric Schmidt, former executive chairman of Alphabet Inc., "Flexner and Dijkgraaf argue that basic research is a proven and essential seed for the revolutionary technologies that fuel the economy, transform society, and provide solutions for the world's problems". The First and the Second World War and the war in Vietnam proved that science is very *useful*—for killing people and destroying their habitat. Ágnes Heller, who taught political theory for 25 years at the New School for Social Research in New York City (https://www.newschool.edu) and drowned in her beloved Lake Balaton, once said: "If I were in the position to devise the curriculum for students of a gymnasium, I would take care not to include any 'useful' subjects, but only 'useless' ones such as Greek, Latin, philosophy, art history and world literature". These were certainly not the subjects Dijkgraaf, the IAS's self-styled director who is never too shy to appear on camera, had in mind.[32]

"The age of chivalry is gone; that of sophists, economists, and calculators has succeeded", Edmund Burke famously wrote. Very little *education* takes place at "educational institutions". Nobody seems to take John Newman's university idea ("education for intellectual and spiritual nobility") seriously anymore. It tends to be forgotten that the root meaning of the word education is "leading out" (what is inside). Education is formation, cultivation, *Bildung*. The French word "*élève*" is cognate with "*élever*" (to elevate, to raise, to uplift), hence *higher* education. In *Wirtschaft und Gesellschaft*, Max Weber (1864–1920) bemoaned the university output of *Fach-leute*, professionals locked up in their discipline and unable to see the bigger picture. In his book *The Aims of Education* (1929), process philosopher Alfred North Whitehead, who distanced himself from Bertrand Russell (one of the founders of analytical philosophy), struck a similar note, asserting that intellectuals lack balance, seeing this set or that set, but not both sets together. The true "philosopher" (lover of wisdom), who lives for philosophy and its applications in daily life, has been replaced by the pseudo-philosopher, who lives *by* it. Higher education has become uninspiring and mind-numbing.

Criticism (the openness which Karl Popper declared to be the chief merit of democracy) is tolerated; it is even stimulated, provided the critic abides by the sacrosanct rules of the game set by God knows who. Whistleblowers, debunkers, mythbusters, or whiners are not welcome. *Fundamental* critique is forbidden. Pointing at a system error or to a flaw in the fabric of pervasive thought is anathema, and whoever dares to do so will be branded as an outlaw, troublemaker, apostate, dissident, traitor or heretic. Galileo Galilei is certainly not the only one having experienced this kind of ostracism. Don't rock the boat! Don't ask who, in the last resort, are at the buttons! Don't look for the producer(s) of the show that must go on, whatever it takes! Never say that the emperor is wearing no clothes or that the current university idea is wrongheaded! Masters, surrounded by opportunists, lackeys and sycophants, are always right, no matter what they say or how stupid they behave. Specialization, for the sake of economic growth (if need be, at the cost of social well-being), is valued everywhere, but a vision of the general direction of the specializations taken together is missing. Aboard the ship of fouls, every sailor fulfills the duty of his/her profession, but the captain, the helmsman is nowhere to be seen. No mariner knows, or seems to care about, where the buoyant watercraft is going to. The Titanic is supposed to be on the right course. Some people raise the alarm, but their voice gets lost in the wind. The "tragedy of the commons" seems to be inevitable. Few people realize that Gaia will be—nay, is already taking revenge, mercilessly.[33]

Undervaluing the importance of character formation and overemphasizing the significance of measurement and quantification,higher education has "impoverished the souls of students" (Allan Bloom). Statistics alone are not enough. Graduates are ideologically straitjacketed by the academic institution they are affiliated with and dependent on. Unaware of the demise of *Paideia*,[35] they have become illiterate about the Great Ideas of the ancients, not to speak of the wisdom of non-Western civilizations. Universities, often rocked by accusations of sexual or racial harassment committed by a professor (with the management standing idle by as students or employees were victimized), have degenerated into places where the student gets confused by, if not totally lost in, the information being provided; into knowledge factories dangerously detached from the immensely complex world; into intellectual, business-oriented department stores where many things are for sale, at ever more rising prices. Universities are under the dictate of the economy (read: money).[36] They have become "undisciplined bazaars of miscellaneous studies" (Barzun), bastions of power rather than beacons of light. Whoever may think that universities are serene locations where the search for wisdom is the one and only selection criterion and where intellectual integrity prevails is badly mistaken. Universities, Colleges, Schools and research institutes are vulgar places of ambition, conceit, arrogance, jealousy, resentment, gossip, slander, chicanery, scheming, prejudice, fraud, abuse of power, and intellectual inbreeding. That is the reason why I "do not long for the company of professors and find peasants more interesting" (Heidegger). It is no coincidence that famous Chinese paintings often feature a reclusive scholar totally absorbed by nature.

The rot sets in when first-year students have (and are readily willing) to undergo hazing, that bizarre, silly, humiliating and often dangerous ritual "to create an *esprit de corps*" (fraternity). In fact, hazing is a kind of brainwashing, a form of menticide reminiscent of what happens in religious sects or political parties. It paves the way to flattery, favoritism, conformism, docility, sheep-like behavior and conservatism, some of the shortcomings of today's higher education. Hazing has proven disastrous, not only for the freshmen concerned (because, like army recruits, they have been turned into willing 'soldiers' who do not dare to march out of step) but also for the society on which these brainwashed academics will be released.

The ongoing discussions about the diseases of our universities are wrongheaded and will remain sterile as long as the bull is not grabbed by the horns. The admission requirements of prestigious universities are opaque, if not arbitrary, and their tuition fees are prohibitively high. On the other hand, the massive influx of youngsters into higher education poses unprecedented challenges. To be able to cope with the crowd that also wants to be highly educated, managers (differentiated in layers) have taken over the role that was traditionally reserved for university or college teachers. The massification of higher education (a concomitant of democratization) has resulted in a bureaucratization afflicted by the *rabies complicatoria* and characterized by the despotism of inertness. Another problem, stemming from mass education and conducive to mediocrity, is the adopted government policy: the more students enrolled, the more subsidy received. Teachers are instructed not to be too demanding in order to minimize the number of dropouts. In addition, the allocation of scarce resources can be seriously questioned. The principle that those who pay for the lecturing shall also determine or control its content causes scientists to "adjust" the outcome of their research, compromising their integrity.[37] There is also a tendency towards grade inflation: progressively higher academic grades for work that would have received lower grades in the past. Furthermore, despite all that has been done to improve the guidance of doctoral studies and the supervision of dissertation work, horror stories keep coming. The rules and regulations on writing a master's or doctoral thesis remain vague. Corruption and nepotism are rampant where wisdom and integrity are supposed to reside. Professors (particularly those who acquired academic tenure) behave like half-Gods. Their position is almost untouchable. Woe betides the student who ventures to criticize or speak against his teacher. Professors anxious to cultivate their business contacts press overworked Ph.D. candidates to perform research for them, ignoring the academic needs of postgraduates. Moreover, the constant pressure to publish in order to sustain or further one's career ("publish or perish") prevent the scientist from engaging in long-term research. Haste breeds waste. Finally, pre-publication peer-review, supposed to be a quality assurance system, is frequently anti-innovatory, easily abused, likely to suffer from lack of evenhandedness and largely a lottery. Superficially carried out by reviewers who are far too busy to read a manuscript carefully, it has become a blockage rather than a filter. The chance for those who "submitted" (what's in a name!) a manuscript to be thwarted by a diligent doorkeeper or silly secretary is high, too high. At the 6th World Conference on Research Integrity, in June 2019, in Hong Kong, all forms of research and reviewing misconduct were discussed—of course without naming and shaming any of the perpetrators present.

> There is much discussion about the peer-review system, and understandably so. Peer-review does not necessarily mean that a submitted manuscript will pass, or has passed, a serious quality control, because the quality of the review varies greatly and the expertise of the reviewers is not always in the subjects brought up by the author of the book or article. Reporting under cover, some reviewers give only their general opinion on the manuscript; other reviewers only check whether the statistics are correct or pass their judgment on the readability of the text; a third group tends to hair-splitting and pays attention to the smallest detail. The anonymity of peer-reviewers may be defensible as being fair and reasonable, it has also a downside: there is no transparency. Quality of research and quality of the reviewing of it are essential for science. To guarantee the reliability and accountability of the quality controller, transparancy is thus of vital importance. I am sympathetic to the suggestion made by Cecile Janssens (research professor at Emory University, in Atlanta) that the reviews and the author's reaction to them should be added, like an appendix, to the publication. Readers could then control the quality of the reviewing, the reviewer's proposal to improve the quality of the work, and whether, or to what extent, the author took the reviewer's advice. In this way, the reviewer cannot safely hide behind a veil of anonymity, which is imbued with power and breeds irresponsibility; he/she has to make his/here judgment on, and argument about, the manuscript public. Many talents have been nipped in the bud by the present peer-review system.

Another trouble with universities is their hopelessly outdated worldview. The mindset of virtually all teachers is reductionistic. They have the ingrained habit of thinking in parts rather than in systems. Lectures are given by blinkered academics on a module-by-module basis. The approach is linear and little attention is given to the interconnections between the components making up a complex system. Problems are dealt with one-by-one, not as being related. The world is conceived of as compartmentalized, not as a wonderful whole. Economic, financial, legal, social, psychological, geographical, political or ecological issues are pigeonholed and, wherever the occasion arises, labeled as subjects to be covered by "colleagues" of other departments/faculties. Developments in systems –, network –, and information science, cybernetics, mathematical biology, nonequilibrium statistical mechanics, chaos-theory, catastrophe theory and synergetics, disciplines which are clearly pointing in the direction of holism and nonlinearity rather than reductionism and linearity, are rarely or never taken into account. The well-paid managers of our universities seem not to have heard of the silent nonlinear-science revolution of the 1970s. Worse still, collaborative research is considered a too complicated matter; compartmentalization is the easiest way to rule (*divide et impera*!).

The ill of grade inflation is aggravated by the proliferation of professorships. In the grey past, one could be a professor of, say, mathematics, logic, physics, chemistry, astronomy, geology, biology, demography, geography, economics, sociology, political science, law, linguistics, psychology, anthropology, or statistics. Their "community" was later joined by professors of physical mathematics, astrophysics, biochemistry, geophysics, economic geology, economic geography, legal economics, biogeography, psycholinguistics, sociolinguistics, social geography, social anthropology, physical anthropology, social psychology, socioeconomics, mathematical economics, mathematical statistics,[38] biostatistics, political economy, legal psychology and political psychology, but nobody asked, or dared to ask, how one person could possibly ride two (in the case econometrics three) horses at the same time. With the subsequent shift in focus from kinds of studying things to kinds of things being studied (a shift probably related to the interference of business with higher education), an explosion in the number of assistant, associate, adjunct, full, visiting, distinguished and endowed chair professors occurred.

This shift, accelerated by the advent of computer science, is of great importance. Instead of taking a physical, chemical, astronomical etc. point of view, or the "viewpoint" of physical mathematics, astrophysics, biochemistry etc., one could become professor of marketing, management, logistics, rail transport, road transport, labor law, cardiovascular imaging, rock music, jazz studies, corporate group liability, agriculture, public health, concept design, sports management, engineering, divinity, ecclesiastical history, ancient history, medieval history, early modern history, contemporary history, international relations, bibliography, the 21st century, Chinese poetry, Russian literature, higher education, military strategy, American history, Russian history, Chinese history, European history, oenology, arbitration and dispute resolution, practice, nutrition and dietetics, information security, artifact conservation, prestidigitation, Indian dance, anti-Semitism, colonial literature, feminism, gender studies, gay studies, queer studies, or postmodernity, to name but a few professorships.[39] According to the Jain philosophy and Nietzsche, anything can be investigated from different points of view.

The reader will notice that the professors I level criticism against belong to the last category. Professors of Turkology, Iranology, Tibetology, Sinology, Koreanology, or Japanology have not been appointed on account of their distinctive and characteristic *explanans* but on account of their presumed knowledge of the country concerned. Kremlinologists were once dealing with soviet affairs, but are now concerning themselves with post-soviet affairs. They do not distinguish themselves by studying these affairs differently (*i.e.* by advancing and testing a distinctive *explanans*), but by studying different affairs; only their *explanandum* is different. *Mutatis mutandis*, the same applies to Pekinologists and Pyongyangologists. Some people may think that I am exaggerating or overstate my case. Whoever thinks so can rest assured: without the fundamental flaw of area studies being properly addressed, the Nobel Prize will never be awarded to an "area or country expert".

Our universities need not only integrity but also integration, cross-disciplinary integration, that is. Genuine interdisciplinary collaboration could cure them from what Barzun aptly called the "gangrene of specialization". In the age of scientific convergences and technological breakthroughs, we need computer-aided collaboration that includes, but is not limited to, the social and human sciences, collaboration that is both international and interdisciplinary in character. Truly collaborative research, however, is loathed by country connoisseurs. These polymaths shrink from changing their course, refuse to stop pretending (tacitly) to be all-purpose researchers, and decline any suggestion either to stick to that humble, yet important job of translating or to devote their time and energy to what they might be well-educated in, that is to say, to what they might be really good at, leaving the subjects that lie outside their discipline to other experts to handle but keeping in touch with them in order not to lose sight of the context of things. Truly collaborative research is also abhorred by trained economists, sociologists, psychologists, lawyers, political scientists, demographers, geographers, geologists or ecologists who are interested in a particular country. In discussions about the country, these specialists are inclined to veer in the direction of another discipline or to drift out to areas of research that do not fall within their purview. Thinking to be free to roam around wherever they please, they do not consider it their duty to stay on their own turf. Staying in touch with colleagues who study another aspect of the same country is recommendable, but overstepping the boundaries of one's discipline is a bad habit. The constant blurring of disciplinary distinctions is the plague of our time. Journalists (who are called "hoernalists" in my home country) may be allowed to blur, but academics shall not be pardoned for committing such crime. Let the greengrocer sell vegetables and fruits, the dairyman sell butter and cheese, and the owner of a sweet shop sell candies and chocolates! Blurring of boundaries leads to chaos. The cobbler shoud stick with his last.

I am certainly not claiming that this book provides the panacea for all the problems universities are currently coping with. Their lack of integrity is a matter of morality and I shall not be in the business of telling leaders of higher education what they are morally obliged to do. They have a conscience, I presume; deep down, they know what to do (or not to do). Integration, however, is a different cup of tea. I am suggesting how universities could get things straight by guarding against scientific devaluation and fragmentation (which is not the same as disciplinary specialization). A plethora of professorships is no guarantee for quality education. My denouncement of area or country studies as conducted in the past and at present will probably not be greeted by tenured professors with great enthusiasm, but universities properly so called may embrace the view that this kind of studies are fundamentally flawed and computer-aided, systems science-based interdisciplinary and international collaboration is the only way out of their predicament. The institutes of higher education are in big trouble. The challenge they face is formidable and can only be taken on by intelligent and brave people.

Adopting a critical second-order stance and arguing for a radically new method with respect to the study of countries, I do not mean to give anybody a hard time or to raise somebody's hackles, but what is conveniently glossed over or deliberately concealed must be exposed—frankly, openly and without regard to any personal interest. To lay bare the basic weakness of country studies all over the world (something nobody has cared or dared to do thus far), I went to the belly of the beast, believing that the subject is of utmost importance for understanding international or foreign affairs. On campuses in Europe, the USA, Japan, Australia and elsewhere, thousands of professors fill the pages of hundreds of journals, the shelves of numerous libraries, and the minds of countless students with their "theories" about a country. When important events occur or a crisis looms in the country, adjunct, associate, assistant or tenured professors, who are deemed to be a valuable source of information, are mobbed by the panting media seeking their opinion. The Federal Government in Washington subsidizes more than a dozen academic centers devoted to studies of the Middle East. Yet for many decades, Middle East Institutes in the US have been schools of error. Those affiliated with them miserably failed at the time when understanding the area was crucial. Western academics, blinded by their ideological prejudices and clinging on to their *modus operandi*, failed to anticipate any of the major developments at the "crossroad of civilizations". The carrousel of opinions keeps on turning. Who foresaw the first oil crisis! Who saw the Arab Spring coming! Who predicted the events in Egypt, Syria, or Iran! Within the field, hardly a voice dared to protest. In an iconoclastic *exposé*, Martin Seth Kramer surveyed the ruins of Middle Eastern studies, asking how and why they went wrong. His message has gone unheeded or has been (deliberately) distorted. In the coterie of "Middle East experts", or in any other clique of country connoisseurs for that matter, nothing has changed fundamentally. Like the Middle East Institutes, the institutes for African, Asian or Latin American Studies, in or outside the US, have been and–*rebus sic stantibus*–will remain "schools of error". No Sinologist anticipated the "Great Proletarian Cultural Revolution", no "China expert" had prescience of the Economic Revolution that Deng Xiaoping would launch and no "China pundit" is able to gauge the impact of Xi Jinping's "Belt and Road Initiative". There has been no shortness of *afterward*-analyses, no lack of swinging after the bell. In November 1919, the conference "Next China: The Future of China and Why It Matters for You" was organized in New York. It was crowded by eager beavers who wanted to know where China would be going. Special sessions were organized where the next billion-dollar trends were promised to be identified. None of the keynote speakers foresaw the coronavirus breakout in Wuhan!

Country Projects

Cross-disciplinary and international country studies should be planned and executed on project basis. A project is always a journey into the unknown, full of risks, because the manager of a project cannot escape from judging what is likely to happen in the future, based on the necessary incomplete information he has. The literature on subjects like project analysis,—benefits,—boundaries,—control,—costs,—design,—evaluation,—format,—implementation,—life cycle,—management,—model,—objectives,—organization,—participants,—planning,—preparation,—selection and—types is overwhelming. See Paul Dinsmore and Jeannette Cabanis-Brewin (eds.), *Handbook of Project Management* (HarperCollins, 2018). Leading journals are the *International Journal of Project Management* (1983ff), published by Elsevier in collaboration with the Association for Project Management (APM) and the International Project Management Association (IPMA), the *Project Management Journal* (1998ff), published by the Project Management Institute (PMI) and "featuring state-of-the-art research, techniques, theories, and applications", and *The Journal of Modern Project Management* (2013ff), which "promotes a balance between theory and practice, highlighting case studies on different sectors". The International Centre for Complex Project Management (ICCPM) is a promising organization that "seeks to translate research on effective management of complex, dynamic systems into practical solutions in different domains".[40] The World Bank Group may be another source of assistance (visit https://www.worldbank.org and search for project management). A list of megaprojects and dozens of project management software packages or tools are available on the Internet.

The planning and executing of country-related collaborative research projects should be done with the following distinctions in mind:

- **Computer-Supported Cooperative Work (CSCW)**

CSCW addresses the question as to how cooperative activities can be supported by computer systems. It is a design-oriented academic field that brings economists, organizational theorists, educators, social psychologists, sociologists, anthropologists and computer scientists together. The expertise of researchers in different disciplines helps to find the avenues for possible development. Despite the variety of disciplines, the focus is on understanding the characteristics of interdependent group work with the object of designing computer-based technology that is adequate for supporting cooperative work. Though CSCW researchers who benefit from the progress made in computer science should and could more involve systems scientists in their work, I am confident that country-students willing to engage in interdisciplinary and international research will learn a lot from them.

- **Computer-Supported Collaborative Learning (CSCL)**

Though CSCL is related to CSCW, its approach is pedagogical. With CSCL, learning takes place via social interaction using a computer network. The knowledge of participants is computationally and interactively shared. Visualization of knowledge plays an important role and the potential of digital libraries is increasingly deployed.

- **Collaborative Information Seeking (CIS)**

CIS focuses on the study of situations, motivations and methods of people working together in information seeking projects, and on the building of a system that supports their activities. CIS can be extended to collaborative information *integration* and sense-making. If scientists from different countries and disciplines collaborate to map the universe (CERN), and if they collaborate to map the human brain (Human Connectome Project), why shouldn't they be able to collaborate to map something in between the two and focus their attention on the situation and developments in a country, taking advantage from the insights gained at both sides!

CSCW, CSCL and CIS may benefit from the advances made in data compression: "the process of modifying, encoding or converting information so as to use fewer bits in order to reduce the consumption of expensive resources such as disk space or transmission bandwidth".[41] Data compression is particularly powerful when combined with the techniques of data visualization. In a recent online article, entitled "Information Compression as a Unifying Principle in Human Learning, Perception, and Cognition", Gerard Wolff gives ample evidence of the important idea that data and information compression are central to human uncerstanding.

By almost general consent, the organization of society and the regulation of production that dominates the modern era have brought mankind to the brink of self-extinction. The root cause of this problem seems to be that modern man is able to distinguish and separate things but unable to see the whole and put Humpty Dumpty together again. "*Die Welt ist aus den Fugen geraten*", Germany's President Frank-Walter Steinmeier once said, and he is not alone. Universities have degenerated into *multi*versities estranged from each other, into isolated faculties where knowledge is further and further compartmentalized until the cat cannot find its own kittens anymore. Present university-managers, the running dogs of captains of industry, pose a major threat to society because they allow their sponsors to keep on "creating wealth by providing goodies". To survive, we need to cross disciplinary borders. Extreme individualism will be fatal. If we do not hang together, we will all be hanged by our own doing, one by one, and human extinction will be inevitable (see Christos Lynteris, *Human Extinction and the Pandemic Imaginary* [Rouledge, 2019] and Note 33 of this chapter; visit http://somatosphere.net/2020/covid–19-forum-introduction.html). At the political, legal, economic, financial, or social level, no expert or specialist will do. We need all the help we can get before it is too late. The road ahead indicated in this book may be long, bumpy and slippery, but the panorama will be breathtaking. Those who abandon their narrow and dim room will see the world wide and bright. In the upward thrust, details important at low level become less important. At high altitude, long-held convictions are disestablished. True and beautiful is, therefore, the dictum of Zhuangzi: "Unification of differences is what is called greatness" (*bùtóng tóngzhì zhī wèi dà*). Surely there is a way, if and when there is a will; if and when "country experts" are willing to kick their bad habit of trespassing on fields of research that belong to sciences they can only borrow from; if and when they are prepared either to dedicate themselves to translation or–should their job be the practicing of a discipline–to listen to the loud call to collaborate and team up; if and when they embark on a cooperative joint venture; if and when they play joyfully together. The stakes are high, higher than many may care to think. See Gennady Shkliarevsky, *The Civilization at a Crossroads* (Glasstree, 2017).

Whenever the subject of interdisciplinarity is brought to the carpet, three questions are posed: What does it mean? Is it necessary? Is it possible? See appendix A. Let me briefly deal with the questions:

(1) **What is meant by "interdisciplinarity"?**

More than half a century ago, in September 1970, the Organization for Economic Cooperation and Development (OECD) hosted a seminar on interdisciplinarity, defining it in her final report as follows:

> Interdisciplinarity is a noun describing the interaction among two or more different disciplines. This interaction may range from simple communication of ideas to mutual integration of concepts, methodologies, procedures, data, terminologies and the organization of research and education in a fairly large field. An interdisciplinary group consists of persons differently trained but organized into a common effort to solve a common problem by means of continuous communication among the participants.

This definition has been cause for much discussion, but consensus is growing that genuine scientific collaboration is not achieved until the individuals involved (*a*) develop a new, systemic way of thinking by listening not passively but actively to one another, (b) venture beyond the borders of their discipline (with due respect for experts) and (*c*) explore the possibility of finding a common ground and unifying knowledge, the possibility of putting into perspective what might at first sight have looked different. Unless the participants change their attitude and way of thought, their "collaboration" remains just a division of labor, useful but not conclusive; the alleged interdisciplinarity is nothing but multidisciplinarity.

(2) **Is interdisciplinarity necessary?**

Mono-disciplinary research is based on unrealistic assumptions, of which *homo economicus* (hero of mainstream economists) is the most conspicuous example. Many scientists have come to the conclusion that the solution of present-day problems requires the pooling of disciplinary knowledge and the combining of analytical skills. A single person cannot master two disciplines, because it is hard enough to be master of just one. However, two persons, each of them being expert in a different discipline (*e.g.* chemistry and biology, law and economics, or psychology and sociology), can inspire each other and co-produce something of great value. In a recent article entitled "Science and Technology Advance through Surprise", Feng Shi (University of North Carolina) and James Evans (University of Chicago), drawing on a wealth of data, write:

> Breakthrough discoveries and inventions involve unexpected combinations of *contents* including problems and methods. Most breakthroughs occur when problems in one field are unexpectedly solved by a research method in another field.

Interdisciplinary collaboration has added value in more than one respect: (*a*) scientific, because it minimalizes duplication, lights up the blind spots of participants, and leads to cross-fertilization, reconceptualization and new research questions or projects, (*b*) personal, because it stimulates creativity and enriches the lives of the scientists involved and (*c*) social, because the best way to understand, and to deal with, complex, society-related problems is to approach them not one-sidedly but in a way that highlights the commonality of different approaches. No forest without trees, however. The input of each and every discipline should be respected; if they cease to exist, there cannot be collaboration between disciplines. The best performance of a symphony is executed by professional players of different instruments.

(3) **Is interdisciplinarity possible?**

Conceptual confusion within or between the natural, social and human sciences is sometimes brought forward as an impediment to interdisciplinarity. This point is to be taken seriously, but the importance of it should not be exaggerated. Conceptual analysis combined with *Begriffsgeschichte* may clear things up. To further this process, one should be aware of (*a*) the difference between words and concepts, (*b*) the power of analogy and metaphor, (*c*) the pitfalls of binary logic as exposed by the advocates of non-classical logics and (*d*) the advantages that can be taken from emerging technologies (an impressive list of which is provided on the Internet). Experts in "argumentation theory" (not to be confused with logic) could officiate as midwives, helping to bring about understanding among differently trained but intellectually open-minded scientists.

The supposed incompatibility of quantitative and qualitative research methods in the "*erklärende Naturwissenschaften*" and the "*verstehende Kulturwissenschaften*" respectively may be regarded as another obstacle to interdisciplinarity. Whereas quantity is a matter of counting and measuring, quality concerns essential and accidental properties that can only be described. What tends to be forgotten, however, is that quantity and quality are like the wings of a flying bird: the one cannot do without the other; the wing of quantity will not be well-functioning without the wing of quality working well, and vice versa. In addition, the so-called hard sciences are softer than their practitioners are willing to acknowledge, and the list of unsolved problems in physics, mathematics, chemistry, astronomy, geoscience, biology, medicine, neuroscience, statistics, computer science and information theory is longer than the representatives of hard science, haunted by the theorems on the limitations of formal systems put forward by Kurt Gödel, Alfred Tarki, Alonzo Church and Alan Turing, would care to admit. Facts are extremely difficult to establish and the issue of causality has been unsettled by nonlinear analyses. A rigorous definition of "point" is not given (a twinkling star in the sky looks like a dot or point but is a huge 3-dimensional object). Theorems are deduced from axioms based on belief, and nobody knows whether mathematics is a consistent body of knowledge. On the other hand, the social sciences, even the humanities, are getting harder. Their foundations are being scrutinized and refurbished. The statistician's toolkit can now be used by natural, biological, social, and human scientists. Unification of large bodies of knowledge (interdisciplinarity) is possible, provided scientists from different disciplines are ready to engage in a genuine dialogue, in a conversation where the concern is not "Who is right?" but "What is right?", provided thus "the willingness to see the other fellow's point of view" (Kenneth Boulding). All disciplines, different though they might be, sprang from the same source: bewilderment. Gazing at the universe (world outside) or looking into the human mind (world inside), we should acknowledge that wondering lies at the heart of scientific query, which originated in religious questioning, this questioning first resulted in philosophical inquiry and then led to scientific research in different domains. In this evolution, which took thousands of years, the questioning remained, but the answers having been given became increasingly circumscribed. Alas, at present, scientists worldwide, having lost the view of the whole and not seeing the intricate connectedness of sciences, do not address the original query of the Ancients anymore.

However, there is some hope glimmering on the horizon. In July 2010, the journal *Biocosmology—neo-Aristotelism* was launched by the Biocosmological Association (http://en.biocosmology.org). The journal aims at the study of life processes in all spheres—biological, ecological, anthropological, sociological and psychological". In June, 2020, the International Conference on Thermodynamics 2.0 was organized in Massachusetts (USA). It was "an opportunity for academics and researchers to learn about historical and contemporary efforts to connect the natural and social sciences" (http://iaisae.org/index.php/about-ict20). In September, 2020, a workshop about "Reconfiguring History beyond Disciplinarity" was held in Jena, Germany. Its aim was "to map the different conceptualizations and uses of the notion of history in order to highlight how they are informed by the changed interdisciplinary settings in which they are rooted". Its organizers, Juliane Tomann (Imre Kertész Kolleg, Jena), Joanna Wawrzyniak (University of Warsaw) and Stéphane Van Damme (European University Institute, Florence), intended "to explore how historical inquiry is (re)shaped when it is taken out of the history department and carried out by those who are not necessarily historians." It seemed to them that, by taking into account the vast variety of research fields, interdisciplinarity "evolved from a trendy catch phrase and a standard buzz word in grant applications (often perceived of by sceptics as mere lip-service which causes more conceptual and practical problems than it solves) to a well-established research practice".

Alluding to a well-known saying of Karl Marx, I exhort all experts from different disciplines interested in a country to unite. Having tuned up their instruments, different musicians, skilled professionals in their own right, play together as one ensemble. They perform a symphony, under the *bâton* of a conductor, who is attentive of the harmony, tempo, rhythm, loudness, timbre and melody. To those wanting to arrive at the edge of country knowledge, my advice would be: seek out the most sophisticated country-interested minds from different disciplines, "put them in a room and have them ask each other the questions they are asking themselves".

Today's digital revolution is not only a matter of technology. It will inescapably transform all sectors of the economy, will have an irreversible impact on the polity, and will fundamentally change the organization of society, including the organization of sciences. Data is said to be the new oil for the twenty-first century. Though this oil cannot be found underground, the main questions are: What can be distilled out of it and how shall it be refined? As information about countries is contained in data collections we today call conventional and tomorrow will call digital libraries, I strongly believe that such information, generated by different disciplines and in different countries, can be computationally mined, processed, integrated and eventually organized into the country knowledge we urgently need for better international understanding and pursuing a foreign policy that is more sensible.

Should the complaint be that these words are only the menu but not the meal, I would reply: "Nobody can justifiably require that I alone prepare and serve the dishes which only a team of scientists using the latest in computer technology can make and table. That is to say, in principle. For the team, that will make steady progress incrementally by connecting the dots, can approach its aim (comprehending the complexity of countries completely) only asymptotically. Only God is said to be omniscient and almighty. Will many cooks spoil the broth? Not when they work together in close harmony. Visit https://www.thealternativesproject.org.

Notes and References

1. See Lokenath Debnath, *The legacy of Leonhard Euler* (Imperial College Press, 2010, Chap. 5) and "A brief historical introduction to Euler's formula for polyhedra, topology, graph theory and networks', *International Journal of Mathematical Education in Science and Technology*, 41:6 (2010), pp. 769–785.
2. Compare with Davide Crippa, *The Impossibility of Squaring the Circle in the 17th Century* (Birkhäuser, 2019).
3. It is important to bear in mind that area/country studies started to grow in the era of overseas discoveries and explorations, *i.e.* in an era related to the period of the rise of European states and–via colonialism and imperialism–leading to the age of globalization, currently the subject of heated debates. In 2020, the journal New Area Studies was launched. It still confuses the study of a country with the study of an area, because an area or region is a certainly a *set* of countries but not necessarily a *system* of counties.
4. See Catherine Cramer et al. (eds.), *Network Science in Education* (Springer, 2018) and visit https://netscisociety.net.
5. There is much to be learned from the art of designing. See Jack Ganssle, *The Art of Designing Embedded Systems* (Elsevier, 2008), Robert Curedale, *Design Thinking* (Design Community College, 2013) and Andrew Pressman, *Designing Relationships* (Routledge, 2014). Visit https://www.aiga.org. Designing looks like, but is not the same as modeling. See end of Chap. 6.
6. Visit http://www.iiisci.org/journal/sci.
7. See Ellis Cashmore and Chris Rojek (eds.), *Dictionary of Cultural Theorists* (Arnold, 1999, pp. 1–13).

8. See Charles Lumsden and Edward Wilson, *Promethean Fire: Reflections on the Origin of Mind* (Harvard University Press, 1983), Carel Germain and Martin Bloom, *Human Behavior in the Social Environment* (Columbia University Press, 1999), Uichol Kim et al. (eds.), *Indigenous and Cultural Psychology: Understanding People in Context* (Springer, 2006), Marion Blute, *Darwinian Sociocultural Evolution* (CUP, 2010), Goldhaber *op.cit.*, Annita Kerr D'Amico and Rene Rochester, *The Culturally-Wired Brain* (Redemption Press, 2015), Kevin Laland, *Darwin's Unfinished Symphony* (PUP, 2017) and Joan Chiao, *Philosophy of Cultural Neuroscience* (Routledge, 2018), José Bermúdez, *Cognitive Science* (CUP, 2020) and Anil Seth, *Bing You: A new science of consciousness* (Faber & Faber, 2021). Browse the journals *Mind, Culture, and Activity* (1994ff), *Mind & Society* (2000ff), *Journal of Cognition and Culture* (2001ff), *International Journal of Brain and Cognitive Sciences* (2012ff) and *Culture and Brain* (2013ff). Visit https://www.iep.utm.edu/embodcog, https://plato.stanford.edu/entries/embodied-cognition and http://www.sphs.info. Lists of cognitive scientists, neuroscientists, biologists, social scientists and philosophers of mind can be found on the Internet. Also visit https://cognitivesciencesociety.org and see Rafael Núñez et al.,"What happened to cognitive science?", *Nature*, June 10, 2019.
9. Visit https://plato.stanford.edu/entries/relativism.
10. See Thomas Benjamin (ed.), *Encyclopedia of Western Colonialism since 1450* (Macmillan, 2006) and Nicholas Dirks, *The Scandal of Empire: India and the Creation of Imperial Britain* (Harvard University Press, 2008). For the foreign aggression against China, see Immanuel Hsü, *The Rise of Modern China* (OUP, 2000, Chaps. 7–10). For the Portuguese series of outrages in Africa, read António Lobo Antunes's trilogy: *Os Cus de Judas* [the Ass of Judas] (1979), *Fado Alexandrino* (1983) and *O Manual dos Inquisidores* [Handbook of the Inquisitors] (1996).
11. See Note 16 in Chap. 2.
12. Look through the pages of *Cross-Cultural Research* (1966ff) and *Journal of Cross-Cultural Psychology* (1970ff), and visit https://sccr.org.
13. Go to http://www.enowkincentre.ca. For the often overlooked difference between debate and dialogue, see Hans-Georg Gadamer, *Wahrheit und Methode* (Mohr Siebeck, 1960).
14. For the difference between tolerance and respect, go to https://plato.stanford.edu/entries/toleration and https://plato.stanford.edu/entries/respect.
15. The literature of the "(New) Silk Road" is large and still growing. There are dozens of books, numerous articles and more than 200 websites on the subject. I would recommend the reading of volume II and IV of Christopher Baumer, *History of Central Asia* (I.B. Tauris, 2012–2018).
16. Hägerstrand's testament and masterpiece, *Tillvaroväven* (the fabric of existence, 2004), has not yet been translated into English. Compare with Michel de Certeau, *L'invention du quotidian*(1980).
17. See Eric Nelson, *Chinese and Buddhist Philosophy in Early Twentieth-Century German Thought* (Bloomsbury, 2017, Chaps.4–6 and 8) and Harald Seubert (ed.), *Neunzig Jahre 'Sein und Zeit'* (Karl Alber, 2019).

18. See Hérodote, 117:2 (2005).
19. See Henry Veltmeyer (ed.), *New Perspectives on Globalization and Antiglobalization* (Routledge, 2008).
20. Churnalism is a form of journalism in which pre-packaged material, supplied by PR agencies, is used to churn out second-hand articles. It paves the way for spreading fake news.
21. See Michael Stanford, *Companion to the Study of History* (Blackwell, 1994, pp. 230–241), Lloyd Kramer and Sarah Maza (eds.), *A Companion to Western Historical Thought* (Blackwell, 2006, Part III), Aviezer Tucker (ed.), *A Companion to the Philosophy of History and Historiography* (Wiley, 2008, Part II), Daniel Woolf et al. (eds.), *The Oxford History of Historical Writing* (OUP, 2011–2012), Nancy Partner and Sarah Foot (eds.), *Handbook of Historical Theory* (Sage, 2013), Rolf Torstendahl, *The Rise and Propagation of Historical Professionalism* (Routledge, 2015), Georg Iggers et al., *A Global History of Modern Historiography* (Pearson, 2017) and Zoltán Simon, *History in Times of Unprecedented Change* (Bloomsbury, 2019), as well as *Zeitschrift für Geschichtswissenschaft* (1953ff), *History & Theory* (1960ff), *The Journal of Interdisciplinary History* (1970ff), *Rethinking History* (1997ff) and *História da Historiografia* (2008ff). Visit https://www.inth.ugent.be/about.
22. For historical sociology, which is concerned with how societies develop through history, see Volker Kruse, "Von der historischen Nationalökonomie zur historischen Soziologie", *Zeitschrift für Soziologie*, 19:3 (June 1990), pp. 149–165, International Social Science Journal, 133, August 1992, Gerard Delanty and Engin Isin (eds.), *Handbook of Historical Sociology* (Sage, 2003), Richard Lachmann, *What is Historical Sociology?* (Polity Press, 2013), Frédérick-Guillaume Dufour, *La sociologie historique: Traditions, trajectoires et débats* (Presses de l'Université du Québec, 2015) and *InterDisciplines: Journal of History and Sociology*, 1:1 (2010), 3:1 (2012), 7:2 (2016) and 9:2 (2018). Browse *Journal of Historical Sociology* (1988ff) and Social Evolution & History (2002ff). Visit https://www.isa-sociology.org/en (search for historical sociology). The literature on (socio) cultural evolution is large.
23. See John Tolan, *Faces of Muhammad: Western Perceptions of the Prophet of Islam from the Middle Ages to Today* (PUP, 2019). Isma'īl Rājī al-Fārūqī and Lois Lamyā al-Fārūqī, *The Cultural Atlas of Islam* (Macmillan, 1986) is an authoritative book, clearly presenting the world view of the Islam.
24. See Shahid Yusuf et al., *Development Economics through the Decades: A Critical Look at 30 Years of the World Development Report* (The World Bank, 2009) and Shahrukh Khan, *A History of Development Economics Thought: Challenges and Counter-challenges* (Routledge, 2014).
25. The literature on the current state of development economics is very large and still growing. For an introduction, see Dwight Perkins et al. (eds.), *Economics of Development* (W.W. Norton, 2013), Michael Tadaro and Stephen Smith, *Economic Development* (Pearson, 2015), Adam Szirmai, *Socio-Economic Development* (CUP, 2015) and Aparajita Mukherjee and Saumya Chakrabarti,

Development Economics: A Critical Perspective (PHI, 2016). Visit https://tinbergentoday.nl and see the previous note.

26. See David O'Connor (ed.), *A Critical History of Western Philosophy* (Free Press, 1964, p. 123), John Shand, *Philosophy and Philosophers* (Acumen, 2002, Chap. 3) and Brian Davies and Eleonore Stump (eds.), *Handbook of Aquinas* (OUP, 2012, Chap. 25). Visit https://www.iep.utm.edu/faith-re.
27. See also Karl Ameriks, *Interpreting Kant's Critiques* (Clarendon Press, 2003) and Henry Allison, *Essays on Kant* (OUP, 2012).
28. Go to https://ir3s.ifi.u-tokyo.ac.jp/en and https://www.routledgetextbooks.com/textbooks/sustainability. Visit also the website of Frontiers in Environmental Science (2015ff).
29. Visit http://positivists.org and https://plato.stanford.edu/entries/hermeneutics.
30. See Ansgar Nünning (ed.), *Lexikon Literatur- und Kulturtheorie* (J.B. Metzler, 2013), Gilbert Rosman, *Why Cultural Studies?* (Wiley, 2015) and Chris Barker and Emma Jane, *Cultural Studies: Theory and Practice* (Sage, 2016). The book series *Radical Cultural Studies*, published by Rowman & Littlefield, examines "the culturopolitical, sociopolitical, aesthetic and ethical implications of international cultures". A list of cultural studies journals can be found on the Internet. See also Note 23 in Chap. 4.
31. The German Bildung tradition is reminiscent of the Chinese *Daxue* (great learning) tradition, for which the interested reader may be referred to Wing-tsit Chan, *A Source Book in Chinese Philosophy* (PUP, 1963, pp. 84–94) and Huaiyu Wang, "On *gewu*: Recovering the Way of the Great Learning", *Philosophy East & West*, 57:2 (April 2007), pp. 204–226. In 2008, *Basic Thought of The Great Learning* (Chinese), written by Jin Yaoji, was published in Hong Kong. In 2012, *The Great Learning and The Doctrine of the Mean Explained* (Chinese), written by Fu Peirong, was published in Taiwan. Context of the German Bildung ideal is provided by Konstanze Baron and Christian Soboth (eds.), *Perfektionismus und Perfektibilität* (Meiner, 2018). Visit http://dx.doi.org/10.1590/2175-623683003 and see Notker Hammerstein and Ulrich Herrmann (eds.), *Handbuch der deutschen Bildungsgeschichte*, Vol. II (Beck, 2005).
32. Dijkgraaf is prime example of the physicist who has not the faintest idea what the humanities are concerned with. He is unaware of the many strains of thought that arose in opposition to the Enlightenment. Dijkgraaf should read C.P. Snow, *The Two Cultures and the Scientific Revolution* (OUP, 1959), Otto Bird, *Cultures in Conflict* (University of Notre Dame Press, 1978), Ed Wilson, *Consilience: The Unity of Knowledge* (Knopf, 1998), Gerasimos Santas, *Goodness and Justice: Plato, Aristotle and the Moderns* (Blackwell, 2001), Jens Høyrup, *through History and Philosophy* (SUNY Press, 2000), Stephen Jay Gould, *The Hedgehog, the Fox, and the Magister's Pox: Mending the gap between Science and the Humanities* (Harmony Books, 2003), Jonathan Gottschall, *Literature, Science, and a New Humanities* (Palgrave Manmillan, 2008), Jerome Kagan, *The Three Cultures: Natural Sciences, Social Sciences, and the Humanities in the 21st Century* (CUP,

2009), Martha Nussbaum, *Not for Profit: why democracy needs the humanities* (PUP, 2010), Edward Slingerland and Mark Collard (eds.), *Creating Consilience: Integrating The Sciences and the Humanities* (OUP, 2012), Joseph Carroll et al. (eds.), *Darwin's Bridge: Uniting the Humanities and Sciences* (OUP, 2016), Christian Madsbjerg, *Sensemaking: The Power of the Humanities in the Age of the Algorithm* (Hachette, 2017) or Richard Miller and Kurt Spellmeyer (eds.), *The New Humanities Reader* (Cengage, 2019). He could also visit https://evolution-institute.org/commentary/the-gap-between-the-humanities-and-the-sciences. Dijkgraaf may widen his mind (*a*) by glancing at the selected works of Wilhelm Dilthey edited by Rudolf Makkreel and Frithjof Rodi (PUP), (*b*) by taking note of the famous Davos disputation between Martin Heidegger and Ernst Cassirer in 1929, or (*c*) by reading about culture (and the varied forms of it studied by cultural anthropologists). Nobody should be surprised that Dijkgraaf has little interest in the humanities: he lives in a country where matter has been prioritized above mind.

33. See James Lovelock's books *The Revenge of Gaia* (Basic Books, 2006) and *The Vanishing Face of Gaia: A Final Warning* (Allen Lane, 2009). Lovelock's thesis has not gone uncriticized. See Toby Tyrrell, *On Gaia* (PUP, 2013). However, the reader should also see: Hicham-Stéphane Afeissa, *La fin du monde et de l'humanité* (PUF, 2014), Herman Daly, *From Uneconomic-Growth to a Steady-State Economy* (Edward Elgar, 2014), Marin Schaffner, *Un Sol commun: Lutter, habiter, penser* (Wildproject, 2019), Nathaniel Rich, *Perdre la Terre* (Seuil, 2019), David Wallace-Wells, *The Uninhabitable Earth: Life after Warming* (Tim Duggan, 2019) and Pierre Charbonnier, *Abondance et liberté* (La Découverte, 2020). In March 2019, the alarming sixth edition of the *Global Environment Outlook* (GEO-6) was published. The reader should further see: Simon Levin (ed.), *The Princeton Guide to Ecology* (PUP, 2009) and the journals *Frontiers in Ecology and the Environment* (2003ff) and *Frontiers in Environmental Science* (2013ff). He/she may also visit https://ensia.com and https://www.umwelt-journal.at.
34. See Marcel Schweiker et al., *Messen und Verstehen in der Wissenschaft* (Springer, 2017). In *The Tyranny of Metrics* (PUP, 2019), Jerry Muller "uncovers the damage that metrics are causing and shows how we can begin to fix the problem". The words of Einstein should be written above the main entrance of every university: "Not everything that counts can be counted and not everything that can be counted counts.".
35. Few are those who read Werner Jaeger's masterpiece *Paideia: Die Formung des griechischen Menschen* (Walter de Gruyter, 1934–1947). See Victor Hanson and John Heath, *Who Killed Homer?* (Free Press, 1998).
36. See Jochen Krautz, *Ware Bildung: Schule und Universität unter dem Diktat der Ökonomie* (Diederichs, 2011), Hans Pechar, *Bildungsökonomie und Bildungspolitik* (University of Oldenburg Press, 2013) and Bas van Bommel, "Between Bildung and Wissenschaft: The 19th-Century German Ideal of Scientific Education", December 2015, online. For research universities, go to http://www.iaruni.org and https://www.leru.org.

37. See Tricia Gallant, *Academic Integrity in the 21st Century* (Jossey-Bass, 2008), Francis Macrina, *Scientific Integrity* (Wiley, 2014), Nicholas Steneck et al. (eds.), *Integrity in the Global Research Arena* (World Scientific, 2015) and Tracey Bretag (ed.) *A Research Agenda for Academic Integrity* (Edgar Elgar, 2020). See *Journal of Academic Ethics* (2003ff) and visit https://www.academicintegrity.org, http://integrity.emory.edu and https://plato.stanford.edu/entries/integrity.
38. Econometrics, attempting to understand the quantitative relations of a modern economy, combines the viewpoints of economists, mathematicians and statisticians. Visit https://www.econometricsociety.org. Antidotes are the article of Denise Cummins, "What Happens when you believe in Ayn Rand and modern economic theory" (https://evonomics.com, February 2016) and Steven Klees, "Neoclassical economics is dead" (*ibid*, October 2020).
39. A list of professorships at the University of Cambridge can be found on the Internet.
40. For project leadership, explore https://iccpm.com. At the next ICCPM meeting, the idea of a project-based country study may be discussed as a roundtable topic. For interdisciplinary research projects, see Steph Menken and Machiel Keestra (eds.) *An Introduction to Interdisciplinary Research* (AUP, 2016). See Appendix A.
41. See Boukari Souley et al., "A Comparative Analysis of Data Compression Techniques", *International Journal of Applied Sciences and Engineering*, 2:1 (April 2014), pp. 63–82 and Khalid Sayood, *Introduction to Data Compression* (Morgan Kaufmann, 2018).

Chapter 9
Reductionism or Holism?

Abstract As the question may arise where to begin the collaborative study of a country (first investigating the whole and then researching its parts, or the other way around), the problem of the nature of, and relationship between, holism and reductionism needs to be addressed. The opportunity will be taken to dwell upon the fundamental difference between the Western (analytic) and Chinese (correlative) way of thought. The relationship between nature and culture is expounded on and the mutual dependency of the sciences of nature and the studies of culture is emphasized.

Since there is confusion about the words reductionism and holism, it might be useful trying to be clear about them. In regard to reductionism, two questions may be raised: (*a*) what is (or should be) reduced to what and (*b*) how is the reduction (supposed to be) carried out? As to holism, it may be asked what the size, contents, structure and borders of a whole are. My answers to these questions may convince the reader that holism is not the negation of reductionism, or vice versa. They are complementary, like the two sides of a coin or like two clapping hands. It is a matter of both-and.

Reductionism has different faces.[1] The most common one, indicated as *methodological* reductionism, concerns the technique of analyzing a system by studying its elements, hence "elementary study" or "basic research". It holds that explanation based on the smallest possible entities (e.g., subatomic particles, molecules, or individuals) is better than, or should precede, explanation at a higher level of organization (atoms, cells and societies, respectively). The strategic maxim of this technique is "divide a problem into smaller problems and solve them one by one before trying to solve the problem". This technique has undoubtedly yielded a rich harvest of scientific discoveries, but it has also contributed to compartmentalization and fragmentation. Methodological reductionism leads to the wearing of blinkers. It locks people up in a room where the world is investigated by peering through a disciplinary straw. Methodological reductionism leaves one with agoraphobia or myopia, for academic specialization requires that each decision to focus on one thing means the decision to disregard or neglect a myriad of other things that are relevant. Politically expedient (*divide and impera*!), methodological reductionism has turned universities into a collection of stovepipes. It is congenial to methodological *individualism*[2] and does not do justice to the complexity and fullness of reality. In addition to

H. Kuijper, *Comprehending the Complexity of Countries*,
https://doi.org/10.1007/978-981-16-4709-3_9

bottom-up explanation (from parts to wholes), there is need for top-down explanation (from wholes to parts).

René Descarte (1596–1650) argued that the world is like a machine, which could only be understood by taking its parts apart and studying them. Giambattista Vico (1668–1744), whose masterpiece *Scienza Nuova* (new science) attempts to organize the humanities, disagreed. He was dissatisfied with the scholastic hairsplitting of the Jesuits and criticized the rationalism and reductionism triggered by Descartes. Given the *Zeitgeist*, Vico was very courageous in doing so. He was the first expositor of the fundamentals of social and human sciences, indeed the precursor of thinking about complexities. Semioticians like Mikhail Bakhtin, Yuri Lotman, Roland Barthes, Algirdas Greimas, Umberto Eco, Roland Posner, Thomas Sebeok, Winfried Nöth, Susan Petrilli and Marcel Danesi are heavily indebted to this poor and tragically misunderstood self-teacher from Naples who was a man of genius. His work is currently revalued.[3]

Descartes' statement was expanded on by Newton, whose *Philosophiæ Naturalis Principia Mathematica* (mathematical principles of natural philosophy), published in 1687, laid the foundations of classical mechanics, which would enable man to set a foot on the moon.[4] But methodological reductionism *à la Descartes* cannot account for life. Biology does not deny chemistry, but chemists are unable to explain life. Monica Anderson, researcher in the field of artificial general intelligence (aka strong AI), once said: "When the reductionist is cutting a living frog to see what makes it tick, the ticking disappears, but all the pieces of the frog are still there." Life cannot be explained by decomposing it. The piecemeal approach that physicists are said to excel in is inadequate. To know life, we must study the living; but living is continuously changing across time and space, and continuity is by definition the opposite of discontinuity or discreteness, the state of being discrete, separated or distinct.[5] The limitations of methodological reduction (breaking down the thing to be explained [the *explanandum*] into discrete parts thought to be more suitable for explaining) become patently obvious when the human mind (or its manifestation: culture) is object of scientific research.

Understanding the mind presupposes a mind, the hallmark of human beings. The self (the subject) cannot be indicated; it is and will always remain ineffable and unknowable, because to know a thing is being able to indicate the thing, i.e., to consider it to be an object for the mind to know.[6] The self cannot be considered to be an object. Scientists who scrutinize somebody's brain delude themselves into believing they are objectifying his/her "me of me". "To BE" shall not be confounded with "to HAVE", as Gabriel Marcel (1889–1973) penetratingly argues in his almost forgotten book *Être et Avoir* (1935). Having an active mind is not the same as being able to explain it. The mind cannot be grasped or known, because the knower (holder of a mind) cannot know himself (hold his mind), which makes "γνῶθι σεαυτόν" (know thyself) a task impossible to fulfill (I am not talking about somebody who is nobody, because he/she is living like a cow). For the same reason, modeling the mind, something Leonid Perlovsky at Harvard University is trying to do, is impossible. The human brain is made up by billions of interconnected neurons. Their firings are electrochemical events, happenings that can be investigated and may be explained, but this cannot be done without the human *mind*. The active brain, being an intricate and fine-meshed network of firing neurons, cannot explain its workings. The brain only functions. This functioning has not been and probably cannot be explained. A robot, even a robot that makes more capable robots, will always need the mind of an engineer who, being able to construct the robot-making robot, outsmarts the thing he ingeniously constructed. Methodological reductionism, glorified in the sciences of *nature*, has been miserably failing in the sciences of *culture*.

Miranda Chih-Ning Cheng (1976) is a Taiwanese-born physicist-mathematician known for formulating the "umbral moonshine conjecture" (https://www.quantamagazine.org/moonshine-master-toys-with-string-theory-20160804). She was doctoral student of Eric Verlinde (a Dutch theoretical physicist and string theorist) and works at the renowned Korteweg-de Vries Institute for Mathematics of the University of Amsterdam. I sent her a letter, asking whether she thought her own mind could be mathematically described. I never got an answer.

Theoretical reductionism is another face of reductionism. The claim of this kind of reductionism is that a number of theories can be derived from an underlying or overarching (more general) theory. The basic idea is that the fundamental laws of physics are universal in scope, always and everywhere valid. Since this idea has not been and cannot be tested *everywhere*, it is nothing but a kind of belief. A textbook example of theoretical reductionism is the relation between mechanics and thermodynamics, two theories long thought to explain aspects of the physical world. Mechanics accounted for the ways physical bodies affect each other, thermodynamics for the working of heat. In the nineteenth century, these theories were unified. Thermodynamic phenomena could be explained in mechanical terms. Thermodynamics was subsumed under, incorporated by and reduced to mechanics. Another example is the relation between chemistry and physics. The long-standing problem as to how atoms are holding together in molecules was finally solved by quantum mechanics. Chemistry was reduced to physics, heart of the physical sciences, which can be defined as the sciences of matter, motion and energy. The ultimate aim of physics, which branched off into fields such as mechanics, statistical mechanics, quantum mechanics, relativistic mechanics, electromagnetism, acoustics, optics, photonics, condensed-matter physics, plasma physics, molecular physics, atomic physics, nuclear physics and particle physics,[7] is to find a unified set of laws governing everything, at micro-scale, at the human scale of everyday life and at macro-scale.[8] Whether this ambitious goal will ever be reached is a matter of heated debate. A theory of everything (considered by physicist Ad Lagendijk to be the arrogance of physicists) has not been found yet. Worse still, it cannot be found. As the word everything, rigorously taken, refers to everything in the universe, such a theory will always remain a guess, a conjecture, not a theory.

In a little book entitled *What Is Life?* (1944), Erwin Schrödinger (1887–1961), who reconceived the "old quantum theory", wrote:

> What I wish to make clear is, in short, that from all we have learnt about the structure of living matter, we must be prepared to find it working in a manner that cannot be reduced to the ordinary laws of physics. And that not on the ground that there is any 'new force' or what not, directing the behavior of the single atoms within a living organism, but because

> the construction is different from anything we have yet tested in the physical laboratory. In biology we are faced with an entirely different situation. (Chap. 7)

Schrödinger's profound statement has been rephrased many times. In *Of Molecules and Men* (1967), molecular biologist, biophysicist and neuroscientist Francis Crick (1916–2004) wrote:

> In one way, you could say all the genetic and molecular biological work of the last sixty years could be considered as a long interlude. Now that the program has been completed, we have come full circle – back to the problems left behind unsolved. How does a wounded organism regenerate to exactly the same structure it had before? How does the egg form the organism? (Second lecture)

These interesting words flowed from the pen of the man who, together with James Watson and Maurice Wilkins, had been awarded the 1962 Nobel Prize in Physiology or Medicine "for their discoveries concerning the molecular structure of nucleic acids and its significance for information transfer in living material". In *The Astonishing Hypothesis: The Scientific Search for the Soul* (Simon & Schuster, 1994), Crick shows himself to be in favor of reductionism, which he describes as "the idea that it is possible, at least in principle, to explain a phenomenon in terms of less complicated constituents", adding that it is "the main method of explanation used by the exact sciences" (pp. 277–278). However, Crick's book is a *hypothesis*, not about the mind but about the *soul*. "Mind" and "soul" are not interchangeable concepts.[9] Scientists may *hypothesize* about the mind, tentatively assuming that man is more than a fleshy machine, more than "the behavior of a vast assembly of nerve cells and their associated molecules" (Crick, p. 3). The discussion is convoluted and may easily lead to misunderstanding.[10] In June 2019, the 14th World Congress of Biological Psychiatry was held in Vancouver, Canada. The participants were probably unaware or insufficiently aware of it that the World Phenomenology Institute, in Hanover (USA) aimed at critically inquiring the ideas of Edmund Husserl, not particularly a defender of materialism.

Sidney Brenner (1927–2019), one of the world's pioneers in genetics and molecular biology (https://www.webofstories.com/play/sydney.brenner/1), once said:

> It may be wrong to believe that all the logic is at the molecular level. We may need to get beyond the clock mechanism.

Alwyn Scott (1931–2007), a theoretical physicist with strong interest in nonlinear phenomena and biology (http://www.consciousness.arizona.edu/alwyn.htm), finishes his magnificent book *The Nonlinear Universe: Chaos, Emergence, Life* (Springer, 2007) by writing:

> Generalizing the concept of emergence from the nonlinear dynamics at a single level of description to include organizational networks – comprising both downward and upward causations among many levels of a being's biological, cognitive, and cultural hierarchies – leads to a very complicated organizational *thing*, the noun that we call *Life*. Far greater than the sum of its parts, life is not a mechanism. Nested within a nonlinear complexity that is unrivaled in the known Universe, life combines all aspects of a creature's dynamics; its physiological, conceptual, motivational, and motoric activities in an unsimulatable autopoietic network.

> Nonlinear science is not yet close to describing this phenomenon, but here, in my view, is where to seek answers to the remaining riddles of biology. (p. 301)

The insight of Scott, who also edited the *Encyclopedia of Nonlinear Science* (Routledge, 2005), is further developed by theoretical biologist and complex systems researcher Stuart Kauffman in his recent book *A World beyond Physics: The Emergence and Evolution of Life* (OUP, 2019). Four years earlier, he co-authored (with Arran Gare) an article entitled 'Beyond Descartes and Newton: Recovering Life and Humanity' (Progress in Biophysics and Molecular Biology, 119:3, pp. 219–244).

By defining the best possible decision support system for farm management, precision agriculture (https://www.ispag.org) aims to optimize the *conditions* for growth, but for growth to happen, it has to rely on Mother Nature. "Life is not a mechanism" and "cannot be explained by decomposing it". "Pulling the rice shoots to help them to grow" will be counterproductive, Mencius (372–289 BC) warned already. See Nobel laureate Paul Nurse, *What is Life?* (W.W. Norton, 2021).

As life is irreducible to the properties of an organism's constituent parts, an organism is more than just a collection of atoms or molecules; it has "emergent properties", qualities none of its constituents has. These properties emerge at a *collective* level and are utterly meaningless at elementary level. However, a description of emergent properties does not contradict a description of constituent parts. The points of view are complementary, each valid at their own level. Zooming-in and zooming-out should not exclude each other; to get the picture right, they should, in the course of a research process, be alternated, not one time but many times. The man who maintains that organisms are nothing but bunches of cells, which in turn are only a great number of molecules made up by atoms consisting of particles, concluding from it that life is not fundamentally different from matter, must be, quite literally, out of his mind. Saying that life does not defy the laws of physics is not the same as saying that the laws of physics explain life.

Admittedly, holistic laws have not been found yet, and how life, being a holistic phenomenon, has emerged is still a mystery. How is a whole to be defined? What are its size and structure? What are its contents and borders? In systems biology, sociology, anthropology and linguistics, disciplines concerned with different systems, structuralism is predicated on the double assumption that (*a*) the elements of, respectively, an organism, society, culture or language are to be understood by way of their interrelationship and (*b*) these interrelationships are related to a broader, overarching system. The premise of structuralism (a subject I will cover in the next chapter) is thus essentially holistic, which does not mean there is *ein freischwebender Geist* or *eine Seele unabhängig vom Körper*. However, structuralists prioritize the whole,[11] mereologists pay more attention to the mutual relationship between the whole and its parts.[12] Perhaps the concept of law is playing tricks on scientists who are engaged in the study of holistic phenomena. To explain what "to be" means is far from easy; more difficult is it to explain what "ought to be" means. Is law to be taken descriptively or prescriptively, in a non-evaluative or in a normative way? Is "law" referring to an order of things ("*Sachverhalt*", or "*Sachverlauf*") or to how things should be ordered, ideally? See Chap. 6.

Going *from matter to life*, one takes a giant step. Going *from life to mind*, however, one takes a step at least as big. Well-informed about the latest developments in relativity theory and quantum mechanics, physicists are of little help to virologists and microbiologists who, facing the COVID-19 pandemic, clearly sit with the hands in their hair, not realizing that their discussion of this deadly disease must be broadened and widened far beyond just biological or medical matters; they are not aware of it that the Anthropocene wreaks havoc on little anthropos. See Erle Ellis, *Anthropocene: A Very Short Introduction* (OUP, 2018) and visit https://populationmatters.org (search for anthropocene). In my view, the disease that affects people all over the world and might be the harbinger of human extinction is the result of a full-blown *mental/psychic* pandemic. Something is wrong with human beings, not with the other animals, their companions. Yet political leaders are relying on "experts" in virology and microbiology. The literature on animal and human rights is voluminous and ever more expanding. Sadly, little effort has been made to critically investigate the biblical basis of our horrible dealings with animals. Our "*Mitwelt*" (social environment) is intimately interrelated with our "*Umwelt*" (natural environment). This biconditional relationship is insufficiently investigated. I leave it to priest, pastors, rabbis and mullahs to inform us about the "*Oberwelt*".

Is the world nothing but matter pushing matter in a void? Are humans nothing but mindless machines operating for the survival of genes? Or is man more than a fleshy machine, more than a chemical factory, more than an input–output mechanism? Is the mind completely reducible to the brain, as Jack Smart, Patricia Churchland, Terrence Sejnowski, Margaret Boden, and Dick Swaab—following in the footsteps of Julien de La Mettrie (1709–1751) and Ludwig Büchner (1824–1899)—would have it?[13] Or should we believe scientists like Francisco Ayala, Michael Gazanniga, Andy Clark, Vincent Descombes, and Dan Zahavi who—following the lead of Immanuel Kant (1724–1804) and Edmund Husserl (1859–1938)—do not completely reduce mental phenomena to physicochemical processes and leave some room for what is *not* determined or computational? Is society merely the sum total of autonomous individuals (individualism), like a heap of sand consisting of separate grains of sand, or are holistic forces at work? Is culture nothing but nature? Or is culture in man's nature? Are we only a part of nature (matter in motion), or also and at the same time apart from it? Is it "matter over mind", "mind over matter", or "matter, mind and meaning"? These are burning questions to which nobody seems to be able to offer watertight answers. I think that mind (*Geist*) and body (*Körper*) are inseparable from each other, but cannot be identified with, or reduced to, one another. We are more than a number of atoms, even more than a number of *organized* atoms. But what makes atoms to be organized? Did they organize themselves, spontaneously, or is the organization imposed on them, from the outside?

Everybody (except the obdurate theist) will agree that without animal life there cannot be such a thing as mind. The materialist who claims that mental phenomena are the result of what happens at atomic or subatomic level will first have to deal with the problem of life. Without life above the level of plants there cannot be mind, but life cannot be completely accounted for by matter. Insects differ from pebbles; insects die, but pebbles do not die. Life is said to be superimposed on matter and mind is thought to be superimposed on life. It seems safe to assume that there is no mind without life and no life without matter. The three should not be muddled up. The human body is a complex system of extremely intricate systems (organs) allegedly consisting of 37.2 billion cells, the interrelations of, or pathways between, which are far from clear. See Bill Bryson, *The Body* (Doubleday, 2019). The mind is not a system but the function of a system. This functioning is a mystery. What makes it happen? Is it happening by itself or on command?

If the active brain is a complex and dynamic system (network) of firing neurons and the mind is a function of it, the branch of mathematical analysis called functional analysis comes in sight, for functional analysis, which is divided into several subbranches, is concerned with functions of sets or spaces of functions. "Holomorphy" (derived from the Greek ὅλος [whole] and μορφή [form]) is its central concept. The literature on functional analysis is large and only accessible to mathematicians. A list of functional analysis topics can be found on the Internet.

The mind abstracts, unifies, forms ideas and conceptualizes (from the Latin *com* + *capere* [to take different things together, to summarize]); we find the prefix *con* also in the words consciousness and conscience). An idea is thus something given form by the mind. The word information stems from *informare* (to give form to). No mind, no idea. The human brain, central organ of the human nervous system, is susceptible of all kinds of measuring, but the mind, being a function of the way in which an active brain works or operates, is not measurable, at least not in the same manner; perhaps in another way, but not in the way a brain can be measured. In his latest book, *From Bacteria to Bach and Back: The Evolution of Minds* (W.W. Norton, 2017), Daniel Dennett builds on ideas from computer science and biology to show how a comprehending mind arose from a mindless process of natural selection. He explains how a crucial shift occurred when humans developed the ability to share memes and found a way of doing things not based on genetic instinct. However, what Dennett exactly means by "a comprehending mind" remains unclear. The noun phrase sounds like "wet water", or "green gras". Furthermore, what are memes? Richard Dawkins (Dennett's slightly younger colleague), Susan Blackmore and Robert Wright have been trying hard to explain the meaning of "meme", and *the Journal of Memetics* (1997ff) provides those interested in the subject with a forum. Memetics is the study concerned with information and culture based on an analogy with Darwin's evolution theory. For analogy, see Chap. 2. Like "consciousness" and "mind", "meme" is an elusive concept. It escapes us as soon as we think we got it. See Jürgen Mittelstraß (ed.), *Enzyklopädie Philosophie und Wissenschaftstheorie*, Vol. 5 (2013), pp. 318–324 and go to https://ice.dartmouth.edu/blog/sean-carroll-and-alan-wallace-on-the-nature-of-reality.

In *The International Dictionary of Psychology* (Crossroad Classic, 1995), experimental psychologist Stuart Sutherland (1927–1998) wrote:

> Consciousness is a fascinating but elusive phenomenon; it is impossible to specify what it is, what it does, or why it evolved. Nothing worth reading has been written on it. The term is impossible to define except in terms that are unintelligible without a grasp of what consciousness means.

In the past two decades or so, much brain research has been done, many brain-books (and articles) have been published, and major technological breakthroughs in the field of neuroscience have been made. See the open-access journal *Network Neuroscience* (2017ff). The list of neuroscientists (available on the Internet) is impressive, and in September 2019, the International Brain Research Organization (IBRO) held its 10th World Congress. Yet Sutherland's remark is still apposite. Consciousness, over the nature and scope of which Indians, Tibetans, Chinese and Japanese have been pondering, is still an elusive, extremely difficult to describe phenomenon.

Einstein's formula "energy equals mass times the speed of light squared" is the very powerful shorthand for expressing or describing the relationship between energy, mass, and the speed of light. For billions of years, nature has been exploiting the relationship, without which the sun, making life on Planet Earth possible, would not burn. However, *living beings* are more than stardust, and the human *mind*, operating within a community of human beings, is more than a pack of firing neurons.[14] "How much more and in what way?" are the as of yet unanswered questions. Descartes's *Cogito ergo sum* is a fallacious argument, for the conclusion to be proved (*sum* [*je suis*]) is validated in the claim (*cogito* [*je pense*]). Dick Swaab also begs the question. The argument put forward in his best-selling book *We Are Our Brains*, translated by Jane Hedley-Prôle (Spiegel & Grau, 2014), seems plausible only through a sleight of hand. "Our" implies "we" and this "we" is made the subject of a sentence about "we". By transforming the sentence "we are our brains" into "we are the brains we have", the nonsense is immediately revealed: as if we are not the arms, the legs, the heart, the liver, the lungs, the bladder, the skin, the hair, the eyes and the kidneys we have! Swaab offers no *explanans* for his *explanandum*, that is, for the "we" of the sentence and for the "we" implied by "our". He explains diligently the workings of the brain at every stage of life (from the womb to what happens when we fall in love or get Alzheimer's), but leaves us in the dark on the "we", man's individuality and identity, the central *locus* which great novelists have been writing about and on the nature of which leading psychologists still disagree. "We have brains" and "*we* have brains, but they have not" are sentences that perfectly make sense, like "we have legs" and "*we* have legs, but they have not". "We *are* our brains", however, is a sentence that makes no sense at all. "Are our brains" is the predicate of a subject ("we") on which materialist Swaab has given absolutely no information. Ergo, the well-paid professor of neurobiology at the University of Amsterdam is talking sheer nonsense.

I have to bypass the voluminous literature on the notorious mind-body problem, but mention should be made of Richard Warner and Tadeusz Szubka (eds.), *The Mind-Body Problem: A Guide to the Current Debate* (Blackwell, 1994), David Armstrong, *The Mind-Body Problem* (Westview Press, 1999), Tim Crane and Sarah Patterson (eds.), *History of the Mind-Body Problem* (Routledge, 2000), Hilary Putnam), *The Threefold Cord: Mind, Body and World* (Columbia University Press, 2001), Alexander Moreira-Almeida and Franklin Santos (eds.), *Exploring Frontiers of the Mind-Brain Relationship* (Springer, 2012), Vincent Descombes, *Les embarrass de l'identité* (Gallimard, 2013), Stanislas Dehaene, *Consciousness and the Brain* (Penguin, 2014), Paavo Pylkkänen, "Fundamental Physicsand the Mind—Is There a Connection?", *Lecture Notes in Computer Science*, 8951 (February 2015), pp. 3–11,

Jonathan Westphal, *The Mind-Body Problem* (MIT Press, 2016), Anna Marmodoro and Sophie Cartwright (eds.), *A History of Mind and Body in Late Antiquity* (CUP, 2018), François Kammerer, "Does the Explanatory Gap Rest on a Fallacy?", *Review of Philosophy and Psychology*, 10:4 (December 2019), pp. 649-667 and Anil Seth, *Being You: A new Science of Consciousness* (Faber & Faber, 2021). See also Note 6 of this Chapter. In his article "Is Identity a Predicate?" (2013), Géza Kállay comes close to the answer given in the *Chandogya Upanishad* (6.8.7): "The self–in its original, pure, and primordial state–is identical with the ultimate reality". The mind-body problem has not been unknown in China. See Walter Liebenthal (1886-1982), "The Immortality of the Soul in Chinese Thought", *Monumenta Nipponica*, 8:1/2 (1952), pp. 327-397, Huiyuan (334-416), "The body will be lost, but the mind does not perish" (Chinese), in Erik Zürcher and Stephen Teiser, *Buddhist Conquest of China: The Spread and Adaptation of Buddhism in Early Medieval China* (Brill, 2007, pp. 204-253), Whalen Lai, "Emperor Wu of Liang on the Immortal Soul", *Journal of the American Oriental Society*, 101:2 (1981), pp. 167-175, and Ming-Wood Liu, "Fan Chen's 'Treatise on the Destructibility of the Spirit and its Buddhist Critics", *Philosphy East and West*, 37:4 (1987), pp. 402-428. Reading Plato's *Phaido*, one cannot be unmoved. Freedom seems to be the mind's essence; its history is described in Annelien De Dijn, *Freedom: An Unruly History* (HUP, 2020). See also Alan Ryan, *The Making of Modern Liberalism* (Princeton University Press, 2012, pp. 45-62 and visit https://iep.utm.edu/freewill, https://plato.stanford.edu/entries/freedom-ancient and https://plato.stanford.edu/entries/liberty-positive-negative.

The scientist who thinks he or she has unraveled the matter-life-mind riddle should expound on the way in which he succeeded in disentangling the knot of all knots (Schopenhauer's *Weltknoten*). Trying to do so, he may find himself to be "a strange loop".[15] Modern physics has improved on but not replaced Newtonian physics. Einstein's formula admirably describes the universe outside, but not the universe inside (without which knowing and mathematizing the universe outside would be impossible), not to speak of the way into which the two universes are interrelated. Einstein (and Sir Roger Penrose, who followed him in his book *The Road to Reality*)[16] completed only half the job and the relatively easy half to boot, for "*l'univers stellaire n'est pas si difficile à comprendre que les véritables actions d'autres personnes*" (Marcel Proust); in the words of his contemporary compatriot Paul Claudel (1868–1955): "*L'universel n 'est pas au-dessus, il est au-dessous*".

Roget's Thesaurus of English words and Phrases (Penguin, 1984) is organized into six classes, which are subdivided into 39 sections. Each section deals with a particular aspect of the class within which it is found. The sections are further subdivided into 990 heads, which themselves are divided into paragraphs grouped together according to their part of speech. The first three classes cover the external world, with sections such as quantity, number, matter, space, time, change and causation. The last three classes deal with the internal world, with sections like reasoning, communicating, volition, emotion, action, morality and religion. Confronting this thesaurus with the International System of Units (SI), the reader will notice that the basic units of this system (s [second], m [meter], kg [kilogram], A [ampère], K [kelvin], mol [mole] and cd [candela]) must be subsumed under the first three classes (Heads 1–446). None of the system's basic units is applicable to the sections of the last three classes (Heads 447–990). Words like action, anger, art, authenticity, awe, beauty, benevolence, boredom, bureaucracy, catharsis, charisma, *charme*, compassion, contempt, courage, courtesy, creativity, cruelty, curiosity, cynicism, decadence, democracy, devotion, diligence, disappointment, disgust, *duende*, ecstasy, elegance, emancipation, embarrassment, Enlightenment, enthusiasm, experience, faith, fear, freedom, friendship, goodness, greed, habit, happiness, harmony, *Heimweh*, holiness, honor, hope, humanism, humor, *iki*, imagination, individualism, inspiration, jealousy, joy, justice, kindness, laziness, loneliness, love, loyalty, meaning, modernization, morality, panic, passion, pathos, patience, poetry, popularity, poverty, pride, prudence, *rasa,* regret, religion, reputation, respect, Romanticism, sadness, satire, *saudade*, secularism, shame, shyness, simplicity, sincerity, *sobornost*, solidarity, style, sublime, suffering, sympathy, taste, temperance, tragedy, trust, truthfulness, value, vulgarity, *wabi-sabi*, westernization, wisdom and *yūgen* cannot be defined in terms of time, length, mass, electric current, temperature, amount of substance or luminous intensity. Belonging to the cultural vocabulary, they stand for manifestations of the mind, for phenomena (appearances) of the νοῦς. Art, beauty, etc., are neither measurable nor quantifiable. They cannot be counted, but certainly count. They cannot be reduced to matter, but do matter. The mind, without which the external world would be unknown, is essentially ungraspable and unknowable.

By saying that the most incomprehensible thing about the universe is that it is comprehensible, Einstein seems to imply that his formula is only applicable to *a part of reality*: the world outside. Yet, an accurate one month weather forecast for all places around the world cannot be made by Einstein's followers and the famous formula cannot be applied to life, which is clearly more than the sheer aggregation of stardust particles (extremely small piece of matter). To be a description or mathematical expression of the *whole* universe (the *ultimate* reality referred to by Śaṃkara), Einstein's formula needs to be completed (see Hanoch Gutfreund and Jürgen Renn, *Einstein on Einstein* [PUP, 2020]). In other words, the world so beautifully described in the (alas, not yet translated) book written by China's most prominent philosopher Tang Junyi (1909–1978), 生命存在與心靈境界 (shēngmìng cúnzài yǔ xīnlíng jìngjiè [the being of life and the boundaries of the mind]), has to be taken into account.[17] Like *Yīn* and *Yáng*, in and out are interrelated (how could there be understanding of "out" without understanding of "in"!). Received wisdom notwithstanding, they cannot be decoupled. On a higher level, the two-valued logic, the logic that satisfies the principle stating that every declarative sentence is either true or false, should be completed with a logic not restricting the number of truth values to only two (visit https://plato.stanford.edu/entries/logic-manyvalued).

This should not be taken to mean that in the West (the nature of) *man* has not been studied. According to Alexander Pope (1688–1744), "the proper study of mankind is man". Western research into (the nature of) man comes in different varieties:

- Anthropology (the study of man): a fast expanding field of *scientific* research;
- Philosophical anthropology: a branch of *philosophy* concerned with the nature of man;
- Psychology: the science of mind and behavior including the study of consciousness, unconscious phenomena, feeling and thoughts;

- Personality psychology: a subfield of psychology focusing on personality and its variation among individuals;
- Evolutionary psychology: the biologically-informed study of human behavior;
- Personalism: an intellectual movement stressing the importance of persons;
- Humanism: a philosophical stance emphasizing the value of human beings, individually and collectively;
- Humanities: academic disciplines studying aspects of the human society (the humanities are closely related to the social sciences).

It would be interesting to investigate whether the basic tenets of Western scientists concerned with the study of man can stand in the face of Confucianism, which has been the main stream of thougt in China, emphasizing the importance for human beings to be truly human by correct social behavior. In *Human Nature: A Critical Reader* (OUP, 1997, p. 17 [note 49], and see note 24 in Chap. 2), Laura Betzig, the editor of this informative book (50 pages of references!), writes that she is neither interested in culture nor in ethics, subjects she regards as "largely non-empirical". However, Edmund Husserl (1859–1938) is the first Western philosopher who was unique in going seriously about the nature of man without appealing to the empty concept of God. See Herbert Spiegelberg, *The Phenomenological Movement* (Kluwer Academic Publishers, 1994, pp. 69–165), Yushun Huang, "On 'viewing things' and 'viewing nothing': A dialogue between Confucianism and phenomenology", *Frontiers in Philosophy in China*, 3:2 (June 2008), pp. 177–193, and Andrea Staiti, *Husserl's Transcendental Phenomenology* (CUP, 2014). In addition, visit https://plato.stanford.edu/entries/husserl.

If, generally, (the nature of) *man* has been the subject of wide-ranging studies, the number of studies about *great* men like Homer, Plato, Aristotle, Alexander the Great, Caesar, Aquinas, Dante, Columbus, Leonardo da Vinci, Luther, Shakespeare, Galilei, Descartes, Rembrandt, Spinoza, Locke, Newton, Bach, Rousseau, Adam Smith, Kant, Goethe, Mozart, Napoleon, Beethoven, Hegel, Gauss, Darwin, Marx, Frege, Freud, Husserl, Max Weber, Einstein, Picaso, Keynes, Bohr, Wittgenstein and Gödel (I restrict myself to the West) is almost uncountable. Scientists and philosophers studying man/mankind tend to overlook this important fact.

In Chap. 2, I maintained that a theory is an explanation of how a piece of reality is the way it is or works the way it works. Natural scientists in general and physicists in particular are said to explain a piece of reality. However, to answer the question whether the fundamental building blocks of reality are particles, fields or both, one has to think beyond physics (see Charles Sebens, *AEON*, October 24, 2019). Furthermore, how big the piece is that physicists are supposed to explain remains to be seen and depends on how these scientists would care to interpret reality. If quantification is the essence of science, and scientific theories have to be empirically tested, a huge part of the *whole* reality is yet to be explained. It would, therefore, be better for natural scientists to get off their high horse. Cultural scientists have something to tell, something worth to listen to. Natural and cultural scientists have an important job to do—together.

Yīn and Yáng

My reference to *Yīn* and *Yáng* shall not be taken as a flight into the unknown. The words, standing for complements, have a ring of modern science (quantum theory). According to the Chinese, black and white, positive and negative, north and south, east and west, interior and exterior, entrance and exit, input and output, life and death, above and under, high and low, even and uneven, hard and soft, hot and cold, short and long, wet and dry, open and closed, foul and fragrant, empty and full, small and big, day and night, dawn and dusk, light and darkness, demand and supply, debit and credit, space and time, one and many, universal and particular, same and other, must (necessity) and may (contingency), male and female, part and whole, or individual and community are not antagonistic toward, but complementary to each other, the one going well with the other. Though clearly distinguishable, ship and harbor, train and station, airplane and airport, electrical connector and power point, right hand and left hand, on and off presuppose, and are dependent on, each other. *Yīn* and *Yáng* are not hostile against, but implicative of, and to be seen as relative to, each other. This bi-conditionality is prevalent in nature and should thus be the guide of scientific research. As indicated in the preface, *Yīn* and *Yáng*, connected by *qì*, constitute the Way (*Dào*) that, according to the *Dàodéjīng*, cannot be spoken of.[18]

Western scientists adhere to the principle of bivalence stating that every declarative sentence is either true or false. They neglect the obvious fact that ordinary language is not restricted to declarative sentences. A declarative sentence does not ask a question, give an order, or express an emotion. The principle of bivalence is also called the law of the excluded middle, stating that there is no other possibility for declarative statements than being either true or not-true. *Tertium non datur*. In terms of set theory, set A and set B are either disjoint (*e.g.*, {1, 2, 3} and {4, 5, 6}) or not disjoint (*e.g.*, {1, 2, 3} and {3, 4, 5}). In the latter case, A is partly or wholly a subset of B, or the other way around. For example, set {1, 2, 3} overlaps partly with set {3, 4, 5} and set {3, 4, 5} is wholly a subset of set {2, 3, 4, 5}.

The Chinese, taking their cue from Mother Nature, think differently: gray is partly white and partly black, but white is in no way a subset of black, and vice versa. The black and the white are to be distinguished, but shall not be separated. They are not mutually excluding but complementing each other, like male and female, winter and summer, high tide and low tide. "If *Yīn* and *Yáng* are obliterated, there would be no means of seeing change" (*Yìjīng*).

The importance of bi-conditionality or mutuality is not to be underestimated. It does equal justice to both sides and shows no preference, is thus less forceful, more objective and impartial. A disjointed speech or piece of writing, having parts that are not connected with each other, is hard to follow, but a coherent one is easy to understand.[19] Clearly, not only the parts but also their interconnectedness is important for comprehending, which is an act of intelligent behavior. A team that performs disjointedly is not a team worthy of its name. Mountains are different from, but not hostile against, or irreconcilably opposite to, valleys. Once a common ground is found, even declared enemies can shake hands and make it up to each other. Compromising (said to be the essence of Chineseness) is the art of reaching mutual agreement, a subject that will interest game theorists. Having impregnated his pencil with ink (two seemingly totally different things), the calligrapher draws a particular line. Following the Way and attuning to the *Dào*, he creates order out of chaos. In China, calligraphy (the art of giving form, of in-forming) is considered to be the queen of arts.[20] It would be interesting to compare Chinese, Japanese, Korean, Mongolian, Tibetan, Indian, Persian, Arabic and Western calligraphy, and to investigate what calligraphy, writing (poetry), painting, architecture, sculpture, dance (ballet) and music have in common. The Chinese regard poetry as invisible painting and painting as visible poetry, the former focusing on process, the latter highlighting pattern (configuration). Dance and music are performing arts, the former is a purposefully selected sequence of human movements, the ordering in space; the latter is sound organized in time. Pattern formation or system building in space and time (subjects I elaborated on earlier) is the very essence of art.[21]

Roads can be found everywhere, but it is impossible to find only one of their sides. In the *Great Appendix* to the *Yìjīng*, a work generally denigrated by Western scientists but of crucial importance for understanding China, it is written and emphasized: 一陰一陽之謂道 (*Yīn* and *Yáng*: that is called the *Dào*).[22] The complementarity view belongs to the ontology of relations rather than to the ontology of things. The so-called Internet of Things (IoT), which is an extension of Internet connectivity, seems to lean to the latter, while neglecting the former. The crux of relational or correlative thinking (as distinct from causal thinking), is its focus on in-betweenness. Chinese culture (*e.g.*, cuisine and medicine) and Chinese art (*e.g.*, painting and poetry) are imbued with it. The Chinese believe that reality originates in nothingness (*wú*), that there is no movement without motionlessness, and that being, in whatever form, rises from non-being. In their view, it is *not something but nothing* that connects and generates everything. It is nothingness or chaos on which order is predicated. The spokes of a wheel come together in, and concentrate on, its center, and yet it is the motionlessness of the hub that makes the turning of the wheel possible. Unity is the very nature of diversity; it is "the great interdependence of all differences" (*Yìjīng*); everything, everybody begins and ends in it. That is why we are speechless at the birth of a baby and at the passing away of a human being.

The Gospel of John opens with the famous declarative sentence: "Ἐν ἀρχῇ ἦν ὁ λόγος, καὶ ὁ λόγος ἦν πρὸς τὸν θεόν, καὶ θεὸς ἦν ὁ λόγος" (in the beginning was the word, and the word was with God, and the word was God). The Chinese do not think so. In their view, *nothingness, silence* was the beginning, the absolute silence broken by the word. As written in the *Dàodéjīng*, Chap. 1: "無名天地之始, 有名萬物之母" (without-name is the beginning of heaven and Earth [the universe], with-name is the mother of ten thousand [all] things). Or in chapter 40: "天下萬物生於有; 有生於無" (all things under heaven spring from being; being emerges from non-being). For example, pots are made of clay, but their usability resides in the not-being of clay. *Wú* (non-being), the unnamable is the source; *yǒu* (being), the namable is the manifestation. Existence is preceded by non-existence. *Le voilà*: the fundamental difference between the Christian and Chinese way of thought! The Chinese view was not totally alien to Westerners, however. *Paradise Lost* (1667), the epic poem of John Milton, opens with the verse: "*In the Beginning how the Heav'ns and Earth Rose out of Chaos*". Whether in the forthcoming second edition of his famous book *Naming and Necessity* (Blackwell) Saul Kripke concurs with the Chinese view remains to be seen.

Augustine (354-430) considered *esse* to be God's innermost nature ("*id quod Deus est*"). In *Sein und Zeit* (1927), Heidegger raises "*die Frage nach dem Sinn von Sein*" which, he mistakenly thought, had been neglected for more than two thousand years. In 1943, *L'Être et le Néant*, a philosophical essay about ontology, was published. Its author was Jean-Paul Sartre (1905-1980), who, like Heidegger, was a disciple of Husserl (1859-1938). Sartre, who was awarded the 1964 Nobel Prize in literature and has influenced sociology, critical theory, literary studies and post-colonial theory, was the progenitor of existentialisme, a philosophical current that takes human *experience* as its starting point.

China's philosophy of nothingness is different from what some modern physicists conceive of as the void.[23] In the Chinese view, reality cannot be grasped; it cannot be caught by any formula whatsoever; it cannot be named (put into words). Trying to do so is like grasping dry, loose sand: it escapes between the fingers. Imagine a straight line connecting the points A (on the left side) and B (on the right side). Somewhere on the line, it is impossible to say whether one leans more to any of the two sides, whether one gravitates more toward A or toward B, but it is after (not: at) this special, absolutely neutral point (called *wú* [nothingness] or *wújí* [ultimateless, unlimited, indefinite] by the Chinese) that both A and B *start* being endpoints, *begin* to culminate in what they really are: ending (or in the opposite direction: starting) points. When the line between A and B is hanging, like a necklace, the very lowest point is where the line does not go up and is absolutely neutral to either A or B.

If the reader has difficulty in understanding this, he should realize that beginning-to-be is, rigorously taken, not the same as not-being. He may think of a spinning potter's wheel and ask himself whether its center is moving or not. It is this seemingly trivial question that has haunted me for many years since I was a freshman at Leiden University. The whole wheel is spinning, including its midpoint, right? Well, not exactly, because a center that turns around is not yet the very center of the wheel. What Westerners would call a *dead* center is for the Chinese the decisive and most *vital* point, the essence and root of all things, the hinge of the door, the ridge of the roof, the pivot around which everything else turns, the hair crown, the "point" where the parts, coming together, end up being parts, the "point" where the parts, not being parts yet, begin to be parts. The Emperor ("Son of Heaven") is the mainmast, second to none, the Serene Highness and August Majesty to be obeyed. Leaning to absolutely no side but being the focal point, the confluence of influences from all sides, he is the Sovereign to whom "everybody under the sky" is supposed to bow like grass in the wind. Lu Ji (261–303), who probably took a leaf out of the book of Liezi (fifth century BC), put it poetically and poignantly: "*kè xū wú yǐ zhēn yǒu; kòu jì mò ér qiú yīn*" (exact being from non-being; knock at silence for sound).

When Westerners think they face, like Buridan's ass, the horns of a dilemma, the Chinese see an opportunity, the possibility of action, the beginning of being truly creative. *Wú* should be taken rigorously. It is not anything whatsoever, goes beyond what is thinkable, conceivable, imaginable, namable or definable. It is probably the answer to the key question in modern physics: how to unite the biggest and the smallest. *Wú* is indefinite and unlimited. It reminds us of Anaximander's ἄπειρον.[24] *Wú* is that from which everything emerges and into which everything will disappear. What is born is bound to die. *Xū* (empty) and *shí* (full) are terms frequently used in Chinese art and literary theory.[25] They relate to non-being and being, respectively. They relate also to each other, for non-being and being are complementary to each other; though different, they go together.

In-betweenness is the zero-ness Westerners are so afraid of (*horror vacui*) but Buddhists use to call *śūnyatā* (emptiness). The Tibetan *mandala* (circle) represents the Cosmos, the Universe. In Zen Buddhism, the *ensō* (circle) is the symbol of completeness. The power of vacuity (ceasing to point at or name) is illustrated in Japanese art forms such as the arrangement of flowers called *ikebana* (literaly: making flowers alive) and the writing of a *haiku* (a very short form of poetry). The distinctive (but from the *Yìjīng* derived!) notion of the *haiku* is the "cut" (*kire*), the articulation point where the antecedent and the subsequent gain in expressive power. Both *Ikebana* and *haiku* invite us to see what is invisible.[26] Only in the context of nothingness, the individual flower fully expresses its beauty, allowing the other flowers of the arrangement to express theirs.

There is affinity between process philosophy[27] and the Chinese (correlative) way of thought, but it needs to be stressed that they are not congruent with each other, as John Cobb Jr., co-founder and current co-director of the Center for Process Studies in Claremont, CA, mistakenly believes. The former posits becoming/functioning over against being/*esse*, giving priority to becoming/ functioning; the latter sets non-being (*wú*) and being (*yǒu*) up in contrary "opposition" to each other, prioritizing non-being. Non-being, however, is not the same as becoming/functioning. Non-being is like non-duality (*bùèr*), which precedes manifestations, also the manifestation called becoming/functioning. The not-yet being (the becoming) is constituent of the being. A ripening apple is an apple in the process of becoming ripe; it is an apple (a being) *in statu nascendi*. Similarly, a baby is an adult (a human being) *in statu nasendi*. Non-being, however, is like zero; it is the unnamable or incalculable from which all naming or calculating starts; it is the beginning of all beginnings, but not the constituent of any being; it is not being whatsoever.

The Chinese, who were originally peasants, distinguish between *běn* (root) and *mò* (end), between *tǐ* (foundation) and *yòng* (function). *Tǐ* and *yòng* relate to each other like pattern and parts. In the Chinese view, *tǐ* (e.g., wood) and *yòng* (e.g., firewood, wood for making fences, wood as building material, wood for making furniture, wood for making [wooden] musical instruments, carving wood, etc.) are correlative, interconnected, not excluding each other. Branches grow out of roots; roots grow into, and end up as, branches. *Pǔ* (the raw, unprocessed wood) is the foundation. According to Ge Hong (283–343), man's calling is to go for the foundation and "to embrace simplicity" (*bàopǔ*). The usual translation of *tǐ* into "substance" is confusing, because the Latin *substantia* stems from the Greek οὐσία (being), words that have triggered heated debates among theologians. Thinkers in ancient China attempted to conceptualize what is earlier than being. Their answer was: non-being (*wú*), the complete darkness, from which light emerges. Without darkness, there can be no light, and without light, the word darkness would be meaningless.

The reader of this book may have noticed that the gist of it hinges on the point discussed in the previous paragraphs. Whereas modern scientists appear to be imprisioned by binary thinking, only considering two possibilities ("yes" or "no", "on" or "off", "true" or "false", "black" or "white"), the ancient Chinese (and the majority of their modern descendants) think correlatively, complementarily, not exclusively but inclusively, thus more harmoniously and more realistically. They make room for a third possibility (*tertium datur*!). While Western scientists are interested in the application and practical use of natural, negative, whole, rational, irrational, real, complex and even hypercomplex numbers, their Chinese counterparts are inclined to pay attention to the root of all roots, to zero, to the numberless and nameless, to the ineffable, to the Absolute—to *wú*. Nothing is, by definition, not anything (in Hinduism: *neti, neti*), "*wú*" is a word referring to the character 無, which is a graphic symbol representing the concept of nothing, *i.e.*, standing for something that is invisible, unsayable and inconceivable. Nothingness is referred to by "*wú*", but cannot be conceptualized, or thought about. Talking about nothingness is thus essentially impossible. *The Chinese think relationally and see things pattern-wise as well as process-wise.* Rather than considering things to be pieces that may or may not fit together, they think of fragments constituting a completeness, or parts belonging to a whole. The Enlightenment, which marked the transition from the dark, Christianity-dominated Middle Ages to Modernity, has been much eulogized in the West, but the light may still have to come—from the East? If that will happen, the old saying *Ex Oriente Lux* will prove to be right. The words spoken by Stephen Hawking on January 23, 2000 are hope-giving: "I think the twenty-first century will be the century of complexity

String theory is fashionable. It is about the point-like particles of particle physics being replaced by one-dimensional objects called strings. See Elias Kiritsis, *String Theory in a Nutshell* (PUP, 2019). An impressive list of string theorists can be found on the Internet. But string theory is speculative, because nobody has empirically proven that "strings" indeed exist. Since they are supposed to be in a state of vibration, it would be interesting to know what the goniometric functions or properties of vibrating strings are. Trigonometry is a branch of mathematics studying the relationships between side lengths and angles of triangles. The subfield also mathematizes objects that are vibrating, which is a kind of turning around or circling, and this – ancient Chinese argued – is a movement that presupposes stillness (*wú*). String theorists are, therefore, advised by Lu Ji "to exact being from non-being". The reader should also visit https://doi.org/10.1103/PhysRevLett.125.181301 and go to note 8 of this chapter.

Interestingly, "ontology" (a word referring to the epistemology-related study of *being*) has been rendered into Chinese as *béntǐlùn* (discussion about root and foundation). In other words, the expression *béntǐlùn* refers to a discourse on the primary source of a being, that is, "*im Grunde genommen*", on its *non*-being. Few, if any, of those parading as experts in Chinese philosophy have pointed out the subtle but crucial difference. Considering *béntǐlùn* the same as ontology is thus glossing over a fundamental difference between the Chinese and Western way of thought. Chinese thinkers, who tried to get to the bottom of things and scrutinized the very concept of being, went deeper than their Western counterparts. The waiting is for a comparative study of Heidegger's distinction between *Sein* and *Dasein* and the Chinese *tǐ-yòng* complementarity.[28] In the second half of the nineteenth century, when China tried to respond to the Western challenge and struggled with her own past, the *tǐ-yòng* complementarity reverberated in the famous slogan *Zhōng tǐ, Xī yòng* (applying Western knowledge for its practical uses, while keeping Chinese values as the core). Xi Jinping has not forgotten this important catchphrase. One almost hears him thinking when he smilingly faces his American counterpart: "How superficial you are!". Ontology dealing with the most general properties of beings, is a subfield of metaphysics, a very broad field attempting to answer questions about how the world is (as distinct from questions about knowing the world [epistemology]). Visit https://plato.stanford.edu/entries/metaphysics. The Chinese translation of "metaphysics" is *xíng ér shàng xué* (the study of what is above forms). "*Xing ér xià xué*" refers to empirical sciences or reasearch.

There are no relations without *relata* (things that are related), but there is said to be nothing that does not change, the meaning of which is that all relata are changing. This makes things to be investigated maddening complicated. The Eastern and Western way of thought should be combined. In their book *Solving the Dynamic Complexity Dilemma* (Springer, 2014), Nabil Abu el Ata and Maurice Perks are not concerned with this subject, neither is apparently any of their colleagues in the field of systems science. Chen Lai (1952), however, does give some attention to the matter in his monumental but untranslated book *You Wu zhi Jing: Wang Yangming Zhexuede Jingshen* (the boundaries of being and non-being: the spirit of Wang Yangming's philosophy), published in Taipei, in 2000. After the nineteenth century revolution of logic, incisively described and analyzed by John Passmore in his landmark book *A Hundred Years of Philosophy* (Penguin, 1968, pp. 120–155), relational–as distinct from predicative–thinking has come in the Western spotlight.[29] There is still a vast area for scientists and philosophers to do research on, however. In order to put the complementarity view into perspective, it would be compulsory for me to elaborate on the subject of relational thinking, but, not being qualified to do so, I must leave the matter to professionals. Visit https://doi.org/10.1093/irap/lcy010.

Reductionisme—Holism Again

The ongoing discussion between realists and anti-realists shall not go unmentioned, though (see Christos Kyriacou and Robin McKenna [eds.], *Metaepistemology: Realism and Anti-Realism* [Palgrave, 2018]; visit https://plato.stanford.edu/entries/structural-realism and https://philpapers.org [fill in: structural realism]). Perhaps the best way to the heart of the matter is to highlight the controversy between essentialists (realists) and relationalists (anti-realists). The argument can be traced back to Aristotle, who questioned the Theory of Forms/Ideas of Plato, his illustrious teacher (without being scolded for doing so!). New essentialism, which stems from the failure of metaphysics to provide a solid foundation for theories about the nature of reality,[30] claims to have developed a comprehensive philosophy of nature. Notable representatives are George Bealer, John Bigelow, Nancy Cartwright, Crawford Elder, Brian Ellis, Evan Fales, John Heil, Caroline Liersen, Charles Martin, Nicholas Maxwell, George Molnar and Sydney Shoemaker. New essentialism distinguishes between individual essence and kind essence, the former being the set of qualities in virtue of which a thing is the individual thing it is, the latter being the set of qualities in virtue of which a thing is a thing of the kind it belongs to. Relationalists, on the other hand, maintain that the identity of an object is constituted by its (constantly changing) relations with other things, and that it is impossible to give an accurate account of any object or person in abstraction from its/his place within a context.[31] All reality is relative. As the Spanish philosopher José Ortega y Gasset (1883–1955) reportedly put it in 1914: "*Yo soy yo y mi circunstancia*" (I am myself and my circumstance). Relationalism reminds us of the message of the great Buddhist thinker Nāgārjuna (c. 150–c. 250 CE) and of the main modern challengers to metaphysical realism, Werner Heienberg (1901–1976), Michael Dummett (1925–2011) and Hilary Putnam (1926–2016). Visit https://plato.stanford.edu/entries/nagarjuna and https://plato.stanford.edu/entries/realism-sem-challenge. The reader of my book will not be surprised by learning that relationalism is congenial to the way of thinking in ancient China.[32]

Neurofication, the act of augmenting a science with the recent discoveries of neuroscience by adding the prefix "neuro" to the moniker (e.g., neurophysics, –chemistry, –biology, –anatomy, –physiology, –psychology, –psychiatry, –sociology, –anthropology, –linguistics, –politics, –economics, –marketing, – ethics, –esthetics, –theology, –informatics and –mathematics), has hardened into neuro*fixation*, to the delight of the pharmaceutical industry, which holds cohorts of scientific researchers hostage. Fixation is a kind of mental illness in which people stop thinking. Reductionism, the credo of neuroscientist, prevails in academia.[33] That's why I think that our universities are sick.

To see the limitations of reductionism, one may look at the important difference between causes and reasons; important, because it shows the gaping gap between the sciences of nature (matter) and the sciences of culture (mind). Some verbs can be transitive or intransitive, as exemplified by the following sentences: "Could you move your car please?" versus "The trees are moving in the breeze", or "She sang a popular song" versus "The kettle was singing on the stove". A man's moving his finger is an action; it is to be distinguished from the mere motion of that finger. Raising your arm is an action, for which you must have a reason, but your arm's rising is not an action; it is an event, having causes. Dying (like sneezing, falling, colliding, raining or flooding) is an event, explained in the natural sciences, but killing (like writing, smoking a cigarette, buying, stealing or migrating) is an action, which cultural scientists try to explain.[34] When dealing with a country, both the discoveries of geologists, biologists and physical anthropologists and the findings of psychologists, sociologists and cultural anthropologists shall thus be taken into account. Geologists are needed for explaining earthquakes or volcanic eruptions, but it would be silly to appeal to them for understanding environmental pollution, the result of human activities. Nature does not polute itself. In nature, there is no waste; she regenerates herself. Events can be analyzed and reduced, but actions are less susceptible to analysis and reduction, because they are linked to cognition, conation and emotion. In contrast to events, actions are more than motions of particles. Interestingly, change (the key concept of the *Book of Change*) relates to both events and actions. Whereas physicists, mathematicians and biologists are familiar with concepts such as change, process and evolution, cultural scientists often use the words development, evolvement and history. Dealing with change, (natural and cultural) scientists should not overlook, neglect or ignore its structure, and vice versa. Processes and patterns are—it is no unnecessary luxury to repeat it—inseparably interconnected. They are the two sides of one and the same coin. This is the reason why I called the problem of transition a vexed problem. See note 5 of this chapter.

Reductionism and holism can and shall be combined. On close investigation, they turn out not to be antagonistic but correlative to each other. It is a matter of *both-and*, not of *either-or*. Generally speaking, while Western scientists tend to lay stress on partition and analysis, their Chinese colleagues are inclined to emphasize blending and synthesis. The strength of both parties must be united. The reductionist who likes to drill deeply has to go hand in hand with the holist who prefers to cast a wide net. "East is East, and West is West; never the twain shall meet", Rudyard Kipling (1865–1936) wrote in the heyday of modern imperialism. In the Age of globalization, I would rather say: it is time for *rapprochement*. Referring to the epigraphs of this book, I therefore finish this chapter by quoting Descartes' contemporary, Blaise Pascal (1623–1662):

> Donc toutes choses étant causées et causantes, aidées et aidantes, médiatement et immédiatement, et toutes s'entretenant par un lien naturel et insensible qui lie les plus éloignées et les plus différentes, je tiens impossible de connaître les parties sans connaître le tout, non plus que de connaître le tout sans connaître particulièrement les parties. (Pensées, Section II, p. 359)

More than 1.400 years earlier, Wang Bi started his famous *Zhouyi Lüeli* (short introduction to the Book of Changes) by writting:

> The many cannot govern the many. The reason why the many hold together is because a master preserves their unity. The movements cannot contol the movements. The reason why the movements accomplish a common rotation is because there is no duality.

Parts and whole are correlated and mutually dependent on each other. The meaning of the former cannot be stated without referring to the latter, and vice versa; each member of the pair draws upon the other for its meaning. A part is a part of a whole, which is a whole of parts. The meaning of parents involves the notion of child, and the meaning of child the notion of parents. Parts that do not belong to a whole are not parts and a whole without parts is not a whole. There is no inside/interior without outside/exterior, no periphery without center. While periphery without center would result in anarchy, center without periphery would result in tyranny. The ruled imply a ruler, and a ruler implies people being ruled. So, in the final analysis, the question is: autocracy (rule by one) or democracy (rule by many)? Thinkers in ancient China have opted for the former, provided the ruler, following his true nature, is a paragon of virtue.

Those who balk at just the prospect of an autocrat, even if he were a benevolent ruler caring for his people, are requested to consider the following:

- The problem of sovereignty is far from being solved.
- The chance of one man being virtuous is higher than the chance of many people being reasonable. A crowd can be defined as multiplied ignorance. "*Panem et circenses*", not "knowledge and wisdom", is what the mob wants. Today: "Hosanna!" tomorrow: "Crucify!" According to Tacitus (56–120 AD), the masses have neither a judgment nor a principle ("*nec principium habere iudicium vulgo*"). Differently put, to be "*beaucoup*" is to be "*bête*".
- The spinning wheel of a potter has one and only one middle point, which arguably is not a point at all and—counter-intuitively—does *not* move. A turning wheel needs its no-turning. An ellipse is a plane curve that surrounds two focal points, but the sum of the two distances to them from each point on the curve is a *constant*.

Asked about his choice between a vicious ruler and democracy, Voltaire (1694–1778), trailblazer of the French Revolution and famous for his advocacy of freedom of speech, reputedly responded: "*Il vaut mieux être mangé par un lion que par mille rats*".

It would be making a mistake to think that my writing about the coherence of things ("*tout tient à tout*" and "*Alles ist vernetzt*"!) only applies to the *study* of countries. In spite of the technical possibilities of fully (not: foolishly) informing people, the general public is (deliberately?) kept ignorant of the wholeness of things that have been happening, of the real situation in countries such as Russia, China, North Korea, Iran, Cuba, Israel, Hungary, Venezuela, Brazil, Egypt and Nigeria. It seems to me that we would live in a better world when the following conditions are met:

(1) If TV items like the melting of the polar icecap, financial statements of troubled companies, fashion shows, sports events, strikes, parades, demonstrations, weddings, funerals, award ceremonies, crimes, accidents, amusement parks, refugees, festivals, traffic congestions, obituaries, horse racing, New Year addresses, casinos, slums, State visits, floods, droughts, tourist resorts, wars, opening ceremonies, the *Oktoberfest* in Munich, animal abuse, carnivals, landslides, the *Tour de France*, court hearings, red light districts, auctions, kickboxing, mass celebrations, cooking competitions, the crowning of a king, forest fires, parliamentary discussions, human trafficking, lottery draws, air and soil pollution, interviews, presidential elections, garbage dumps, buying sprees, sickness and old age, agricultural or fishing problems, gala dinners, the exorbitant wealth of few and the misery of many people, military operations, dilapidated buildings, bankruptcies, mergers and acquisitions, arms waving crowds, hungry children, beauty contests, crying mothers, volatile financial markets, auto racings, decadent behavior, drugs or alcohol abuse, exploitation, cases of the perpetual show business, anorexia or obesity, processions, open-air concerts, abandoned factories and the coronavirus pandemic were not separated by, or interspersed with, misleading, distracting or intrusive advertisements.

(2) If, in case necessary, some of the TV items were shown and commented on not as completely independent and fully autonomous but as fundamentally interrelated and intimately interconnected subjects.
(3) If only "the truth, the whole truth and nothing but the truth" were revealed.
(4) If we are fully and frankly told what is going on behind the scenes or in the corridors of power.
(5) If the *real* reality of everyday life were shown, that is to say, not a reality arbitrarily broken down into pieces of information or infotainment which are carefully edited, arranged, measured out and broadcast.

People are horribly fooled by the media, the hiding and well-to-do tycoons of which shamefully keep on lining their pockets. Dictators in distant countries could not improve on those who find the smartest AI experts at their side. I am not only referring to immoral figures who are providing many types of disinformation (fake news being only one of them), but also and primarily to the "*divide et impera*" policy persistently pursued by media bosses, who only serve their commercial interests. Essentially, man is a moral being, having a conscience. He/she should not be ruled by Mammon. For the desire of money is the root of all evil.

Notes and References

1. The literature on this subject (related to mereology, the theory of parthood relationships) is large. See Andras Angyal, "The Structure of Wholes", *Philosophy of Science*, 6:1 (1939), pp. 25–37, Paul Davies, *God and the New Physics* (J.M. Dent, 1983, pp. 58–71), Christopher Lawrence and George Weisz (eds.), *Greater than the Parts* (OUP, 1998), Hale and Wright *op. cit.* pp. 227–247, Rick Looijen, *Holism and Reductionism in Biology and Ecology* (Springer, 2000), Richard Jones, *Reductionism: Analysis and the Fullness of Reality* (Bucknell University Press, 2000), Marc van Regenmortel and David Hull (eds.), *Promises and Limits of Reductionism in the Biomedical Science* (Wiley, 2002), Vieri Benci et al. (eds.), *Determinism, Holism, and Complexity* (Springer, 2003), Karen Gloy, *Die Geschichte des ganzheitlichen Denkens* (Komet, 2005), Ji Zhang, *One and Many* (University of Hawaii Press, 2012), Katharine Farrell et al. (eds.), *Beyond Reductionism* (Routledge, 2013), Sergio Chibbaro et al., *Reductionism, Emergence and Levels of Reality* (Springer, 2014), Julie Zahle and Finn Collin (eds.), *Rethinking the Individualism-Holism Debate* (Springer, 2014), Alfred Abächerli, "Holism and Reductionism: How to Get the Balance Right", *Open Access Library Journal*, 3:5 (May 2016), pp. 1–15, Eric Kandel, *Reductionism in Art and Brain Science: Bridging the Two Cultures* (Columbia University Press, 2016), Anne Whitehead et al. (eds.), *Companion to the Critical Medical Humanities* (Edinburgh University Press, 2016, Chapter 3), Gerard Verschuuren, *The Holism-Reductionism Debate* (CreateSpace, 2017) and Len Pismen, *The Swings of Science: From Complexity to Simplicity and Back* (Springer, 2018). The word holism was coined by Jan Smuts (1870–1950) in his book *Holism and Evolution* (Macmillan, 1926). Claudius van Wyk wrote an article on the book entitled "Holism and Evolution by Jan Christian Smuts:

a re-evaluation after 90 years". Go to https://holos-earth.org and see Note 26 in Chap. 5.

2. See Lars Udehn, *Methodological Individualism* (Routledge, 2001). I consider methodological reductionism and methodological individualism to be rooted in individualism, the emergence of which is probably related to the rise of Christianity. See Colin Morris, *The Discovery of the Individual 1050–1200* (University of Toronto Press, 1972), Daniel Shanahan, *Toward a Genealogy of Individualism* (University of Massachusetts Press, 1992) and George Smith and Marilyn Moore (eds.), *Individualism: A Reader* (Cato Institute, 2015). The pros and cons of individualism and collectivism (read: capitalism and communism) should be impartially compared.
3. See Giorgio Tagliacozzo et al., *Vico and Contemporary Thought* (Macmillan, 1976), Marcel Danesi (ed.), *Giambattista Vico and Anglo-American Science* (Mouton de Gruyter, 1995) and Joseph Mali, *The Legacy of Vico in Modern Cultural History* (CUP, 2012).
4. See Eduard Dijksterhuis, *The Mechanization of the World Picture*, translated by C. Dikshoorn (OUP, 1961, pp. 403–418), Bruce Brackenridge, *The Key to Newton's Dynamics* (University of California Press, 1995, pp. 17–24), Richard DeWitt, *World Views* (Wiley, 2010, pp. 79–204) and Lars Jaeger, *Die Naturwissenschaften* (Springer, 2015, pp. 89–104). For the shift from the Cartesian-Newtonian to the New Physics paradigm, see e.g., Fritjof Capra, *The Turning Point* (Simon & Schuster, 1982, pp. 63–89).
5. In addition to Erwin Schrödinger's classical book on the subject, see Josef Seifert, *What is Life?* (Rodopi, 1997) and Addy Pross, *What is Life?* (OUP, 2012). Visit https://plato.stanford.edu/entries/life. The philosophy/science of mechanics does not solve the vexed problem of transition encountered in physics, chemistry, biology, geology, politics, sociology, linguistics, economics, mathematics and cybernetics, Visit https://philpapers.org/s/transition. The obscure Chinese Book of Change(s) might also be called the book of transition(s). "Transition" refers to the *Strukturwandlung* or dynamics of a (complex) system; another word, with basically the same meaning, is "metamorphose". Read Ovid, *Metamorphoses* (first published in 8 AD).
6. For attempts to comprehend consciousness, see Juleon Schins, *Empirical Evidence for the Non-Material Nature of Consciousness* (Edwin Mellen Press, 2004), Carlo Penco et al. (eds.), *Explaining the Mental* (Cambridge Scholars Publishing, 2007), Eric Schwitzgebel, *Perplexities of Consciousness* (MIT Press, 2011), *Internationales Jahrbuch des Deutschen Idealismus* (volume 11, 2013), Georg Northoff, *Unlocking the Brain* (OUP, 2014, volume II), Steven Sevush, *The Single-Neuron Theory: Closing in on the Neural Correlate of Consciousness* (Palgrave, 2016), Stephen Leach and James Tartaglia (eds.), *Consciousness and the Great Philosophers* (Routledge, 2017), Susan Schneider and Max Velmans (eds.), *Companion to Consciousness*

(Wiley-Blackwell, 2017), Robert Kirk, *Robots, Zombies and Us* (Bloomsbury, 2017), Philip Goff, *Galileo's Error* (Pantheon, 2019) and "the paradox of self-consciousness", an Edge conversation with Markus Gabriel, November 11, 2019 (online). Go to https://plato.stanford.edu/archives/win2019/entries/qt-consciousness, https://consciousness.arizona.edu and https://journals.sagepub.com/doi/full/10.1177/2398212818816019. Browse the leading Journal of *Consciousness Studies* (1994ff). The basic postulate of the Yogācāra school of Mahayana Buddhism is that consciousness is the only reality. See John Makeham (ed.), *Transforming Consciousness: Yogacara Thought in Modern China* (OUP, 2014). Interesting is also Alan Wallace (ed.), *Buddhism & Science* (Columbia University Press, 2003, pp. 87–279) and the website https://www.mindandlife.org (search for consciousness, self, and neuroscience). "Consciousness" and "conscience" are cognates. See Patricia Churchland, *Conscience: The Origins of Moral Intuition* (W.W. Norton, 2019).

7. Notice that no physicist has mastered all the subfields of physics. Interdisciplinarity is inescapable. See Appendix A.
8. See Steven Weinberg, *Dreams of a Final Theory* (Hutchinson Radius, 1993), Stephen Hawking, *The Theory of Everything* (Jaico, 2006) and Udayashankar Bhat, *An Envisioned Theory of Everything* (Notion, 2018). String theory is supposed to be a step on the road to a theory of everything. A list of string theory topics can be found on the Internet. See Elias Kiritsis, *String Theory in a Nutshell* (PUP, 2019).
9. See Edward Reed, *From Soul to Mind* (Yale University Press, 1998).
10. See Richard Swinburne, *The Evolution of the Soul* (Clarendon, 1997), Sandro Nannini, *Seele, Geist und Körper* (Peter Lang, 2006), Michael Trimble, *The Soul In the Brain* (Johns Hopkins University Press, 2007), *idem* and Mark George, *Biological Psychiatry* (Wiley, 2010), Gerhard Medicus, *Was uns Menschen verbindet* (VWB, 2013), Robert Spitzer, *The Soul's Upward Yearning* (Ignatius Press, 2015), Danijela Kambaskovic (ed.), *Conjunctions of Mind, Soul and Body from Plato to the Enlightenment* (Springer, 2014) and Jenny Teichman, *The Mind and the Soul* (Routledge, 2015). The gift of Buddhism to the world has been its relentless probing of the human mind. The similarities of, and differences between, Buddhism and Vedanta are matters of dispute. For the mind–body problem, see the literature referred to in chapter 9, and for consciousness, see Note 6 of this Chapter.
11. In France, structuralism rose to prominence in the wake of existentialism. For mathematical and physical structuralism, see Geoffrey Hellman and Stewart Shapiro, *Mathematical Structuralism* (CUP, 2018) and go to https://plato.stanford.edu/entries/physics-structuralism, respectively. Structural realism is currently considered the most defensible form of scientific realism, which is a reaction to logical positivism. For the concept of structure and related concepts, see the next chapter.
12. Go to https://plato.stanford.edu/entries/mereology and https://plato.stanford.edu/entries/lesniewski. See Hans Burkhardt et al. (eds.), *Handbook of Mereology* (Philosophia, 2017). Readers interested in mereology and its related

subjects may scroll through the valuable online list compiled by Frederick Rickey, "An Annotated Leśniewski Bibliography", February 1, 2014.

13. See Jack Smart, "Sensations and Brain Processes", *Philosophical Review*, 68:2 (April 1959), pp. 141–156, Margaret Boden, *Mind as Machine* (OUP, 2006), Dick Swaab, *We are our Brains* (Penguin, 2015) and Patricia Churchland and Terrence Sejnowski, *The Computational Brain* (MIT Press, 2016). They are essentially followers of Thomas Hobbes (1588–1679), according to whom there is nothing but matter in motion. There are scores of critical studies about Hobbes, harbinger of "modernity". Marcus Adams (ed.), *A Companion to Hobbes* (Wiley-Blackwell) is forthcoming and will perhaps be supplementing *The Cambridge Companion to Hobbes*, edited by Tom Sorell (1996), *Hobbes*, written by Aloysius Martinich (Routledge, 2005), and *The Bloomsbury Companion to Hobbes* edited by Sharon Loyd (2012). See also *Hobbes Studies* (1988ff). Visit https://philpapers.org/s/hobbes and https://plato.stanford.edu/entries/hobbes.
14. See Stuart Judge, "Nothing than a Pack of Neurons?", *The Faraday Papers*, 16 (March 210). For critical appraisals of Einstein's ideas, see Paul Schilpp (ed.), *Albert Einstein: Philosopher-Scientist* (Open Court, 1949, pp. 199–241, 555–562). See also Note 1 in Chap. 1.
15. See Douglas Hofstadter, *I Am a Strange Loop* (Basic Books, 2007) and Chap. 5.
16. See also David Deutsch, *The Fabric of Reality* (Allen Lane, 1997), Bernard d'Espagnat, *In Search of Reality* (Springer, 1983), John Gribbin, *Schrödinger's Kittens and the Search for Reality* (Back Bay Books, 1995), Barry Stroud, *The Quest for Reality* (OUP, 2000), Hilary Putnam, *Representation and Reality* (MIT Press, 2001), Brian Greene, *The Fabric of the Cosmos: Space, Time, and the Texture of Reality* (Knopf, 2004), Max Tegmark, *Our Mathematical Universe: My Quest for the Ultimate Nature of Reality* (Knopf, 2014), Frank Wilczek, *Fundamentals: Ten Keys to Reality* (Penguin, 2021) and John Heil, *Appearance in Reality* (OUP, forthcoming). Explore the website of SFI's Collective Computation Group, co-directed by Jessica Flack and David Krakauer. Penrose, Deutsch, D'Espagnat, Gribbin, Stroud, Putnam, Greene, Tegmark, Wilczek, Flack and Krakauer fail to define reality. Ignoring, or forgetting about, what prominent ontologists had/have to say, they look outward, not inward. The philosophy/science of mind has not yet matured. For an "atlas of reality" (!), see Robert Koons and Timothy Pickavance, *The Atlas of Reality: A Comprehensive Guide to Metaphysics* (Wiley, 2017). In *Farewell to Reality* (Pegasus, 2013), Jim Baggott argues "how modern physics has betrayed the search for scientific truth" and in *Irrationality: A History of the Dark Side of Reaso*n (PUP, 2019), Justin Smith shows "how history reveals that any triumph of reason is temporary and reversible". See Note 6 of this Chapter. Basically, there are two things more perplexing than reality: (I) the knowledge of and (II) the language about it. The nature of, and relationship between, reality, thought and language has been a conundrum for both Eastern and Western philosophers. As natural scientists rely on exact measurement, they can only grasp *part* of reality, and—given the wholeness of reality—even that capability

is questionable. Natural scientists seem to ignore that reality became highly problematic in the second part of the nineteenth century, particularly among symbolist authors such as Stéphane Mallarmé, Paul Verlaine, Arthur Rimbaud, Maurice Maeterlinck, Paul Valéry and Rainer Maria Rilke. Though the use of symbols can be traced back to Plato, Plotinus, medieval mystics and German Romanticists, the symbolists used them to *suggest*, rather than to stand for, reality. See Note 29 in Chap. 2.

17. Tang's two-volume book, published in 1977, is the key text of modern philosophy in China. It should be compared with Augustine's *Soliloquies* (inner dialogues). Anja Steinbauer (Hamburg University) and Ady van den Stock (Ghent University) wrote scholarly articles *about*, but did not translate any work of, Tang Junyi. Thomas Fröhlich (Hamburg University) wrote even a thick book about him. Apart from some translated quotations, Tang himself does not get the floor. In 1990, *Universe of the Mind: A Semiotic Theory of Culture*, the magnum opus of Yuri Lotman (founding father of the Tartu-Moscow School) was published.
18. The concept of *Dào* is central to both the *Dàodéjīng* and the *Yìjīng*. The relationship between these seminal works is debated among scholars. For a translation of the *Dàodéjīng*, see D.C. Lau, *Tao Te Ching: A Bilingual Edition* (The Chinese UniversityPress, 1963 [2001]). Also visit http://www.tao-te-king.org. For the *Yìjīng*, visit https://www.biroco.com/yijing. Ming-che Lee, "The early transmission and rendition of the *Yijing*", *Compilation and Translation Review*, 13:1 (2020), pp. 71–114, is a must-read.
19. A consistent speech is free from any contradiction but can still be incoherent. Having badly organized his material, the speaker is difficult to follow.
20. Chiang Yee's *Chinese Calligraphy* (Methuen, 1955) is probably the unsurpassed book on the subject. See also Alfreda Murck et al. (eds.), *Words and Images* (Metropolitan Museum of Art, 1991), Michael Sullivan, *The Three Perfections: Chinese Painting, Poetry and Calligraphy* (George Braziller, 1999) and Ouyang Zhongshi and Wen Fong (eds.), *Chinese Calligraphy* (Yale University Press, 2008).
21. The Chinese (and Indian) literature on art is very large.
22. See Bo Mou (ed.), *History of Chinese Philosophy* (Routledge, 2009, pp. 71–106) and Gudula Linck, *Yin und Yang, Die Suche nach Ganzheit im chinesischen Denken* (Karl Alber, 2017). Readers interested in philosophical thinking in China should visit https://www.iep.utm.edu/chin-pho, https://plato.stanford.edu (search for Chinese philosophy), https://plato.stanford.edu/entries/chinese-translate-interpret, https://warpweftandway.com, http://www.crvp.org/publications/Series-III/Series03-1.html and https://philpapers.org/browse/contemporary-chinese-philosophy. They could possibly subscribe to *Journal of Chinese Philosophy* (1973ff), *Contemporary Chinese Thought* (1997ff), *Dao: A Journal of Comparative Philosophy* (2002ff), *Frontiers of Philosophy in China* (2006ff), *Modern Chinese Philosophy* (2010ff) and/or *Journal of Chinese Humanities* (2014ff), or turn to the book series *Chinese Philosophy and Culture* (SUNY Press, 1988ff), *Dao Companions to Chinese philosophy*

(Springer, 2010ff) or *Key Concepts in Chinese Thought and Culture* (Palgrave, 2019ff). A useful website is http://cjfraser.net/links.

23. See Wilczek *op.cit.* pp. 118–124 and p. 393. Compare this with Carlo Rovelli, *Reality Is Not What It Seems* (Allen Lane, 2016) and Achille Varzi, "The Magic of Holes", in Giuseppina Marsico and Luca Tateo (eds.), *Ordinary Things and Their Extraordinary Meanings* (IAP, 2019, pp. 21–33). See note 8 and 16 of this Chapter, and visit https://plato.stanford.edu/entries/nothingness. "To be, or not to be—that is the question" is the famous opening phrase of a soliloquy given by Prince Hamlet in the so-called "nunnery scene" of William Shakespeare's play *Hamlet*. Looking around in this depraved world, I would rather say: to fool, or to be fooled—*that* is the question.
24. See Carlo Rovelli, *The first Scientist: Anaximander and his legacy* (Westholme, 2011, Chap. 5) and Radim Kočandrle and Dirk Couprie, *Apeiron* (Springer, 2017).
25. For their relationship, see Liu Xie, *The Literary Mind and the Carving of Dragons* (Chinese University Press, 1983, pp. xli-xlviii) and Zong-qi Cai (ed.), *A Chinese Literary Mind* (Stanford University Press, 2002).
26. See Maurice Merleau-Ponty, *Le visible et l'invisible* (Gallimard, 1979). *The Arts of Japan* is an informative and downloadable publication of the Smithsonian Institution. See also JeeLoo Liu and Douglas Berger (eds.), *Nothingness in Asian Philosophy* (Routledge, 2014) and Brook Ziporyn, *Emptiness and Omniscience* (Indiana University Press, 2016). In addition, go to https://plato.stanford.edu/entries/japanese-aesthetics.
27. See Jenny Helin et al. (eds.), *Handbook of Process Philosophy and Organization Studies* (OUP, 2014). Visit https://plato.stanford.edu/entries/process-philosophy, https://plato.stanford.edu/entries/chinese-change and https://samim.io/p/2019-12-28-the-yijing-whitehead-and-time-philosophy-by-yih-hsi.
28. See Graham Parkes (ed.), *Heidegger and Asian Thought* (University of Hawaï Press, 1987), David Leahy, *Novitas Mundi: Perception of the History of Being* (SUNY Press, 1994, section A and B, and Chap. 12), Reinhard May, *Heidegger's Hidden Sources: East Asian Influences on His Work* (Routledge, 1996), Herman Philipse, *Heidegger's Philosophy of Being: A Critical Interpretation* (PUP, 1998), Charles Guignon (ed.), *Companion to Heidegger* (CUP, 2006), Mark Wrathall, *Heidegger and Unconcealment* (CUP, 2011), John Richardson, *Heidegger* (Routledge, 2012), Eric Meyer, *Questioning Martin Heidegger* (University Press of America, 2013, pp. xi–xix), Lin Ma (2016), "On the Paradigm Shift of Comparative Studies of Heidegger and Chinese Philosophy", *Confluence: Journal of World Philosophies*, 4 (2016), pp. 81–98, Mahon O'Brien, *Heidegger's Life and Thought: A Tarnished Legacy* (Rowman & Littlefield, 2019) and Harald Seubert, *Heidegger: Ende der Philosophie oder Anfang des Denkens* (Karl Alber, 2019). See also Note 17 in Chap. 8. For "relational ontology", see Wesley Wildman, "An Introduction to Relational Ontology", May 15, 2006 (online) and Andrew Benjamin, *Towards a Relational Ontology* (SUNY Press, 2016).

29. See Daniel Merrill, *Augustus De Morgan and the Logic of Relations* (Springer, 1990), Didier Debaise, "What is Relational Thinking?", *INFleXions*, 5 (March 2012), pp. 1–11, Andrew Benjamin, *Towards a Relational Ontology* (SUNY Press, 2015), Anna Marmodoro and David Yates (eds.), *The Metaphysics of Relations* (OUP, 2016), Anna Marmodoro, *Everything in Everything* (OUP, 2017) and Joanna Golińska-Pilarek and Michał Zawidzki (eds.), *Ewa Orłowska on Relational Methods in Logic and Computer Science* (Springer, 2018, pp. 3–24). Also visit https://plato.stanford.edu/entries/relations and https://philpapers.org/browse/relations. See end of previous note.
30. Heidegger claims to be the first philosopher who since the Presocratici was seriously interested in the question of being, but he forgot about Augustine. See Søren Overgaard, *Husserl and Heidegger on Being in the World* (Springer, 2004). See also Note 28 of this Chapter.
31. See Brian Ellis, *The Philosophy of Nature: A Guide to the New Essentialism* (Acumen, 2002), David Oderberg, *Real Essentialism* (Routledge, 2007), Joseph Kaipayil, *Relationalism: A Theory of Being* (JIP, 2009) and Mark Young, "Relevance and Relationalism", *Metaphysica*, 12:1 (May 2011), pp. 19–30.
32. See Kwang-Kuo Hwang, *Foundations of Chinese Psychology: Confucian Social Relations* (Springer, 2012, Preface and pp. 187–217, 297–326) and Leah Zhu, *The Power of Relationalism in China* (Routledge, 2018). Carlo Rovelli, *Helgoland* (Penguin, 2021) must not be overlooked. The book is about quantum mechanics and its *relational* interpretation. The key question is whether or not relations are human constructs. See Chap. 2.
33. See Nikolas Rose and Joelle Abi-Rached, *Neuro: The New Brain Sciences and the Management of the Mind* (PUP, 2013), Sally Satel and Scott Lilienfeld, *Brainwashed: The Seductive Appeal of Mindless Neuroscience* (Basic Books, 2013) and Gary Marcus and Jeremy Freeman (eds.), *The Future of the Brain: Essays by the World's Leading Neuroscientists* (PUP, 2015). For neuroscience, see Mark Bear et al., *Neuroscience: Exploring the Brain* (Lippincott Williams & Wilkins, 2007) and Stephen Casper, "Neuroscience needs some new ideas", *Nature*, vol. 580 (2 April 2020), pp. 23–24. Go to https://plato.stanford.edu/entries/neuroscience, http://ctcn.utexas.edu and http://www.humanconnectomeproject.org. There are more than hundred journals on neuroscience, and a long list of neuroscientists can be found on the Internet.
34. Committing suicide is not an event but an action, in which social factors are involved. It is thus not to be explained by physicians but by sociologists and psychiatrists. See Émile Durkheim, *Le Suicide* (Félix Alcan, 1897).

Chapter 10
General Conclusions

Abstract In this chapter, general conclusions are drawn from the premises that collaborative country studies are scientifically and technically feasible. Some concepts are further clarified, and the complementarity of interdisciplinary and international collaboration is emphasized. The book's thesis is buttressed by some publications, and a cue is taken from astronomy, space science and geography. If the question regarding the fundamental problem posed is whether it can be computationally solved or whether the complexity of a country can be mathematically modeled and if this question has to be answered negatively, there is a second-best option not be rejected. Bridging the chasm between the sciences of nature and the sciences of culture is shown to be necessary and possible, and the concept of structure, being essential to the study of systems, is explored.

This book is critical of the work of so-called area or country experts, but it is critical in a constructive way. Showing the way out of their predicament (the way ahead), I add the sweet to the sour. *Nourri dans le sérail* (Leiden University), I cannot be content with what is happening at institutions devoted to area or country studies. The studies carried out there are beset with a fundamental problem: their practitioners, having an *explanandum*, do not have an *explanans* that makes them different from colleagues in, e.g., the human or social sciences. Saying that their study *objects* are different would be stating the obvious. Of course, the study object of Iranologists (Iran) differs from what Japanologists are supposed to explain (Japan). The issue is that, while meticulously describing what has been going on in the country attracting their attention, country pundits do not have a theory of their own, thereby exemplifying what the Romans already said: "*Praxis sine teoria sicut currus sine* via" (see Chap. 2).

The *praxis* of **area** students is questionable, because areas or regions such as the Middle East, (Sub-Saharan) Africa, (Central, South, East, or Southeast) Asia, (North, Middle or South) America and (North, East, South, West or Central) Europe are heterogeneous groups of juridically demarcated, militarily guarded and in many ways considerably different countries which only immortals can dream of grasping intellectually.[1] A country is certainly a geographic area, but an area or region is

H. Kuijper, *Comprehending the Complexity of Countries*,
https://doi.org/10.1007/978-981-16-4709-3_10

not a country. Middle East pundits, for example, (should) know that talking away about an area encompassing Egypt, Iran, Turkey, Saudi Arabia, Syria and Israel cannot be taken seriously. No musician masters the violin, the cello, the bassoon, the trombone, the oboe and the clarinet. Decision makers in Washington, Brussels or Tokyo who deliberate on, say, the strategy to be pursued with regard to China's "One Belt One Road Initiative" would do well to realize that the interests of dozens of different countries are involved and no scholar is, or will ever be, able to compare and weigh all these historically, geographically, socially, economically, politically and culturally molded interests.

Last year, the director of the International Institute for Asian Studies (IIAS) to whose infinite wisdom I already referred in the Introduction launched a "project of interdisciplinary research aimed at studying the Belt and Road Initiative, with special attention to the impact of the 'New Silk Road' on countries, regions and peoples outside of China". In order to promote interest in the project, which is directed by Richard Griffiths, IIAS and Leiden University released a MOOC (massive open online course), entitled *Building the New Silk Road* and hosted by the Stanford-based Coursera, which is, with 33 million enrolled users, the largest educational platform of its kind. I wish emeritus professor Griffiths and his colleagues all the luck in the world, but I hope that they will realize the importance of having an *explanans* concerning their enormous *explanandum*, of having a *praxis cum teoria*. If they do not, their chariot will be without a road. The Dutch would call this "*luchtfietsen*" (to cycle in the air).

In 2016-2017, following the "successful" completion of the three-year project "Rethinking Asian Studies in a Global Context", IIAS received another major grant from the New York-based Andrew W. Mellon Foundation to continue its "innovative work" to develop new research and teaching paradigms that seek "to move beyond traditional area studies, towards what can best be defined as globally connected humanities". The four-year grant supports work to create an alternative humanistic pedagogy curriculum. The new program is entitled "Humanities across Borders: Asia and Africa in the World" (https://humanitiesacrossborders.org) and aims at "expanding the scope of the humanities by mobilizing the knowledge in everyday life practice that have remained largely unrepresented in contemporary academia". It seems to me that the IIAS and its sponsor are totally lost. Instead of moving beyond traditional area studies, they should restrain themselves and think about how the complexity of a country is best to be comprehended. For the human sciences, see chapter one (note 9) and seven (note 12).

The study of a country, let alone the study of a heterogeneous area or region, is still in its infancy. If astronomical research institutes had been satisfied with only employing people with above average knowledge of celestial objects and phenomena, the development of astronomy would have been greatly retarded and little would have come of its spin-off, spaceflight (search the Internet for "James Web Space Telescope"). The knowledge of "country experts" is essentially nothing but above average knowledge of a country, knowledge of country-related things worth knowing. These pundits are not disciplined; their knowledge is not scientifically grounded. At best they have directly or indirectly perceived a lot of things, but they do not holistically comprehend the things having been perceived. For the crucial difference between perceiving and understanding, see Erwin Panofsky, *Studies in Iconology* (OUP, 1939, pp. 3-17) and *Barry Stroud, Seeing, Knowing, Understanding* (OUP, 2018).

The *praxis* of **country** students is also questionable, as long as there is no repeatedly tested *country* theory, so long as there is only sundry borrowing or stealing from natural, social or human sciences. So-called country experts, who regularly appear in the media, must be distrusted, because—taken as *country* experts—they do not explain what they are justifiably supposed to explain, the carefully cultivated appearance of the opposite of it notwithstanding. These pseudo-experts are sorely mistaken in forgetting, if not attempting to forget, the winged words of Pliny the Younger (*multum, non multa*), in believing that the sheer exposition of things to be known in regard to a country makes them scientists, in naively assuming that the opening of a cabinet of curiosities is equivalent to the advancement and development of a scientific and overarching theory in regard to the things being exposed. By definition, a generalist is not a specialist and an expert is not an omnivore. Science is not a mass of information or a great number of unconnected dots. Science is "organized knowledge" (Herbert Spencer), the systematic classification of things that are known or knowable. Science implies a refutable theory. Only displaying curiosities may lead to fury, to the accusation of being an overturned bookcase. The accumulation of "facts" that are not put into a scientific perspective is nothing but the piling or juxtaposing of bricks for building a house. "A scientist without imagination is like a butcher with dull knives and out-worn scales" (Khalil Gibran). A genuine scientist has a vision; a great scientist has a great vision.

The illusion of being expert on a country probably stems from the capability to understand and speak the language of the people living there and to read their literature, one of the products of the elusive mind of that people. However, both language and literature have become the subject of a scientific discipline the scope of which is widening, the refinement of which is increasing and the analysis of which is deepening. Linguistics is the scientific study of (a certain) language; the discipline is concerned with the investigation of its form, meaning and use. "Literary science" is a *portmanteau* word for all scientific studies relevant to literature, including literary theory sensu stricto, literary criticism, history of literature, sociology of literature and comparative literature. Languages, which are "the pedigrees of nations" (Samuel Johnson), are usually divided into language families, a list of which can be found on the Internet. The literature of a country is written in the language of its inhabitants. It is a beautiful corner in the vast and richly variegated "*jardin d'imagination fleurie*" called literature, i.e., a particular case of an enormous system in many ways related to other systems, such as the linguistic, social, political and economic system of the country. Surprisingly few "experts" on a country are really fluent in the language spoken by the people living there. The "expert" who utters some standard phrases in, say, Chinese will promptly produce expressions of awe and admiration on the faces of his/her audience. One can almost hear the interviewer thinking: "The interviewee in front of me *must* be an expert on China". An old and favorite trick of "country experts" is to put up a linguistic smoke screen to conceal their lack of disciplinary expertise.

This being the case, there is good reason even to question the professionality of "country experts" who claim to be knowledgeable about a certain aspect of the country but cannot show their scientific credentials. For example, asked at which university they got their PhD in literary science, many professors of Turkish, Arabic, Persian, Indian, Hebrew, Russian, Mongolian, Chinese, Korean or Japanese literature will fall by the basket, fancying themselves they can lecture on a topic in a foreign literature without being schooled in the basics of literary science. By saying that university education in literary science is absolutely required before hurling oneself at a particular literature, I certainly do not mean to say that literary theories developed at a Western university are universally valid. As explained in Chap. 2, in order to be a scientific theory and not just an opinion, a theory has to be tested. In science, a theory is considered an explanation of something when based on repeatedly tested and widely established ideas.

Self-respecting scientists have a journal devoted to the history of their discipline. Examples are: *American Journal of Legal History* (1957ff), *Journal of the History of Biology* (1968ff), *Terrae Incognitae* (1969ff), *Journal for the History of Astronomy* (1970ff), *History of Education* (1972ff), *History and Anthropology* (1973ff), *Studies in the History of the Language Sciences* (1973ff), *Historia Mathematica* (1974ff), *Journal of the History of Economic Thought* (1980ff), *Earth Sciences History* (1982ff), *Language & History* (1984ff) and *History of Psychology* (1998ff). Dozens of *books* on the history of science or a particular discipline have been published. Looking at the history of one's discipline is not a case of senseless navel-gazing. It is the best way to see when and why the discipline branched out, how the discipline has matured, which lacunas have yet to be filled, in which direction the discipline's investigations are or should be going, and whether connections with other disciplines have been or might be made. Investigating the history of a discipline is a golden opportunity to consider the possibility of embarking on a cross-disciplinary research project.

Social scientists like Adam Smith, Karl Marx, Herbert Spencer, Vilfredo Pareto, Émile Durkheim, Max Weber, John Maynard Keynes, Bronisław Malinowski, Karl Polanyi, Pitirim Sorokin, Karl Mannheim, Norbert Elias, Alfred Schutz, Leslie White, Eric Voegelin, Talcott Parsons, Karl Popper, Theodor Adorno, Gregory Bateson, Claude Lévi-Strauss, Karl Deutsch, Robert Dahl, Reinhard Bendix, Harold Berman, Kenneth Arrow, Seymour Martin Lipset, Reinhart Koselleck, Shmuel Eisenstadt, Claude Lefort, Clifford Geertz, François Furet, Zbigniew Brzezinski, Charles Tilly, Neil Smelser, Pierre Bourdieu, Mancur Olson, Pierre Clastres, Giovanni Arrighi, Michael Mann, Theda Skocpol, Philip Gorski and James Mahoney have put forward impressive theories, on which "country experts" have been non-professionally drawing as it suits them. However, it needs to be emphasized that these theories have not been tested in a wide range of countries and, if they are tested in a particular country, do not take all the aspects or dimensions of the country into account. Historical sociology and comparative history are thriving branches of research, but the close relationship between the patterns and the processes of natural events and sociocultural actions is a problem still haunting their practitioners.

A radically new kind of country study is needed. Collaborative research based on computational systems science is the one and only viable way—viable in view of the growing literature on e-science, e-research and human–computer interaction.[2] Scientific collaboration across disciplines means: working jointly rather than independently to accomplish a task or project, but it must not be confused with consensus. Collaboration does not mean that everyone must agree before any decision can be made. Nor does it mean that there is no room for individual creativity. Cross-disciplinary research based on computational systems science means collaboration whereby the deepening insight of complex systems science is used in combination with the increasing power of connected computers; it means relying on an infrastructure that consists of the components or operations that somehow play a role in information technology. *Collaborative* computing is a set of information technologies designed to support work between individual researchers. Like *cognitive* computing, it is a new branch on the living tree of computer science. In the future, collaborative and cognitive computing may pool their resources.[3] Scientific collaboration *across disciplines* should go hand in hand with scientific collaboration *across national boundaries* to safeguard against parochialism and ethnocentrism.

Country studies as currently conducted are not only incomplete.[4] Their practitioners are also wrong-headed. They (tacitly) claim to be knowledgeable about a whole country. In actual fact, however, they can, *at best*, only explain part of it. To be well informed about the economy of a country, we turn to economists who, graduated from a reputable university, tested their theories on that economy, and for reliable knowledge about the literature of a country, we turn to literary scientists who, graduated from a renowned university, attempted to apply their theories to the country's literature. Alas, "country experts", who are not graduated in any scientific discipline whatsoever (or are graduated in only one discipline) but talk away about whatever is on demand, are the people television viewers, radio listeners or newspaper readers have to content themselves with and put their confidence in. Having an *explanandum* sine *explanans*, country studies are seriously flawed. County pundits should have the guts to acknowledge this.

My thesis that so-called country experts have no theory of their own and consequently cannot be considered scientists bears the character of a null hypothesis. This is necessarily so. Since the literature on countries is overwhelmingly large (see Appendix C), the thesis is not proven or established. However, *pace* Popper, it might be disproved. In other words, the thesis advanced stands as long as it has not been refuted. My bold and perhaps unwelcome claim is to be upheld so long as my accusations and allegations have not been rebutted by mentioning only one single practitioner of country study who really and demonstrably has a repeatedly tested theory concerning his/her *explanandum*.

Some years ago, I attempted to breathe fresh life into a discussion on "Chinese Studies and the Disciplines". The proceedings of this discussion had been published in *The Journal of Asian Studies* (23: 4 [August 1964], pp. 505–536), but the American editor-in-chief of the journal did not reply to the email in which I tried to revive the discussion. Even a polite reminder remained unanswered, as if I had made a dishonorable proposal or had been trampling the American flag. In this book, the writing of which was instigated by that uncivilized chief editor, I do not only stress the need of country studies being less incomplete by taking the expertise of the greatest possible variety of scientists into account; I also indicate how country studies could be changed fundamentally, i.e., could have an *explanandum* cum *explanans* by cross-disciplinary and cross-national research based on computational systems science.

The reader may have noticed that the book is also foundational to International Relations Studies, as none of these studies can be carried out successfully without giving due attention to the individual countries themselves, or to the constituent parts of a particular area. Not only the *relations* between countries but also the *relata* (the countries themselves) should be studied. The study and planning of the *economic* development of a country involve consideration of its initial conditions, internal factors and external elements. In other words, they take into account (*a*) its natural resources, (*b*) its demand and supply, (*c*) its saving and investment, (*d*) its agricultural, industrial and service sector, (*e*) its governmental revenue and spending, (*f*) its production pattern, (*g*) its inflow and outflow of goods, services and capital, and (*h*) all sorts of institutions. Many factors are thus included in the equation.[5] Students of a country should take into consideration its geographic, geological, biological, demographic, ecological, political, legal, military, social, anthropological, educational, medical, economic, financial, religious and cultural aspect, being brave enough to appeal to the expertise of (other) scientists interested in the country. The study should not only involve the pairwise comparing of different things, domestically or internationally; it should also mean a serious attempt to unify (not: equalize) all the things investigated, to put them into perspective and to indicate how an overarching and testable country theory may look like. Being able to comprehend the complexity of a country depends on the level of abstraction one can achieve, on the capability to see the (nature of the) relations and to connect the (clusters of) dots.

A society is a group of people who share a particular territory, have largely the same culture, and are politically or otherwise organized. Societies are characterized by leveled and connected patterns and processes. In other words, societies can be identified as distinctive complex-and-dynamic networks of networks. A country, we saw in chapter three, is a bordered stretch of land with politically or otherwise organized people who, to some extent, share a history and have a sense of belonging. So the words society and country are virtually interchangeable. A society is essentially a ὅλον (whole, *Ganzheit*), not a πᾶν (total, *Gesamtheit*). Words like "trust", "solidarity", "concord", "harmony", "balance" and "order" are inherent in a society (from Latin *societas* [association, community, fellowship, union]), in contrast to words such as "distrust", "disagreement", "discord", "disharmony", "imbalance" and "chaos". A country or society is basically an order, although there might be temporary disorder. The case I am arguing for hinges on the subtle difference between ὅλον and πᾶν. The book's main thesis that the current study of countries is fundamentally flawed and their future study should take all the dimensions of these *explananda* into account by engaging in collaborative research based on computational systems science is thus premised on the idea that each and every country is a typical example of order. What these different, possibly constitutionally based orders or systems have in common and how they are or could be ordered into a world order is for complex systems scientists to investigate.[6] May the United Nations become a real and peaceful "Family of Nations"!

Should anybody have doubts about the need for, and possibility of, truly collaborative country research, he/she should read *Group Genius: The Creative Power of Collaboration* (Basic Books, 2017), a book in which psychologist Keith Sawyer tears down some of the most popular myths about creativity, revealing that creativity is always collaborative. Alternatively, the doubtful reader could learn something from the book written by Geoff Mulgan and published, in 2018, by Princeton University Press, *Big Mind: How Collective Intelligence Can Change Our World*. Gathering insights from diverse disciplines (an activity one should always be suspicious of), the author reveals how collective intelligence can guide corporations, governments and universities to make the most of human brains and digital technologies. Looking at recent experiments and groundbreaking technologies, he emphasizes that collective intelligence has to be organized and orchestrated in order to harness the intellectual and digital power available. "A smart group is not the same as a group of smart people", Mulgan writes, not only considering why organizations full of intelligent people can make foolish mistakes, but also showing how to avoid these mistakes. Informed by the latest work on big data, Web platforms and artificial intelligence, he cogently argues how a smart combination of humans and computers could cope with the challenges that governments, universities and companies are facing. For that to happen, however, we need new professions, new institutions and new ways of thinking, something the academic establishment may stubbornly resist to. According to Howard Gardner, celebrated author of *Multiple Intelligences: New Horizons* (Basic Books, 2006), Mulgan convincingly shows how collective intelligence can be mobilized to deal effectively and wisely with the urgent problems on our planet. In my view, the misconceptions about countries are the root cause of most of the Earth's problems. The venerated "country experts" are to blame for those misconceptions.

With advanced digital techniques available, it should be possible to mine, collect, process and integrate information from different sources and to approach the ideal of understanding countries. Data can already be stored on a single atom and quantum computing, predicted to dwarf conventional computing, is said to be just around the corner. In 2019, IBM announced its ability to build universal quantum computers for business and science, bringing the systems approach to quantum computing. Under the heading "Quantum Computing Takes Flight" (*Nature*, vol. 574, October 24, 2019), William Oliver, computer scientist at the Massachusetts Institute of Technology, writes:

> A programmable quantum computer has been reported to outperform the most powerful conventional computers in a specific task—a milestone in computing comparable in importance to the Wright brothers' first flights.

Fine days later, Andreas Baumhof, Vice President Quantum Technologies at QuintessenceLabs, uploaded the article "Quantum leap: why the next wave of computers will change the world". For the possibilities of digital communications, the reader should explore the Web site of the World Economic Forum.

Al we need is the will and determination to *apply* available technology to the collaborative study of countries. It will cause a sea change in and far beyond academia. If the USA or the EU delays the decision to take the necessary steps, forward surging China may take advantage of the situation and build a decisive lead. In a provocative article published under the heading "The big idea: one change that could transform universities" (*Times Higher Education*, November 29, 2018), the challenging question was raised as to what single change of university practice would have the most positive and far-reaching consequences. Only a few academics cared to offer their views or succeeded in getting their ideas published in the journal, but none of them suggested that country studies should be interdisciplinary and international exercises rooted in systems science and based on the latest in information technology.

EAI Chinacom 2021, an international conference on communications and networking, presenting "original and fundamental research in the field of communications, networks, visualization and Internet applications", is scheduled for November this year, in Beijing. It would be a matter of great importance, if the West had to give up its strategic leadership in this field.

The advantage of interdisciplinary over multidisciplinary (country) research can be easily demonstrated. Imagine the equilateral triangle ABC, with the length of each edge being 1 km. C can be reached by first going from, say, A to B, and then from B to C. The total distance walked is then $1 + 1 = 2$ km (the distance from B to C is added to the distance from A to B). However, there is a considerably shorter way: going from A to C. At each point on line AC, both a part of the way (line) AB and a part of the way (line) BC will be covered, which—*nota bene*—is not a matter of addition. Arriving at C, the distance walked is only 1 km. Generally, in a triangle ABC, any edge is shorter than the total length of the other edges. This is not a trivial remark—unless one admits that scientists working together are more capable than scientists working separately. Composition goes beyond juxtaposition.

Working this book, I was inspired by the development of interferometry, an important technique in fields such as astronomy, metrology, oceanography, seismology, remote sensing, particle physics and quantum physics. A modern interferometer radio-telescope processes and integrates signals received from different types of antennas, enabling astronomers to look deeper into the universe and further back in time. Huge projects are planned or already in progress. If the knowledge gained and accumulated in different disciplines is regarded as various ways of knowing (*modi sciendi*), multimodality could help to achieve the objective of comprehending a country in and of itself. It could promote information and knowledge integration based on semantic interoperability.[7]

Space science, which encompasses all the disciplines involving space exploration, was also inspiring. Looked at from a geostationary satellite, planet Earth can only be studied one-sidedly, that is to say, partly. However, a *network* of carefully calibrated and intercommunicating satellites can observe the globe as a whole. A single satellite could do this too, but not at one or another time. Interferometry and space science prove that interdisciplinary country research is technically possible.

Cartography is an extensive and growing body of theoretical and practical knowledge used to construct maps as a distinct mode of visual representation (modeling). According to the International Cartographic Association, founded in 1959, cartography is "the discipline dealing with the conception, production, dissemination and study of maps". The *Deutsche Gesellschaft für Kartographie* defines the discipline as "*die Wissenschaft, Technik und Kunst der Darstellung und Kommunikation raumzeitlicher Informationen*". Maps are fascinating, because they tell a story by means of a mix of colors, lines, drawings, symbols, words and figures; they are our eyes on the world and stimulate our imagination. The person who has no historical map gets lost in the past. Since the 1960s, computer-assisted cartography (digital mapping) and geographic information systems (GIS) have been growing by leaps and bounds, enabling users to notice and assess relationships between different kinds of information.[8] There are different kinds of maps. A unitary thematic map shows data of the same kind, e.g., population density, annual rainfall or the rate of a disease. A bivariate map displays two variables on a single map by combining two different sets of graphic symbols or colors. For example, a map showing both annual rainfall and the rate of a disease may be used to explore a possible correlation between the two phenomena. More than two sets of data leads to multivariate mapping. For example, a single map showing population density, annual rainfall and the rate of a disease. Multivariate mapping is a method used by collaborating scientists from different disciplines. Country students could learn a lot from modern mapmakers.[9]

The key question is: can or will software be designed for scientists from different disciplines and countries ready to embark on a collaborative country research project? Given the drive towards mathematical expression and computational solution of complex problems, my hunch is that engineers are, or will soon be, able to design such software. If they think they are not, and will never be, able to do so, I would remind them of René Thom's remark: "*Pour atteindre les limites du possible, il faut rêver l'impossible*". However, whether they are willing, or expected to be encouraged by superiors or outside parties, to design and develop the software, is a matter about which I feel less confident. The computer-aided country research project I envision requires a team that works on a software program which may consist of hundreds of thousands, if not millions, of code lines. Technically, I believe the project is feasible, but I am not so sure whether it is also economically and politically possible. Competition from financially resourceful China may force the West to follow suit.

My belief that a computer-supported and collaborative country-research project is feasible may be ill-founded, however. There is an extensive study of which mathematical problems are computable or solvable and which are not. Hilbert (1862–1943) thought that all mathematical problems are solvable, but in 1930s, Gödel, Turing and Church showed that this is not the case (https://plato.stanford.edu/entries/computability). So the question basically is: Can the complexity of a country be mathematically modeled and is it calculable? If the answer to this question is negative, we do not have to satisfy ourselves with the subjective opinions ventilated by "country experts", because there is a second-best option not to be sneezed at. In my view, countries, having many, evolutionarily or historically molded dimensions, can be better (though not completely) understood through real interdisciplinary and genuine international research, which requires the willingness to cooperate sincerely and integrate scientifically. However, if the answer is positive, the important question rises why no effort whatsoever has been made thus far to model a country's complexity mathematically. Is it because no "country expert" knows what a scientific model is and how powerful it can be (https://plato.stanford.edu/entries/models-science)? See Chap. 2.

Each and every country, irrespective of its size, is a dynamic whole consisting of parts that are intimately related, not only to each other but also to the whole, which itself is part of a constantly changing whole: the turbulent and eventful world. A country is an astonishingly intricate and delicate web of webs with numerous nonlinear relations. Each of its dimensions affects, and is affected by, at least one of the web's other dimensions. No gluing together of partial studies can give a perfect idea of the whole and yet the whole country (including its foreign relations) should be studied, however, crudely that may be. In a globalizing world troubled by many conflicts, an interdisciplinary and international country approach based on general systems science and aided by a fast developing computer science is urgently needed. As the Persian proverb goes, "The ant cannot perceive the pattern of the whole carpet". Andrew Carnegie (1835–1919) put it differently:

> No person will make a great business who wants to do it all himself or to get all the credit.

"Area experts" or "country connoisseurs" are horrible misnomers, because nobody has a poly-ocular vision and each and every one of us is nothing but an ant, though some people may believe otherwise. Whoever lectures on the affairs of a country (not to mention the affairs of an area that encompasses a number of different countries, or Global/International Affairs) runs the risk of being accused of biting off more than he/she can possibly chew, of not realizing how many unknowns are involved, of being way too vague about his/her model,[10] or about the key concepts underlying his/her lecture; in short, of being shamelessly pretentious.

Two kinds of scientific collaboration are conceivable:

(1) **Cross-cultural or <u>international</u> collaboration**

This type of scientific collaboration is fashionable in the era of globalization. Physicists, chemists, astronomers, cosmologists, geologists, geographers, biologists, ecologists, demographers, political scientists, jurists, economists, sociologists, anthropologists, psychologists, linguists, literary scientists, cognitive scientists, neuroscientists, mathematicians, logicians, statisticians, computer scientists and systems scientists: each group has its international association or society that publishes a flagship journal, and members of each group travel to, and participate in, international conferences, colloquia or congresses at touristic places all over the world, of course not at their own expenses but at the expense of the subsidized university or research institute they are affiliated with.

(2) **Cross-disciplinary or <u>interdisciplinary</u> collaboration**

This type of scientific collaboration is more difficult to establish, because it is less glamorous. Coming home with stories about "important" meetings you had in a faraway country is more impressive than telling you had a good conversation with a scientist from another institute on the campus of your university. However, cross-disciplinary collaboration is considerably cheaper and potentially more profitable. Contrary to popular belief, truly interdisciplinary studies are not a step backward to generalization or speculation. The studies try to go beyond the dichotomy between the general and the particular; they attempt to combine width and depth. In other words, scientists engaged in such activities attempt to fully utilize their intelligence, their capacity to "make sensible choices and synthesize". The word intelligence comes from *inter* (between) + *legere* (to choose, to pick out).[11] Cross-disciplinary research based on computational systems science (the discipline where computer science and systems science flow together) falls within the purview of science and technology (S&T). It profits from the strength at both sides. The barriers to such research are not insurmountable, as exemplified by the new climate science. Climate change is recognized as posing one of the most urgent societal challenges of our time, demanding the integration of natural, social and human sciences. Climate science is now subsumed under sustainability science.[12] Computer-aided interdisciplinary research is not only the concern of climate/sustainability scientists. Natural, social and human scientists increasingly realize that cross-disciplinary research strengthened by powerful and interconnected computers is required to be able to cope with the complex problems encountered by humanity. "Country scholars" should be awakened to this fact, that is staring them in the face. Requested to fund their activities, financial institutions should think twice before opening their lock gates. Trusts and foundations should exhort "country experts" to engage in truly interdisciplinary and genuinely international country research. See appendix A.

In October 2016, the International Council for Science (ICSU) and the International Social Science Council (ISSC) agreed to pursue a merger between the two umbrella organizations. A year later, a joint and official announcement was made, saying that in the first half of 2018 the **International Science Council** (ISC), serving as the global voice for science, would be inaugurated. As intended, the Council was inaugurated and is headquartered in Paris. However, the science for which ISC claims to be serving as the global voice is only a part of the science that encompasses the natural, social, human and formal sciences. Whereas the *human sciences*, including, but not limited to, such widely ramified sciences as cultural anthropology, linguistics and literary science, are concerned with the human mind and its manifold manifestations, the *formal sciences*, including, but not limited to, such widely ramified sciences as mathematics, computer science and systems science, are concerned with the relations, patterns and forms of non-formal sciences. The one-time leaders of ICSU and ISSC are to be highly praised for pooling their resources, but the current leadership of ISC seems to have no intention to follow through. A horse which jumps the first fence successfully and then starts nibbling the grass is unlikely to win the race. International Science Council writ large should be speaking on behalf of *all* sciences and not only in the name of some of them. If such a merger were realized, it would be truly momentous for the study of countries. In July 2018, I took the liberty of drawing the attention of ISC's president and president-elect to this point, but the response I received was—well, not very enthusiastic.

In October 2017, the secretary general of the European University Association delivered a keynote speech at the 52nd Annual Assembly of the Latin American Council of Management Schools taking place at California State University, in San Bernardino. She had flown first class and stayed in a first class hotel. With the usual dose of superiority, she reflected on how the wealth of European experience with interuniversity collaboration and internationalization over the past 30 years could help to envisage and develop new models and opportunities for North and Latin American collaboration with European universities. "One way of addressing the uncertainty of globalization is for universities to further develop partnership that enable them to internationalize their programs, enhance their research capacity and strive to help students to become truly global citizens," she said. The secretary general, who graduated from the University of Glasgow and the Institute of Higher European Studies at the University of Strasbourg, was right, but she was not telling the *whole* truth (see Note 4 of this Chapter). Apparently unaware of the groundbreaking work on complexity and interdisciplinarity of Edgar Morin (well known in Latin America),[13] she forgot to tell her audience that international collaboration will only be fully effective when connected with *interdisciplinary* collaboration. For the benefit of the whole world, such a linkage would change universities to what they originally were and hopefully will be soon again: universities.

In *Convergence of Knowledge, Technology and Society* (Springer 2013), Mihail Roco (senior advisor for nanotechnology at the USA National Science Foundation), William Bainbridge (sociologist and co-editor of the online report *Managing Nano-Bio-Info-Cogno Innovations*), Bruce Tonn (international known for his research in the fields of energy and environment policy) and George Whitesides (recipient of the Priestley Medal, the highest honor conferred by the American Chemical Society) present a new model of convergence, suggesting the necessity for a proactive government. Going beyond the convergence of nano –, bio –, info – and cognitive technologies, they draw attention to (*a*) the accelerating and transformative interaction between seemingly distinct scientific disciplines, (*b*) the desirability and feasibility of mutual compatibility and (*c*) synergism and stigmergic collaboration. Their study identifies international opportunities "to develop and implement a program aimed at creating a vibrant community of scientists with different backgrounds, but willing to actively listen to, and learn from, each other". Though the authors of this timely, relevant and readable book do not address the deep divide between (the sciences of) nature and (the sciences of) culture,[14] I hope that their message will not fall on deaf ears among practitioners of area or country studies.

Last year, on May 24, Nobel laureate Murray Gell-Mann, who revealed symmetry and order in the world of subatomic particles, passed away, at the age of 89. Though he was best known for his contributions to particle physics, this eminent scholar-scientist wanted to understand the chain of relationships that connects the universal laws of physics with the complexities of economies, polities, societies, languages and cultures. In *The Quark and the Jaguar* (W.H. Freeman and Company, 1994), he describes these two extremes of interest as "two aspects of nature: on the one hand, the underlying physical laws of matter and the universe; on the other hand, the rich fabric of the world that we perceive directly and of which we are a part". To enable rigorous study of the latter aspect of reality, Gell-Mann co-founded in 1984 the now world-famous Santa Fe Institute, where he closely collaborated with physicists, biologists, economists, political scientists, linguists, sociologists, mathematicians, statisticians and computer scientists who shared his passion for finding the fundamental principle that governs the whole world. Whether Professor Gell-Mann also collaborated with scholars who drew his attention to thinkers in ancient China, I do not know. My experience with other Distinguished Fellows at the Santa Fe Institute frightened me off asking him, but I strongly believe that the *Great Appendix* of the *Yijing* would have interested Gell-Mann. See Note 18 in Chap. 9.

On 11 February last year, Doyne Farmer (director of the Complexity Economics program at the Institute for New Economic Thinking at the Oxford Martin School), Fotini Markopoulou (CEO of Empathic Technologies), Eric Beinhocker (professor of public policy practice at the Blavatnik School of Government, the University of Oxford) and Steen Rasmussen (professor of physics and director of the Center for Fundamental Living Technology at the University of Southern Denmark) published in *AEON* a provocative article under the title "Collaborators in Creation" in which they write:

> While we are fascinated and sometimes frightened by the pace of evolution of physical technologies, we experience the evolution of social technologies differently. Our values, laws and political organisations define and shape our identities. We often regard those who use different social technologies – people from different cultures, regions, nations, religions or those with different values and beliefs – as 'others'. When social technologies change too quickly, we experience a loss of identity, a collective confusion about who we are and how we distinguish ourselves from others. But when social technologies change too slowly, this can create tensions too – for example, when political institutions fail to keep pace with wider changes in society. Physical and social technologies co-evolve all the time, pushing and pulling each other. The influence is in both directions. Physical and social technologies are so entangled that it can be hard to separate them.

Nature and Culture

"Philosophy" was once thought to stand for *libido sciendi* (the knowledge drive dwelling in every human being); for the desire to have a cosmic, all-embracing view, to gather all information and organize or systematize all knowledge and to unify all sciences (*universitas scientiarum*); for "the science of sciences" (Kant's *scientia scientiarum*); for the will to grasp the container of the contained, the background of backgrounds, the setting of settings and the context of contexts; for the yearning "to know what the world in its innermost core holds together" (Goethe, *Faust*), to find "the hidden connections" (Fritjof Capra) and "the hidden pattern" (Ben Goertzel), to see "nature's deep design" (Frank Wilczek). It was supposed to aim for comprehending the elusive *Unum e Pluribus* and for understanding that "strange loop" called man, who is entangled in "*selbstgesponnene Bedeutungsgewebe*" (Max Weber).

After reading Jan Smuts' *Holism and Evolution* (1926), Albert Einstein wrote: "Two mental constructs will direct human thinking in the next millennium: the constructs of relativity and holism." Philosophers in Ancient China conceived of the *Unum*, the "Great Interdependence (of all differences)", as the grand-scale 道理 (*dàolǐ* [process-pattern, flowing order, dynamic structure]). They visualized this magnificent concept in the famous unity-in-diversity, diversity-in-unity, infinity and nonlinearity suggesting black and white *Yīn-Yáng* symbol, the origin of which is shrouded in mystery. The symbol featured on the coat of arms of Niels Bohr, but by coloring it Bohr overlooked that black and red are not the "*contraria*" he rightly thought to be "*complementa*".

Unlike the Western cross (†), the Chinese symbol or diagram on Bohr's coat of arms (☯) cannot be dichotomized. Though it has a bilateral configuration, the symbol is chiral (asymmetrical), like a pair of shoes or a pair gloves. It has no axis of reflective symmetry. The two interrelated and inseparable components of the symbol are isomorphic, but not inverse. While the circle has the highest possible degree of symmetry for a two-dimensional object (the number of symmetry planes around its diameter being infinite) and, in contrast, the left side of the Christian cross is symmetrical with its right side, transformation or mapping by turning the Chinese diagram in space around any imaginary straight line drawn through the center of the diagram plane (*Umklappung* by 180°) does not bring about a mirror image. Though *Yīn* and *Yáng* are complementary, they are not reducible to each other. The *Yīn-Yáng* symbol is significantly different from the cross, the symbol which is crucial in understanding Christianity, indeed in understanding the Western mindset.

There is another and related reason for looking closely at the Chinese diagram. Mathematical logic (the combination of two formal sciences!) is predicated on the semantic principle of bivalence stating that a declarative sentence is either true or false, meaning that there is no other possibility. It is precisely this principle that is violated in the undichotomizable Chinese symbol. When a child is born, it starts dying. Deatch is the *sine qua non* for life, as sunrise is for sunset. Though the mix varies greatly, a man is partly a woman and a woman is partly a man. That's why *Yáng* (the white) contains a bit of *Yīn* (the black dot) and the other way around. The West has no "the A of not-A" (or "the not-A of A"). On the other hand, China has few "either A or not-A". The Chinese way of thought is correlative and relational. It differs fundamentaly from the Western way of reasoning, which tends to be binary (yes/true, or no/untrue, *tertium non datur*). The Chinese, who know that A has not-A as an integral part and not-A has A as an integral part, are more flexible than Westerners. In the Chinese way of though, contrary rather than contradictory reasoning prevails. In his well-written and thought-provoking book *On Chinese Body Thinking: A Cultural Hermeneutic* (Brill, 1997), Kuang-Ming Wu argues that this kind of circularity may be maddening at first, but could benefit both East and West. Read also Kuang-ming Wu, "China as Culture", Philosophy Study, 9:7 (July 2019, pp. 377–386.

Seeing more than two possibilities, the Chinese are both difficult and easy to negotiate with. The growing problems faced by mankind are extremely complex and do not have an "either-or" solution. Sticking to the principle of mathematical logic will not help us. Genuinely interdisciplinary and truly international country research allowing for more than two possiblities is thus the only way-out of the predicament of our own making.

Dàolǐ is (or should be) the most basic concept in category theory, which takes a central position in contemporary mathematics and theoretical computer science by formalizing the concept of structure.[15] In *Great Circle of Mysteries* (Birkhäuser, 2018), Misha Gromov, a world-famous Russian-French mathematician known for his contributions to various mathematical subjects, suggests how the development of mathematical concepts in the spirit of category theory may lead to the unraveling of the mysteries of the world (including the human mind and its manifestations).

Introducing his marvelous, well-documented book, *Patterns in Nature: Why the Natural World Looks the Way It Does* (University of Chicago Press, 2016), Philip Ball, a British science writer who was, for over twenty years, editor of the journal *Nature*, writes:

> The world is a confusing and turbulent place, but we make sense of it by finding order. We notice the regular cycles of day and night, the waxing and waning of the moon and tides, and the recurrence of the seasons. We look for similarity, predictability and regularity, which have always been the guiding principles behind the emergence of science. We try to break down the complex profusion of nature into simple rules, to find order among what might at first look like chaos. This makes us all pattern seekers. (p. 6).

The book (to be read in conjunction with Ernst Gombrich, *The Sense of Order* [Phaidon, 1994]) summarizes and updates Ball's trilogy, *Nature's Patterns: A Tapestry in Three Parts* (OUP, 2009), which elaborated on his book *The Self-made Tapestry: Pattern Formation in Nature* (OUP, 1999). Inspired by the classic study of pattern formation by the Scottish biologist and mathematician D'Arcy Wentworth Thompson (1860–1948), taking a leaf out of Alexander von Humboldt's massive *Kosmos* (1845–1862), but apparently not realizing that it was the Russian chemist Dmitri Mendeleev (1834–1907) who had put everything in place by creating the periodic table of elements, Ball explains why "the windblown ripples of desert sand resemble the stripes of a zebra or marine fish", why "the trellis-like shells of microscopic sea creatures show the same geometry as the bubble walls of foam" and why "the forks of lightning mirror the branches of a river network or tree". He reveals how nature weaves its tapestry without any blueprint, without an orchestrator or mastermind, without a Creator or benevolent "Father in Heaven". Instead, the interactions between the parts—whether they are the subatomic particles, the grains of sand, molecules or living cells—give rise to spontaneous order, a planless system that is at the same time complex and beautiful. Many patterns, he maintains, are universal, everywhere recurring again and again. By describing "the curvature of shells and horns, the hexagons of honeycombs, the leopard's spots, the arrangement of leaves on a plant, the oscillations in populations of predators and their prey, the turbulent eddies in fast-flowing streams, the swirl of rising smoke, the flight of flocks of geese, the branching of trees, a snowflake, cracks and coastlines", Ball reveals how nature forms patterns. In "Living Robots" (https://doi.org/10.1038/s41563-020-0627-6), the author takes the same position. Had he broken out of his intellectual bubble, however, he would have taken the man-made world into account and attempted to explain why the *cultural* universe looks the way it does See Chap. 6.

Ball is not alone in only paying attention to *nature* and forgetting about *culture*, "*das Gewebe von Bedeutungen die es zu interpretieren gilt*".[16] Morphogenesis, the biological processes that cause an organism to develop its shape, is a subject vigorously investigated at many centers, but the cultural factors are disregarded.[17] In his book *The Physics of Life: The Evolution of Everything* (NY: St. Martin's Press, 2016, Chaps. 9 and 11), Adrian Bejan, a Romanian American scientist who has made important contributions to thermodynamics, explains how what he calls the "constructal law" accounts for all forms, structures, configurations, systems and designs in nature, but he leaves out of consideration, or forgets all about, the *cultural* patterns and processes.

On the other hand, the Design History Society, established in London and "welcoming scholars from disciplines such as anthropology, art history, business history, craft history, economic and social history, cultural studies and management studies", is concerned with objects of design in their historical and stylistic contexts. The Society emphasizes *cultural* artifacts *without paying attention to pattern formation in nature.*[18]

The question thus arises whether nature and culture are related to each other and, if they are, how this relationship is to be defined. The question is of great important, because it bears on the nature of, and relationship between, *the sciences* of nature and *the sciences* of culture. I venture the opinion that (*a*) nature and culture are inextricably intertwined and intimately related to each other, (*b*) ontology (the study of being) shall not be dissociated from epistemology (the study of knowing) and, consequently, (*c*) scientists of nature and scientists of culture shall actively listen to each other, because the gap between them is only imaginary.[19]

Modern physics is the study of sensible or instrumentally detectable things that change.[20] The Latin word *natura* stems from the Greek word *φύσις*, primarily signifying *change*. Since *change* presupposes *being* (for there can be no change without being),[21] the sentence "Nature exists independently of the mind" can be said to contain a tautology, because the words nature (the being that is prerequisite for change) and existence have the same meaning. Whatever naturalists or (structural) realists maintain, nature cannot be spoken of, or known about, without mind. At the same time, without nature the mind cannot function. Without certain natural products there would be no Stonehenge, Egyptian pyramids, Taj Mahal, Alhambra, Potala or palace of Versailles, and without sound (air vibration) there would not be the art form called music. One may quibble over what came first or what is more important (nature or culture), but in a trailblazing book published in 1966, Peter Berger and Thomas Luckmann strongly argued for "the social construction of reality".[22] Claude Lévi-Strauss (1908–2009) bemoaned the chasm between the sciences of nature and the sciences of culture, a chasm that—*nota bene*—has been established by culture (the mind) and which culture (the mind) should bridge.

> It is not easy to explain what "*Kulturwissenschaften*" are, or to say what the concern of "*kulturwissenschaftler*"[23] is (see Chap. 1). In a book edited by Heinz Dieter Kittsteiner and entitled *Was Sind Kulturwissenschaften?* (Wilhelm Fink, 2004, pp. 7–23), thirteen answers are given, but it remains unclear what the difference is between culture (singular) and cultures (plural). "*Kulturwissenschaften*", "cultural studies" and "*études culturelles*" are not interchangeable concepts. For "cultural *studies*", see Rivkin and Ryan *op.cit.* pp. 1023–1115. Oddly enough, *The Syntopicon* (1952), a two-volume collection of topics which the editors, Mortimer Adler (1902–2001) and his team, deemed to be the fundamental ideas contained in the famous *Great Books of the Western World*, does not contain an article on culture. The literature on *Kulturwissenschaften*, cultural studies/analysis and *études culturelles* is very large and rapidly growing. See Friedrich Jaeger and Burkhard Liebsch (eds.), *Handbuch der Kulturwissenschaften* (J.B. Metzler, 2004), Tony Bennett and John Frow (eds.), *Handbook of Cultural Analysis* (Sage, 2008), Anne Chalard-Fillaudeau, *Les études culturelles* (*Presses Universitaires de Vincennes*, 2015), Jerry Moore, *Visions of Culture: An Introduction to Anthropological Theories and Theorists* (Rowman & Littlefield, 2018), Peter Trifonas (ed.),

Handbook of Theory and Research in Cultural Studies and Education (Springer, 2020) and the culture-related publications of he UNESCO.[24]

"Experience", "language", "action", "value", "history" and "identity" are the core concepts of cultural studies, *Kulturwissenschaften* and *études culturelles*.[25] Experience is the foundation of both the natural and cultural sciences: no natural science without experience (experiments) and no cultural science without knowledge, the carrier of experience. To reconcile them, one should be clear about the meaning of the word experience. What exactly is experience? Can it be measured? Does it involve more than sense perception? To what extent is the human mind involved? Is experience a kind of representation? What is the scope of experience"? Does it occur in degrees and are there varieties of experience? "Speech is the expression of human experience; writing is the notation of speech", the Chinese say. The difference between actions and events and the discrepancy between ontology (the study of *Sein*) and axiology (the study of *Sollen*) have been pointed out above. Tradition, the way of acting that people in a particular society or group have been following, is interwoven with history. The word identity stems from *identitas* (sameness). It refers to the relation of each thing to itself and indicates the difference with other things.[26] Wearing a uniform, i.e. relinquishing one's identity, is thus not a token of civilized behavior. Animals have no, or only a rudimentary form of, identity sense.[27]

Cultural psychology, which shall not to be confused with cross-cultural psychology, is based on the principle that culture and mind are closely interconnected and mutually constitutive. Culture enables and constrains individual behavior. Human beings are culture-building as well as culture-bound. Cultural constructs are human constructs made possible by Mother Nature. People are shaped and restricted by the culture of their own making. Communication (the social act of conveying meanings by means of mutually understood signs and symbols) and culture (crystallization of the workings of the mind) are entangled with each other. Psychology, sociology and the science (and philosophy) of the human mind form a triangle, from which the natural sciences shall not be far away.[28]

In *Zhēn Yuán Liù Shū* (six books on purity and primacy), a not yet translated philosophical book published during World War II, Feng Youlan (1895–1990) expounds on 道 (*dào* [way, process]), 天 (*tiān* [heaven, nature]) and 大全 (*dàquán* [the great whole]), emphasizing that all is one and one is all. The book, said to be "the most original Chinese philosophy work of the twentieth century", is to be read in close conjunction with *Xīn Wéishílùn* (a new treatise on consciousness-only), a profound book written by Xiong Shili (1885–1968) and translated and annotated by John Makeham.[29] These masterpieces of Chinese thinking elaborate on the central theme of Zhu Xi (1130–1200) and Wang Yangming (1472–1529), the main protagonists of Neo-Confucianism (*lǐxué* [the study of systems]), a subject about which the reader may visit https://www.iep.utm.edu/neo-conf and https://warpweftandway.com (search for Zhu Xi, and for Wang Yangming). "Taiwan New Confucianism" (see Umberto Bresciani, *Reinventing Confucianism: The New Confucian Movement* [Ricci Institute, 2001]) is to be distinguished from "Mainland New Confucianism" (see Ruiping Fan [ed.], *The Renaissance of Confucianism in Contemporary China* [Springer, 2011]). Their central theme, the importance of which has been emphasized by all Chinese philosophers (lately by Shiying Zhang, Shenzhi Li, Ying-shih Yu, Zehou Li, Shu-hsien Liu, Chung-ying Cheng and Chen Lai), is epithomized in the crucial saying 天人合一 (*tiān rén hé yī* [heaven and man correspond]). See Chuang Liu, "*Tian Ren He Yi*: An Ontology for the Quantum World", *Journal of East–West*

Thought, 1:2 (2012), pp. 123–133, Philip Ivanhoe, *Three Streams: Confucian Reflections on Learning and the Moral Heart-Mind in China, Korea, and Japan* (OUP, 2016) and Mike Peng et al., "*Tian-ren-he-yi* Strategy: An Eastern Perspective", *Asia Pacific Journal of Management*, 33:3 (September 2016), pp. 695–722.

> "*Tiān*" refers to the sky, to the firmament and, more generally, to the pristine nature. It should be stressed that "*tiān*" does not denote or mean the realm where people would be going to after death if they had lived a virtuous life. For *Shintō*, the Japanese way of thought which emphasizes the harmonious relations between human beings and natural forces and teaches "to live within nature rather than attempting to dominate it", see Motohisa Yamakage, *The Essence of Shintō: Japan's Spiritual Heart* (Kodansha, 2006). For the concept of *tiānxià* (all under heaven), see Chap. 4, under urban economics. Chinese thinkers knew that laws are needed to keep people in line, but they have never thought man to be totally depraved or utterly sinful and they have never fancied a cleavage between the City of God and the Earthly City.
>
> When Benedict of Nursia (480–543) and Francis of Asisi (1182–1226) spoke about the mutual involvement of nature and man, they were actually speaking about "*tiān rén hé yī*". See Ian Simons, *Changing the Face of the Earth* (Blackwell, 1996, pp. 420–427). In times of environmental degradation and a COVID-19 pandemic, what else does man have to learn than living in harmony with nature again! Nature supports man, so man must support nature in turn. In the words of Teddie Potter (university of Minnesota): "We need to move from domination of to partnership with nature". See Kenneth Gregory et al., *Environmental Sciences: A Student's Companion* (Sage, 2009) and Debora MacKenzie, *COVID-19: The Pandemic that Never Should Have Happened* (Bridge Street Press, 2020). See also Note 33 in Chap. 8.

While Feng Youlan stressed the importance of 天理 (*tiānlǐ* [the pattern/design of nature]), Xiong Shili emphasized the significance of 人心 (*rénxīn* [the human heart/mind]). Zhu Xi, Wang Yangming, Xiong Shili and Feng Youlan struggled with the fundamental problem of reconciling man with nature., as Zhuangzi (339-286 BC) and Dong Zhongshu (179-104 BC) had been doing. They restated the controversy over being (*yǒu*) and nothingness (*wú*) in which great thinkers like Wang Bi, Dharmarakṣa, Guo Xiang, Lu Ji, Dao'an, Zhidun, Huiyuan, Kumārajīva, Daosheng, Kou Qianzi and Sengzhao had been involved in the third, fourth and first half of the fifth century AD, at the time when China was beginning its great Buddhist adventure and Indian texts on the "*prajñāpāramitā*" (perfection of wisdom) were translated into Chinese.[30]

In my study hangs a scroll featuring the beautiful calligraphy of a Chinese friend of mine. It reads: 天理人心 (*tiānlǐ rénxīn* [the pattern/laws of nature and the heart/mind of man]). The characters summarize my lifelong quest for the reconciliation of matter (nature) and mind (culture). Though Western philosophers and scientists are largely and fatefully neglecting this problem, it cannot be denied that man is both a part of and apart from nature; that he is a two-in-one, a mysterious combination of body and mind; that man (with his *pied-à-terre*) and nature can be distinguished but shall not be dissociate from each other; and that the natural and cultural sciences are inextricably intertwined. In the words of Jean Racine (1639–1699):

> Je trouve deux hommes en moi: l'un tout esprit, tout celeste; l'autre par son poids funeste me tient vers la terre penché.

In our time, "being is exalted" (*guìyǒu*) and "non-being is belittled" (*jiànwú*). The body, the perceivable outwardness is exposed in many ways, but the mind, the invisible inwardness is neglected. The tragedy is that those who, like the pre-Socratic philosophers, are gazing at the starry sky or, to quote Oscar Wilde, "who, being in the gutter, are looking at the stars" are rapidly diminishing in numbers. Astronomers and astrophysicists (a long list of their names can be found on the Internet) probe and research *the world outside*, but do not (even try to) fathom *the world inside*, the bottomless self without which no research on heavenly bodies is possible. Urged by profit-seeking university managers who have lost their moral compass, they scrutinize the branches, but neglect to look for and to cultivate the roots. They are reaching out at the stars without wondering about the invisible universe inside, leaving this fascinating subject to colleagues from whom they are estranged and who forget about "looking at the stars".

The controversy (stirred up by C.P. Snow in 1959) over the allegedly unbridgeable gulf between natural sciences and humanities, is not new. The fatal split in the intellectual life of the (Western) world endures; it even endangers the human species. In a brilliant flash of insight, Bertrand Russell wrote in 1961:

> The sciences have developed in an order the reverse of what might have been expected. What was most remote from ourselves was first brought under the domain of law, and then, gradually, what was nearer: first the heavens, next the earth, then animal and vegetable life, then the human body, and last of all (as of yet very imperfectly) the human mind.[31]

Though study of the human mind falls at the end of the list of Western intellectual challenges, it forms the heart of the thriving cognitive sciences.[32] The not yet mended gap between the natural sciences and humanities, as described by the paleontologist Stephen Gould in his final book, *The Hedgehog, the Fox, and the Magister's Pox* (Harmony Books, 2003), is revealed in the ongoing debate between "analytic" and "continental" philosophy. Whereas analytic philosophers such as Frege, Russell, Moore, Wittgenstein, Ayer, Carnap, Quine, Davidson, Dummett, Dretske, Goodman, Hintikka, Parfit, Putnam, Kripke and Williamson (*a*) aim for argumentative clarity and precision, (*b*) draw freely on the tools of logic and mathematics and (*c*) rely on the sciences of nature, continental philosophers like Hegel, Dilthey, Nietzsche, Husserl, Cassirer, Heidegger, Gadamer, Adorno, Merleau-Ponty, Levinas, Bauman, Foucault, Habermas, Serres, Bourdieu and Latour are more concerned with the human condition and its historical dimension. Analytic philosophers accuse their continental counterparts of being insufficiently clear. The latter accuse the former's philosophy of being insufficient (https://againstprofphil.org/2020/02/17).[33] Julian Young, an American philosopher known for his expertise in post-Kantian philosophy, once remarked:

> The continental tradition contains most of the great European thought of the past 200 years. While analytic philosophy has proved of little or no interest to the humanities, the impact of continental philosophy has been enormous. But there is also a great deal of humbug in the continental tradition. That is why there is a powerful need for philosophers equipped with the analytic toolkit to work within the continental tradition — to sort out the gold from the humbug. (*Times Literary Supplement* [July 10, 1998, p.17])

For such things to happen, however, analytic philosophy (driven by logical positivism) has to be renovated, returning to Charles Sanders Peirce (1839–1914), "the greatest American thinker ever" (Bertrand Russell), "one of the greatest thinkers of his time" (Umberto Eco) and founder of semiotics, the field under which he subsumed logic. Peirce was never appointed as professor, because he dared to speak his mind, something that, I experienced firt-hand, is still anathema to many so-called intellectuals. Focusing on the relations between the parts of, and on their relations to, a system, he conceived of semiotics as possibly being the missing link between natural sciences and cultural studies, between the *mundus sensibilis* and the *mundus spiritualis*.

For semiotics, see Roland Posner et al (eds.), *Handbook on the Sign-Theoretic Foundations of Nature and Culture* (Walter de Gruyter, 1997) and explore the websites of the Semiotic Society of America, the International Association for Semiotic Studies and the International Conference on Informatics and Semiotics in Organizations. Paul Cobley and Litza Jansz, *Introducing Semiotics* (Icon, 2012) is an amusing introduction to the "science of signs", but the two-volume book co-edited by Robin Fawcett, *The Semiotics of Culture and Language* (Bloomsbury, 2015) should not be overlooked. Peter Trifonas (ed.), *International Handbook of Semiotics* (Springer, 2015) "brings together many otherwise fragmented academic disciplines and currents, uniting them in the framework of semiotics".

Frdinand de Saussure (1857–1913), Peirce's contemporary and founding father of structural linguistics (Roman Jakobson, Nikolái Trubetzkói, Louis Hjelmslev, Algirdas Greimas, Leonard Bloomfield, Charles Hockett, Robert Hall, Zellig Harris and many other linguists), argued that linguistics is but a part of semiology ("*la science qui étudie la vie des signes au sein de la vie sociale*"). Semiology is itself part of social psychology, an immense field of research that, interfacing with the natural sciences, increasingly encompasses the social and human sciences. In chapter three, I give an example of the interconnectedness of sciences[34] and plead for applying the science of systems to the system of sciences, an exercise that will clear up many problems and wiil be highly profitable for the radically new country study advocated in this book. The reader should realize that linguistics, logic and mathematics constitute *trois champs qui s'entrecoupent*. They form a complex scientific system that is deeply embedded in semiotics and could possibly be tackled by the precision instruments of category theory, which is becoming the hub of mathematics.[35]

The relationship between numbers, words and images is fascinating. Numbers are the concern of mathematicians. Algebra, a generalization of arithmetic, is the study of mathematical symbols and the rules that govern their manipulations (see Chap. 2, symbolic models). According to the mathematician Israel Hersein, algebra is "the unifying thread which interlaces almost all of mathematicsgeometry, number theory, analysis, topology, and even applied mathematics" (*Topics in Algebra* [Wiley, 1975, p. 1]). Words consist of morphemes, which are the smallest meaningful units of a language. Words and their combinations are the concern of semantics, an extended subfield of linguistics, which is a discipline encompassing phonetics, phonology, morphology, syntax, lexicography, discourse and text analysis and pragmatics, subjects that are closely interrelated. See Gillian Ramchand and Charles Reis (eds.), *Handbook of Linguistic Interfaces* (OUP, 2007). Words refer to, stand for, or represent something. Images, finally, are also representations. They are similar to signs and symbols (important in symbolism and symbolic logic). Images are portraits, *Abbildungen, Chiffern* (Karl Jaspers), not the real things. Numbers, words and images dominate our lives; they are nothing but simulacra. See Plao's allegory of the cave in the *Republic* (514a–520a) and Paul Devanandan, *The Concept of Maya* (Lutterworth Press, 1950).

For the possible connection between the sciences of nature and cultural studies, the reader should turn to Robert Brandom, who, in his landmark study, *A Spirit of*

Trust: A Reading of Hegel's Phenomenology (Harvard University Press, 2019), draws on the work of Peirce and attempts to forge a bridge between analytic and continental philosophy. Perhaps one day we will see the confluence of systems science, network science, cybernetics, informatics, cognitive science and semiotics, with the concept of *dàolǐ* (flowing order) being the common denominator. Close collaboration between these disciplines may lead scientists to realize what "ORDER" means and to see what that almost forgotten Vedic word *Ṛta* actually stands for.[36] The scientist's calling is, after all, to create order. May present and future scientists be faithful to that calling!

The Concept of Structure

My argument for computer-aided collaborative country-research rooted in general systems science is based on the concept of structure, which is the core idea of structuralism, an intellectual movement put on the defensive in the social and human sciences, but gathering momentum in mathematics and the natural sciences. The concern of Alexander Grothendieck (1928–2014), the reclusive French mathematician who invented modern algebraic geometry, was not the resolving of particular problems, but the understanding of general structures.

The word structure stems from Latin *struere* (to place or join together, to arrange, to order). It is related to PIE **stere-* (to spread). Structure is essential to systems; a system without structure is no system at all. In mathematics, a structure is a set endowed with some additional features. "*La structure révèle l'analyse interne d'une totalité*" (Jean Pouillon).The word structure underlines the systemic approach to an object or situation. It indicates the presence of at least two constituents, the existence of relations between the elements making up a thing and the ordering of these relations to a system. No element of the *Ganzheit* (as distinct from *Gesamtheit*) of a study-object (e.g., a country) may be considered isolated. Use of the word structure serves to clarify not only the ordering of the parts of a whole but also the permanence of this ordering. "*Yīn <u>and</u> Yáng: that is called the Dào*". "Structure" (viewed synchronically or diachronically), "system" and "organization" are thus essentially interchangeable. Structuralism is the scientific approach emphasizing the importance of structures. Postmodernism, a broad movement that started with the French school called "post-structuralism", is a radicalized version of structuralism.[37]

Formalism (the analysis of forms as a means of evaluating a work of art) has to do with forms, *i.e.* with structures. The word form refers to the internal organization and shape of a piece of art, particularly a text; it stems from the Latin *forma* (form, contour, figure, appearance, likeness, outline, model or pattern) and is cognate with the Greek εἶδος (originally: "the look of something by means of which it is recognizable"). On account of its εἶδος, a thing is known or intelligible as a distinct entity. This cognitive connotation of "form" has been a recurrent theme in Western aesthetics. The book to be read is Ernst Gombrich, *The Sense of Order: A Study in the Psychology of Decorative Art* (Phaidon, 1994). The basic meaning of εἶδος is the internal relationships, the constitution, ordering or organization of something.

"Internal organization" suggests the richer and more varied relationship of a thing to its parts than does the word shape alone. For Aristotle, the form of a thing is what constitutes it as the entity it is. Without form, a thing cannot be known or conceived of; without form, there is only chaos. The form of a work of art is therefore the set of relationships among its parts or aspects that constitutes it as an artistic entity.

For centuries, students of literature had been concerned with everything about literature except the literary *text*. In the first half of the 20th century, that situation changed. To Viktor Shklovsky, Yury Tynyanov, Vladimir Propp, Mikhail Bakhtin, Roman Jakobson and Grigory Gukovsky, the main representatives of *Russian Formalism*, the literary *text* was most important. Influenced by cubism, which itself was a reaction to the idealist movement called symbolism, Russian Formalism began in two groups: (*a*) the Moscow Linguistic Circle, founded in 1915, and (*b*) the OPOJAZ (society for the study of poetic language), founded in 1916 at Saint Petersburg, thus shortly before the Russian Revolution of 1917. Using techniques developed by structuralist Ferdinand de Saussure (1857–1913), the Russian formalists gave priority to form over content or ideas. In 1923, the merger of the groups was dissolved by the upholders of socialist realism. However, the formalist movement continued in Czechoslovakia, where the Prague Linguistic Circle was founded in 1926 by Roman Jakobson, Vilém Mathesius, Nikolai Trubetzkoy and Jan Mukařovský. Eventually, the formalism was adopted throughout Europe and the USA. Jacobson's statement that "the object of study in literary science is not literature but literariness (*literaturnost*)" became the guiding principle not only of new formalism (a late 20th century development in American poetry drawing attention to the verse in terms of meter, rhyme and symmetry), but also of related movements such as structuralism (see above), deconstruction (a form of text analysis largely derived from the work of Jacques Derrida) and semiotics (the science of signs and sign systems). Following his emigration to the United States, in 1941, Jakobson became a naturalized citizen and taught at Harvard University, the Massachusetts Institute of Technology and a number of other universities. Claude Lévi-Strauss (French anthropologist and ethnologist) and Noam Chomsky (the father of modern linguistics) are some of the scientists who acknowledged being deeply indebted to him. Jakobson famously distinguished between the "referential", "poetic", "emotive", "conative", "phatic" and "metalingual" function of a language. Communication scholars such as Theodor Adorno, Gregory Bateson, Karl Deutsch, Marshall McLuhan, Claude Shannon, Thomas Sebeok, Niklas Luhmann and Jürgen Habermas would elaborate on the distinction. Shannon (1916–2001) is known as the founding father of information theory, which should not be confused with information *science* (aka information studies).

To John Crow Ransom, Allen Tate, Robert Penn Warren, Cleanth Brooks, William Empson and William Wimsatt, the main representatives of *New Criticism*, poetry was a special kind of discourse, a particular means of communicating feelings and thoughts. Though poetry differs qualitatively from the language of scientists and philosophers, it conveys meanings that are equally valid. The protagonists of New Criticism, which has dominated the teaching of literature at American universities after the Second World War, also broke with the previous practice of literary study. They tried to formalize poetry, paying special attention to both the structure of a text (close reading) and the associative values of words. In *Theory of Literature* (1949; rev.ed. 1956), a book which was for many years the standard text in higher literature-education, René Wellek and Austin Warren attempted to combine Russian Formalism and New Criticism, but pointed out their neglect of literary history. Monroe Beardsley (1915–1985), a renowned philosopher of art, took it on him to provide New Criticism with a philosophical foundation. For a survey of Russian formalism and New Criticism, see Michael Ryan (ed.), *Literary Theory: A Practical Introduction* (Wiley-Blackwell, 2017, Chap. 1) or Raman Selden et al., *A Reader's Guide to Contemporary Literary Theory* (Routledge, 2017, Chaps. 1 and 2).

"Morphology" stems from μορφή (form) and λόγος (discourse). It refers to the study of word form or structure. Morphology is a branch of linguistics, although the word is also used in astronomy, biology, physical geography, archeology, urbanization studies and mathematics. Linguistic morphologists are mainly concerned with morphemes, the smallest meaningful units in a language. Morphology interfaces with phonetics (the study of speech sounds), phonology (the science of linguistic sound systems), syntax (the study of word combinations and sentence structure), semantics (the science of linguistic meaning) and pragmatics (the study of how a context contributes to meaning). Each of these branches has been the subject of extensive studies. See Ramchand and Reiss, *op.cit.* For synergetic linguistics, which tries to integrate different linguistic laws into a complex model, see Tetiana Dombrovan, *An Introduction to Linguistic Synergetics* (Cambridge Scholars Publishing, 2018). I already dealt with morpho*genesis* and pattern *recognition* (see Notes 17 and 18 of this Chapter).

I am leaving the perennial problem of the relationship between form (*Gestalt*) and content (*Gehalt*), or between ὕλη (matter) and μορφή (form), out of consideration. The interested reader may see Roman Ingarden, "*Das Form-Inhalt Problem im literarischen Kunstwerk*", *Helicon*, 1–2 (1938), pp. 51–67, Georgios Anagnostopoulos (ed.), *A Companion to Aristotle* (Wiley-Blackwell, 2013, pp. 162–185) and Kathrin Koslicki, *Form, Matter, Substance* (OUP, 2018, Part 1). In addition, he/she may visit https://plato.stanford.edu/entries/form-matter and https://philpapers.org/s/hylemorphism. The matter-form problem is the *Leitmotiv* running through the controversial work of the juridical scholar Herman Dooyeweerd (1894–1977), who, challenging Karl Barth (1886–1968), is said to be the Thomas Aquinas of Calvinist theologians. See Herman Dooyeweerd, *A New Critique of Theoretical Thought* (Presbyterian and Reformed Publishing, 1953–1958).

I also bypass the vexed problem of the relationship between form and function. Readers interested in this subject are referred to Horatio Greenough, *Form and Function: Remarks on Art* (University of California Press, 1958), Saunders Mac Lane, *Mathematics, Form and Function* (Springer, 1986), Susan Lambert, *Form Follows Function?* (Trafalgar Square, 1995), Frederick Newmeyer, *Language Form and Language Function* (MIT Press, 2000), Robert Hillenbrand, *Islamic Architecture: Form, Function, and Meaning* (Columbia University Press, 2004), Christopher Breward, *The Suit: Form, Function and Style* (Reaktion Books, 2016) and Magdalena Walenta, *Form-Function Mapping in Content-Based Language Teaching* (Springer, 2019).

In Chap. 9, I made already reference to structural realism, currently taken to be the most defensible form of scientific realism. For the Chinese character that refers to structure or pattern, see Tang Junyi, "The six different interpretations of *lǐ* in the history of Chinese philosophy", *Hsin Ya hsüeh-pao* (the New Asia journal), 1:1 (1955), pp. 45–98, Jeeloo Liu, "The status of the cosmic principle in neo-confucian metaphysics", *Journal of Chinese Philosophy*, 23:3 (2005), pp. 391–407, John Krummel, "Transcendent or immanent? Significance and history of *lǐ* in confucianism", *Journal of Chinese Philosophy*, 37:3 (2010), pp. 417–437, Jana Rošker, "The concept of structure as a basic epistemological paradigm of traditional chinese thought", *Asian Philosophy*, 20:1 (2010), pp. 79–96, idem, *Traditional Chinese Philosophy and the Paradigm of Structure* (Cambridge Scholars Publishing, 2012), Brook Ziporyn, *Ironies of Oneness and Difference* (SUNY Press, 2012), idem, *Beyond Oneness and Difference* (SUNY Press, 2013) and Kai-chiu Ng and Yong Huang (eds.), *Dao Companion to Zhu Xi's Philosophy* (Springer, 2020, pp. 243–263). Also visit https://unizarfilosofia.wordpress.com/2010/10/31/la-notion-de-li-dans-la-pensee-chinoise. I do not know of any study of *lǐ* conducted by a "China expert" who was/is well-informed about the Western concept of structuralism, formalism, morphology, morphogenesis, pattern recognition, or structural realism. Nor do I know of any Western expert in structuralism etc. who is familiar with, and refers to, the crucial Chinese concept of *lǐ*. According to Max Tegmark, "consciousness is a mathematical pattern".

In Chap. 5 (sub 'systems science'), we saw that the *formal* sciences are concerned with the *structure* of physical, biological, psychological, social and cultural sciences. The formal sciences themselves, however, are scrutinized by *formal* ontology and *formal* epistemology. The study of being in so far as it is shared by all entities (ontology) can be traced back to ancient Greece. A list of ontologists can be found on the Internet. Influenced by Alexius Meinong (1853–1920), Edmund Husserl (1859–1938) advanced the idea of *formal* ontology, distinguishing it from *formal* logic. Whereas the latter dealt with the form of scientific theories, he argued, formal ontology would deal with the domain of objects which such theories were supposed to be about. Gottlob Frege and Bertrand Russell tended to blur this distinction, though current analytic philosophers seem to be more aware of it. See Michael Beaney (ed.), *Handbook of the History of Analytic Philosophy* (OUP, 2013) and go to https://plato.stanford.edu/entries/analysis, https://www.iep.utm.edu/analytic and https://philpapers.org (search for "analytic philosophy"). In *Mainstream and Formal Epistemology* (CUP, 2007), Vincent Hendricks demonstrates how mainstream epistemology (the study of knowledge) and formal epistemology "may significantly benefit from one another". See also the *Formal Epistemology Reader* (CUP, 2013), a book he compiled with Horacio Arlo-Costa and Johan van Benthem. In addition, visit https://philpapers.org (search for "formal epistemology").

Systems, or organizations, are inherently patterned and structured. Systems scientists and mathematicians have a keen eye for it. In his famous autobiographical account, *A Mathematician's Apology* (1940), Godfrey Harold Hardy (1877–1947) writes:

> A mathematician, like a painter or a poet, is a maker of patterns. If his patterns are more permanent than theirs, it is because they are made with *ideas*. A painter makes patterns with shapes and colours, a poet with words. A painting may embody an idea, but the idea is usually commonplace and unimportant. In poetry, ideas count for a good deal more; but, as Housman insisted, 'the importance of ideas in poetry is habitually exaggerated. I cannot satisfy myself that there are any such things as poetical ideas. Poetry is not the thing said but a way of saying it.' A mathematician has no material to work with but ideas, and so his patterns are likely to last longer, since ideas wear less with time than words. The mathematician's patterns, like the painter's or the poet's, must be beautiful; the ideas, like the colours or the words, must fit together in a harmonious way. (pp. 13–14)

Category theory (CT) is a fundamental theory in mathematics co-founded by Saunders Mac Lane (1909–2005) and Samuel Eilenberg (1913–1998). Dealing with structures and systems of structures, CT is a powerful conceptual framework, allowing us to discern the relations among the components of a family of structures (e.g., the relations between sciences); it is "unmatched in its ability to organize abstractions, to find commonalities between all kinds of structures and to facilitate communication between different scientific communities". See Chap. 5 and go to https://www.quantamagazine.org (search: category theory). Ralph Krömer, *Tool and Object: A History and Philosophy of Category Theory* (Birkhäuser, 2007) is an accessible monograph devoted to the history of this burgeoning field of research. For a more demanding introduction, see William Lawvere and Stephen Schanuel, *Conceptual Mathematics* (CUP, 2009). For *applied* CT, see Brendan Fong and David Spivak, *An Invitation to*

Applied Category Theory: Seven Sketches in Compositionality (CUP, 2019). For a survey of the field, go to https://plato.stanford.edu/entries/category-theory. In March 2020, the International Centre for Mathematical Studies, in Edinburgh, organized the "Topics in Category Theory" Spring School, where Olivia Caramello, professor at the University of Insubria (Italy), delivered four videoed lectures on Grothendieck's concept of topos (sort of category). Grothendieck's idea and Nishida Kitarō's concept of logic of place seem to be interrelated. It remains to be seen whether category theory can be successfully applied to problems in network science, particularly when a society or country is the subject of study (see Chap. 5 and Note 15 of this Chapter).

For better understanding the constitution, complexity and interdependence of systems, their structures (forming a network of networks) may be considered at different levels interrelating like parts and wholes. The systems are thus analyzed *as if* broken into parts. Each level embraces lower system levels as parts or subparts of a configuration that exhibits emergent properties, which cannot be reduced to the characteristics of any of the lower layers. Each level is an emergence. Biologists distinguish between the structure of a cell, an organ and an organism; sociologists between the structure of a household/family, a community and a society; and political scientists between the structure of a city/municipality, a province/county/*département* and a nation-state (country). Levels of structure can be found in each science and in each country. A science of the complex system of sciences should take this into account, and collaborative country studies should be aware of it. For level differentiation, which can be more or less refined, several criteria may be used that need to be explicitly stated. One should be careful not to make a category mistake, in which something is said of, or believed to be the case on, one level when in fact it belongs to, or is valid for, another. As in map drawing, scale is of paramount importance; its negligence is a guarantee of confusion and misunderstanding.[38] For example, an economic system may be thought of as a hierarchy of layers, each layer or tier having its own processes. The elements of each process form a logical sequence: information → decision → implementation → outcome. The outcome, in turn, may feedback on the information, leading to revision, correction or adjustment of the past course of action.[39]

After the collapse of communism, when the defenders of capitalism claimed victory, the study of economic systems seems to have slipped down on the list of priorities (despite the publication of journals like *Economic Systems Research* and *Economic Systems*), the result being that hardly any Western economist dares to think out of the box (the "hands off" economy). Mao Zedong, who supported the "hands on" economy, was wrong, even terribly wrong, but not totally wrong. His lifetime project, taking the teaching of Confucius a step further and aiming to mold humans into full-blown social beings, tragically derailed, but his (and Aristotle's)idea that man is not an autonomous being or isolated individual should be embraced by anybody who is serious about creating a world *order*. See Andrew Walder, *China under Mao: A Revolution Derailed* (Harvard University Press, 2015), a book reviewed by Dingxin Zhao in *Contemporary Sociology: A Journal of Reviews*, 45:5 (September 2016), pp. 664–665.

Focusing one's mind on the complex and layered structure of a country would not be enough. Due attention should also be paid to its *dynamics*, for there is not only structure, at a certain time; there is also "*Strukturwandlung*", change of structure *through* time, transition and tradition. As explained above, space (the extent in which events or actions occur) and time (their duration) are intimately interrelated; they are, so to speak, the two sides of the same coin. If the conditions (the environmental factors of the system) are favorable, one of the eggs laid on a leaf by a butterfly becomes a larva, which changes into a pupa, which transforms into a butterfly. In the same way, from a fertilized female egg a baby will develop from which a child, growing into an adult, will develop. The adult in turn may find a mate and pass life onto the next generation.

In his *opus magnum*, *Metamorphoses* (book of transformations), the exiled Roman poet Ovid (43 BC–17 AD) narrates the history of the world as a cosmic courtship dance, as an omnipresent, never ending process of change. Repeating Pythagoras, following Heraclitus and anticipating Nietzsche (1844–1900), he proclaims that "*omnia mutantur, nihil interit*" (everything changes, nothing disappears). Essentially the same is written in the first of all Chinese classics, the *Yìjīng* (book of changes). The world is not a mechanism, an aggregate of separable and addable parts, as René Descartes and Isaac Newton mistakenly thought. The world is a dynamic and complex system, the hidden connections of its parts (often systems of themselves) modern scientists are strenuously attempting to discover and explain. What holds for the world as a whole also applies to the individual countries constituting it. The "mechanization of the world picture" (E.J. Dijsterhuis) is now making room for the organic worldview, the view of the ancient Chinese. It would be interesting to lay the masterpiece of Ibn Khaldun (1332–1406), the *Muqaddimah* (introduction chapter of his planned book on world history), alongside Ovid's *Metamorhoses* and the *Yìjīng*, and to compare these great works. The complexity of a whole cannot be comprehended by the sheer aggregation or juxtaposition of the parts it consists of. Complexities can only be comprehended by considering them to be composed, that is to say, by scientists from different disciplines and countries working truly together.

Notes and References

1. Under the auspices of the Council for European Studies (inaugurated in 1970) the 25th International Conference of Europeanists [sic] was held in Chicago, in March 2018.
2. See William Dutton and Paul Jeffreys (eds.), *World Wide Research: Reshaping the Sciences and Humanities* (MIT Press, 2010), Xiaoyu Yang et al. (eds.), *Guide to e-Science* (Springer, 2011), Ned Kock, *Interdisciplinary Applications of Electronic Collaboration Approaches and Technologies* (IGI Global, 2013), Ayse Kok and Hyunkyung Lee (eds.), *Cultural Behavioral, and Social Considerations in Electronic Collaboration* (IGI Global, 2016) and Paul Nikodemus, *Lernprozessorientiertes Wissensmanagement und kooperatives Lernen* (Springer, 2017). Explore the *International Journal of e-Collaboration* (2005ff) and visit https://escience-conference.org and http://www.oerc.ox.ac.uk. For human–computer interaction, see Chap. 7.
3. For collaborative computing, go to http://collaboratecom.org. For cognitive computing, read John Russell, "Mapping Out a New Role for Cognitive Computing in Science", February 25, 2016 (online), go to https://synthexis.com/cognitive-computing and https://www.ibmbpnetwork.com/ibm-cognitive-computing.

4. Being incomplete in providing information is something advertisers and propagandists can be accused of. A half-truth is worse than a lie, because a lie can be detected as being untrue, whereas a half-truth can be made to seem true. A child may come home, cryingly saying that he was beaten by the neighbor's boy, but leaving out that he himself had hit the boy first. Politicians know this kind of tricks, and if they do not know, they surely will be trained how to do such a dirty trick by public relations professionals. One can lie by telling the truth, a curious paradox that many people refue to see.
5. See Gerald Meier and James Rauch, *Leading Issues in Economic Development* (OUP, 2005) and Adam Szirmai, Socio-Economic Development (CUP, 2015).
6. The literature on world order, global society, international law, world peace, utopia and cosmopolitanism is overwhelming. The reader could begin with Reus-Smit and Snidal *op.cit.* pp. 62–83, 259–263, 270–279, and follow up with Frank Sauer and Carlo Masala (eds.), *Handbuch Internationale Beziehungen* (Springer, 2017, pp. 21–49) and Gert Krell and Peter Schlotter, *Weltbilder und Weltordnung* (Nomos, 2018). Evan Luard, *International Society* (Macmillan 1990), Philip Allott, *Eunomia: New Order for a New World* (OUP, 1991), Richard Falk, *The Declining World Order* (Routledge, 2004), Christan Reus-Smit, *American Power and World Order* (Polity Press, 2004), Thomas Metzger, *A Cloud Across the Pacific: Essays on the Clash between Chinese and Western Political Theories Today* (Chinese University Press, 2005), Giandomenico Majone, *Europe as the Would-be World Power* (CUP, 2009), Michael Reisman, *The Quest for World Order and Human Dignity in the 21st Century* (Nijhoff, 2013), Henry Kissinger, World *Order* (Penguin, 2014), John Ikenberry et al. (eds.), *America, China, and the Struggle for World Order* (Palgrave Macmillan, 2015), Bruno Maçães, *Belt and Road: A Chinese World Order* (Hurst, 2018), Michael Mazarr et al., *China and the International Order* (Rand Corporation, 2018), Jeffry Frieden et al. *World Politics* (W.W. Norton, 2018), Ahmet Davutoğlu, *Systemic Earthquake and the Struggle for World Order* (CUP, 2020), Axel Berkofsky and Giulia Sciorati, *Mapping China's Global Future* (Ledizioni, 2020) and the Springer series *Global Power Shift* (2012ff) are recommendable books. The World Federalist Movement advocates "the establishment of a global federal system of institutions subjected to the principles of subsidiarity, solidarity and democracy", but solidarity, the feeling of being connected to others on which justice depends, is a bone of contention and in Chap. 3 we saw that democracy is a hotly debated issue. See also Emilian Kavalski and Magdalena Zolkos (eds.), *Defunct Federalisms: Critical Perspectives on Federal Failure* (Ashgate, 2008), Hans Kuijper, "Wereldfederalisme: Een Wenkend Perspectief" (Dutch), *Eén Wereld*, 43:2 (Summer 2018), pp. 22–25 and Donald Kettl, *The Divided States of America: Why Federalism Doesn't Work* (PUP, 2020). Visit http://www.wfm-igp.org, https://plato.stanford.edu/entries/world-government and https://www.worldgovernmentsummit.org. For the Chinese view about world order, see John King Fairbank, *The Chinese World Order* (HUP, 1968), Kung-chuan Hsiao's book, *A Modern China and a New World* (University of Washington Press, 1975), Yunling Zhang, *Rising*

China and World Order (World Scientific, 2010), Y.T. Kueh, *Pax Sinica* (Hong Kong University Press, 2012), Li Xing (ed.), *The Rise of China and the Capitalist World Order* (Routledge, 2016), Tze-ki Hon and Kristin Stapleton (eds.), *Confucianism for the Contemporary World* (SUNY, 2017, pp. xi-xxix), Ban Wang (ed.), *Chinese Visions of World Order* (Duke University Press, 2017), Yun Zhao (ed.), *International Governance and the Rule of Law in China under the Belt and Road Initiative* (CUP, 2018), John Friend and Bradley Thayer, How China Sees the *World* (Palgrave Mcmillan, 2018), Kai-Fu Lee, *AI Superpowers* (Houghton Mifflin Harcourt, 2018), Xuetong Yan, *Leadership and the Rise of Great Powers* (PUP, 2019), Timothy Brook, *Great State: China and the World* (Profile, 2019), Huiyun Feng et al., *How China Sees the World* (Palgrave Mcmillan, 2019), David Shambaugh (ed.), *China and the World (*OUP, 2020) and Feng Zhang en Richard Ned Lebow, *Taming Sino-American Rivalry* (OUP, 2020).

7. See Claire Lutkewitte, *Multimodal Composition* (Bedford/St. Martin's, 2013), Carey Jewitt et al., *Introducing Multimodality* (Routledge, 2016), Arlene Archer and Esther Breuer (eds.), *Multimodality in Higher Education* (Brill, 2016), Ognyan Seizov and Janina Wildfeuer (eds.), *New Studies in Multimodality* (Bloomsbury, 2017), John Bateman et al., *Multimodality* (De Gruyter, 2017), Markus Höllerer et al. (eds.), *Multimodality, Meaning, and Institutions* (Emerald, 2018) and *Routledge Studies in Multimodality* (2011ff). For semantic interoperability, see Giovambattista Ianni et al. (eds.), *Reasoning Web: Semantic Interoperability on the Web* (Springer, 2017).
8. See the *UN Handbook on Geographic Information Systems and Digital Mapping* (2000), Alexander von Lünen and Charles Travis (eds.), *History and GIS* (Springer, 2013), Bo Huang, *Comprehensive Geographic Information Systems* (Elsevier, 2017) and *International Journal of Geographical Information Systems* (1987ff).
9. See Nicolas Lambert and Christine Zanin, *Manuel de cartographie* (Armand Colin, 2016), Alexander Kent and Peter Vujakovic (eds.), *Handbook of Mapping and Cartography* (Routledge, 2018), Kenneth Field, *Cartography: A Compendium of Design Thinking for Mapmakers* (Redlands, CA: Esri, 2018), Norman Thrower, *Maps and Civilization* (University of Chicago Press, 2018), *Cartography* (1956ff), *The Cartographic Journal* (1964ff), *Cartographica* (1964ff) and *International Journal of Cartography* (2015ff).
10. A model is not a jumble of ideas, but a simplifying representation of a piece of actual or imagined reality. A model can be empirical, theoretical, static, dynamic, discrete, continuous, descriptive, prescriptive, explicative, predictive, deterministic or stochastic. See chapter two.
11. Visit https://arxiv.org/pdf/0706.3639.pdf.
12. Visit https://sustainabledevelopment.un.org, http://unsdsn.org and https://network.futureearth.org.
13. Edgar Morin (1921) is a French philosopher and sociologist whose six-volume *La Méthode* (1977–2004) is highly acclaimed as "an inexhaustible treasure trove of insights". Drawing on systems science, cybernetics and information theory, Morin provides the reader with an alternative to the traditional assumptions and methods of inquiry. He founded and is president of the Association

pour la Pensée Complexe (APC), an organization largely ignored by American complexity scientists. Visit http://www.intelligence-complexite.org.

14. In their book *Von der Quantenphysik zum Bewusstsein* (Berlin, 2016), physicist Thomas Görnitz and psychologist Brigitte Görnitz attempt to bridge the gulf. Mention should also be made by the work of Mark Elliott, founder of Collabforge, a Melbourne-based firm that specializes in "driving collaboration and innovation, both within and across organizations through strategy, facilitation and technology". Visit http://www.stigmergicsystems.com and https://www.edge.org/conversation/seth_lloyd-communal-intelligence.
15. See Stevan Harnad, "To Cognize is to Categorize", in Henri Cohen and Claire Lefebvre (eds.), *Handbook of Categorization in Cognitive Science* (Elsevier, 2017, pp. 21–54).
16. See Leggewie *op.cit.* p. 13 and Jeremy Lent, *The Patterning Instinct* (Prometheus. 2017). For a negative review of the latter book, see Peter Stearns, *Journal of Interdisciplinary History*, 48:4 (2018), pp. 540–541.
17. See Paul Bourgine and Annick Lesne (eds.), *Morphogenesis* (Springer, 2011) and Chaouqi Misbah, *Complex Dynamics and Morphogenesis* (Springer, 2017). Go to http://www.scholarpedia.org/article/Morphogenesis, https://doi.org/10.3389/fphy.2020.00031, www.swemorph.com and http://web.mit.edu/martin-lab.
18. For pattern *recognition*, go to http://www.scholarpedia.org/article/Category:Pattern_Recognition, https://iapr.org, http://www.jprr.org, https://www.worldscientific.com/loi/ijprai and https://www.journals.elsevier.com/pattern-recognition.
19. See Richard Kraut (ed.), *Cambridge Companion to Plato* (CUP, 1992, pp.277–310) and Joerg Tuske (ed.), *Indian Epistemology and Metaphysics* (Bloomsbury, 2017). The Chinese classic, *Book of Change*, is certainly not outdated.
20. Modern physicists would be well-advised to consult (if necessary with the help of those who can read Chinese) the *Zhōnggúo Yìxué Dà Cídiăn* (great Chinese compendium on Yijing studies), a monumental two-volume work compiled by a large team of Chinese specialists and published in Shanghai, in December 2008.
21. The question as to what is primary (being and stasis, or change and flux) was the main bone of contention in the discussion between Parmenides and Heraclitus. Aristotle opted for the view of the latter. See Note 11 in Chap. 4 and visit https://plato.stanford.edu/entries/parmenides and https://plato.stanford.edu/entries/heraclitus.
22. See *Constructivist Foundations* (2005ff). Visit https://plato.stanford.edu/entries/social-construction-naturalistic and https://www.iep.utm.edu/con-ethi. In *Cultural Analysis* (Routledge & Kegan Paul, 1984), a book co-authored by Robert Wuthnow et al., the work of Berger, who was influenced by Émile Durkheim, Max Weber, Georges Gurvitch and Alfred Schütz, is compared with the work of Mary Douglas, Michel Foucault and Jürgen

Habermas. For sociology of (scientific) knowledge, go to http://wissenssozio logie.de and https://plato.stanford.edu/entries/epistemology-social. Giambattista Vico (1668–1744) was a constructivist *avant la lettre*. Karl Mannheim, *Ideologie und Utopie* (1929) is a classic. See Martin Jay, "The Frankfurt School's Criticism of Karl Mannheim and the Sociology of Knowledge", *Telos*, 20 (1974), pp. 72–89.

23. A long list of "Kulturwissenschaftler" can be found on the Internet.
24. See also Note 24 in Chap. 4 and 23 of this Chapter.
25. The Chinese binomes for culture (*wénhuà* [文化] and *wénmíng* [文明]) contain the character *wén* (文), which refers to writing or literacy. Originally, *wén* stood for the stripes of an animal, the eddy or swirling of a fluid, the ripples on the water, or the lines (grain) of wood. It is a cognate of *lǐ* (理), which stood for the veins of jade. The common idea of pattern is thus taken from nature. See preface. Furtheremore, *huà* and *míng* mean transformation and clarity respectively. *Wénlǐ* (文理) means humanities and natural sciences. In the West, culture is distinguished from civilization. See Joachim Gentz, *Keywords Re-Oriented* (Universitätsverlag Göttingen, 2009, pp. 35–40) and Weigui Fang, *Modern Notions of Civilisation and Culture in China* (Palgrave Macmillan, 2019).
26. Go to https://plato.stanford.edu/entries/identity and https://plato.stanford.edu/entries/identity-time.
27. See Kevin Laland and Bennett Galef (eds.), *The Question of Animal Culture* (Harvard University Press, 2009).
28. See David Matsumoto (ed.), *The Handbook of Culture and Psychology* (OUP, 2001, pp. 11–34), Jaan Valsiner (ed.), *The Oxford Handbook of Culture and Psychology* (OUP, 2012), Kevin Laland, *Darwin's Unfinished Symphony: How Culture Made the Human Mind* (PUP, 2017), Young Yun Kim (ed.), *The International Encyclopedia of Intercultural Communication* (Wiley-Blackwell, 2018) and Dov Cohen and Shinobu Kitayama (eds.), *Handbook of Cultural Psychology* (Guilford Press, 2019). Visit https://plato.stanford.edu/entries/culture-cogsci, https://www.ccp.aau.dk and https://www.intercultural-academy.net. In 2001, Brill Publishers launched the journal *Cognition and Culture*.
29. For a review by Jana Rošker, see *Philosophy East and West*, 67:2 (April 2017), pp. 605–611. Makeham does highlight the conjunction between the thoughts of Xiong Shili and Feng Youlan. However, the divergence between their thoughts is remarkable.
30. See Thomé Fang, *Chinese Philosophy: Its Spirit and Development* (Linking, 1981, pp. 157–172), Zenryu Tsukamoto, *A history of early Chinese Buddhism*, translated by Leon Hurvitz (Kodansha, 1985), Paul Demiéville,"Philosophy and Religion from Han to Sui", in Dennis Twitchett and Michael Loewe (eds.), *The Cambridge History of China*, Vol. 1 (CUP, 1986, pp. 808–872), Alan Chan, *Two visions of the Way* (SUNY Press, 1992), Anne Cheng, *Histoire de la pensée chinoise* (Seuil, 1997, pp. 325–393), Erik Zürcher and Stephen Teiser, *The Buddhist Conquest of China* (Brill, 2007), Peter Bol, *Neo-Confucianism in History* (HUP, 2010), Stephen Angle, *Sagehood: The Contemporary Significance of Neo-Confucian Philosophy* (OUP, 2012), Robert Buswell and Donald Lopez, *Dictionary of Buddhism* (PUP, 2014),

Kenneth Ch'en, *Chinese Tranformation of Buddhism* (PUP, 2015) and David Chai (ed.), *Dao Companion to* Xuanxue (*Neo-Daoism*) (Springer, 2020). Go also to https://iep.utm.edu/neo-conf and https://plato.stanford.edu/entries/neo-daoism. For the prajñāpāramitā literature, see Edward Conze, *Perfect Wisdom* (Buddhist Publishing Group, 1973) and Barbara O'Brien, "The Prajñāpāramitā Sutras" (June 25, 2019, online). Visit http://www.chinabuddhismencyclopedia.com (search for Chinese Buddhism and prajñāpāramitā).

31. See Erich Dietrich and Valerie Hardcastle, *Sisyphus's Boulder: Consciousness and the Limits of the Knowable* (Amsterdam: John Benjamins, 2005, p. 6). In the Neo-Confucian treatise *Jìnsīlù* (notes on thoughts about the nearby), it is explained that whoever wants to have knowledge has to search inside his heart, that is to say, at a nearby place, because the laws of nature are the laws written in man's heart. Outside and inside are inseparable; "heaven and man correspond". As student, I did not understand the title of the book, but now I know.
32. See Keith Frankish and William Ramsey (eds.), *Handbook of Cognitive Science* (CUP, 2012), Carolyn Sobel and Paul Li, *The Cognitive Sciences: An Interdisciplinary Approach* (Sage, 2013), Henri Cohen and Claire Lefebvre (eds.), *Handbook of Categorization in Cognitive Science* (Elsevier, 2017), *Cognitive Science: A Multidisciplinary Approach* (1977ff) and Trends in Cognitive Sciences (1997ff). Also visit https://plato.stanford.edu/entries/cognitive-science and https://www.iep.utm.edu/embodcog.
33. For analytic philosophy, see Michael Beaney (ed.), *Handbook of the History of Analytic Philosophy* (OUP, 2013) and visit https://www.iep.utm.edu/analytic, https://plato.stanford.edu/entries/analysis, http://analyticphilosophy.eu, http://sshap.org and https://plato.stanford.edu/entries/latin-american-analytic. For "continental philosophy", see Simon Glendinning (ed.), *The Edinburgh Encyclopaedia of Continental Philosophy* (Edinburgh University Press, 1999), Brian Leiter and Michael Rosen (eds.), *Handbook of Continental Philosophy* (OUP, 2007) and Alan Schrift (ed.), *The History of Continental Philosophy* (University of Chicago Press, 2011). Also visit https://www.ascp.org.au, https://scot-cont-phil.org and https://www.c-scp.org. See further Hao Wang, *Beyond Analytic Philosophy* (MIT Press, 1985), Richard Cobb-Stevens, *Husserl and Analytic Philosophy* (Springer, 1990), Tiziana Andina (ed.), *Bridging the Analytical Continental Divide* (Brill, 2014), Guillermo Haddock (ed.), *Husserl and Analytic Philosophy* (De Gruyter, 2016), Jeffrey Bell et al. (eds.), *Beyond the Analytic-Continental Divide* (Routledge, 2016) and Maria Baghramian and Sarin Marchetti (eds.), *Pragmatism and the Europen Raditions: Encounters with Analytic Philosophy and Phenomenology before the Great Divide* (Routledge, 2017), but first and foremost see the *opus magnum* of Robert Brandom, *A Spirit of Trust* (HUP, 2019). If Brandom's central idea of "objective idealism" can be supported (after critical reviews by seasoned philosophers at both sides of the "analytic-continental divide"), it would provide the possibility of a fruitful dialogue between Anglo-Saxon and Indian philosophers. Even more interestingly, it could possibly pave the way for a triangular dialog between Anglo-Saxon, Indian and Chinese philosophers, as philosophers in ancient China paradoxically thought CHANGE to be the

essence of BEING. A thorough study of the work of Chinese philosophers from the viewpoint of analytic philosophers is a great *desideratum*. See JeeLoo Liu, "Converting Chinese Philosophy into the Analytical Context", October 2009 (online). The first problem to be solved, however, is whether there is such a thing as Chinese philosophy. In *Philosophy East and West*, 51:3 (July 2001), pp. 393–413, Carine Defoort (KU Leuven) offers a not universally shared answer. See Note 22 in Chap 1.

34. See further Alexander Christian et al. (eds.), *Philosophy of Science* (Springer, 2018). The book co-edited by Simon Lohse and Thomas Reydon, *Grundriss Wissenschaftsphilosophie* (Meiner, 2017), "*bietet einen einführenden Überblick über den gegenwärtigen Forschungsstand der verschiedenen Philosophien der Einzelwissenschaften*".
35. See Brendan Fong and David Spivak, *An Invitation to Applied Category Theory: Seven Sketches in Compositionality* (CUP, 2019).
36. See Benveniste *op.cit.* pp. 99–105. Visit http://www.iranicaonline.org/articles/asa-means-truth-in-avestan. The object of mathematics, indeed of all sciences, is to create order out of chaos. In *Natural Law and Natural Rights* (Clarendon, 1980, pp. 136–139), John Finnis defines order, of which he identifies four types, as "a set of unifying relationship". The journal *Order* (1984ff) is concerned with the theory of ordered sets and its applications.
37. See Jean Piaget, *Le structuralisme* (PUF, 1969), Dan Sperber, *Qu'est-ce que le structuralisme?* (Seuil, 1973), Alfred Schmidt, *History and Structure* (MIT Press, 1981), François Dosse, *Histoire du structuralisme* (La Découverte, 1991–1992), Hans-Heinrich Lieb (ed.), *Prospects for a New Structuralism* (John Benjamins, 1992), Wolfgang Balzer and Carlos-Ulises Moulines, "Structuralist Theory of Science", *Erkenntnis*, 51:2–3 (November 1999), pp. 353–356, Maxime Parodi, *La modernité manquée du structuralisme* (PUF, 2004), Jonathan Culler (ed.), *Structuralism* (Routledge, 2006), Kathrin Koslicki, *The Structure of Objects* (OUP, 2008), Rivka and Robert Oxman (eds.), *The New Structuralism* (Academy Press, 2010), Alisa Bokulich and Peter Bokulich (eds.), *Scientific Structuralism* (Springer, 2011), Peter Marti, *Theory of Structures* (Wiley, 2013), Johannes Angermuller, *Le champ de la théorie: Essor et déclin du structuralisme en France* (Hermann, 2013), Steven French, *The Structure of the World* (OUP, 2014), John Burgess, *Rigor & Structure* (OUP, 2015), Herman Hertzberger, *Architecture and Structuralism* (Nai010 Publishers, 2016), Edith Kurzweil, *The Age of Structuralism* (Routledge, 2017) and Karl-Eugen Kurrer, *The History of the Theory of Structures* (Wiley, 2018). For postmodernism, see Lawrence Cahoone (ed.), *From Modernism to Postmodernism* (Blackwell, 2003, Part III), Steven Connor (ed.), *Companion to Postmodernism* (CUP, 2004, pp. 1–19), Gerhard Hoffmann, *From Modernism to Postmodernism* (Rodopi, 2005, pp. 13–32), Tim Woods, *Beginning Postmodernism* (Manchester University Press, 2010), Stuart Sim (ed.), *Companion to Postmodernism* (Routledge, 2011, pp. vii-xiv), Brian McHale, *Introduction to Postmodenism* (CUP, 2015, Introduction) and Godfrey Ozumba et al. (eds.), *Critical Essays on Postmodermism* (ERIC, 2017, pp. 1–19).
38. See Geoffrey West, *Scale* (Penguin, 2017).
39. See Václav Holešovský, *Economic Systems: Analysis and Comparison* (McGraw-Hill, 1977, pp. 22–25).

Chapter 11
Summary and Suggestion

Abstract For the reader's convenience, the subjects covered in the book are listed. A practical suggestion is made to solve the basic problem I noticed. The need to apply the science of systems to the system of sciences is re-emphasized. Remarks are made and reasons are given for interspersing the text with references to the (ancient) Chinese way of thought.

This book is about comprehending the complexity of countries. "Country", "complexity" and "comprehension" have been investigated, and the subjects have been found to be interrelated. Countries are complex and connected parts of a changing reality (world), but this reality is certainly not limited to what natural scientists are dealing with. Countries are complex units/systems, consisting of many, intricately related complex subsystems and having numerous dimensions. Only when this is taken into account and indigenous experts are involved in the research, countries can be comprehended. For this purpose, connected computers could be deployed (digitized libraries). Subsumable under the category of philosophy of sciences, the book is primarily meant for academics concerned with study of (the geological, geographical, biological, demographic, ecological, political, legal, military, economic, financial, social, psychological, medical, educational, linguistic, literary, religious, philosophical, paleontological, archeological or cultural aspect of) a country. In case my meandering argument has not retained the reader's attention, it may be useful to provide a summary and to spell out the gist of the book. The main subjects covered are the following.

How Countries Have Been Studied

Turkologists, Japanologists, Koreanologists, Sinologists, Tibetologists, Mongolists, Russianists, Indologists and similar so-called country experts have, by definion, committed themselves to explaining a country (their *explanandum*), but they have failed to develop and come out with a distinctive tested theory that does the explaining. They do not have an *explanans* of their own. All they can do is to pick and choose

H. Kuijper, *Comprehending the Complexity of Countries*,
https://doi.org/10.1007/978-981-16-4709-3_11

ideas by trespassing on disciplinary domains and to show off other people's feathers. "Country experts" have the bad habit of "borrowing" a theory from one of the social or human sciences to do the explaining they miserably fail to do themselves. In doing so, they make a big mistake. For example, without being schooled in economics, sinologists are inclined to resort indiscriminately to an economic theory—not only to explain the economic development of China, but to explain the state of affairs in China *tout court*. Worse still, they often use a heady brew of theories advanced in the social and human sciences to picture and supposedly explain the situation in China, that is to say, theories that are not or insufficiently tested in the country. Their language skill often serves as a screen concealing this scientifically shameful behavior from a critical look.

Self-styled country experts are not the only culprits of such behavior. They can also be, and are in fact often, found among scientists such as economists, political scientists, sociologists, psychologists, legal experts, geographers, demographers, statisticians and ecologists who leave their home turf and feel called to let the world know about their "expertise" in regard to a country on which they never tested their own or borrowed theories.

The change from "Turkology", "Japanology", etc., to "Turkish Studies", "Japanese Studies", etc., is misleading on two accounts. By definition, "Turkish Studies" are studies carried out by *Turkish* scientists or scholars on some subject (not necessarily a subject related to Turkey). If "Turkish Studies" is a substitute for "Turkology", the question arises what kind of study in regard to Turkey is meant: a geographic, demographic, linguistic, literary, economic, political, legal, military, social or another kind of study? "Professor of Chinese, or Russian, Studies" sounds like "Professor of Social, or Human, Sciences", almost empty designations, for there are many and widely different social, or human, sciences. Nobody is or will ever be a consummate and seasoned professional in all of them.

An expert is someone who is educated and has prolonged or intensive experience in a particular field—"educated" meaning: has acquired well-organized knowledge. A person who knows all kinds of things in regard to a country is not necessarily an expert. What he/she misses is scientifically based, well-grounded knowledge, a tested or at least testable theory, an *explanans*. If knowing many things concerning a country would be a sufficient condition for being rightly called country expert, every visitor of, tour guide to, or newspaper correspondent stationed at, the place would qualify.

What is a Country?

A country is a geologically formed, politically organized, juridically defined, militarily defended, economically regulated and socio-culturally ordered populated stretch of land in an area or region of the world. For example, Syria is a country in the area called the Middle East, Bolivia is a country in the region Latin America and Korea is a (divided) country in the area usually indicated as East Asia. A country is a geographical area, but a region consists of different countries. The differences of countries in the same region are noticeable. Both states and nations are indicated by the word country. The different stages preceding, or leading to, the formation of a state have been left out of consideration. The literature on this subject is substantial and heterogeneous.

Countries and cities have more in common than usually realized. They are both, and to a great extent, human constructs with different aspects and interrelated dimension. Their identity is undeniable. Bolivia is not the same as Botswana, and Peking will not be mistaken for Paris. The history of a city and the history of a country are interconnected. The city was the seat of ancient civilizations, and cities came into being with the country or culture they belonged to. Cities are the focal points of a country and carry its most salient characteristic. Spanish cities are first and foremost Spanish cities, quite distinct from, say, Japanese cities, which epitomize Japan's individuality. Countries consist of a number of cities or towns and their countryside (land outside of them). These surrounding rural areas are dotted with villages and hamlets connected to the urban areas. Like countries, cities are complex and layered systems of systems, not to be studied by a single scientist (who can only be disciplined in one science) but by a team of scientists from various disciplines. Like countries, cities vary in size, site and character.

A country is a woven web, a fabric with a warp and a weft, a system of traditions, a diachronic and synchronic whole, a kind of organism with interrelated cells (human individuals having different ways or styles of living). Each aspect of the country affects, and is affected by, all other aspects. No partial study can give a perfect idea of the layered whole. "Part of a country" may refer to one of the country's geographical or administratieve areas; it may also stand for one of its aspect or dimensions. A country's part must be situated in its context: the whole country, including its ever-evolving foreign relations. All parts should be studied as parts of a whole; no part, aspect or dimension should be investigated or analyzed in isolation. A doctor investigating a little finger without being interested in the functioning of the whole body is not a good doctor.

A country is a constantly changing composite of codes to be deciphered. It is a complex, dynamic and adapative system of systems, which are geological, ecological, geographic, demographic, political, legal, military, economic, financial, social, psychological, educational, medical, linguistic, literary, artistic or religious in character. A country is a *system*, because, surrounded by an environment (foreign countries); it is composed of interrelated parts or subsystems; it has a network structure, with nodes or dots that are and should be connected. A country is an *adaptive* system, because the members of its population are learning, and consequently changing their behavior. Actions have *reasons*, which are difficult to explain exhaustively. In contrast, events have *causes*, which are difficult to map comprehensively. Actors have one or more than one motive, often unknown or insufficiently known to the actors themselves. A motive (reason for doing something) should not be confused with a motif (idea, subject, theme or image regularly repeated). For example, "The suspect had no motive for the crime" vs. "Freedom is a central motif of the novel".[1] While events are *natural*, actions are *cultural* happenings. Nature and culture are closely connected and inextricably intertwined. To treat actions like events or the other way round is making a "category mistake" (Gilbert Ryle). A country is also a *dynamic* system, because it changes all the time. In addition, a country is a *complex* system, because it has many different parts that are nonlinearly related. "Complex" is not the same as "complicated". Trees and human beings are complex, cars and clocks are complicated systems.

A country is an *individuum*, something that cannot be divided into parts without losing sight of its identity. It is not a *Gesamtheit* (aggregate) of elements just to be added up for being understood, but a constantly changing, living *Ganzheit* (whole) that needs to be looked at from some distance by abstracting from concrete cases and particular instances. Each country has a unique form, not only meaning that it has a particular physical configuration (to be investigated by natural scienists). It also implies that the country is a distinct human construct (to be studied by cultural scientists). If anywhere, it is especially here that natural scientists and cultural scientists have to collaborate with each other. A country without physical configuration does not exist, and a country without culture is inconceivable. Myths or symbols are as real as mountains or rivers. Each country has a face and every part of the country "*ritiene in sé la natura del tutto*" (Leonardo da Vinci).

Being largely a human construct related to its physical appearance, a country is a highly intricate mix of different kinds of arrangements. Some of them are spontaneous (dictated from the *inside*); others are man-made (dictated from the *outside*). Organisms[2] are spontaneous, natural orders, but countries, like companies, are founded or managed by man. Blades of grass, worms, bees and butterflies are not made; casinos, stethoscopes, rifles and robots are. The ruler or government of a country may be well advised to be fully aware of this difference and to take it into account when policies are implemented or enforced. The view often prevails that everything is, or will be, arrangeable and fixable. Science and technology (the application of scientific knowledge) are believed to solve all problems sooner or later. Some people believe that a theory of everything is just around the corner, that technological singularity is coming, that computers will replace man, and that there will be no place for wondering anymore. In short, man (the *Rechner*) will be master of the universe. Erik Brynjolfsson and Andrew McAfee, respectively, director and co-director of the MIT Initiative on the Digital Economy (IDE), want us to believe that "in the coming decades, exponential improvements in artificial intelligence, robotics, network analysis and digitization will increasingly affect the economy and society". I am not convinced that all these things are going to happen, at least not very soon. Human development has been proceeding in a way far beyond anybody's imagination. Human beings often act illogically. Who predicted the COVID-19 pandemic and its devastating effects on economy and society! When the present leadership, advisory board, fellows and staff members of IDE are all dead and gone, and when they cannot be called to account anymore, the world may look quite different from what they imagined.[3] A computer capable of tying my cravat or the laces of my shoes is yet to be invented. What I do believe, however, is that the ancient Daoist idea of spontaneity is worth of being thoroughly investigated by collaborating scientists.

Sinologists are unable to do this, because—their language skill notwithstanding—they are not scientifically qualified; their philological approach is old-fashioned and inadequate.

How Countries Should Be Studies

I take sides with the linguist interested in the Chinese language, not with the Sinologist who lectures on it, even though he/she is assumed (!) to speak and read Chinese fluently. I support the literary scientist taking an interest in the Persian literature, not the Iranologist who elaborates on it. I prefer to listen to the economist who has special knowledge about the Russian economy, rather than to be invited to attend a meeting where a prominent "Russia expert" will cover the subject. Assuming that the linguist, literary scientist and economist know when to give the floor to a colleague from another discipline.

Each and every country has aspects or facets that have been historically evolved. For practical purposes, these aspects can be distinguished, but they shall never be isolated. Among the potentialities of a nation-state, there are "*Bedingtheiten*" (Jacob Burckhardt). These mutual dependencies ought to be carefully investigated. Walking is moving along by putting one foot in front of the other, allowing each foot to touch the ground before lifting the next. By investigating the movement of only one foot of a walking person, one never understands the process of walking. A country's geology, geography, demography, polity, economy, society and ecology can only be *conceived of as* being separated from each other. They can be analytically distinguished, but they are inseparable, because they are intimately interlinked. Separating or dissecting the *explanandum*, breaking it down into bite-sized chunks would be like killing an organism to understand life, dividing a dance into different steps to value dancing or breaking a melody up into notes in order to enjoy the music. The complexity of, and the convoluted relationships between, a country's internal (domestic) and external (foreign) interactions can thus not be ignored or neglected. Its mentally distinguishable but physically inseparable dimensions are conjoined. Thirty two panels sewn together make a football. Eleven players trained together form a football *team*. It takes two to tango, two parties to make war or peace, two people "to make love", a seller and a buyer to negotiate. "*E Pluribus Unum*" is the motto of the USA. The reader should judge for himself how seriously these words are taken in time of COVID-19. To be is not to-be-divided but to-be-united. Distinctions are nothing but mental conceptions, human constructs which are culture-specific. In science, no division or subdivision would be meaningful if things are taken out of context. Distinctions are to be compared; they ask for being abstracted from, for being generalized. This requires researchers who are open-minded, scientists who are willing and able to think out of their little box, scholars who are ready to truly collaborate with each other.

Each country is a member of the group of nations said to have, under international law, equal status. A country is a multifaceted whole, having many aspects, which can be intellectually distinguished but shall never be separated. A country is a distinctively articulated unit, having many dimensions and maintaining all kinds of relations with other countries, which have also numerous interrelated dimensions and foreign relations. For example, the United States of America, the Republic of India, the Federative Republic of Brazil, the People's Repubic of China, the Commonwealth of Australia and the Republic of the Marshall Islands are widely recognized sovereign countries. Yet their differences are unmistakable. It is deeply regrettable that, at the UN Climate Change Conference, in November 2021 (COP26), the participants, who flocked to Glasgow by polluting aeroplanes, were insufficiently aware of this basic fact. Don't eat cherries with great lords—they'll throw the pits in your face!

The reality of a country is more than what natural scientists are concerned with. It also includes the realm of culture, which emerges from, but is irreducible to, nature. The human face (culture) of a country is its "*speculum mentis*". A country's culture is the manifestation, expression or crystallization of its mind, which cannot be measured or quantified, because measuring and quantifying presuppose somebody having a mind, a "*Lebenswelt*" (Edmund Husserl).[4] Since the mind (culture) is part of reality and cannot be measured or quantified, I disagree with Max Planck, who reportedly said: "Reality is only what is measurable". The mind of the measuring person is as real as his or anybody else's measurable body. The AI-community, currently the center of worldwide attention, seems to forget that machine learning is ultimately based on machine *teaching* (programming), which is not an event, but the action a human being. An algorithm is inherently amoral, not immoral; if it leads to something morally wrong, the writer of the algorithm is to be blamed, because he behaved immorally. Computers, i.e., their tangible, dead parts or components called hardware, run on software, the set of instructions or orders (given by man) that tells a computer how to do a particular job and controls what the machine is doing. Robots (will) never *act*, because they are mindless. An atomic bomb is not randomly dropped by a bomber, but purposefully dropped by a trained and brainwashed pilot, that is, by the uniformed commander of the pilot, that is, by the political leader of the commander, that is, by people having chosen and following the leader, who is skilled in the art of lying by telling the truth ("politics is poly tricks"). So who is responsible for the killing and destroying? Certainly not the bomb or the aircraft carrying it! We are still far from the realization of Leibniz's dream: the *characteristica universalis* and the *calculus ratiocinator*.[5] The extent to which epistemic access to the world is possible has long been and will probably always remain an unsolved problem. The reality that scientists attempt to model is an impoverished reality. Indeed, it is an emasculated and castrated reality, a reality devoid of spirit. The *real* reality is and will remain elusive. Things can be pointed at or indicated, but reality cannot be shown or measured. Understanding understanding is on top of the list of unresolved problems. In his beautiful novel, *Machines Like Me* (Penguin, 2019), Ian McEwan asks "whether a machine can understand the human heart". If the present is both where the past ends (without being the past) and the future begins (without being the future), it is extremely difficult, if not impossible, to know, understand and describe the structural features of the present, wherever one imagines the present to be.[6]

Regrettably, physicists tend to forget the words of Blaise Pascal: "*Le cœur a ses raisons que la raison ne connaît point*".

A country cannot be explained by a single person, as the pattern of a carpet cannot be seen by an ant. Whoever attempts to explain a country single-handedly bites off more than he/she can chew. He/she is bound to paint an incomplete, if not distorted picture. Foreign policies based on such a picture are doomed to fail. Wrong policy based on poor country scholarship can cause widespread sufferings and bring about great costs! Since the Afghanistan scandal was revealed, "Afghanistan experts" have been wrapping themselves in deafening silence.

A country must be studied interdisciplinarily, by a team of scientists, each of them focusing, or zooming in, on the aspect about which he/she, schooled in a scientific discipline, could possibly say something meaningful, while keeping things in perspective and "taking a crude look at the whole" (Murray Gell-Mann). The team members listen *actively* to, and *regularly* compare notes with, each other. Interdisciplinary and multidisciplinary research should not be muddled up. *Multi* (various and of many kinds) is not the same as *multum* (comprehensive and thorough). With multidisciplinary research, the parts are not considered as parts of a whole. The equation variables are not provided with a coefficient. Whereas interdisciplinary studies result in a composition, multidisciplinary studies will end up with delivering nothing but a juxtaposition of different views. A cathedral is not a pile of stones, a symphony is not a number of notes, and a river is more than an aggregate of drops of water. In contradistinction to what late professor and Nobel laureat Paul Samuelson wanted us to believe, an economy is not the sheer aggregate of individual economic activities. The whole is more than the sum of its parts. Country studies should thus be based on general systems theory.

When studying a country, it is important to keep in mind that the whole and its parts belong together, because they are interdependent and mutually complementary or reinforcing. Countries being studied are like novels being written. A sentence consists of, but is more than the sum of its, words and the novel consists of, but is more than the aggregate of its, sentences. Coherence, pattern and style confer character and individuality upon the novel. The close connection between what seem to be different things is mysterious. Like *Yin* and *Yang*, on and off, day and night, winter and summer or mountain and valley, "*les extrêmes se touchent*". At the ridge of the roof, the two different sides unite, covering the whole house.

Science must be distinguished from philosophy. Whereas scientists theorize about a *part* of reality, leaving everything that lies outside of their *explanandum* for other researchers to explain, philosophers are concerned with reality (everything), or the epistemic relationship to it, being anxious that their argument meets the requirements of logic (which or what kind of logic is still a mater of dispute), and being aware that informal logic often prevails in real life. In contrast to specializing scientists, who study something (a particular thing) and hope to find a theory of everything, philosophers attempt to provide a plausible view about the whole of things (reality) under which anything (any one thing) can be subsumed. The unity of science is the greatest challenge faced by scientists. Taking a point of view outside of the world of people (including philosophers) who are *in* it is the awesome challenge lying ahead of philosophers. While scientists are on the way from the many to the one, philosophers try to reconcile diversity/difference and unity/similarity by walking in opposite direction.[7] "Classicism", Oscar Wilde once pointed out, "is the subordination of the parts to the whole; decadence is the subordination of the whole to the parts". However, there is no subordination but complementarity. Scientists and philosophers need each other; they can learn something of great importance from each other. The book to be read is Michael Sidiropoulos, *The Mind of Science: From Aristotle to Einstein* (CreateSpace, 2015). Its author explores the intersection of science and philosophy in the pursuit of knowledge.

Philosophy of science, under which this book is to be subsumed, is a thriving field of research.[8] Science is an ever more fragmentizing kind of business with specific rituals, rules and regulations. It is divided into natural, social, human and formal sciences. Whereas the natural, social and human sciences are concerned with substance, the formal sciences (including but not limited to mathematics, logic, computer science and systems science) pay attention to the formal side of the other (non-formal) sciences, being stimulated by them and urging them to be as exact as possible. There is a growing tendency in the natural, social and human sciences to work together (see Appendix A), but collaboration in the formal sciences leaves much to be desired. Systems scientists should know their restrictions and team up with mathematicians, logicians, statisticians, decision theorists and computer scientists. I consider a study of the-system-of-sciences-and-the-science-of-systems to be a *desideratum*. There is a way to make the general systems theory, as outlined by Ludwig von Bertalanffy (1901–1972), the unifying framework for all sciences. Indeed, there is a road to realize the great dream of Aleksandr Bogdanov (1873–1928), as described in his massive groundbreaking but marginalized (by Lenin) masterpiece *Tektologiya* (general organization science), the assumption being that the difference between natural and cultural sciences is adequately taken into account.

A country must not only be the object of study by a team of scientists disciplined in different areas of research; it must also be studied *internationally*, by scientists from various countries, in particular by scientists from the country being, or to be, studied. Interdisciplinary and international studies must be interconnected, because they can reinforce each other. For example, if Russia is the country to be studied, American economists who focus their mind on the Russian economy should collaborate with internationally renowned sociologists, geographers, demographers, political scientists, legal experts, military scientists and ecologists who have an interest in Russia from *their* point of view. International research prevents the study of a country from becoming ethnically or nationally prejudiced.

The study of a country should be analytic as well as holistic. It should pay attention to both the parts and the whole, because parts belong by definition to a whole and a whole always includes parts. Differentiation and integration are mutually complementary. Zooming in asks for zooming out, and vice versa. A fully rounded picture is the result. As the ancient Chinese said: 一陰一陽之謂道 (*Yīn* and *Yáng*: that is called the *Dào*). Like the road and the river, a coin has two distinguishable but unseparable sides. Birds have two wings to fly, and "port" and "starboard"refer to, respectively, the left and right side of a ship. "*Audi alteram partem*" is the judicial principle. We need two hands to clap, two legs to walk, two chopsticks to eat, two jaws to chew. It takes two parties to clinch a deal. Like space and time, subject and object are mutually related, but irreducible to each other.

The *method* of study is the structural, systemic (and systematic) approach, whereby the whole and its constituent parts are alternately paid attention to. Over-emphasizing the whole may provoke the complaint of vagueness; laying too much stress on the parts may result in the accusation of being incoherent. Globally connected computers linked up with digital libraries could be profitably used to process and integrate the growing amount of country-related information, that is, "to merge information from heterogeneous sources with differing conceptual, contextual or typographical representations".[9]

Assyriologists, Egyptologists, Indologists, Sinologists or Mongolists do not need to be familiar with the workings of the New York Stock Exchange, the common law system, the mechanisms of institutional conflicts in the European Union, the voting system of the Security Council or the intricacies of ecosystems. However, able to read Sumerian inscriptions, hieroglyphs, Sanskrit texts, Chinese characters and the Mongolian script, respectively, they should be willing to seriously engage in an in-debth conversation with linguists, economists, sociologists, political scientists, geographers, demographers, ecologists and other experts. This would be the way to prevent them from making statements on matters they do not understand. It would also be the way for the linguists, economists etc., to find out whether or not their theories are universally valid.

Writing an interdisciplinary and international study of countries, one should organize the information about them like the contents of a general encyclopedia, which is not only a storehouse of facts and bare information (like a telephone book), but also and primarily a systematic survey of all departments of knowledge. However, while a general encyclopedia represents the state of affairs at a certain time, an electronically accessible country-encyclopedia should provide information and knowledge related to the country on continuous basis. The knowledge provided should be well-reasoned, and its organization should be all-rounded. This can only be done when the principles of general systems science are respected. The requirement that the knowledge provided is well-reasoned leaves the door open to future, deeper insights; the obligation of the knowledge organization to be all-rounded makes room for the context without which nothing is understandable. Partial information or knowledge provided by some but not all departments to be involed is by definition incomplete. An encyclopedia is an en-cyclo-paedia, a circle of learning, an ouline of knowledge.[10] The organized and integrated knowledge of a country consists of the knowledge gained by the physical, biological, social and human sciences as far as these sciences or their numerous subdisciplines are related to the country. The developments of the physical, biological, social and human sciences will be stimulated by the advancement made by the formal sciences. This advancement will in turn be furthered by progress in the non-formal sciences.

A country study is essentially the study of a system of systems delineated by its boundaries, surrounded by its environment (with which the country interacts) and characterized by both its structures and processes. It must be based on the principles of the general systems science envisioned by Ludwig von Bertalanffy (1901–1972) and currently revisited by the Bertalanffy Center for the Study of Systems Science (BCSSS), in Vienna. In combination with this requirement, country studies will hugely profit from the fast developing computer science, which is branching off into ever more subdisciplines. Practitioners of Iranology, Turkology, Egyptology, Japanology, Koreanology, Sinology, Tibetology, Indology, (Latin) American–, Caribbean–, Russian–, Slavic–, European–, African–, Arabic–, Middle Eastern–, Central Asian–, Mongolian– or Southeast Asian studies will greatly benefit from taking careful note of, and giving due attention to, the proceedings of:

(1) The annual meeting of the International Society for the Systems Sciences, and
(2) The annual International Symposium on Frontiers of Combining Systems.

Country studies should be planned and executed on project basis. There is plenty of software for project management. The list of megaprojects in domains such as science, aerospace, energy, transportation, sport, urban planning and water management is very long and can be found on the Internet. What all these projects have in common is the large-scale pooling of resources, the joint and concerted effort of differently educated and trained people under the leadership of a project manager, who has the accomplishment of a specific task in mind and makes use of sophisticated computer hardware and software. A technically, economically, financially and politically feasible project is carefully planned. Before it is embarked on, a decision must be made after many in-depth analyses have been carried out and comprehensive discussions have been held. Whoever is willing to design and complete a country-study project should follow the example of, and learn something from, those who have practical experience in planning and executing a megaproject. Should he/she be wondering how the study-project might be organized and managed, a visit to the Web site of the International Centre for Complex Project Management would be recommendable. Not everything will be plain sailing. Labor pains will be unavoidable, but, as written on the coat of arms of my hometown: "*PER ASPERA AD ASTRA*" (through hardships to the stars). In the beautiful phrase of Baruch Spinoza: "*Omnia præclara tam difficilia quam rara sunt*" (all things excellent are as difficult as they are rare).

The team of scientists embarking on the study of a country will soon realize that structure (the synchronic) cannot be decoupled from process (the diachronic), for space and time are not to be seen as independent from each other.[11] The ancient Chinese knew it already: space and time are inextricably interconnected. Their compression is the essence of globalization, the variety of aspects of which is bewildering. Globalization affects all countries, some of which are key players, the main protagonists in a humanitarian drama of worldwide proportions. Whether the rulers of these countries really understand their counterparts and with them the countries they represent will determine the fate of humanity. So the weight of the burden on the shoulders of country experts cannot be overestimated. There is a path that few historians are willing to follow: the trajectory from colonialism via imperialism to today's globalization. The world government which some people are dreaming of may be a pipe dream, impossible or very unlikely to come true. There will never be, I surmise, one captain on spaceship earth - a captain recognized by all people on board.

Realizing that structure (space) and process (time) are connected may lead to the realization that **area** (country) and **era** (history) studies are intimately interlinked and firmly interlocked. Historians also have an *explanandum* (an era to be explained), but—like "country experts"—they have no distinctive *explanans* (a theory of their own that does the explaining). So my critique of *country* studies might as well apply to *history* studies, particularly to the history of countries. Like so-called country experts, historians can do nothing but to show the feathers of other people and to flaunt what is not theirs. Without history's auxiliary sciences, historians would be completely helpless; with them, their narrative is essentially a chronicle, the listing or enumerating of what happened at a certain place in consecutive order. Anything more would be fantasy, not the result of scientific, collaborative research. If "era experts" as well as "area experts" do have a last, they definitely do not stick to it.

The human drama is unfolding on the territory of countries. This should make scientists realize that geography is nothing but history in space and history is only geography in time. Ideally, geographers and historians would thus have much to tell each other. In other words, the geology-, ecology-, demography-, polity-, economy- and society-related geography of a country cannot be disconnected from its geology-, ecology-, demography-, polity-, economy- and society-related history. They are just two sides of the same coin.

Although the sciences have been branching off, they should all be involved in the study of a country. Natural, social, human and formal sciences are more interrelated than their practitioners care to admit. The study of a country shall not be the prerogative of philologists who have no schooling in any of these sciences; it must not be monopolized by browsing amateurs.

Country studies have to be placed on a totally new footing. When they are, their impact on the humanities and social sciences, hence on higher education at large, will be profound. The reverse will be true as well. Radical new country studies will profit tremendously from using the latest edition of the multi-volume *International Encyclopedia of the Social and Behavioral Sciences*, which comprises over 3,900 articles (commissioned by dozens of section editors) and includes more than 90,000 bibliographic references. However, whether the study of countries *will* be thoroughly revised is a question I cannot answer. Much will depend on the will and determination of the leadership of universities to take the necessary measures, and this, in turn, will depend on the will and determination of politicians, who are prone to conservatism, I am afraid. "Don't rock the boat" will be the establishment's moto. It will be hard to convince those who, lurking in the wings, exercise their influence in the political arena. Mental sclerosis is said to be incurable, but the choice to be made is crystal clear: do we allow practitioners of country studies to continue their old way of doing things, or do we urge them to have the courage to change course by shifting the helm?

I am not pleading for a kind of academic fusion cooking, for undoing scientific differentiation, for a situation in which all scientists look grey. To gain a complete and coherent, an articulate and fully rounded country picture, however, interdisciplinary and international research based on computational systems science is indispensable.

Seeing both the trees and the forest is an absolute prerequisite, not only for comprehending the complexity of a country but also for pursuing a sound and sensible policy toward it. Being a unique, multilayered, multifaceted and ever-changing reality, a country can only be comprehended by seasoned natural, social and human scientists who are seriously interested in it and are—in association with the *formal* scientists—willing to collaborate, to connect all the relevant dots, to see the whole elephant and to change from what Ulrich Weinberg aptly calls "Brockhaus thinking" to systems or network thinking. The different fingers of a hand belong together and none of them is as strong as the hand. A single person may be knowledgeable about one part or aspect of the country, but it is absolutely impossible for him or her to be knowledgeable about all the closely connected parts/aspects of it. It is utterly ridiculous to consider oneself to be, or to be called, a country expert, not to speak of an *area* specialist. The talking head who shamelessly expresses his or her opinion in the mass media without the caution and modesty befitting a scientist must be exposed. "Middle East experts", "Far East experts"—please don't make me laugh.

The complexity of a country can only be comprehended by taking things together (the hallmark of intelligence), by engaging in truly interdisciplinary and genuinely international work, by embarking upon a country-project that uses the latest in computer engineering techniques and computer software. I am not arguing for width at the expense of depth, but I am neither pleading for the particular at the cost of the general. The reader will notice that I attempted to go beyond what only appears to be a trade-off. For without parts (depth) there can be no whole (width) and without a whole (general) there can be no parts (particulars). In other words, depth/parts and width/whole are fundamentally interconnected, like *Yīn* and *Yáng*. Country studies should be wide as well as deep. The *uomo universale* concerning a country does not, and will never, exist.

To be taken seriously, country studies should be interdisciplinary as well as international exercises based on computational systems science, the discipline where computer and systems science converge. The new and integrative country studies I imagine could get inspiration from systems biology, health science, systems ecology and computational social science. If they meet the requirements spelled out in this book, country studies will have finally come of age, having been uplifted onto a higher, scientific level. Though needed for better international and intercultural understanding, alas, the rquirements are flatly rejected by tenured country-professors who populate our universities and stick to their old way of doing things. *Sapienti sat*!

In sharp contrast to Western scientists, who are inclined to analysis and reductionism, Chinese researchers think correlatively and complementarily, seeing things pattern-wise and process-wise at the same time. They know what matrix-thinking really is. Their relational way of thinking is deeply rooted in the main concepts of great thinkers in ancient China: the concepts of 道 (*dào* [way, process]) and 理 (*lǐ* [structure, pattern]) referred to in the preface. The Western reductionistic and the Chinese holistic way of thought should be combined. If the parts are important, so is the whole to which they belong. Seeing the many in the one and the one in the many is the key to comprehending the complexity of countries.

Suggestion

An **international conference** on the current and future state of area or country studies is needed. The subjects to be deeply thought about and seriously dealt with should include:

(1) The difference between area/country studies conducted in Europe and the USA;
(2) The fundamental problem with these studies;
(3) The relevance of the problem in a time of rising international tensions;
(4) The definition of "area" and "country";
(5) The definition of "complex and dynamic system";
(6) The need for genuinely *interdisciplinary* country studies;
(7) The need for genuinely *international* country studies;
(8) The *scientific* feasibility of interdisciplinary and international country studies;
(9) The *technical* feasibility of such studies;
(10) The difference between spontaneous and man-made organizations;
(11) The nature of the chasm between natural and cultural studies;
(12) The origin and development of this divide;
(13) The possibility of bridging the gap;
(14) The requirements to refurbish, even to redesign universities.

If it would turn out at the meeting that a *country* theory (general or particular) does not exist, "country experts" should be expelled from universities without delay, because their "theories" are nothing but a bunch of opinions with no compelling argument or empirical evidence to back them up. Universities are meant to be places where testable theories are advanced or developed, not safe havens for people who only ventilate opinions.

The conference I envisage may result in the founding of a **Center for Interdisciplinary and International Country-studies** (CIIC), a forum where scientists from various disciplines and nations meet, and discuss with, each other (physically or virtually), a platform where scientists have unhindered access to centrally collected and country-related data. In other words, the conference may result in the creation of a global hub of knowledge about countries, which have many interrelated dimensions and are uniquely formed by both nature and culture over a certain period of time. The regularly updated **home page** of CIIC should not only give an introduction to the Center; it should also contain links to books, articles, lectures, (contributors to) research discussions, and (past and forthcoming) conferences, symposia or workshops about a particular country.

The conference may also result in the launching of an open access journal (possibly entitled ***Frontiers in Country Studies***) to which scientists from different quarters interested in a particular country may (be invited to) contribute, and in which recently published country studies are rigorously reviewed, with the review electronically accessible and the name of the reviewer mentioned! See Chap. 8.

In addition, the conference may lead to the compilation of an electronically accessible **country-encyclopedia** as explicated above.

The congress will cause an earthquake in higher education and its impact could prove to be as far-reaching as the still reverberating effects of the Apollo project. Should the organizational requirements seem to be too difficult to fulfill, one could possibly appeal to the chief librarian of national libraries or to the chief executive of ICT companies. A beautiful task for the Director-General of UNESCO!

The possible objection that such a conference is not needed has to be rejected, for isn't it remarkable that, while mathematicians, physicists, chemists, astronomers, geologists, geographers, biologists, demographers, economists, sociologists, psychologists, anthropologists, political scientists, linguists and literary scholars of different stripes and persuasions have their international conferences, an international meeting attended by scholars engaged in the study of different countries has never been convened! The relentless stream of sundry works wrought by self-styled country experts of different stripes and persuasions is concrete proof of it that I am not setting up a straw man which I am wasting my time fighting. Should the reader want additional proof, he/she is invited to look at book series like *Key Issues in Asian Studies*, *Asia Past & Present* and *Asia Shorts* [sic], published under the auspices of the Association for Asian Studies, a "scholarly, non-political and non-profit professional association open to all persons interested in Asia and the study of Asia". Asia is Earth's largest and most populous continent, reaching from 26°4'E to 169°03'W and from 81°17'N to11°00'S, and comprising fifty countries. May the real "Asian expert" stand up!

Richard Feynman once hit the nail on its head by saying that science is imagination within a disciplinary straitjacket. This would mean that Egyptology, Iranology, Indology, Turkology, Japanology, Koreanology, Sinology, etc., are not sciences, because the study of Egypt, Iran, India, Turkey, Japan, Korea, China, etc., are not cases of disciplined imagination. Not only does this book expose country pundits and, *a fortiori*, area experts as lacking a disciplinary straitjacket; it also challenges them to be more imaginative. Arguing for genuine collaboration (interdisciplinary as well as international), the book is a plea for pushing back scientific frontiers, not unlike the pushing back of frontiers on the American continent in the past, but this time without bloodshed and humiliation.

In *Fundamentals of Comparative and Intercultural Philosophy* (SUNY Press, 2016), Lin Ma and Jaap van Brakel discuss the conditions on which comparative and intercultural philosophy would be possible. Claiming that it is not necessary to share a common language to engage in comparison, they posit that the controversial and much-publicized issue of universalism versus relativism can be resolved by replacing the concept of ideal language with the idea of family resemblance, an idea made popular by Ludwig Wittgenstein in his posthumously published book *Philosophical Investigations* (1953). Whether the claim can be sustained is debatable.[13] It would have been interesting and perhaps more fruitful if Ma and Van Brakel had added the wider perspective of interdisciplinary and intercultural collaboration.

Stating that my ideal of a radically new kind of country study cannot, on purely practical grounds, be achieved does not legitimize the work currently done by so-called area/country experts. Aware of the risks being taken, I claim that cross-disciplinary and cross-national collaboration, combining depth and width of research, is the one and only way out of the predicament into which country connoisseurs have brought themselves.[14] The fast developing computer science could be helpful in carrying out interdisciplinary research, which is to be distinguished (but by no means separated) from international scientific collaboration. The mystery of all mysteries is that differentiation and integration, the discrete and the continuous, the many and the one implicate each other. "*Tat Tvam Asi*" (that thou art), the *Chāndogya Upanishad* says, meaning that the Self, in its original, pure and primordial state, is identical with, and the same as, the Ultimate Reality, the ground and origin of all different looking phenomena. The crux of the matter is to see the one as well as the many, to see both the whole and its parts, for parts are always parts of a whole and a whole has always parts, as the ancient Indian and Chinese thinkers intuitively knew already. Parts in turn can be, and are often, composite wholes. Doing computational research on the system of sciences (by means of the science of systems) does not mean that no attention should be paid to the interrelated subdivisions of a science and to the interconnected subparts into which these subdivisions may be divided. Whereas parts are conceivable, the ultimate whole is certainly not. The fomer can be described or put into words; the latter finally not, because it is ineffable.

The text of the book is interspersed with remarks on, or references to, the Chinese way of thought. Given my education, the reader may forgive me these escapades, but the asides should not be mistaken. I think that the West could learn something of great importance from wise men such as Confucius, Mengzi, Xunzi, Laozi, Zhuangzi, Zhu Xi and Wang Yangming, but I am not an admirer or fellow traveler of the brutal and oppressive regime in the PRC. I got hooked on, but not by, China.

> The name "People's Republic of China" is a pleonasm ("republic" comes from *res publica*!). Many, if not most, universities, think tanks and political or economic institutions in the Western world employ Chinese academics, whose hidden dog loyalty to the government in Beijing should be a cause for great concern in their host country. Though unnoticed by the general public, it is a plain fact that there is hardly any organization of importance in the world where the tentacles of the Chinese octopus have not been reaching. Nearby or far away, at high or at low level, Beijing's fifth column is active everywhere outside the PRC. China is certainly not weak on soft power, as Joseph Nye would have it.[15] Unlike Samuel Huntington, who in *The Clash of Civilizations and the Remaking of World Order* (Simon and Schuster, 1996)[16] argued that future wars would be fought not between countries, but between cultures, and that Islamic extremism would become the biggest threat to world peace, I think that China is the greatest challenge the West has to meet. To meet this challenge, however, the first thing to do is to put oneself in the position of China, who, rightly or wrongly, considers the West since the fateful First Opium War (1839–1842) as a decadent, complacent and intrusive existenial threat. While the Muslims adhere to one of the Abrahamic religions and a common ground for discussion could thus be found, the difference between the Chinese and Western ways of thought is more fundamental.[17]

Although I am proud that my cradle stood in Europe, "*le petit cap du continent asiatique*" (Paul Valérie) where Homer, Pythagoras, the Greek tragedians,

Plato, Aristotle, Euclid, Archimedes, Ovid, Virgil, Aquinas, Dante, Columbus, Leonardo da Vinci, Erasmus, Copernicus, Michelangelo, Luther, Cervantes, Shakespeare, Galilei, Hugo Grotius, Descartes, Rembrandt, Milton, the three great playwrights of seventeenth-century France, Pascal, Spinoza, Locke, Newton, Vico, Bach, Montesquieu, Voltaire, Hume, Rousseau, Adam Smith, Kant, Watt, Goethe, Mozart, Beethoven, Hegel, Ritter, Balzac, John Stuart Mill, Darwin, Kierkegaard, Marx, Baudelaire, Dostoevsky, Pasteur, Ibsen, Tolstoy, Maxwell, Dilthey, Mendeleev, Nietzsche, Cantor, Frege, Freud, Hertz, De Saussure, Joseph Conrad, Émile Durkheim, Husserl, Max Weber, Curie, Proust, Thomas Mann, Rilke, Schweitzer, Einstein, Fleming, Picasso, James Joyce, Kafka, Keynes, Malinowski, Bohr, Wittgenstein, Von Bertalanffy, Gödel, Claude Lévi-Strauss and Turing were born,[18] I have a lot of sympathy for old China, where the "scholar official" (*shìdàfū*) stood high on the social ladder (in contrast to Japan, the country of shōguns and sword-brandishing samurai) and merchants were frowned upon (in contrast to America, the country of businessmen and moneymakers). Whereas Moses, the most important prophet in Judaism, is believed to have received, at Mount Sinai (Sînāy), the Ten Commandments from "somebody" speaking through a burning bush (*səneh*); whereas the *Iliad* is a story of rage and resentment (Achilles' *μῆνις*); and whereas the *New Testament* is the weird but persistent narrative of a man from Nazareth who, fancying himself to be "the son of God", felt called (by nobody but his "Father in Heaven") to be deeply humiliated and finally crucified in order to redeem us ("through the grace of God"!) from "our sins", the Chinese Classics have none of this. There is no hocus-pocus or party trick and there are no theophanies (divine manifestations) or stories of rancor and bad blood in these marvelous books, noted for their relentless pleading for humaneness, that is to say, for really being (and not, in one way or another, *simulating* to be) man and acting on the principle of "togetherness" (仁). What the Chinese have cherished are the worldly ideals of social order and harmony, great learning and the ruling of a country, not by law, which can be immoral, but by wise men.[19] Their ideals have not been realized thus far, but in his classic book *La pensée chinoise* (Albin Michel, 1999 [1934]), Marcel Granet (1884–1940), member of the Durkheimian school of sociology and never surpassed by any "China expert", writes:

> S'il faut, pour conclure, essayer d'indiquer les traits les plus remarquables de la civilisation chinoise, la formule la moins imprudente sera celle qui présentera un aspect negative. Insistant sur le fait que les Chinois ne subissent volontiers aucune contrainte, je me bornerai à caractériser l'esprit des moeurs chinoises par la formule: ni Dieu, ni Loi. (pp. 475–476)

The guiding thought of the promising journal *Ancient Philosophy Today* (2019ff), published by the Edinburgh University Press and "providing a forum for the mutual engagement between ancient and contemporary philosophy", is that ancient theories are not obsolete or out-of-date and irrelevant to current philosophical debates. "Instead of being primarily approached in a philological fashion, they should be taken seriously as philosophical texts, which can be assessed for their philosophical merits as well as more modern texts can be". In much the same vein, the *basso continuo* of this book is that the reading and interpreting of Chinese Classics, and of the *Yìjīng* first and foremost, cannot be left to philologists who are not educated in any scientific discipline whatsoever and, due to lack of sufficient command of the source language, have to rely on the language proficiency of Chinese colleagues. The study of China shall be committed to a team of seasoned scientists who are seriously interested in the country. The Chinese Classics, the correct reading of which is indispensable to understanding the Chinese mind, were translated into English by James Legge (1815–1897), a representative of the London Missionary Society in Malacca and Hong Kong. His translation has become standard, but Legge was not a scientist.

Remembering the words of one of my teachers, Erik Zürcher (1928–2008), who in his inaugural speech, in 1962, lamented the "dialogue of misunderstandings" but did not seem to see the subtle difference between debate and dialogue, I share the view of François Cheng (1929), who believes that cultures could engage in a fruitful dialogue. I also concur with the opinion of Léon Vandermeersch (1928), who thinks that by engaging in a real conversation and exchanging views we may learn something from the children of the Yellow Emperor and from the offspring of Yao and Shun.[20]

It has been said that, to know a person, one has to walk at least a mile in his shoes, and that, to know a town, one has to walk there many miles. The idea advanced and developed in this book is that, to really comprehend a country, a scientific *team* is needed, with each member presenting his/her professional view on an aspect of it, while *actively* listening to what other team members have to say. For no single scholar will ever be able to see the historically evolved complex structure of a country. In a constantly changing, hypercomplex world, no problem can be dealt with by the logical tribunal that decides on true or false (two-valued logic). Real life cannot be modeled in black and white, because there are infinitely more possibilities. Reality is to be represented in hues of grey. Nobody knows where to draw precisely the boundary between raining and drizzling, or when many grains of sand change to a heap of sand.[21] Real life has many shades of color, but Western scientists, claiming to be real scientists, can only paint in black and white. Miguel de Unamuno (1864–1936) opens his masterpiece *Del Sentimiento Trágico de la Vida* (about the tragic sense of life) by drawing attention to "*el hombre de carne y hueso, el que nace, sufre y muere–sobre todo, muere*" (the man of flesh and bone, the one who is born, suffers and dies–above all, dies). Safely navigating the incalculable sea of a community living in a particular area requires a special compass.

At the first national conference of sinologists convened in November 1971 by the new Far Eastern Institute (*Institut Dal'nevo Vostoka*) of the Academy of Sciences of the USSR, its director, Dr. Mikhail Sladkovsky, said: "China will undoubtedly play a great role in the world. Study of the country demands, therefore, an all-round and deeper approach. It must be an integrated science, studying all aspects, the internal material and spiritual life of the Chinese people as well as its intercourse with the the world outside". Currently, the Far Eastern Institute, a leading research institute of the Russian (!) Academy of Sciences, is headed by professor Sergey Lusyanin, successor to Mikhail Titarenko (1934–2016), who was chief editor and co-author of the famous *Dukhovnaya Kultura Kitaya Entsiklopediya* (encyclopedia of the spiritual culture of China), a six-volume online work financed by the Chinese Development Bank and published from 2006 to 2010.[22] The Institute should be distinguished from the 200 hundred years old Institute of Oriental Studies of the Russian Academy of Sciences. Without committing myself to the ideology of communists, I could not more agree with Dr. Sladkovsky.

"*Ceterum censeo Carthaginem esse delendam*", Cato the Elder is reported to have been repeatedly saying in the Roman Senate. My less destructive *censeo* is that experts from different disciplines and from all over the world should unite and form, according to the country they are interested in, a scientific team. They should have the courage to embark upon a country research project, because scientists working together in a project are more capable than scientists working separately. In that case too, the whole is more than the sum of its parts. Unity means strength! So I hope that at the 11th International Conference on Intelligent Computing and Applications, in Chongqing,[23] the discussion will also be about the intelligent application of systems research and computer science to the study of countries.

I am realistic enough to expect that well established, high-profile and celebrated "country connoisseurs" will find it extremely difficult, if not impossible, to openly agree with my whistle-blowing.[24] They may think that their very livelihood is at stake and that they will lose their bread-and-butter by concurring with my views. However, indignant about my rebellion against their authority, they may overlook what is valuable in the book, which is meant to be an exercise in "creative destruction" (Joseph Schumpeter). Upset by the destructive side of it, they may conveniently forget about its *constructive* side, pushing aside a palatable dish that is *sour and* sweet. Nobody could blame me for asking for a long overdue online discussion or international congress regarding an issue that affects all of us, and increasingly so. There are really no country experts such as Americanists, Russianists, Iranologists, Turkologists, Japanologists and Sinologists, not to speak of *area* specialists with regard to Europe, the Middle East, Latin America, The Caribbean, sub-Saharan Africa or Southeast Asia. I am not afraid of being refuted—far from it. As the old saying goes, "the truth emerges from a clash of opinions". However, the meaning of this saying can only be that without an opportunity or possibility of opinions to clash, the truth has no chance of coming out and the confusion pointed out in Chap. 1 will endure. Being refuted, rebutted or rightly contradicted, I will immediately recant and promptly acknowledge that I was wrong in making allegations against so-called area or country experts at universities, research institutes or think tanks. I will sincerely and publicly apologize for having done so, even though I may still hear Galileo Galilei whispering: "*E pur si muove*". What I fear, however, is to be misunderstood, inadvertently or deliberately.

Is comprehending a country based on general systems science and aided by computers unnecessary or unneeded? I don't think so. A new kind of country study is not only necessary and urgently needed; it is also technically feasible. In an age of globalization and global strife between superpowers, mutual understanding at different levels is of vital importance. Is such a comprehending impossible or very unlikely to happen? Again, I don't think so. Admittedly, the differences are great and should not be glossed over. The natural sciences differ from the cultural sciences, and the concern of sociologists is certainly not the same as what biologists are dealing with. The *explanandum* of formal sciences is not the substance but the form or structure of natural, social and human sciences. However, the natural, social, human and formal sciences are all sciences, and scientists are searching for patterns. They try to explain, to make something clear or easy to understand by describing or giving information about it. The pursuit of knowledge or the search for a pattern (the ancient Chinese called it *lǐ*) is what all scientists have in common. A truly intelligent man *com*prehends, takes things together, locally and, if possible, more widely. The book to be read is Bruce Robertson, *The Pattern Paradigm* (Xlibris, 2012).

Notes and References

1. See James Milne, "Motives—Grothendieck's Dream", April 24, 2012 (online). Motives and motifs have a common denominator, which links nature to culture. A motive is a need, want or desire, possibly/usually acquired through the influence of culture. Motivation is man's direction to behavior; it drives or "causes" man, or gives him/her reason, to repeat a behavior and to move forward in life. Motivation results from the interaction of conscious and unconscious factors; it governs the choices or decisions being made. Motifs are related to fractals (recurrent patterns). Visit https://plato.stanford.edu/entries/moral-motivation and https://philpapers.org (fill in: motive and motivation). For causes, see Yemima Ben-Menahem, *Causation in Science* (PUP, 2018).
2. See Kristi Lew, *Taxonomy: The Classification of Biological Organisms* (Enslow, 2018).
3. See Erik Brynjolfsson and Andrew McAfee, *The Second Machine Age* (W.W.Norton, 2014) and Andrew McAfee and Erik Brynjolfsson, *Machine, Platform, Crowd: Harnessing Our Digital Future* (W.W. Norton, 2017).
4. See Andrea Staiti, *Husserl's Transcendental Phenomenology* (CUP, 2014), Sara Heinämaa et al. (eds.), *Phenomenology and the Transcendental* (Routledge, 2014, pp. 1–19) and Kwok-Ying Lau, *Phenomenology and Intercultural Understanding* (Springer, 2016). The word *Lebenswelt* (lifeworld) refers to the state of affairs in which the world is experienced, to the world of *Erlebnisse*. Read the enthralling "noetic trilogy" of Karel Čapek, *Hordubal* (1933), *Meteor* (1934) and *An Ordinary Life* (1934), and Jaeger and Liebsch, who elaborate on the concept of *Erfahrung* (see *op.cit.* Vol. 1, pp. 1–91). Interestingly, empiricists, who usually distance themselves from Husserl and his followers, maintain that real knowledge comes from experience, but they do not tell what "experience" (*Erlebnis* or *Erfahrung*?) exactly means and how it relates to "perception", a highly problematic notion. See Stathis Psillos and Martin Curd (eds.), *Companion to philosophy of Science* (Routledge, 2013, Chap. 15) and visit https://philpapers.org/s/experience. "Experience" is also a core concept in theology, the study of religious beliefs, places and practices at which empiricists turn their nose up. Visit https://plato.stanford.edu/entries/scheler (Sect. 5). I borrow the phrase "*speculum mentis*" from Robin Collingwood (1889–1943). See Ray Monk, "How the untimely death of RG Collingwood changed the course of philosophy forever", September 5, 2019 (online).

5. See the *Annual Review in Automatic Programming* (1960–1994), the *Journal of Automated Reasoning* (1985ff) and Robert Arp et al., *Building Ontologies with Basic Formal Ontology* (MIT Press, 2015). Also visit https://plato.stanford.edu/entries/reasoning-automated.
6. For computational constraints, read the article of Artem Kaznatcheev in the March 2019 issue of the *Journal Genetics* and Melanie Mitchell, *Artificial Intelligence: A Guide for Thinking Humans* (Farrar, Straus and Giroux, 2019).
7. There is a large literature on the related subjects "one vs. many", "universal vs. particular" and "same vs. other". Visit https://plato.stanford.edu/entries/scientific-unity, https://www.iep.utm.edu/universa, https://plato.stanford.edu/entries/mereology and https://philpapers.org (search for "one many", "universal particular" and "same other"). Read Maria Keet, *An Introduction to Ontology Engineering* (College Publications, 2018) and Achille Varzi, *Ontologia* (Laterza, 2019).
8. See Note 11 in Chap. 1. For the reverse, see Bruce Robertson, *The Pattern Paradigm: The Science of Philosophy* (Xlibris, 2012) and Frank George, *The Science of Philosophy* (CRC Press, 2017). Metaphilosophy (aka philosophy of philosophy) is "the investigation of the nature of philosophy". See *Metaphilosophy* (1970ff) and visit https://philpapers.org/browse/metaphilosophy.
9. See Norman Anderson (ed.), *Contributions to Information Integration Theory* (Psycholoy Press, 1914). Visit http://www.iiwas.org. See also end of Note 10 in Chap. 7.
10. See *Encyclopædia Britannica, Propædia*, fifteenth edition, 1990, pp. 5–8.
11. See Stephen Kern, *The culture of Time and Space, 1880–1918* (Harvard University Press, 2003) and David Carr and Chan-Fai Cheung, *Space, Time and Culture* (Springer, 2004). Visit https://www.iep.utm.edu/time, http://www.spacetimesociety.org and http://www.qiss.fr/consortium.html. In February 2019, the "Space Time Matrices" conference was held at the Institut des Hautes Études Scientifiques, in Paris, with 12 speakers from 8 different countries working on the structure of space and time.
12. Visit https://plato.stanford.edu/entries/world-government, https://worldgovernmentsummit.org and http://www.wfm-igp.org.
13. See Steven Hales (ed.), *A Companion to Relativism* (Wiley-Blackwell, 2011), István Lakatos, "Thoughts on Universalism versus Cultural Relativism", *Pécs Journal of International and European Law*, 1 (2018), pp. 6–25, and the journals *Cross-Cultural Research* (1966ff) and *Journal of Cross-Cultural Psychology* (1970ff). Also visit https://sccr.org, https://plato.stanford.edu/entries/comparphil-chiwes, https://www.iep.utm.edu/comparat, https://plato.stanford.edu/entries/relativism and https://www.iep.utm.edu/relativi.
14. Sinologists who are afraid of getting unemployed due to my scathing attack on their work should realize that there are many Chinese books crying out to be translated and annotated, *e.g.*, the invaluable books about *Dào*, *Lǐ* and *Qì* compiled by Liwen Zhang and published by the Chinese People's University Press.

15. See Joshua Kurlantzick, *Charm Offensive: How China's Soft Power Transforms the World* (Yale University Press, 2007), Michael Barr, *Who's Afraid of China? The Challenge of Chinese Soft Power* (Zed Books, 2011), Marshall Sahlins, *Confucius Institutes: Academic Malware* (University of Chicago Press, 2014) and "Are Confucius Institutes Good for American Universities?", *ChinaFile*, April 4, 2019 (online).
16. Huntington wrote this book in response to *The End of History and the Last Man* (The Free Press, 1992), written by his former student Francis Fukuyama. See Davide Orsi (ed.), *The 'Clash of Civilizations' 25 Years On* (E-International Relations, 2018).
17. See the preface. For different views, see Arnold Toynbee, *A study of History* (Thames and Hudson, 1972, pp. 52–72), Derrick Fiedler and Bjørn Thomassen, "Civilization on Trial – Again", *Historical Processes and Peace Politics*, 7:8 (2009), pp. 71–104, John Blair and Jerusha McCormack, *Comparing Civilizations: China and the West* (Global Scholarly Publications, 2013) and David Wilkinson, "The comparative study of Civilizations and its Relation to China", *Comparative Civilizations Review*, 81:2 (2019), pp. 6–8. Ashley Tellis et al. (eds.), *Strategic Asia 2020: U.S.-China Competition for Global Influence* (NBR, 2020) is an aggressively marketed book, but does not provide an essentially new vision of the difference between the Chinese and Western ways of thought. The boundaries of a civilization are difficult to delineate. The French speak of "*le rayonnemen de la civilisation*".
18. Europe started two world wars causing endless suffering and widespread devastation, yet the long list of its great minds cannot be matched by any list of prominent intellectuals elsewhere, including the USA.
19. The relationship between law and morality is intriguing. The question as to whether law is independent of, or identical with, morality has not yet been settled. The literature on the subject is large. See Kaila Morin, "Bridging the Conceptual Gap between Law and Morality", *UCL Journal of Law and Jurisprudence*, 1:1 (2012), pp. 120–136, Philip Huang, "Morality and Law in China: Past and Present", *Modern China*, 41:1 (January 2015), pp.3–39 and Emima Alistar, "The Relation between Law and Morality", April 25, 2019 (online). See further Chap. 6 (*nomos* vs. *physis*).
20. See Madeleine Bertaud and Cheng Pei, *François Cheng à la croisée de la Chine et de l'Occident* (Droz, 2014) and Léon Vandermeersch, *Ce que la Chine nous apprend* (Gallimard, 2019).
21. There are borderline cases. Visit https://plato.stanford.edu/archives/sum2018/entries/vagueness. What Anaximander (c. 610 – c. 546 BC) exactly meant by *ἄπειρον* is unclear.
22. For Titarenko, go to https://larouchepub.com/other/2016/4312titarenko_obit.html. For the state of Russian Sinology, visit http://www.unz.com/akarlin/russian-sinology.
23. Go to http://www.icica.org.
24. The University of California Press claims to be disciplined in publishing African, American, Asian, Latin American, Middle Eastern and Global Studies

(https://www.ucpress.edu/disciplines), but turned down my request to publish a book exposing the fundamental flaw of all kinds of country studies and attempting to advance them. The prestigious company does not seem to be interested in publishing a book that digs a bit deeper, pleads for interdisciplinary and international country research and encourages big-picture thinking.

Chapter 12
The Main Message

Abstract Today, even reading a book's summary seems to be asking too much. So, finally, a double-distilled drink made from diligence and resignation is offered.

- Geographical areas such as Europe, Africa, the Middle East, Central Asia, South Asia, Southeast Asia, East Asia and Latin America are sets of significantly different countries, which are wholes the complex constituent parts of which are intimately related. Being examples of a *Ganzheit* (not: *Gesamtheit*), countries need to be studied as *individua*, as things that can be "analyzed" but shall not be split up. When studying a country, distinctions must be made, but it shall not be conceived of as being able to be broken down into scientifically manageable chunks. Distinction (seeing differences) is not the same as separation (making independent). The geographically, socially, economically, politically or otherwise distinguished parts of a country cannot exist autonomously or independently, without relations to the other parts. If they could, they would not really be parts or aspects of the country. Each and every country can be defined as a huge set (*ensemble*) of interrelated elements and is part of a still larger whole, this ultimately being part of the universe. As its parts are intimately interconnected, a country is more than a set: it is a *system*; indeed, it is a complex and dynamic system, because the parts are nonlinearly interrelated. A set is a collection or group of things; a system is a set of *connected* things; and in a complex systems the things are related in a nonlinear way, making the system unfit to be studied conventionally. Systems science focuses on the relationships between parts, rather than reducing an entity to its parts. Analysts (splitters) tend to overlook this, accusing systems scientists of blurring boundaries by lumping things together. While the systems scientist fully acknowledges that the trees are different, he looks at, and studies, the wood that consists of different kinds of trees. In a complex system, the parts are interconnected in a nonlinear way.
- The solar system, Planet Earth, plants and animals are natural, self-organizing systems, dictated from the inside; no human activity is involved. Assuming that they are created by God is naively believing in a *deus ex machina*. Ecological problems are the result of man's tinkering with, and interfering in, natural processes.

H. Kuijper, *Comprehending the Complexity of Countries*,
https://doi.org/10.1007/978-981-16-4709-3_12

On the other hand, economies, companies, governments, political parties, universities, hospitals, trade unions, advocacy groups, alliances, religious institutions, armies and sports clubs are *not* natural, *not* self-organizing systems. They are dictated from the *outside*; human activity is involved. The difference is crucial and should not be overlooked,[1] Natural, self-organizing systems can perfectly do without man, but man, or the systems of his making (the not natural and not self-organizing systems), cannot do without man/nature; man is both a part of and apart from nature. Physicians and engineers are able to do a great job, but the healing of a wound is something they have to leave to Mother Nature. *Homo Faber* (man the maker), celebrated in the West and increasingly so in the East, has designed and manufactured many things (from cigarette lighters to condoms and from musical instruments to spacecraft), but he cannot make himself, as a human being, as a unity of body and mind. *Homo Faber* has organized numerous organizations (from birthday parties to multinationals and from sports events to festivals), but he is unable to design and build a single member of any of them.

- The elements of the system called country can be physical, biological, or social in character. Social elements (i.e., the things social and human scientists are concerned with) are dependent on, but not completely reducible to, biological elements; and biological elements are, in turn, dependent on, but not entirely reducible to physical elements. Physical and biological elements and processes are the concern of physicists and biologists, respectively. Mathematization and computation of their theories are possible, and division of their labor is an ongoing process, risking becoming uncontrollable. However, their separation from, indifference to, if not animosity toward social and human scientists is deeply worrisome, because of its profound negative effects on society, particularly on future generations. Things that are by nature connected shall not be separated or tried to separate. Heart and lungs need each other, are dependent on each other; a good cardiologist or pulmonologist know this, but the cardiologist will not pretend to be a pulmonologist, and vice versa.
- A system can be very complicated, involving a lot of different parts. Cars, watches and airplanes are just a few examples. If the relations between the elements are nonlinear, the system is not only complicated, it is *complex*. Examples are biological, social and cultural systems. Network science tries to exploit the properties that make a system complicated, but it is not very successful yet in the field of *social* sciences, that is, as soon as human beings (i.e. beings with an incalculable and unpredictable mind) are involved. The study of sociocultural networks, so relevant to the study of countries, is still in its infancy, but its future is said to hold much in store.
- When the mind focuses on, or shifts to, (the relations between) sociocultural elements,[2] further distinctions can be made between geographic, demographic, political, legal, military, economic, financial, cultural, linguistic, literary, medical, psychological, educational and environmental phenomena. Their study is becoming increasingly subdivided and fragmented. No scientist seems to take the interdependencies into account or "to take a crude look at the whole". Sociocultural elements can be considered either at a certain time (synchronically) or

over a shorter or longer period of time (diachronically), but the two distinguishable points of view shall not be separated.

- Science is more than having an opinion, or a philosophy; it is organized, systematized knowledge (see Chap. 2). "Science" comes from the Latin word *scientia* (knowledge) and is related to the Proto-Indo-European root **skei-* (to cut, to split). However, what exactly is meant here by "to cut" or "to split"? Is it reality arbitrarily cut into pieces, or is it a way of knowing reality? Any way of knowing reality is arguably limited and partial. Though truth (i.e., being in accord with facts or reality) is what scientists are said to be searching for, it is a hotly debated issue and its relationship to value is enigmatic. I take science not to be entirely value-free. See Chap. 2. Science and morality (defined as behavior based on the realization that humans are social beings somehow connected and obligated to each other) are mutually complementary, like mountain and valley (no mountain without valley and no valley without mountain). Science and morality shall not be driven apart by scientists who only focus their attention on what is (the world outside), while attempting to keep out questions as to what ought to be (the world inside). Unaware of their biases or hidden agenda, scientists in general and physicists in particular abstain from value judgments, leaving it to politicians to make decisions based in their own value judgments, that is to say, to people who are unreliable, because they are notoriously mendacious (not telling the whole truth). On the world inside, which is inextricably related to the world outside, see Morton Hunt, *The Universe Within* (Simon & Schuster, 1982), Patricia Alexander, *Psychology: Exploring Our Universe Within* (Pearson, 2012) and Neil Shubin, *The Universe Within* (Penguin, 2013). In addition, visit https://plato.stanford.edu/entries/mind-indian-buddhism, https://plato.stanford.edu/entries/computational-mind and https://philpapers.org/s/mind. The human brain is divisible and calculable, but the human mind is, and will probably always remain, indivisible and unfathomable.
- The true scholar is not only smart; he/she is also sincere and open-minded, which are, according to ancient Chinese, the prerequisites for investigating things and extending knowledge. The true scholar or scientist strives, therefore, to overcome prejudice and steers clear of partiality. "*Jūnzǐ bù dǎng*" (the noble man [not: nobleman] is not a party member), the Master said. A sentence, or a set of sentences, may be cognitively meaningful if and only if either it is analytic or empirically falsifiable,[3] but there can be no question that meaning, also the meaning of (the distinguishable parts of) a country, only exists in context, contemporary (crosswise) as well as historical (lengthwise). To see the meaning of something is to comprehend its coherence and to be able to put the thing into perpsective; it is nothing but seeing a pattern, seeing both the whole and its parts, seeing the dynamic order, seeing "the great interdependence of all differences" (*Yìjīng*).
- A necessary, but certainly not sufficient, condition of comprehending the complexity of a country is to command its *language*, which is—I am paraphrasing a saying from Heidegger—"*das Haus des Seins eines Volk*". Another necessary condition is the application of mathematically grounded and computer-supported

general systems theory. The implication of this condition is collaboration, cross-disciplinary as well as worldwide. Multidisciplinary research will not do; what we need is genuinely interdisciplinary and truly international research. Juxtaposition, or accumulation, is not enough; composition, integration, scientific collaboration are the things needed. A cathedral is more than a pile of stones and a novel is more than a number of words.[4] If it is true that, to some degree, all men are intelligent and have a sense of order (*Ṛta, Dào, Lǐ*),[5] the study of countries, which are essentially intricate combinations of self-regulating and not self-regulating systems, should be carried out interdisciplinarily as well as internationally. The basic and fundamental pattern of a country, its wiring has to be seen, but this can never be done by a single scientist. Only a tight *team* of scientists who are well-informed about the advances made in complex systems and computer science will be able to perceive the pattern of the carpet, the order of things in a country.

- Country studies such as Turkology, Iranology, Indology, Tibetology, Sinology, Koreanology and Japanology are fundamentally flawed. They are usually taken to be scientific studies. However, while in some parts of the world (predominantly Europe) practitioners of such studies do not provide a distinctive *explanans* (a tested or testable theory of their own) about their *explanandum* (a country), in other parts (mainly the USA), scientists like economists, political scientists, sociologists and ecologists who are interested in a foreign country have the bad habit of trespassing on the domain of colleagues at other university departments without notification and without assurance that their own (or borrowed) theory applies in the country concerned. Briefly, country experise as defined in this book can nowhere be found. Country studies could be lifted to a higher level by shifting one's attention to the SYSTEMS PARADIGM (https://systemsinnovation.io/systems-paradigm and https://necsi.edu/iccs-2018), that is, by changing from one-man operation to teamwork, to the joint and concerted efforts of groups of scientists who love working with other scientists and try to forge bridges between the perhaps not so exact "exact ciences" and the social and human sciences. One could possibly begin with a collaborative study of Singapore, Hong Kong, or Qatar. If the results of these pilot projects are successful and promising, i.e., if Singapore, Hong Kong, and Qatar can be coherently viewed from different sides instead of being one-sidedly and unevenly, hence distortedly described, one could proceed and embark upon the collaborative study of countries such as Israel, Hungary, Cuba, North Korea, Poland, Venezuela and Iran, ending up with collaboratively studying India, China, Russia and the USA, perhaps the most complex country on the Planet (not to speak of the EU).[6]
- Books or articles about, say, China and written by so-called China experts (Sinologists) sorely lack a tested or testable theory that is different from theories about a part or aspect of China advanced by, e.g., political scientists, legal scientists, economists, sociologists, linguists, literary scientists, ecologists or demographers. My critique does not, at least not primarily, concern the study about countries taken together or about the whole world. It concerns the study about countries taken individually and allegedly belonging to the "Family of Nations". Sinologists, Japanologists, Koreanologists, Indologists, Iranologists, Turkologists and Russianists,

to name but a few so-called country experts, are to be blamed for the enduring confusion about the respective countries. They should therefore be removed from universities, places where science (properly so called) is promoted and developed. Though there is no lack of opinion about country X or Y, there is absolutely no theory on any country whatsoever. If there would be a country expert having a distinctive and tested or testable theory on a certain country (his self-declared study object), I will openly recant. If the author only focuses his/her mind on a part of the country concerned (e.g., its literature, economy, or political system), has tested (or seriously tried to test) his theory on it and has not overstepped the boundaries of his *discipline*, his or her work falls outside the purview of my criticism.

- A country is an extremely complex and dynamic system interacting with other, so-called foreign countries and changing structurally (*Wandlungsstruktur*). Collaborating and computer-aided scientists will probably never be able to draw a country's changing structure as finely meshed as Richard Feynman's diagrams, which are pictorial representations of the mathematical relations between subatomic particles, but they will certainly be convinced that, in regard to the country under investigation, nothing is isolated and "*tout tient à tout*", or, in the memorable words of Leonardo da Vinci, "*D'ogni cosa la parte ritiene in sé la natura del tutto*" (the part preseves always the nature of the whole). Parts are, by definition, parts of a whole and there is no whole without parts. The one and the many are mutually implicated. The taste of the soup prepared by a master chef is closely related but not completely reducible to the taste of its indispensable ingredients taken apart. It is the melding or *mélange* that matters. "*Yīn and Yáng: that is called the Dào*".[7]
- Since natural, social as well as human scientists are increasingly listening to scientists engaged with the study of complex systems, it would not be outlandish to suggest that the science of complex systems should be applied to the complex system of sciences. Inasmuch as all the parts of this system (the individual sciences) are pertinent to the study of a country, it is thus time to change and to explore whether the science of complex systems should and could be applied to *countries*, complex and dynamical systems (of systems) *par excellence*. Given the dire international situation, I am arguing that it *should* be and, with the astonishing growth of human–computer interaction research, *could* be applied. The insights of complex systems scientists could be profitably applied to the study of countries, if the massive and fast-developing potential of computer and telecommunication science were harnessed for the benefit of it.
- I have been bold enough to attempt to advance country studies and to lift them to a higher, scientific level, claiming and demonstrating that the country pundits currently associated with universities, research institutes or think tanks only ventilate opinions. Having themselves no tested or testable theory about the country they maintain to investigate, these connoisseurs are not scientists properly so called. Scientific and international teamwork planned and executed on project basis, I think, would create a more balanced view and provide readers insight into

all the interrelated aspects of a country. Such teamwork would take the justified call to comprehend the complexity of a country more seriously.

- Whoever wants to frustrate my attempt should realize that an ant is unable to see the pattern of the whole carpet and that a single individual scholar will never succeed where a team of scientists may be more successful. "*Nourri dans le sérail et connaissant ses détours*", I can only exhort "country experts" to change their dead-end course as soon as possible.
- The main conclusion of this book is that computer-aided interdisciplinary and international collaboration based on the science of systems is the key to comprehending countries, the dynamic and extraordinary complex parts of a worldwide whole. Though a country can not (yet) be modeled mathematically (represented by a system of mathematical equations), it can be scientifically modeled all the same. The tetrahedron suggests the lines according to which a country can be studied analogically.[8] In order to comprehend the dynamic complexity of the whole world (the heavy task of those engaged in world politics), it is of utmost importance to comprehend the dynamic complexity of individual countries in the first place. Talking about the general (the whole) without knowledge of the particular (its parts), which is often a composite whole by itself, is talking like a headless chicken.
- Articial Intelligence may hold immense promise in enabling scientific breakthroughs and discoveries in diverse areas, as professor Anima Anandkumar of the California Institute of Technology recently maintained in a lecture delivered at the prestigious Simons Foundation, but not unless and only when the mind exactly knows what the mind is, how it works, how it differs from matter, and how far or deep it can reach.
- The claims made in this book are bold. They may upset or even enrage many professors of, or lecturers on, Asian, African, South American, Middle American, Middle Eastern, Russian, European, American, Canadian or Australian studies. Yet inconvenient truths should be exposed for correction without delay, because they are the ugly reality which casts a shadow on our universities. I hope to have got my point across. I am neither a physicist nor a neuroscientist, but a social scientist greatly admiring the ancient Chinese, who–hardly any Western scientist is aware of it–were complex systems thinkers "*avant la lettre*" (see my article "The Foundation of Chinese Systems Thinking" in *Systems Research & Behavioral Science* [forthcoming]). It would be great if an international conference concerning the current and future study of countries were organized and if the conference were presided over by a professional moderator, who knows how to wield the gavel. I am ready for fair criticism and open for constructive suggestions, but if—for one reason or another—the urgently needed (online) conference can not be organized, I will accept it wih resignation. Thomas Kuhn once pointed out that a truly new idea is usually rejected by those who are indoctrinated in the prevailing way of thought and like to be surrounded by yes-men. He was rephrasing what Einstein famously said: "We cannot solve our problems with the same thinking we used when we created them". So-called country experts, in particular those whose rights are legally nailed down (tenured professors), will

undoubtedly find all kinds of reason for not following my unsolicited advice and for sticking to their guns, not being aware of, or conveniently forgetting about, the grave political consequences their narrow-mindedness and stubbornness to fundamental change may have. However, there is a good reason to lift the studies with which they have so long been busy to a higher, that is to say, a genuinely scientific level. May my "casting a brick attract a jadestone" (*pāo zhuān yǐn yù*)!

Notes and References

1. Experts on self-organization tend to overlook this difference.
2. For sociocultural systems, see Frank Elwell, *Sociocultural Systems: Principles of Structure and Change* (AU Press, 2013) and go to https://systemsinnovati on.io (search for sociocultural-systems).
3. See Barry Stroud, *Seeing, Knowing and Understanding* (OUP, 2018, pp. 233–272).
4. A great number of cathedral or castle pictures (outside only) can be found on the Internet, and the list of novels is almost endless.
5. See Ernst Gombrich, *The Sense of Order* (Phaidon, 1979), Christopher Alexander, *The Nature of Order* (Berkeley: Center for Environmental Structure, 2004) and Martino Stierli and David Brownlee (eds.), *Complexity and Contradiction at Fifty* (MoMa, 2019). "*Lǐ*" stands for 理 (natural pattern) and for 禮 (social order). In the ancient Chinese worldview, the social order should reflect, or be modelled on, the natural order. This is the main lesson to be learned from the ancient Chinese, particularly at the time when environmental problems are pressing. See Chap. 6. The relationship between 理 and 禮 (both characters pronounced in the same way, the former meaning order in the [meta]physical, the latter meaning order in the social domain) reminds us of the relationship between the Indian concepts of *Ṛta* (cosmic or natural order) and *Dharma* (ritual, moral or social order). See Patrick Olivelle (ed.), *Dharma: Studies in its Semantic, Cultural and Religious History* (Motilal Banarsidass, 2009), Alf Hiltebeitel, *Dharma: Its Early History in Law, Religion and Narrative* (OUP, 2011) and James Kalomiris, *The Vedic Dharma: Explorations in the Vedic Natural Order* (Balboa Press, 2019). If there is one concept embracing the whole of my book, it is the concept or ORDER. In Chap. 3, I maintained that the idea of order, being of central importance in astronomy (cosmology), physics (entropy), informatics, politics, economics and organizational studies, should be the first subject on any agenda for an East–West Dialogue. Other words for "order" are "organization", "system" and "complexity". Comprehending the complexity of a country is a matter of seeing order, the perennial and recurrent order mankind has been deviating from and should return to before it is too late.
6. By now, the reader may understand why I cannot take a Professor of European Studies/History seriously. Though EU is not congruent with Europe, nobody is, or will ever be, an expert on the European Union.
7. See Robert Allison (ed.), *Understanding the Chines Mind* (OUP, 1989, pp. 236–264) and Kuang-ming Wu, "China as Culture", *Philosophy Study*, 9:7 (2019),

pp. 377–386. For an exploration of China's polity, economy, society, culture, and international relations, see Silvio Beretta et al. (eds.), *Understanding China Today* (Springer, 2017). Emeritus professor Beretta seems to be the ant able to see the pattern of the whole carpet called China.

8. A country is a demarcated piece of land and its organized people. Differently put, a country is a tripartite system consisting of three basic, interrelated elements: its territory, its people and the organization of its people. Countries can thus be represented (modeled) by a triangle, the sides of which are variable in length. Since each and every country has a history, the triangle has to be changed into a tetrahedron.

Appendix A
Literature on Interdisciplinary Research

Nicolai Hartmann, *Der Aufbau der realen Welt* (Walter de Gruyter, 1940).

Hans Lenk, *Handlungstheorien - interdisziplinär* (Brill, 1975).

Joseph Kockelmans (ed.), *Interdisciplinarity and Higher Education* (Pennsylvania State University Press, 1979).

Kunihiro Jojima, *Ökonomie und Physik: Eine neue Dimension der interdisziplinären Reflexion* (Duncker & Humblo, 1985).

Wilhelm Vosskamp, *Utopieforschung: Interdisziplinäre Studien zur neuzeitlichen Utopie* (J.B. Metzler, 1985).

Wilhelm Vosskamp, "From Specialization to the Dialogue between the Disciplines", *Issues in integrative studies*, 4 (1986), pp. 17–36.

Marcel Jollivet (ed.), *Sciences de la nature, sciences de la société, les passeurs de frontier* (CNRS, 1992).

Edgar Morin, "*Sur l'interdisciplinarité*", 1994 (http://ciret-transdisciplinarity.org/bulletin/b2c2.php#).

Liora Salter and Alison Hearn, *Outside the Lines: Issues in Interdisciplinary Research* (McGill-Queen's University Press, 1997).

Jürgen Mittelstraß, *Die Häuser des Wissens* (Suhrkamp, 1998).

Diederik Aerts, "Synthesis and analysis, interdisciplinarity and foundations", *Foundations of Science*, 3:2 (1998), pp. 203–206.

Alex Coles and Alexia Defert (eds.), *The Anxiety of Interdisciplinarity* (1998).

François Kourilsky, *Ingénierie de l'interdisciplinarité: Un nouvel esprit scientifique* (L'Harmattan, 2002).

H. Kuijper, *Comprehending the Complexity of Countries*,
https://doi.org/10.1007/978-981-16-4709-3

Roland Posner, "*Semiotik und andere interdisziplinäre Wissenschaften*", Roland Posner et al. (eds.), *Semiotics: A Handbook on the Sign-Theoretic Foundations of Nature and Culture* (Walter de Gruyter. Vol. III, 2003, pp. 2341–2374).

Harald Mieg, "Interdisziplinarität braucht Organisation!", *Umweltpsychologie*, 7:2 (2003), pp. 32–52.

Jürgen Mittelstraß, *Transdisziplinarität* (UVK, 2003).

Heiko Hecht et al. (eds.), *Looking Into Pictures: An Interdisciplinary Approach to Pictorial Space* (MIT Press, 2003).

Manfred Max-Neef, "Foundations of Transdisciplinarity", *Ecological Economics*, 53:1 (April 2005), pp. 5–16.

Sharon Derry et al. (eds.), *Interdisciplinary Collaboration: An Emerging Cognitive Science* (Psychology Press, 2005).

Wesley Shrum et al., *Structures of Scientific Collaboration* (MIT Press, 2007).

Walter Schweidler, *Postsäkulare Gesellschaft: Perspektiven interdisziplinärer Forschung* (Alber, 2007).

John Parker et al. (eds.), *Collaboration in the New Life Sciences* (Routledge, 2010).

Melissa Anderson and Nicholas Steneck (eds.), *International Research Collaborations* (Routledge, 2011).

Moira Cockell et al. (eds.), *Common Knowledge: The Challenge of Transdisciplinarity* (EPFL Press, 2011).

Diederik Aerts et al. (eds.), *Worldviews, Science and Us: Bridging Knowledge and Its Implications for Our Perspectives of the World* (World Scientific, 2012).

Matthias Bergmann et al., *Methods for Transdisciplinary Research: A Primer for Practice* (Campus, 2012).

Günter Ropohl, *Allgemeine Systemtheorie: Einführung in Transdisziplinäres Denken* (Sigma, 2012).

Stefan Brüggemann and Claudia d'Amato, *Collaboration and the Semantic Web* (IGI Global, 2012).

Ned Kock, *Interdisciplinary Applications of Electronic Collaboration Approaches and Technologies* (IGI Global, 2013).

Andrew Barry and Georgina Born (eds.), *Interdisciplinarity: Reconfigurations of the Social and Natural Sciences* (Routledge, 2013); reviewed by Yves Laberge in *Electronic Green Journal*, 1:38 (Spring 2015).

Gabriele Bammer, *Disciplining Interdisciplinarity* (ANU E-Press, 2013).

Carolyn Sobel and Paul Li, *The Cognitive Sciences: An Interdisciplinary Approach* (Sage, 2013).

Sylvie Catellin and Laurent Loty, *Sérendipité et indisciplinarité* (*Hermès*, Vol. 67, 2013).

Jerry Jacobs, *In Defense of Disciplines: Interdisciplinarity and Specialization in the Research University* (University of Chicago Press, 2013).

Katharine Farrell et al. (eds.), *Beyond Reductionism: A Passion for Interdisciplinarity* (Routledge, 2013).

Michael Jungert et al. (eds.), *Interdisziplinarität: Theorie, Praxis, Probleme* (WBG, 2013).

John Aldrich, *Interdisciplinarity: Its Role in a Discipline-based Academy* (OUP, 2014).

Peter Weingart and Britta Padberg (eds.), *University Experiments in Interdisciplinaritity* (ZiF, 2014).

Andrzej Wlodarczyk, *Academe Demarcated No More: Disciplines and Interdisciplinarity* (AuthorHouse, 2014).

Sönke Bartling and Sascha Friesike (eds.), *Opening Science: The Evolving Guide on How the Internet is Changing Research, Collaboration and Scholarly Publishing* (Springer, 2014).

Divya Jindal-Snape and Elizabeth Hannah (eds.), *Exploring Dynamics of Personal, Professional and Interprofessional Ethics* (Policy, 2014).

Christian Chevandier (ed.), *Travailler ensemble? Des disciplines aux sciences sociales* (*Presses universitaires de Rouen et du Havre*, 2014).

Barry Bozeman and Craig Boardman, *Research Collaboration and Team Science* (Springer, 2014).

Mauricio Fau, *Interdisciplina: vital para la ciencia* (*Colección Resúmenes Universitarios*, Nr. 411, 2015).

Felicity Callard and Des Fitzgerald, *Rethinking Interdisciplinarity across the Social Sciences and Neurosiences* (Palgrave Macmillan, 2015).

Maureen Ehrensberger-Dow et al. (eds.), *Interdisciplinarity in Translation and Interpreting Process Research* (John Benjamins, 2015).

Julie Thompson Klein, *Interdisciplining Digital Humanities* (University of Michigan Press, 2015).

Paolo Diviacco et al., *Collaborative Knowledge in Scientific Research Networks* (IGI Global, 2015).

Harvey Graff, *Undisciplining Knowledge* (Johns Hopkins University Press, 2015).

Maureen Ehrensberger-Dow et al. (eds.), *Interdisciplinarity in Translation and Interpreting Process Research* (John Benjamins, 2015).

Paul Gibbs (ed.), *Transdisciplinary Professional Learning and Practice* (Springer, 2015).

Yoneyuki Sugita (ed.), *Japan Viewed from Interdisciplinary Perspectives* (Lexington, 2015).

Patricia Leavy, *Essentials of Trandisciplinary Research* (Routledge, 2016).

Gerald Hartung and Matthias Herrgen (eds.), *Interdisziplinäre Anthropologie* (Springer, 2016).

Scott Frickel et al. (eds.), *Investigating Interdisciplinary Collaboration* (Rutgers University Press, 2016).

Ben Shneiderman, *The New ABCs of Research: Achieving Breakthrough Collaborations* (OUP, 2016).

Banu Akçeşme et al. (eds.), *Interdisciplinarity, Multidisciplinarity and Transdisciplinarity in the Humanities* (Cambridge Scholars Publishing, 2016).

Steph Menken and Michael Keestra (eds.), *An Introduction to Interdisciplinary Research: Theory and Practice* (Amsterdam University Press, 2016).

Andrew Balmer et al., "Five rules of thumb for post-ELSI interdisciplinary collaborations", *Journal of Responsible Innovation*, 3:1 (2016), pp. 73–80.

Eyup Saritas (ed.), *Interdisciplinary Studies in Turkey: New Ideas New Strategies* (Trafford, 2016).

Frédéric Darbellay et al., *La recherche interdisciplinaire sous la loupe* (Peter Lang, 2016).

Bianca Vienni and Bianca, *Key readings about interdisciplinarity* (October 25, 2016).

Tanya Augsburg, *Becoming Interdisciplinary: An Introduction to Interdisciplinary Studies* (Kendall Hunt, 2016).

Mihai Nadin (ed.), *Anticipation across Disciplines* (Springer, 2016).

Robert Frodeman et al. (eds.), *Handbook of Interdisciplinarity* (OUP, 2017).

Mark Burgin and Wolfgang Hofkirchner (eds.), *Information Studies and the Quest for Transdisciplinarity* (World Scientific, 2017).

Victoria Martin, *Transdisciplinarity Revealed: What Librarians Need to Know* (Libraries Unlimited, 2017).

Kristóf Fenyvesi and Tuuli Lähdesmäki (eds.), *Aesthetics of Interdisciplinarity: Arts and Mathematics* (Birkhäuser, 2017).

Frédéric Darbellay et al. (eds.), *Creativity, Design Thinking and Interdisciplinarity* (Springer, 2017).

Franz-Josef Kahlen et al. (eds.), *Transdisciplinary Perspectives on Complex Systems* (Springer, 2017).

John Parker et al. (eds.), *Collaboration in the New Life Sciences* (Routledge, 2017).

Uskali Mäki et al. (eds.), *Scientific Imperialism: Exploring the Boundaries of Interdisziplinarity* (Routledge, 2017).

Mohammed Nasser Al-Suqri et al. (eds.), *Promoting Interdisciplinarity in Knowledge Generation and Problem Solving* (IGI Global, 2018).

Agnieszka Olechnicka et al., *The Geography of Scientific Collaboration* Routledge, 2018).

Thomas Boyer-Kassem et al. (eds.), *Scientific Collaboration and Collective Knowledge: New Essays* (OUP, 2018).

Roy Bhaskar et al., *Interdisciplinarity and Wellbeing* (Routledge, 2018).

Azad Madni et al. (eds.), *Disciplinary Convergence in Systems Engineering Research* (Springer, 2018).

Antonietta Di Giulio and Rico Defila (eds.), *Transdisziplinär und transformativ forschen: Eine Methodensammlung* (Springer, 2018).

David Rousseau et al., *General Systemology: Transdisciplinarity for Discovery* (Springer, 2018).

Robert Axelrod, "*Collaboration and the Evolution of Disciplines*", July 1, 2019 (online).

Olivia Bina et al. (eds.*), Intrepid Knowledge: Interdisciplinary & Transdisciplinary Research and Collaboration* (COST, 2019).

Allen Repko and Rich Szostak, *Interdisciplinary Research: Process and Theory* (Sage, 2020).

Jens Allwood a.o. (eds.), *Controversies and Interdisciplinarity: Beyond Disciplinary Fragmentation for a New Knowledge Model* (John Benjamins, 2020).

See also:

- The special issue of *Revue Europeénne des Sciences Sociales* on "L'interdisciplinarité existe-t-elle?" (43:132 [2005])
- *Transdisciplinary Journal of Engineering and Science* (2010ff)
- *InterDisciplines: Journal of History and Sociology* (2010ff)
- *Journal of Interdisciplinary Studies in Education* (2012ff)
- The special issue of *The Historyof the Behavioral Sciences* on "The social sciences in a cross-disciplinary age" (51:1, winter 2015)
- The online *Survey Report for the Global Research Council 2016 Annual Meeting*, May 2016
- The open access interdisciplinary journal *New Trends in Social and Liberal Sciences* (2016ff)

- The proceedings of the 7th World Conference on Metaphysics, October 2018, in Salamanca
- In "A Nobel opportunity for interdisciplinarity" (*Nature Physics*, 14:11 [November 1, 2018], pp. 1075–1078), the authors argue that "despite the growing interdisciplinarity of research, the Nobel Prize consolidates the traditional disciplinary categorization of science". On the other hand, in partnership with the British Academy, the Royal Academy of Engineering and the Royal Society and with generous support from the Leverhulme Trust, the APEX award 2019 scheme offers "established independent researchers with a strong track record in their respective area an exciting opportunity to pursue genuine interdisciplinary and curiosity-driven research to benefit wider society".
- Roger Waldeck (ed.), *Methods and Interdisciplinarity* (Wiley, 2019) is "a follow-up to the 2018 winter school's presentations entitled '*Rencontres interdisciplinaires sur les systèmes complexes naturels et artificiels*'", annually held in Rocheburne, France.

In addition, visit:

https://www.nature.com (search for "interdisciplinarity")
https://www.nsf.gov (search for "interdisciplinarity")
https://www.sciencemag.org (search for "interdisciplinarity")
https://interdisciplinarystudies.org
http://ciret-transdisciplinarity.org
https://www.eolss.net
https://doi.org/10.1177/1368431018823140
https://www.bu.edu/cgs/citl
http://intrepid-cost.ics.ulisboa.pt
https://www.shapeid.eu
https://www.vub.be/CLEA
https://inidtd.org
https://www.ubc.ca (search for "interdisciplinary studies")
https://www.nss-journal.org (search for "interdisciplinarity")
https://warwick.ac.uk/fac/cross_fac/cim/research/inventing-indicators-of-interdisciplinarity
https://www.isko.org
http://www.protosociology.de
https://www.parmenides-foundation.org/research (4)
https://www.academia.edu (search for interdisciplinarity)
https://www.master-anthropologie.uni-freiburg.de
https://collabforge.com and
https://i2insights.org (this is probably the best website for quick orientation to the subject).

For the related subject of **unity of science**, see:

Brian McGuinness (ed.), *Unified Science* (Springer, 1987), Gerhard Schurz, *Philosophy of Science: A Unified Approach* (Routledge, 2014), Harmke Kamminga and Geert Somsen (eds.), *Pursuing the Unity of Science* (Routledge, 2016), David Lowenthal, *Quest for the Unity of Knowledge* (Routledge, 2018) and the book series *Logic, Epistemology, and the Unity of Science* (2004ff). The *International Encyclopedia of Unified Science*, an ambitious project conceived of by the Vienna Circle, was never completed. An incomplete list of encyclopedias of science can be found on the Internet.

- Go also to:

  https://plato.stanford.edu/entries/scientific-unity and
  http://atlas-disciplines.unige.ch.

- *Interdisciplinary Science Reviews* (1976ff) is a quarterly journal aiming "to explore the social, philosophical and historical interrelations of the natural sciences, engineering, mathematics, medicine and technology with the social sciences, humanities and arts".
- It should be borne in mind that the outcome of ongoing interdisciplinary or interfacultary research may also be useful to: (*a*) Heads of intergovernmental organizations having to take into account the particulars of different countries (a list of such organizations can be found on the Internet), (*b*) Heads of governments facing the problem of coordinating regional/local interests or activities, (*c*) Ministers wrestling with the problem of interdepartmental consultation and (*d*) CEO's of multinational corporations confronted with the fundamental choise between organizing their enterprise sectorally or according to the countries they are doing business with.
- Whoever wants to create or maintain a **whole** cannot escape from looking at, and taking care of, its **parts**. For a whole/system is an arrangement of interrelated parts subject to changes and a whole/system without interrelating parts is not a whole/system at all. Though the one and the many differ from each other, they are mutually involved. Order means unity and the human mind is, by its very nature, receptive to order. Indeed, it is man's ordering organ, his pattern seeking or forming capacity. Understanding the nature of each and every element of a system means understanding the relationships among them forming the structure of a whole, and vice versa. There is no priority. The single one conceals, and is pregnant with, the many, which are oriented to oneness, wholeness and unity. It is possible to temporarily consider the parts of a system in semi-isolation, but to really understand the nature of the parts one has to put them into perspective, and to really understand the wholeness or *Ganzheit* of things one has to recognize the interdependencies of the pieces and to see how they fit together. "*Tout tient à tout*" and "*qui n'entend qu'une partie n'entend rien*". In other words, Heads of an intergovernmental organizations, Heads of governments, Heads of state departments (ministers) and Chief Executive Officers of multinational corporations wanting to give specific guidelines should have an overview, which requires insight into particular or local issues and problems. Similarly, scientists concerned with the

study of a country should have an overall, uniting, inclusive and integrated view—a view of both the general and the particular. Overcoming their deep-seated mutual animosity, the generalist and the specialist should shake hands. Deployment of computers and ICT will be helpful.

- The parts-whole relationship can be transposed to the sociopolitical domain. Whereas individualists and liberalists tend to emphasize the importance of the parts, collectivists and socialists are inclined to prioritize the whole. However, it should not be forgotten that a community is constituted by its individual members, who belong to the community. May the champion of liberalism and the supporter of socialism, therefore, realize that parts and whole are epistemologically distinguishable but ontologically not separable, the liberalist considering society to consist of individuals, who cannot be taken to be completely autonomous, and the socialist conceiving of society as a whole, made up by individuals to be respected!

Appendix B
Literature on Higher Education

The Journal of Higher Education (1930ff)

The Chronicle of Higher Education (1966ff)

Times Higher Education (1971ff)

New Directions for Higher Education (1973ff)

Forum for the Future of Higher Education (founded in 1984)

Higher Education: Handbook of Theory and Research (1985ff)

Academic Questions (1988ff)

RoutledgeFalmer Studies in Higher Education (1994ff)

International Higher Education (1995ff)

Quality in Higher Education (1995ff)

Journal of Studies in International Education (1997ff)

Jahrbuch für Universitätsgeschichte (1998ff)

The Internet and Higher Education (1998ff)

Higher Education Dynamics (2002ff)

Global Perspectives on Higher Education (2005ff)

Higher Education Research in the 21st Century (2009ff)

Core Concepts in Higher Education (2010ff)

International Journal of Educational Technology in Higher Education (2010ff)

H. Kuijper, *Comprehending the Complexity of Countries*,
https://doi.org/10.1007/978-981-16-4709-3

Higher Education Studies (2011ff)

The Changing Academy (2011ff)

Global Perspectives on Higher Education (2011ff)

International Journal of Higher Education (2012ff)

Innovations in Higher Education Teaching and Learning (2014ff)

Higher Education for the Future (2014ff)

Theory and Method in Higher Education Research (2015ff)

Palgrave Critical University Studies (2016ff)

Higher Education Horizons (2018ff).

- The *Daxue* (great learning), one of the celebrated foundational texts of Confucianism, begins as follows: 大學之道 在明明德 (the way of great learning consists in showing [one's] bright virtue).
- Informative websites on higher education are:

https://www.whed.net

https://en.unesco.org/themes/higher-education

https://www.acls.org (search for higher education, and university)

https://philpapers.org/s/university

https://www.ilr.cornell.edu/cheri

http://en.ichei.org

http://www.echer.org

https://www.ashe.ws

https://www.srhe.ac.uk

https://ec.europa.eu/education/events/european-universities-initiative_en

http://www.sussex.ac.uk/education/cheer

https://www.powerlineblog.com (search for higher education)

http://www.insidehighered.com

https://www.nap.edu/read/13417/chapter/9

https://zu.libguides.com/HEA

https://www.greatertexasfoundation.org/research-resources/annotated-bibliography-higher-education-and-the-workforce

https://doi.org/10.1007/s10734-018-0262-1

https://doi.org/10.1057/s41307-017-0073-5

https://doi.org/10.1515/9783110445244

http://www.shanghairanking.com

http://www.shanghairanking.com/WCU

https://www.universite-paris-saclay.fr

https://www.oxfordbibliographies.com (search for higher education)

https://plato.stanford.edu/entries/john-salisbury

https://plato.stanford.edu/entries/vico.

- Interesting publications are:

Immanuel Kant, *Der Streit der Fakultäten* (1798)

John Henry Newman, *The Idea of a University* (1858).

James Bryce, *The functions of a university* (1912).

Thorstein Veblen, *The Higher Learning In America* (1918).

Karl Jaspers, *Die Idee der Universität* (Springer, 1923).

Abraham Flexner, *Universities: American–English–German* (OUP, 1930).

José Ortega y Gasset, *Missión de la Universidad* (Jacobo Muños, 1930).

Stephen d'Irsay, *Histoire des universités françaises et étrangères* (Picard, 1935).

Werner Jaeger, *Paideia: die Formung des griechischen Menschen* (De Gruyter, 1936).

Robert Hutchins, *The Higher Learning in America* (Yale University Press, 1936).

Hastings Rashdall, *The Universities of Europe in the Middle Ages* (Clarendon, 1936).

Nathan Schachner, *The Medieval Universities* (Allen & Unwin, 1938).

Frank Leavis, *Education and the University* (Chatto Windus, 1943).

James Conant, *On Understanding Science* (Yale University Press, 1947).

Paul Kristeller, *Die Italienische Universitäten der Renaissance* (Scherpe, 1953).

Michael Young, *The Rise of the Meritocracy* (Thomas and Hudson, 1958).

Noah Fehl, *The Idea of A University in East and West* (Chung Chi College, 1962).

Helmut Schelsky, *Einsamkeit und Freiheit: Idee und Gestalt der deutschen Universität und ihrer Reformen* (Rowohlt, 1963).

Georges Gusdorf, *L'université en question* (Payot, 1964).

Helene Wieruszowski, *The Medieval University: Masters, Students, Learning* (D. Van Nostrand, 1966).

Jacques Barzun, *The American University: How It Runs, Where It Is Going* (The University of Chicago Press, 1968).

Robert Nisbet, *The Degradation of the Academic Dogma: The University in America, 1945–1970* (Basic Books, 1971).

Immanuel Wallerstein and Paul Starr, *The University Crisis Reader* (Random House, 1971).

Alain Touraine, *Université et société aux États-Unis* (Seuil, 1972).

Talcott Parsons and George Platt, *The American University* (HUP, 1973).

Alan Cobban, *The Medieval Universities: Their Development and Organization* (Methuen, 1975).

Joseph Kockelmans (ed.), *Interdisciplinarity and Higher Education* (Pennsylvania State University Press, 1979).

Pierre Bourdieu, *Homo academicus* (Minuit, 1984).

Stephen Ferruolo, *The Origins of the University: The Schools of Paris and Their Critics, 1100–1215* (Stanford University Press, 1985).

Daniel Gardner, *Chu Hsi and the Ta-hsueh: Neo-Confucian Reflection on the Confucian Canon* (Harvard University Press, 1986).

Ron Eyerman and Lennart Svensson, *Intellectuals, Universities, and the State in Western Modern Societies* (University of California Press, 1987).

Allan Bloom, *The Closing of the American Mind* (Simon & Schusters, 1987).

Clyde Barrow, *Universities and the Capitalist State* (University of Wisconsin Press, 1990).

Henry Rosovsky, *The University: an Owner's Manual* (W.W. Norton, 1991).

Paul Shore, *The Myth of the University: Ideal and Reality in Higher Education* (University of America Press, 1991).

Edward Shils, *The Idea of the University: Obstacles and Opportunities in Contemporary Societies* (Minerva, 1992).

Sheldon Rothblatt and Björn Wittrock (eds.), *The European and American University since 1800: Historical and Sociological Essays* (CUP, 1993).

Miguel Escobar et al., *Paolo Freire on Higher Education* (SUNY Press, 1994).

Jürgen Mittelstraß, *Die unzeitgemäße Universität* (Suhrkamp, 1994).

Peili Wang, *Wilhelm von Humboldt und Cai Yuanpei* (Waxmann, 1996).

Bill Readings, *The University in Ruins* (Harvard University Press, 1997).

Wesley Shumar, *College for Sale: A Critique of the Commodification of Higher Education* (Routledge, 1997).

Jerzy Brzeziński and Leszek Nowak (eds.), *The Idea of University* (Rodopi, 1997).

Thomas Boudreau, *Universitas: The Social Restructuring of American Undergraduate Education* (Praeger, 1998).

Victor Hanson and John Heath, *Who Killed Homer?* (Free Press, 1998).

Herman van der Wusten (ed.), *The Urban University and its Identity* (Springer, 1998).

John Daniel, *Mega-universities and Knowledge Media* (Routledge, 1999).

Ulrich Schneider, *Philosophie und Universität: Historisierung der Vernunft im 19. Jahrhundert* (Felix Meiner, 1999).

Peter Scott, *Higher Education Reformed* (Falmer, 2000).

Benjamin Elman, *A Cultural History of Civil Examinations in Late Imperial China* (University of California Press, 2000).

Clark Kerr, *The Uses of the University* (Harvard University Press, 2001).

Frank Rhodes, *The Creation of the Future: The Role of the American University* (Cornell University Press, 2001).

Jacques Derrida, *L'université sans condition* (Galilée, 2001).

Christine Musselin, *La longue marche des universités françaises* (PUF, 2001).

Stanley Aronowitz, *The Knowledge Factory: Dismantling the Corporate University and Creating True Higher Learning* (Beacon Press, 2001).

Paul Grendler, *The Universities of the Italian Renaissance* (Johns Hopkins University Press, 2002).

Dennis Hayes and Robin Wynyard (eds.), *The McDonaldization of Higher Education* (Praeger, 2002).

Gordon Graham, *Universities: The Recovery of an Idea* (Academic, 2002).

Thomas Tighe, *Who's in Charge of America's Research Universities? A Blueprint for Reform* (State University of New York Press, 2003).

Jean-Hervé Lorenzi and Jean-Jacques Payan, *L'université maltraitée* (Plon, 2003).

David Kirp, *Shakespeare, Einstein, and the Bottom Line: The Marketing of Higher Education* (Harvard University Press, 2004).

Donald Stein (ed.), *Buying In or Selling Out? The Commercialization of the American Research University* (Rutgers University Press, 2004).

Brent Ruben, *Pursuing Excellence in Higher Education: Eight Fundamental Challenges* (Jossey-Bass, 2004).

Hugh Graham and Nancy Diamond, *The Rise of American Research Universities* (Johns Hopkins University Press, 2004).

John Darley et al. (eds.), *The Compleat Academic: A Career Guide* (American Psychological Press, 2004).

Jim Black, *Freefall of the American University* (WND, 2004).

Paul Gibbs, *Trusting in the University* (Kluwer, 2004).

Christopher Brooke et al. (eds.), *A History of the University of Cambridge* (CUP, 2004).

Richard Hersh and John Merrow (eds.), *Declining by Degrees: Higher Education at Risk* (St. Martin's Press, 2005).

David Frank and Jay Gabler, *Reconstructing the University: Worldwide Shifts in Academia in the 20th Century* (Stanford University Press, 2006).

Derek Bok, *Our Underachieving Colleges: A Candid Look at How Much Students Learn and Why They Should Be Learning More* (PUP, 2006).

Mordechai Feingold and Victor Navarro-Brotons, *Universities and Science in the Early Modern Period* (Springer, 2006).

Emilio Ambasz et al., *The Universitas Project: Solutions for a Post-Technological Society* (Museum of Modern Art, 2006).

Jean-Robert Pitte, *Jeunes, on vous ment! Reconstruire l'Université* (Fayard, 2006).

Jennifer Washburn, *University Inc.: The Corporate Corruption of Higher Education* (Basic Books, 2006).

Jochen Krautz, *Ware Bildung: Schule und Universität unter dem Diktat der Ökonomie* (Diederichs, 2007).

James Duderstadt, *The View from the Helm: Leading the American University during an Era of Change* (University of Michigan Press, 2007).

Ted Tapper, *The Governance of British Higher Education* (Springer, 2007).

Daniel Greenberg, *Science for Sale: The Perils, Rewards and Delusions of Campus Capitalism* (University of Chicago Press, 2007).

Mark Allen (ed.), *The Next Generation of Corporate Universities* (Wiley, 2007).

Bill Marsh, *Plagiarism: Alchemy and Remedy in Higher Education* (SUNY Press, 2007).

Sally Baker and Brian Brown, *Rethinking Universities: The Social Functions of Higher Education* Bloomsbury, 2007).

Shahid Yusuf and Kaoru Nabeshima (eds.), *How Universities Promote Economic Growth* (World Bank, 2007).

Andrea Bonaccorsi and Cinzia Daraio (eds.), *Universities and Strategic Knowledge Creation: Specialisation and Performance in Europe* (Edward Elgar, 2007).

Henry Giroux, *The University in Chains: Confronting the Military-Industrial-Academic Complex* (Paradigm, 2007).

Marc Bousquet, *How the University Works* (New York University Press, 2008).

William Adams (red.), *Reflections on the University Scene* (Xlibris, 2008).

Bruce Smith et al., *Closed Minds? Politics and Ideology in American Universities* (Brookings Institution Press, 2008).

Douglas Jacobsen and Rhonda Jacobsen (eds.), *The American University in a Postsecular Age* (OUP, 2008).

Alain Renaut, *Quel avenir pour nos universités ? Essai de politique universitaire* (Timée, 2008).

Chris Lorenz (ed.), *If You're So Smart, Why Aren't You Rich? Universiteit, Markt & Management* (Boom, 2008, Dutch).

Frank Donoghue, *The Last Professors: The Corporate University and the Fate of the Humanities* (Fordham University Press, 2008.

Carmelo Mazza et al. (eds.), *European Universities in Transition: Issues, Models and Cases* (Edward Elgar, 2008).

Sheila Slaughter and Gary Rhoades, *Academic Capitalism and the New Economy: Markets, States and Higher Education* (Johns Hopkins University Press, 2009).

Jussi Välimaa and Oili-Helena Ylijoki (eds.), *Cultural Perspectives on Higher Education* (Springer, 2008).

Christophe Charle and Charles Soulié, *Les ravages de la "modernisation" universitaire en Europe* (Syllepse, 2008).

Alain Caillé, *L'Université en crise: Mort ou résurrection?* (La Découverte, 2009).

Gaye Tuchman, *Wannabe U: Inside the Corporate University* (University of Chicago Press, 2009).

Robert Maranto et al., *The Politically Correct University: Problems, Scope, and Reforms* (AEI, 2009).

Willem Otterspeer, *The Bastion of Liberty: Leiden University Today and Yesterday* (Leiden U.P., 2009).

James Garland, *Saving Alma Mater: A Rescue Plan for America's Public Universities* (Univerity of California Press, 2009).

Mark Taylor, *Crisis on Campus* (Alfred A. Knopf, 2010).

Jürgen Kaube and Johannes Schmidt, *Die Wirklichkeit der Universität* (Lucius & Lucius, 2010).

Christoph Markschies, *Was von Humboldtnoch zu lernen ist* (Berlin University Press, 2010).

Martha Nussbaum, *Not for profit: why democracy needs the humanities* (PUP, 2010).

Michèle Lamont, *How Professor Think: Inside the Curious World of Academic Judgment* (HUP, 2010).

Anthony Nocella et al. (eds.), *Academic Repression: Reflections from the Academic Industrial Complex* (AK Press, 2010).

Hua Shaoxiang and Gregor Kneussel, *Das Große Lernen* (*Verlag für fremdsprachige Literatur*, Beijing, 2010).

Louis Menand, *Marketplace of Ideas: Reform and Resistance in the American University* (W.W. Norton, 2010).

Andrew Hacker and Claudia Dreifus, *Higher Education?* (Times Books, 2010).

Cary Nelson, *No University is an Island: Saving Academic Freedom* (New York U.P., 2010).

Klaus Dörre and Matthias Neis, *Das Dilemma der unternehmerischen Universität* (Sigma, 2010).

Christopher Loss, *Between Citizens and the State: The Politics of American Higher Education in the 20th Century* (PUP, 2011).

Clayton Christensen and Henry Eyring, *The Innovative University* (Wiley, 2011).

Ben Wildavsky et al. (eds.), *Reinventing Higher Education* (Harvard Education Press, 2011).

Benjamin Ginsberg, *The Fall of Faculties* (OUP, 2011).

Victor Ferrall, *Liberal Arts at the Brink* (Harvard University Press, 2011).

William Locke et al. (eds.), *Changing Governance and Management in Higher Education* (Springer, 2011).

Richard Arum and Josipa Roksa, *Academically Adrift: Limited Learning on College Campuses* (university of Chicago Press, 2011).

Ruth Hayhoe et al., *Portraits of 21st Century Chinese Universities* (Springer, 2011).

Janette Ryan, *China's Higher Education Reform and Internationalisation* (Routledge, 2011).

Martin Clarke, *Higher Education in the Ancient World* (Routledge, 2012).

Stefan Collini, *What Are Universities For?* (Penguin, 2012).

Jonathan Cole, *The Great American University* (Public Affairs, 2012).

Charles Clotfelter (ed.), *American Universities in a Global Market* (University of Chicago Press, 2012).

Walter Köhnlein, *Sachunterricht und Bildung* (Julius Klinkhardt, 2012).

Ulrich Sieg and Dietrich Korsch (eds.), *Die Idee der Universität heute* (De Gruyter, 2012).

Yehuda Elkana and Hannes Klöpper, *Die Universität im 21. Jahrhundert: Für eine Einheit von Lehre, Forschung und Gesellschaft* (Körber-Stiftung, 2012).

Andrew Jewett, *Science, Democracy, and the American University* (CUP, 2012).

Jacques Verger, *Les universités au Moyen Age* (PUF, 2013).

Qi Wang et al. (eds.), *Building World-Class Universities* (Springer, 2013).

James Lang, *Cheating Lessons: Learning from Academic Dishonesty* (Harvard University Press, 2013).

Andrew McGettigan, *The Great University Gamble: Money, Markets and the Future of Higher Education* (Pluto, 2013).

Ray Land and George Gordan (eds.), *Enhancing Quality in Higher Education* (Routledge, 2013).

Richard Levin, *The Worth of University* (Yale University Press, 2013).

Jerry Jacobs, *In Defense of Disciplines: Interdisciplinarity and Specialization in the Research University* (University of Chicago Press, 2013).

Benjamin Elman, *Civil Examinations and Meritocracy in Late Imperial China* (Harvard University Press, 2013).

Rui Wang, *The Chinese Imperial Examination System: An Annotated Bibliography* (Scarecrow Press, 2013).

Hao Ping, *Peking University and the Origins of Higher Education in China* (Monkfish, 2013).

George Fallis, *Rethinking Higher Education: Participation, Research, and Differentiation* (Queen's University, 2013).

Paul Kemeny (ed.), *Faith, Freedom, and Higher Education* (Pickwick, 2013).

George Steiner, *Universitas* (Nexus Institute, 2013).

Jeffrey Selingo, *College (Un)Bound: The Future of Higher Education and What it Means for Students* (New Harvest, 2013).

Jim Dator et al., "Campuses 2060: Four Futures of Higher Education in Four Alternative Futures of Society", in Munir Shuib et al. (eds.), *Developments in Higher Education: National Strategies and Global Perspectives* (Universiti Sains Malaysia Press and National Higher Education Research Institute, 2013, pp. 116–162).

Edgar Morin, *Enseigner à vivre* (Actes Sud, 2014).

Glen Sherman, *Refocusing the Self in Higher Education: A Phenomenological Perspective* (Routledge, 2014).

Jos Boys, *Building Better Universities* (Routledge, 2014).

Yuzhuo Cai and Vuokko Kohtamäki (eds.), *Transformation of Higher Education in Innovation Systems in China and Finland* (Tampere University Press, 2014).

Nicholas Maxwell, *How Universities Can Help Create a Wiser World: The Urgent Need for an Academic Revolution* (Imprint Academic, 2014) .

Yves Dupont, *L'Université en miettes: servitude volontaire, lutte des places et sorcellerie* (L'échappée, 2014).

Ramalingam Ponnusamy and J. Pandurangan, *A Handbook on University System* (Allied Publishers, 2014).

William Devane, *Higher Education in Twentieth-Century America* (Harvard University Press, 2014).

Sandra Featherman, *Higher Education at Risk (*Stylus, 2014).

Paola Mattei (ed.), *University Adaptation in Difficult Economic Times* (OUP, 2014).

Marshall Sahlins, *Confucius Institutes: Academic Malware* (Prickly Paradigm, 2014).

Robert Rhoads et al., *China's Rising Research Universities: A New Era of Global Ambition* (Johns Hopkins University Press, 2014).

Heather Brook et al., *Universities in Transition* (University of Adalaide, 2014).

Howard Wiarda, *Universities in Decline: From the Great Society to Today* (UPA, 2014).

Bob Bruner, "Suggested Readings about the Crisis in American Higher Education" (October 2014).

Lene Foss and David Gibson (eds.), *The Entrepreneural University* (Routledge, 2015).

Derek Bok, *Higher Education in America* (PUP, 2015).

Michael Crow and William Dabars, *Designing the New American University* (Johns Hopkins University Press, 2015).

Simon Schwartzman et al. (eds.), *Higher Education in the BRICS Countries: Investigating the Pact between Higher Education and Society* (Springer, 2015).

Karen Kelsky, *The Professor Is In: The Essential Guide to Turning Your PhD into a Job* (Penguin, 2015).

Hans-Jürgen Gralke, *Universitätsmanagement und universitäre Individualität* (Hampp, 2015).

Michael Roth, *Beyond the University: Why Liberal Education Matters* (Yale University Press, 2015).

Christophe Granger, *La destruction de l'université française* (La Fabrique, 2015).

Christophe Charle and Charles Soulié, *La deregulation universitaire* (Syllepse, 2015).

Christophe Charle and Jacques Verger, *Histoire des universities* (PUF, 2015).

Steven Cranfield, *F.R. Leavis: The Creative University* (Springer, 2015).

Michael Bastedo et al. (eds.), *American Higher Education in the 21st Century* (Johns Hopkins University Press, 2016).

James Axtell, *Wisdom's Workshop: The Rise of the Modern University* (PUP, 2016).

Philip Altbach, *Global Perspectives on Higher Education* (Johns Hopkins University Press, 2016).

Laurence Brockliss, *The University of Oxford: A History* (OUP, 2016).

Charles Sykes, *Fail U: The False Promise of Higher Education* (St. Martin's Press, 2016).

Ken Robinson and Lou Aronica, *Creative Schools* (Penguin, 2016).

Jessica Hoffmann Davis (ed.), *Discourse and Disjuncture between the Arts and Higher Education* (Palgrave Macmillan, 2016).

Elizabeth Popp Berman and Catherine Paradeise (eds.), *The University under Pressure* (Emerald, 2016).

James Arvanitakis and David Hornsby (eds.), *Universities, the itizen Scholar and the Future of Higher* (Palgrave, 2016).

Stephanie Blackmon and Claire Major (eds.), *MOOCS and Higher Education* (Wiley, 2016).

Matthias Barth et al. (eds.), *Handbook of Higher Education for Sustainable Development* (Routledge, 2016).

Hans Pechar, *Bildung im Spannungsfeld von Ökonomie und Politik* (Waxmann, 2016).

William Rouse, *Universities as Complex Enterprises* (Wiley, 2016).

Arlene Archer and Esther Breuer (eds.), *Multimodality in Higher Education* (Brill, 2016).

Michael Samuel et al. (eds.), *Disrupting Higher Education Curriculum* (Springer, 2016).

Hong Zhu et al., *Development and Reform of Higher Education in China* (Chandos, 2016).

Louis Vaccaro and Mi Zhou, *East Meets West: Perspectives on Chinese and American Higher Education* (Createspace, 2016).

Sara Goldrick-Rab, *Paying the Price: College Costs, Financial Aid, and the Betrayal of the American Dream* (University of Chicago Press, 2016).

Christopher Newfield, *The Great Mistake: How We Wrecked Public Universities and How We can Fix Them* (Johns Hopkins Un iversity Press, 2016).

Esther Breuer and Arlene Archer (eds.), *Multimodality in Higher Education* (Brill, 2016).

Stephen Kosslyn and Ben Nelson (eds.), *Building the Intentional University: Minerva and the Future of Higher Education* (MIT Press, 2017).

Henry Malden, *On the Origin of Universities and Academical Degrees* (HardPress, 2017).

Bert van der Zwaan, *Higher Education in 2040* (Amsterdam University Press, 2017).

Sharlene Cochrane et al. (eds.*), Culturally Responsive Teaching and Reflection in Higher Education* (Routledge, 2017).

William and Carol Allen, *Habits of Mind: Fostering Access and Excellence in Higher Eduacation* (Routledge, 2017).

Philip Altbach et al., *Responding to Massification: Differentiation in Postsecondary Education Worldwide* (Körber-Stiftung, 2017).

Tilly Miller and Margit Ostertag (eds.), *Hochschulbildung: Wiederaneignung eines existentiell bedeutsamen Begriffs* (De Gruyter, 2017).

Robert Wolff, *The Ideal of the University* (Routledge, 2017).

Rui Yang, *The Third Delight: Internationalization of Higher Education in China* (Routledge, 2017).

Carey Normand and Lorraine Anderson (eds.), *Graduate Attributes in Higher Education: Attitudes on Attributes from Across the Disciplines* (Routledge, 2017).

Hastings Rashdall, *The Universities of Europe in the Middle Ages* (CUP, 2017).

Dennis Hayes (ed.), *Beyond McDonaldization: Visions of Higher Education* (Routledge, 2017).

Frank Furedi, *What's Happened To The University? A Sociological Exploration of Its Infantilisation* (Routledge, 2017).

David Labaree, *A Perfect Mess: The Unlikely Ascendancy of American Higher Education* (University of Chicago Press, 2017).

Kevin Downing and Fraide Ganotice, Jr. (eds.), *World University Rankings and the Future of Higher Education* (IGI Global, 2017).

Joseph Lizier et al., "An interview based study of pioneering experiences in teaching and learning Complex Systems in Higher Education" (https://arxiv.org/abs/1802.02707, February 2018).

Michael Nietzel, *Coming to Grips with Higher Education* (Rowman & Littlefield, 2018).

"*Géopolitique des universities*", *Hérodote*, 168:1 (2018).

Alfred Wu and John Hawkins (eds.), *Massification of Higher Education in Asia* (Springer, 2018).

Debaditya Bhattacharya (ed.), *The University Unthought: Notes for a Future* (Routledge, 2018).

Dennis Ahlburg (ed.), *The Changing Face of Higher Education* (Routledge, 2018).

Adrian Curaj et al. (eds.), *European Higher Education Area: The Impact of Past and Future Policies* (Springer, 2018).

Kristina Powers and Angela Henderson (eds.), *Cultivating a Data Culture in Higher Education* (Routledge, 2018).

Johan Östling, *Humboldt and the Modern German University* (Lund University Press, 2018).

Christopher Hill and Rozilini Fernandez-Chung (eds.), *Higher Education in the Asian Century* (Routledge, 2018).

Brendan Cantwell et al. (eds.) *High Participation Systems of Higher Education* (OUP, 2018).

Cristóbal Salinas and Michelle Boettcher (eds.), *Critical Perspectives on Hazing in Colleges and Universities* (Routledge, 2018).

Steven Brint, *Two Cheers for Higher Education: Why American Universities Are Stronger Than Ever and How to Meet the Challenges They Face* (PUP, 2018).

Jung Cheol Shin and Pedro Teixeira (eds.), *Encyclopedia of International Higher Education Systems and Institutions* (Springer, 2018).

Bryan Caplan, *The Case against Education: Why the Education System is a Waste of Time and Money* (PUP, 2018).

Nancy Gleason, *Higher Education in the Era of the Fourth Industrial Revolution* (Palgrave, 2018).

Kenneth Westphal, "Higher Education and Academic Administration: Current Crises Long Since Foretold", *Social Epistemology Review and Reply Collective*, 7:1 (January 2018), pp. 41–47.

Xiufeng Liu and Wen Ma (eds.), *Confucianism Reconsidered: Insights for American and Chinese Education in the Twenty-First Century* (SUNY, 2018).

Jianmin Gu et al., *Higher Education in China* (Springer, 2018).

Nathan Grawe, *Demographics and the Demand for Higher Education* (Johns Hopkins U.P., 2018).

Marcelo Rabossi et al. (eds.), *In Pursuit of World-Class Universities: A Global Experience* (Studera Press: 2018).

Clyde Barrow, *The Entrepreneurial Intellectual in the Corporate University* (Palgrave Macmillan, 2018).

Stefan Collini, *Speaking of Universities* (Verso, 2018).

Donia Zhang, "Philosophical Influences on Education in China: Different Schools of Thought on Self-Cultivation", *Journal of Contemporary Educational Research*, 2:3 (May 2018), pp. 38–50.

Xi Yang and You You, "How the World-Class University Project Affects Scientific Productivity: Evidence from a Survey of Faculty Members in China", *Higher Education Policy*, 31:4 (December 2018), pp. 583–605.

Fabian Cannizzo, Nick Osbaldiston (eds.), *The Social Structures of Global Academia* (Routledge, 2019).

Jian Li and Xudong Zhu, *Conceptualizing and Contextualizing Higher Education with Chinese Characteristics: Ontological and Epistemological Dimensions* (Springer, 2019).

Roger Geiger, *American Higher Education since Second World War: A History* (PUP, 2019).

Bang Nguyen et al. (eds.), *Strategic Brand Management in Higher Education* (Routledge, 2019).

Paul Tough, *The Years That Matter Most: How College Makes or Breaks Us* (Houghton Mifflin Harcourt, 2019).

Martin Carnoy et al. (eds.), *Higher Education in Federal Countries: A Comparative Study* (Sage, 2019).

Raewin Connell, *The Good University: What Universities Actually Do and Why It's Time for Radical Change* (ZED, 2019).

Daniel Golden, *The Price of Admission: How America's Ruling Class Buys Its Way into Elite Colleges–and Who Gets Left Outside the Gates* (Broadway Books, 2019).

Daniel Markovits, *The Meritocracy Trap* (Allen Lane, 2019).

John Thelin, *A History of American Higher Education* (Johns Hopkins University Press, 2019).

Xudong Zhu and Jinhui Xu, *Faculty Development in Chinese Higher Education: Concepts, Practices, and Strategies* (Springer, 2019).

Chang Zhu and Merve Zayim-Kurtay (eds.), *University Governance and Academic Leadership in the EU and China* (IGI Global, 2019).

James Forest (ed.), *University Teaching: International Perspectives* (Routledge, 2019 [1998]).

Lukas Mitterauer et al. (eds.), *Systeme im Wandel: Hochschulen auf neuen Wegen* (Waxmann, 2019).

Hantian Wu, "Three dimensions of China's 'outward-oriented' higher education internationalization", *Higher Education*, 77:1 (January 2019), pp. 81–96.

Anna Valero and John Van Reenen, "The Economic Impact of Universities", *Economics of Education Review*, 68 (February 2019), pp. 53–67.

Bruce Chapman and Dung Doan, "Higher Education Financing: Student Loans", *Economics of Education Review*, 71 (August 2019), pp. 1–146.

Charles Edel et al., "How Should Universities Respond to China's Growing Presence on Their Campuses?" (*ChinaFile Conversation*, November 4, 2019).

Ant Bagshaw and Debbie McVitty (eds.), *Influencing Higher Education Policy* (Routledge, 2019).

Jian Li, *Comprehensive Global Competence for World-Class Universities: Context, Concept, Model and Evaluation* (Springer, 2020).

Eryong Xue and Jian Li, *Innovating World-Class Thnology-Oriented Higher Education in China: Ideas, Strategies, and Practices* (Springer, 2020).

Jerrid Freeman et al., *Higher Education Response to Exponential Societal Shifts* (IGI Global, 2020).

Floris Cohen, *De Ideale Universiteit* (Prometheus, 2020, Dutch).

Kyle Long, *The Emergence of the American University Abroad* (Brill, 2020).

Christopher Caterine, *Leaving Academia: A Practical Guide* (PUP, 2020).

Khadija Hussain, *Can Elite Universities Justify their Wealth?* (Verso, 2020).

Bryan Alexander, *Academia Next: The Futures of Higher Education* (Johns Hopkins U.P., 2020).

- Martin Heidegger's inaugural lecture "*Die Selbstbehauptung der deutschen Universität*" at Freiburg University (1933) is still worth reading.
- *The Economist* (March 28th–April 3rd, 2015) devoted a special report to universities.
- The most authoritative and comprehensive book on European universities to date is probably Walter Rüegg (ed.), *Geschichte der Universität* (C.H. Beck, 1993–2010).
- The bi-annual volumes *Historyof Universities* (OUP, 1985ff) are an inexhaustible treasure trove.
- For the history of higher education, see further the website of *History of Education Quarterly* (1991ff) and search for "higher education".
- For European higher education, go to http://www.ehea.info, https://www.study.eu/article/the-european-higher-education-area-ehea and https://ec.europa.eu/education/policies/higher-education/about-higher-education-policy_en. The European University Association plays a crucial role in influencing EU policies with regard to higher education; the Association of American Universities is its American counterpart.
- In March 2020, the conference Global Higher Education in 2050: Imagining Universities for Sustainable Societies was held in Santa Barbara (https://uchri.org/events/the-university-after-globalization-are-there-democratic-futures-for-universities).
- On October 28, 2020, the Leiden Asia Center released the report *Towards Sustainable Europe-China Collaboration in Higher Education and Research*, building upon previous publications of the Center on the same topic (in 2017, 2018 and 2019).
- The Boston College Center for International Higher Education created a map of more than 200 higher education research centers/institutes around the world.
- The Center for 21st Century Universities at the Georgia Institute of Technology is a living laboratory for fundamental change in higher education.
- Universitas21 (https://universitas21.com) is a global network of research-intensive universities for the twenty-first century.
- A list of Higher Education Associations and Alliances and a list of University Networks can be found on the Internet.
- Five things should be borne in mind:

(1) The books and articles about *higher* education is a subset of publications on *education*, the organization and process of teaching or learning (especially in schools or colleges) and the study of teaching methods and theories. See Penelope Peterson et al. (eds.), *International Encyclopedia of Education* (Elsevier, 2010, volume 4, pp. 217–627), John Rury and Eileen Tamura (eds.), *Handbook of the Historyof Education* (OUP, 2019) and the *International Journal of Complexityin Education* (2020ff).

(2) *Philosophy* of education is a thriving field of research. See John Dewey, *Democracy and Education: An Introduction to the Philosophy of Education* (Macmillan, 1916), Otto Hansmann, *Die Bildung des Menschen und*

des Menschengeschlechtes: Eine herausfordernde Synopse vom 18. Jahrhundert bis zur Gegenwart (Logos, 2014) and Olivier Reboul, *La philosophie de l'éducation* (PUF, 2018). Visit https://www.philosophyofeducation.org, http://www.ibe.unesco.org/en/document/thinkers-education, https://philpapers.org (search for philosophy of education) and https://plato.stanford.edu/entries/education-philosophy.

(3) Discussions about universitiesor higher education spill easily over into a debate over the unity/philosophy of science. See note 11 in Chap. 1, 22 in Chap. 5 and 8 in Chap. 11, and Appendix A.

(4) Given the torrent of information (the list of publications is by no means exhaustive), it is no exaggeration to say that something is wrong with higher education. Man has drifted far away from what universities were originally meant to be. Paolo Freire (1921–1997), leading Brazilian advocate of crititical pedagogy, once said: "Education should raise the awareness of students so that they become subjects, rather than objects, of the world. Education is an act of knowing, not of memorizing. This act can never be accounted for in its complex totality by a mechanistic theory". Even in translation, the Chinese Classic *Daxue* (see above) is still worth reading. See note 31 in Chap. 8. Like companies only focusing on the interest of shareholders (particularly since the monetary counter-revolution unleashed by Milton Friedman, in the 1970s), universities only serve partial interests. So universities need to be redesigned and refurbished.

(5) Universities are usually the breeding places for heads of intergovernmental organizations, heads of government, ministers and chief executive officers. The requirement to redesign and refurbish universities, research institutes and think tanks along two dimensions (integration of research methods and integrity of mentality, attitude and behavior) is thus urgent.

Appendix C
List (Incomplete) of Associations Related to Area or Country Studies

- The European Association for Chinese Studies
- The American Association for Chinese Studies
- The European Association for Japanese Studies
- The Association for Jewish Studies
- The European Association for Jewish Studies
- The European Association of Israel Studies.
- The European Association for Southeast Asian Studies
- The European Association for Studies of Australia
- The European Studies Association Australia and New Zealand
- The Association of European Studies for the Caucasus
- The European Studies Association of Sub-Saharan Africa
- The Hong Kong Association for European Studies
- The Israeli Association for the Study of European Integration
- The Irish Association for Contemporary European Studies
- The Bulgarian European Studies Association
- The Ukrainian European Studies Association
- The Czech Association of European Studies
- The German Association for East European Studies
- The Southeast European Studies Association
- The Association for European Studies
- Center for European Studies
- Council for European Studies
- The European Studies Asssociatio
- The European Union Studies Association
- The Study Association for European Studies
- University Association for Contemporary European Studies
- The American Studies Association

H. Kuijper, *Comprehending the Complexity of Countries*,
https://doi.org/10.1007/978-981-16-4709-3

- The European Association for American Studies
- The Association for Slavic, East European & Eurasian Studies
- Center for Research and European Studies
- The International Association for Mongol Studies
- The International Association for Tibetan Studies
- The Association for the Advancement of Central Asian Research
- Amsterdam Center for European Studies
- The Japan Studies Association
- The Association for Korean Studies in Europe
- The European Association of Taiwan Studies
- The North American Taiwan Studies Association
- The Middle East Studies Association
- The Association for the Study of the Middle East and Africa
- The British Association for American Studies
- The British Association for Islamic Studies
- The British Association for Japanese Studies
- The British Association for Korean Studies
- The British Association for South Asian Studies
- The British Association for Slavonic and East European Studies
- The North American Association of Islamic and Muslim Studies
- The Australian Association of Islamic and Muslim Studies
- The European Union Studies Association Asia Pacific
- The International Association for Scandinavian Studies
- The German Studies Association
- The Mediterranean Studies Association
- The American Association for Italian Studies
- The Association of Professional Italianists in South Africa
- The Association for the Study of Modern Italy
- The American Portuguese Studies Association
- The Turkish Studies Association
- The Ottoman and Turkish Studies Association
- The Association for Iranian Studies.
- The Irish Association for Russian, Central and East European Studies
- The International Australian Studies Association
- The International Council for Central and East European Studies
- The Latin American Studies Association
- The Caribbean Studies Association
- The Irish Association for American Studies
- The German Association for American Studies
- The French Association of American Studies
- The Hellenic Association for American Studies
- The Russian Association of Baltic Studies
- The Association for the Advancement of Baltic Studies
- The African Studies Association
- African Studiesin Europe

- The Association for Asian Studies
- The International Association for Asia Pacific Studies
- The Asian Studies Association of Australia
- The South Asian Studies Association
- The Association of Southeast Asian Studies
- The Society for French Studies
- The Nineteenth-Century French Studies Association
- The Australian Society for French Studies
- The Spanish Studies Association
- The Hispanic Studies Association
- The Association for Contemporary Iberian Studies
- The Association for Canadian Studies
- The Eighteenth-Century Russian Studies Association
- The Society for Co-operation in Russian and Soviet Studies
- Center for Russian, European and Eurasian Studies
- World Association for Chinese Studies
- International American Studies Association
- The Chinese Studies Association of Australia.

Thousands of articles and hundreds of book reviews have appeared in the journals edited by these associations, societies, or centers. For university-based area studies centers designated by the Chinese Ministry of Education, see Ren Xiao and Liu Ming, *Chinese Perspectives on International Relations in the Xi Jinping Era* (NBR Special Report, June 2020, p. 7)

Author Index

H. Kuijper, *Comprehending the Complexity of Countries*,
https://doi.org/10.1007/978-981-16-4709-3

L

M

Q

R

S

Subject Index

H. Kuijper, *Comprehending the Complexity of Countries*,
https://doi.org/10.1007/978-981-16-4709-3

G

H

I

J

Q

R

T

U

V

The manufacturer's authorised representative in the EU is Springer Nature Customer Service Centre GmbH, Europaplatz 3, 69115 Heidelberg, Germany. If you have any concerns regarding our products, please contact ProductSafety@springernature.com

Printed and bound by CPI Group (UK) Ltd, Croydon, CR0 4YY
30/09/2026
02225296-0016